Perspectives in Animal Ecology and Reproduction

The Editors

Dr. Vijay Kumar Gupta, Ph.D., FLS, London (born 1953-) Chief Scientist, CSIR- Indian Institute of Integrative Medicine, Jammu, India. He did his M.Sc. (1975) and Ph.D. (1979) in Zoology both from University of Jammu, Jammu-India. His research capabilities are substantiated by his excellent work on histopathology, ecology and reproductive biology of fishes, turtles, birds and mammals, which has already got recognition in India and abroad.

Dr. Gupta has to his credit more than 100 scientific publications and review articles which have appeared in internationally recognized Indian and foreign journals. His current areas of interest are histopathology, toxicology, pre-clinical safety pharmacology, reproductive efficacy studies of laboratory animals and biodiversity.

He is the Series Editor of the recently published multi-volume set of books, **"Comprehensive Bioactive Natural Products (Vols. 1-8)"**, published by M/S Studium Press, LLC, USA. He is also Editor-in-Chief of the books, **"Utilisation and Management of Medicinal Plants"**, **"Medicinal Plants: Phytochemistry, Pharmacology and Therapeutics (Vols.1&2)"**, **"Traditional and Folk Herbal Medicine"**, **"Natural Products: Research Reviews"**, **"Bioactive Phytochemicals: Perspectives for Modern Medicine"**, **"Perspectives in Animal Ecology and Reproduction** (Vols.1-8)" and **"Animal Diversity, Natural History & Conservation"**. Dr. Gupta appointed as Nominee for the *Committee for the Purpose of Control and Supervision of Experiments on Animals* (CPCSEA, Govt. of India).

Dr. Anil K. Verma, Ph.D., M.N.A.Sc., FLS, London (born 1963-) Prof. & Head, Department of Zoology, Govt. (P.G.) Degree College Rajouri, J&K State, did his M.Sc. in Zoology (1986) from University of Jammu, Jammu. He has undergone his M.Phil. (1988) and awarded first rank and Ph.D.(1993) in the field of animal reproduction at the same University. Dr. Verma has published about 60 research papers and review articles in reputed journals and books. He is also a member Editorial Board of the book series "Advances in Fish and Wildlife: Ecology and Biology" a Daya Publications, New Delhi.

Dr. Gurdarshan Singh (born 1962), working as Principal Scientist in the PK-PD-Toxicology Division of Indian Institute of Integrative Medicine (CSIR), Jammu. He did his M.Sc. (1987), M.Phil. (1989) and Ph.D. (1993) in Zoology from University of Jammu, Jammu-India.

Dr. Singh has published 16 research papers, both in national and international journals of high repute and impact factor.

Perspectives in Animal Ecology and Reproduction

— Volume 9 —

Editor-in-Chief
Dr. V.K. Gupta

Editors
Dr. Anil K. Verma
Dr. G.D. Singh

2013
Daya Publishing House®
A Division of
Astral International Pvt. Ltd.
New Delhi – 110 002

© 2013 EDITORS
ISBN 978-81-7035-829-9

Publisher's Note:

Every possible effort has been made to ensure that the information contained in this book is accurate at the time of going to press, and the publisher and author cannot accept responsibility for any errors or omissions, however caused. No responsibility for loss or damage occasioned to any person acting, or refraining from action, as a result of the material in this publication can be accepted by the editor, the publisher or the author. The Publisher is not associated with any product or vendor mentioned in the book. The contents of this work are intended to further general scientific research, understanding and discussion only. Readers should consult with a specialist where appropriate.

Every effort has been made to trace the owners of copyright material used in this book, if any. The author and the publisher will be grateful for any omission brought to their notice for acknowledgement in the future editions of the book.

All Rights reserved under International Copyright Conventions. No part of this publication may be reproduced, stored in a retrieval system, or transmitted in any form or by any means, electronic, mechanical, photocopying, recording or otherwise without the prior written consent of the publisher and the copyright owner.

Published by : **Daya Publishing House®**
A Division of
Astral International Pvt. Ltd.
81, Darya Ganj, Near Hindi Park,
Delhi Medical Association Road,
New Delhi - 110 002
Phone: 011-4354 9197, 2327 8134
Fax: 2324 3060
E-mail: info@astralint.com
Website: www.astralint.com

Laser Typesetting : **Classic Computer Services**
Delhi - 110 035

Printed at : **Salasar Imaging Systems**
Delhi - 110 035

PRINTED IN INDIA

Editorial Board

– Editor-in-Chief –

Dr. V.K. Gupta, *FLS, London*
Chief Scientist
CSIR-Indian Institute of Integrative Medicine,
Canal Road, Jammu – 180 001, India
E-mail: vgupta_rrl@yahoo.com

– Editors –

Dr. A.K. Verma, *FLS, London,*
Professor and Head
Department of Zoology, Government Degree College,
Rajouri, Jammu – 180001, India
E-mail: anilvermaverma@lycos.com

Dr. G.D. Singh
Principal Scientist
Pharmacology Division,
Indian Institute of Integrative Medicine (CSIR),
Canal Road, Jammu – 180 001, India
E-mail: singh_gd@rediffmail.com

– Members –

Prof. Clement A. Tisdell
Professor Emeritus,
School of Economics, The University of Queensland, Brisbane, Australia
E-mail: c.tisdell@economics.uq.edu.au

Prof. Tej Kumar Sherstha
Head,
Central Department of Zoology, Tribhuvan University, Kathmandu, Nepal
E-mail: drtks@ccsl.com.np

Dr. Justin Gerlach
Chief Scientist,
The Nature Protection of Seychelles,
133 Cherry Hinton Road, Cambridge CBI 7BX,UK
E-mail: justgerlach@aol.com

Prof. Vanitha Kumari
Head,
Department of Zoology, Bharathiar University, Coimbatore, India
E-mail: regr@bharathiaruni.org

Dr. Sunil Kumar
Deputy Director (Sr. Grade),
Division of Reproductive Toxicology & Histochemistry,
National Institute of Occupational Health,
Meghani Nagar, Ahmedabad – 380 016, India
E-mail: sunilnioh@yahoo.com

Dr. A.K. Pandey
Principal Scientist,
NBFGR (ICAR),
Canal Ring Road, Lucknow – 224 002, U.P. India
E-mail: akpandey_cifa@yahoo.co.in

राष्ट्रीय मत्स्य आनुवंशिक संसाधन ब्यूरो

(भारतीय कृषि अनुसंधान परिषद्)

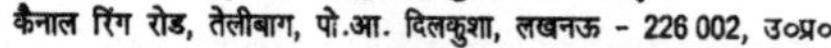

कैनाल रिंग रोड, तेलीबाग, पो.आ. दिलकुशा, लखनऊ - 226 002, उ०प्र०

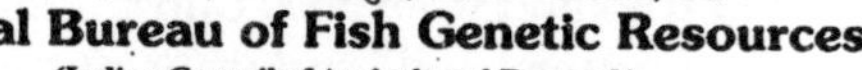

National Bureau of Fish Genetic Resources

(Indian Council of Agricultural Research)

Canal Ring Road, Telibagh, P.O. Dilkusha, Lucknow - 226 002, U.P., India
Ph. : (0522) 2441735, 2440145, 2442440, 2442441 Fax : (0522) 2442403 Gram : Germplasm
E-mail : nbfgr@sancharnet.in; director@nbfgr.res.in; Website : www.nbfgr.res.in

Dr. A. K. Pandey,FAEB,FZS,FZSI,FSB,FZSI,FNC,FISEP
FBRS FAScAW,FAZRA,FEnRA,FIAES,FNEA,FSESc,FSLSc
Principal Scientist

Foreword

Living organisms exist in different forms on this earth from protozoan to the highly evolved human beings. Their mode of reproduction also varies greatly from asexual to highly complex sexual cycle according to the phylogenetic progression. Biodiversity of life is distributed from highest of mountains to the deep seas and are adapted to the ecological niches they occupy. Biodiversity is variations in the genetics and life forms of populations, species, communities and ecosystems which affect the capacity of living systems to respond to changes in the environment essential for providing goods and services from the ecosystems.

Ecology and biodiversity are to be viewed in different perspectives with the changing scenario including anthropogenic interventions in the nature leading to environmental pollution, climate changes, resource depletion and threatening of the species. Owing to the excessive genetic loss during the last few decades, the United Nations Convention on Biological Diversity (Agenda 21) envisaged conservation, sustainable use and equitable sharing of the benefits arising from the biological resources. The Post-Earth

Summit Assessment Report (2002) revealed destruction of 2.4 per cent of the world forests (90 million ha) cover and about 40 per cent of world population faced water shortage during this period. The need to feed rising global population with exacerbated increase in food consumption (from 2100 calories to 2700 calories/day in developing countries and from 3000 calories to 3400 calories/day in developed countries) will certainly put extra pressure on agricultural production which consumes about 70 per cent of the global freshwater supplies. Signs of climate change linked to global warming causing most frequent and intense droughts in parts of Asia and Africa and rising sea levels would be addressed in the dimensions of biodiversity and ecosystem management, productivity enhancement and regulatory measures.

Over the past decades, the field of reproduction has changed drastically. It was known for many years that the gonadotropin-releasing hormone (GnRH), elaborated from hypothalamus of the vertebrates, promotes reproductive activity by stimulating production and release of pituitary gonadotropins (GtH-I and II) but the recent identification of multiple forms of GnRH with wide expression patterns have given a different dimensions to the concept. Molecular biology and genomics have revolutionized the study of reproduction in vertebrates related to the structure of hormones and their receptors, intracellular signaling molecules and enzymes involved in synthesis and release of steroids. Our knowledge on reproductive physiology of crustaceans is still in infancy. Better understanding of physiology of reproduction is having wider applications in the propagation of commercially important species for effective utilization and diversification. Chemical characterization of releaser, primer and imprinter pheromones in the animal kingdom has given new insights into their potential utility in animal husbandry and wildlife ecology.

It gives me immense pleasure to write the *Foreword* for the *Perspectives in Animal Ecology and Reproduction, Vol. 9.* I heartily congratulate Dr. V.K. Gupta, Dr. A.K. Verma and Dr. G.D. Singh for bringing together the eminent scientists with updated reviews and research papers in the rapidly expanding but interlinked disciplines of animal ecology and reproductive biology and doing justice in editing the series meticulously in elegant manner. I am sure that the current publication will provide valuable source of information and serve reference book for students, teachers and scientists in the relevant field.

A.K. Pandey

Preface

Ecology and reproduction have become the focus of attention of the scientists and scholars now-a-days and all the efforts are underway for a better understanding of the complex interaction among various living organisms and the abiotic environment. The first volume of the series *"Perspectives in Animal Ecology and Reproduction"* was published in 2002 when there was tremendous pressure upon the animals and other resources of the earth and the situation has not changed much in the last decade. In India the population has since crossed the one billion mark and the ever increasing population of the world has continued to create pollution and destruction of the resources.

Present volume of the series (Volume 9) comprises the vast body of research on the subject and has been designed in such a way as to serve the purpose of the students and the scientists. The book includes twenty six original research articles written by eminent researchers from within India and abroad. The book thoroughly covers all basic information on the subject of animal ecology and reproduction to make reader updated with the latest developments in this field where propagation of scientific information has been enormous.

The papers incorporated herein represents extremely modest selection of several facts of the breeding ecology and diversity of the animals. The lead articles focus upon the taxa ranging from minor phyla to mammals representing research work, review or overviews. The volume includes topics *viz*. Synthesis of silver nanoparticles using Gerifort drug extract and evaluation of their antimicrobial activities; Behavioural studies on the Hemipteran-*Diplonychus rusticus;* Noval bio-pesticides for mosquito vectors; Use of bio-insecticides for control of Dengue vector; Larvicidal effect of Spinosad and Azadirachtin against *Aedes aegypti;* Nutritional status of molluscs in Nepal; Utilization of feeds on the survivality of Tiger prawn and for sustainable aquaculture productivity.

Other notable inclusions in this volume are: Effect of dye effluents on haematological parameters of Indian major carp *Labeo rohita;* Effect of Doramectin administration on biochemical parameters of live fish *Channa punctatus;* Anti-hepatotoxic effects of vitamins on damaged liver of rats; Role of vitamin A, C and E on the histopathological changes induced in the pancreas of rats; Effects of Triphla and Trikatu on obesity in albino rats; Protective role of Vitamin E on Aroclor 1260 induced oxidative damage in steroidogenic enzymes in TM_3 Leydig cell line and the Anti-oxidant activity of Genistein on Aroclor 1260 induced toxicity in TM_3 Leydig cell line.

The authors have tried to analyze the bio-ecological spectra vis-à-vis breeding strategies adapted by diverse animals *in vivo* and *in vitro* as well. It is hoped that this volume shall generate further database for the researchers and academicians working on this field who shall find this book very useful and indispensable in their relevant research pursuits.

Dr. V.K. Gupta

Contents

Chapter 1

Synthesis of Herbal Drug Mediated Silver Nanoparticles Using Geriforte Drug Extract and Evaluation of their Antimicrobial Activities

V. Maruthappan[1], R.T. Rajendra Kumar[2], M. Ramesh[1] and K. Sakthi Shree[3]*

[1]Department of Zoology, [2]Department of Physics, Bharathiar University, Coimbatore – 641 046, India

[3]Department of Zoology, Government Arts College (Autonomous), Coimbatore – 641 018, India

ABSTRACT

There is an increasing commercial demand for nanoparticles due to their wide applicability in various areas such as electronics, catalysis, chemistry, energy, and medicine. Metallic nanoparticles are traditionally synthesized by wet chemical techniques, where the chemicals used are quite often toxic and flammable. In this work, we describe a cost effective and environment friendly technique for green synthesis of silver nanoparticles from 1mM $AgNO_3$ solution through the extract of geriforte herbal drug as reducing as well as capping agent. Nanoparticles were characterized using UV-vis absorption

* Corresponding author: E-mail: zoomaruthu@yahoo.co.in

spectroscopy, XRD and SEM. X-ray diffraction and SEM analysis showed the average particle size of 15 nm as well as revealed their cubic structure. Further these biologically synthesized nanoparticles were found to be highly toxic against different multi drug resistant human pathogens. This is for the first time that any plant extract was used for the synthesis of nanoparticles.

Keywords: *Silver nanoparticles, Green synthesis, Antibacterial activity.*

Introduction

Nanotechnology is expected to be the basis of many main technological innovations in the 21st century. Research and development in this field is growing rapidly throughout the world. A major output of this activity is the development of new materials in the nanometer scale, including nanoparticles. These are usually defined as particulate materials with at least one dimension of less than 100 nanometers (nm), even the particles could be zero dimension in the case of quantum dots. Metal nanoparticles have been of great interest due to their distinctive features such as catalytic, optical, magnetic and electrical properties (Rassaei *et al.*, 2008; Bar *et al.*, 2009).

Nanoparticles exhibit completely new or improved properties compared with larger particles of the bulk materials and these novel properties are derived due to the ariation in specific characteristics such as size, distribution and morphology of the particles. Nanoparticles present a higher surface area to volume ratio with decrease in the size of the articles. Specific surface area is relevant to catalytic activity and other related properties such as antimicrobial activity of AgNPs (Pal *et al.*, 2007; Gurunathan *et al.*, 2009; Bae *et al.*, 2010). Nanoparticles of noble metals, such as silver, gold and platinum are widely applied in produces that directly come in contact with the human body, such as shampoos, soaps, detergent, shoes, cosmetic products and toothpaste, besides medical and pharmaceutical applications.

A number of approaches are available for the synthesis of silver nanoparticles for example facile method (Nidhi *et al.*, 2009), thermal decomposition of silver compounds (Navaladian *et al.*, 2007), electrochemical (Starowicz *et al.*, 2006), sonochemical (Esau *et al.*,

2010), microwave assisted process (Sreeram *et al.*, 2008) and recently via green chemistry route (Begum *et al.*, 2009). Unfortunately, many of the nanoparticle synthesis or production methods involve the use of hazardous chemicals, low material conversions, high energy requirements, difficult and wasteful purifications. Therefore, there is a growing need to develop environmentally friendly processes for nanoparticle synthesis without using toxic chemicals. Biosynthetic methods employing either microorganisms or plant extracts have emerged as a simple and viable alternative to chemical synthetic procedures and physical methods.

Biological approaches using microorganisms and plants or plant extracts for metal nanoparticle synthesis have been suggested as valuable alternatives to chemical methods (Bhattacharya and Gupta, 2005; Mohanpuria *et al.*, 2007). The use of plant materials for the synthesis of nanoparticles could be more advantageous, because it does not require elaborate processes such as intracellular synthesis and multiple purification steps or the maintenance of microbial cell cultures. Recent research reported that silver nanoparticles have been synthesized using various natural products like green tea (*Camellia sinensis*) (Vilchis-Nestor *et al., 2008*), *neem* (*Azadirachta indica*) *leaf broth* (Shankar *et al.*, 2004), natural rubber (Abu Bakar *et al.*, 2007), *Aloe vera* plant extract (Chandran *et al.*, 2008), latex of *Jatropha curcas* (Bar *et al.*, 2009).

Microorganisms that are exposed to pollutants in the environment such as metal ions have a remarkable ability to fight with metal stress. These metal-microbe interactions have already found an important role in biotechnological applications. It is only recently that microorganisms have been explored as potential bio-factories for synthesis of metallic nanoparticles (Rajesh *et al.*, 2002). Silver nanoparticles act as an antimicrobial, anti-bacterial, anti-viral, antibiotic and anti-fungal agent when incorporated in coatings, nano-fiber, first aid bandages, plastics soap and textiles, in self cleaning fabrics and as conductive filler. It is also used in nanowire and in certain catalyst applications. Silver nanoparticles spectrally designed for selective coating for solar energy absorption and intercalation material for electrical batteries as optical receptors, polarizing filters and catalysts in chemical reaction. These particles also use as staining pigments for glasses and ceramics, transparent conducting coating, electronics, in surface-enhanced Raman spectroscopy, bio-labeling and antimicrobial agents (Peterson *et al.*, 2006). The use of anodic silver as preserving agents in cosmetics was also tested by a challenge test in a set of cosmetic dispersions with the addition of known preservative inhibitors or microorganism growth promoters such as humectants, hydro soluble collagen and vegetable extracts (Rajesh *et al.*, 2002).

The antimicrobial mechanism of AgNPs is also thought to be driven by surface area and the formation of reactive oxygen species. Baker *et al.* (2005) found that complete bacterial death could be achieved at 8 $\mu g/cm^2$ and smaller particles were more efficient antimicrobials. Others support this finding, and found that the amount of chemisorbed Ag^+ and aggregation status influences antimicrobial efficacy (Lok *et al.*, 2007). The formation of reactive oxygen species has been implicated in bacterial toxicity (Choi and Hu, 2008) and is thought to perturb cell membrane integrity (Su *et al.*, 2009).

Ancient medical literature on ayurveda gives a vivid and comprehensive description of this group of plant medicines which are tonic, prevent disease and postpone ageing. The indigenous drug formulation in tablet form, geriforte (M/s. Himalaya Drug Co., Bangalore, India), is a comprehensive compound of rasayana drugs, the principal component of which is chyavanprash. It also contains such nervine tonics as brahmi, ashwagandha, herbal and mineral extracts known for their metabolic and tonic properties and is prepared in the juices and decoctions of various reconstituents, tonics, hepatic stimulants and digestives (Kishore *et al.*, 1983; Upadhyaya *et al.*, 1990). Geriforte, an indigenous herbomineral compound, is a mixture of several major and minor herbal extracts and minerals.

Geriforte induces cellular regeneration (Lobo *et al.*, 1975), increases hormonal utilization (Kishore *et al.*, 1983), regulates enzymes and bioamine metabolism (Bardhan *et al.*, 1985; Upadhyaya *et al.*, 1988), inhibits the formation of lipofuscin in ageing rats (Sharma *et al.*, 1991; 1992) and maintains the cytoprotective and immunomodulating activities (Bansal *et al.*, 2001). The clinical significance of this remedy has already been reported in the management of anxiety and psychosomatic disorders (Singh *et al.*, 1978; Dubey *et al.*, 1984), maintains positive nitrogen balance and cures menopausal symptoms (Bardhan *et al.*, 1985). It is also reported to act on the cardiovascular, sexual and other systems and helps to correct the metabolism of proteins, fat, carbohydrates, serum cholesterol, triglycerides, phospholipids etc. (Singhal, 1978).

Geriforte has the capacity to undergo ion exchange, a property that can be easily exploited for many uses. Previously, the synthesis of a patterned geriforte template containing AgNPs and its antimicrobial efficiency was described (Sabbani *et al.*, 2010). Others have also characterized antimicrobial activities of silver-exchanged geriforte and found that Ag^{+} ions migrate from the substrate and that they are also sufficient to cause bacterial death of both *E. coli* and *S. aureus* (Fernandez *et al.*, 2010). Kwakye-Awuah *et al.* (2008) also demonstrated the antimicrobial power of silver geriforte: complete bacterial death was observed after 1 hour, even when geriforte was retrieved, washed and reused for up to three experiments.

There is growing concern surrounding the increasing use of AgNPs. The use of silver has already proved successful for many biomedical applications and AgNPs may expand these applications as an efficient antimicrobial for disease prevention and treatment. Here in, we report for the first time synthesis of silver nanoparticles, reducing the silver ions present in the solution of silver nitrate by the cell free aqueous extract of geriforte drug. Further these biologically synthesized nanoparticles were found highly toxic against different multi drug resistant human pathogens.

Materials and Methods

Materials

All chemicals used in this experiment were of highest purity and obtained from M/s. Sigma, Bangalore, India and M/s. Merck, Mumbai, India. Geriforte, herbal drug was purchased from M/s. Himalaya Drug Co., Bangalore, India.

Preparation of the Drug Extract

Whole drug was used to make the aqueous extract. Drug was crushed into 100 ml sterile distilled water and filtered through Whatman No.1 filter paper (pore size 25 µm). The filtrate was further filtered through 0.6 µm sized filters.

Synthesis of Silver Nanoparticles

1mM aqueous solution of Silver nitrate ($AgNO_3$) was prepared and used for the synthesis of silver nanoparticles. 10 ml of drug extract was added into 90 ml of aqueous solution of 1 mM Silver nitrate for reduction into Ag^+ ions and kept at room temperature for 5 hours. A Change from brown to reddish colour was observed.

Characterization Method

UV-vis Spectra Analysis

UV-vis spectroscopy measurements (Jasco-UV-vis) were carried out as a function of time of the reaction at room temperature operated at a resolution of 1 nm. The reduction of silver ions was confirmed by qualitative testing of supernatant obtained after centrifugation with a pinch of NaCl.

Structural Analysis

X-ray diffraction (XRD) measurements of the bio-reduced silver nitrate solution drop coated onto glass substrates were done for the determination of the formation of AG by an X'Pert Pro P Analytical X-ray diffractometer instrument with X'Pert high score plus software operating at a voltage of 45 KV and a current of 40 mA with Cu Kα radiation.

Surface Analysis

SEM Analysis Protocol

Scanning electron microscopic (SEM) analysis was done using Hitachi S-4500 SEM machine. Thin films of the sample were prepared on a carbon coated copper grid by just dropping a very small amount of the sample on the grid. Extra solution was removed using a blotting paper and then the films on the SEM grid were allowed to dry by putting it under a mercury lamp for 5 minutes.

Antibacterial Assay

Antibacterial activities of the synthesized Ag nanoparticles were determined using the agar well diffusion assay method (Perez *et al.*, 1990). Approximately 20 mL of molten and cooled media (Nutrient agar) was poured in sterilized petri dishes. The plates were left overnight at room temperature to check for any contamination to appear. The bacterial test organisms were grown in nutrient broth for 24 hour. A 100 mL nutrient broth culture of each bacterial organism ($1x10^5$cfu/mL) was used to prepared bacterial lawns. Agar wells of 5mm diameter were prepared with the help of a sterilized stainless steel cork borer. Two cells were prepared in the agar plates. The wells were labelled as A and B. 'A' well was loaded with 30 μL of positive control drugs (Chlorophenical) was used as positive control. The plates containing the bacterial and Ag nanoparticles were incubated at 37°C. The plates were examined for evidence of zones of inhibition, which appeared as a clear area around the wells. The diameter of such zones of inhibition was measured using a meter ruler and the mean value for each organism was recorded and expresses in millimetre.

Results

UV-vis Spectra Analysis

The reaction mixture, geriforte drug extract with aqueous solution of the silver nitrate, started to change its colour from yellowish brown to reddish brown (Figure 1.1). It indicated the formation silver nanoparticle with the reduction of silver ion. The characteristic surface plasmon absorption bands were observed at 440 nm. Extinctions spectra of silver synthesized from $AgNO_3$ were shown in Figure 1.2.

SEM Study

Figure 1.3 represents the SEM image recording from drop coated films of the silver nanoparticles synthesized with geriforte drug extract. The SEM image showed cubical and relatively uniform shape of nanoparticle formation with diameter range 48-67 nm.

XRD Study

The studies with XRD pattern obtained in Figure 1.4 showed number of Bragg's reflection. There are three intense peaks in the whole spectrum of 2θ values ranging from 10 to 70. A comparison of our XRD pattern with the standard continued that the silver particles formed in our experiments were in the form of nanocrystals, as evidenced by the peaks at 2θ values of 38.18°, 44.18°, 64.10° and integrated intensity values at (111), (200) and (220), respectively, for silver.

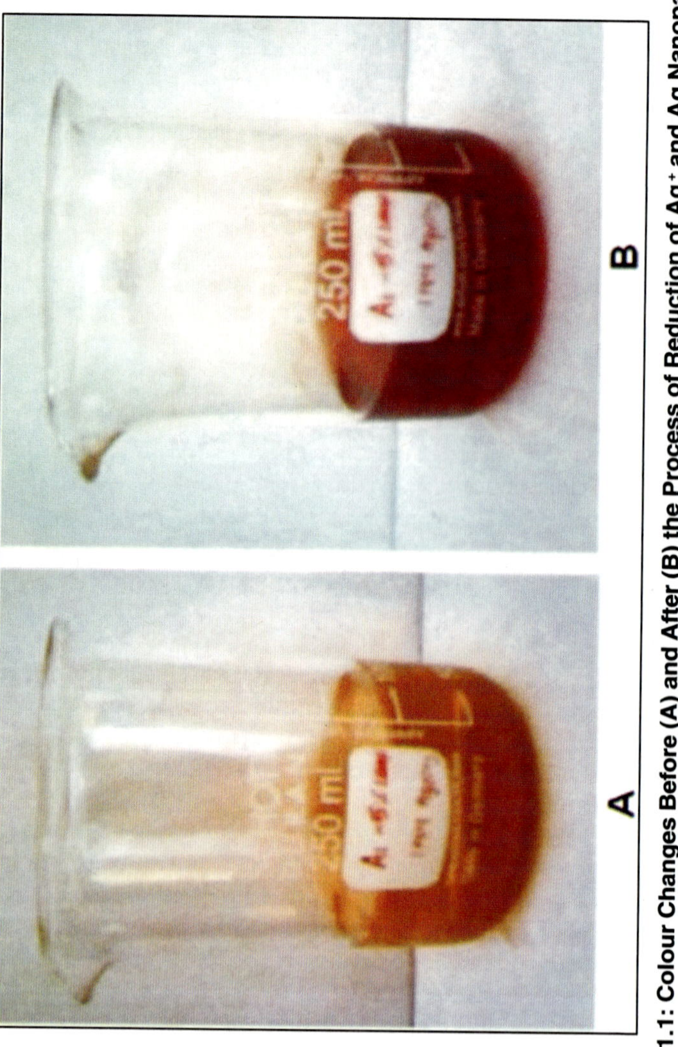

Figure 1.1: Colour Changes Before (A) and After (B) the Process of Reduction of Ag^+ and Ag Nanoparticles

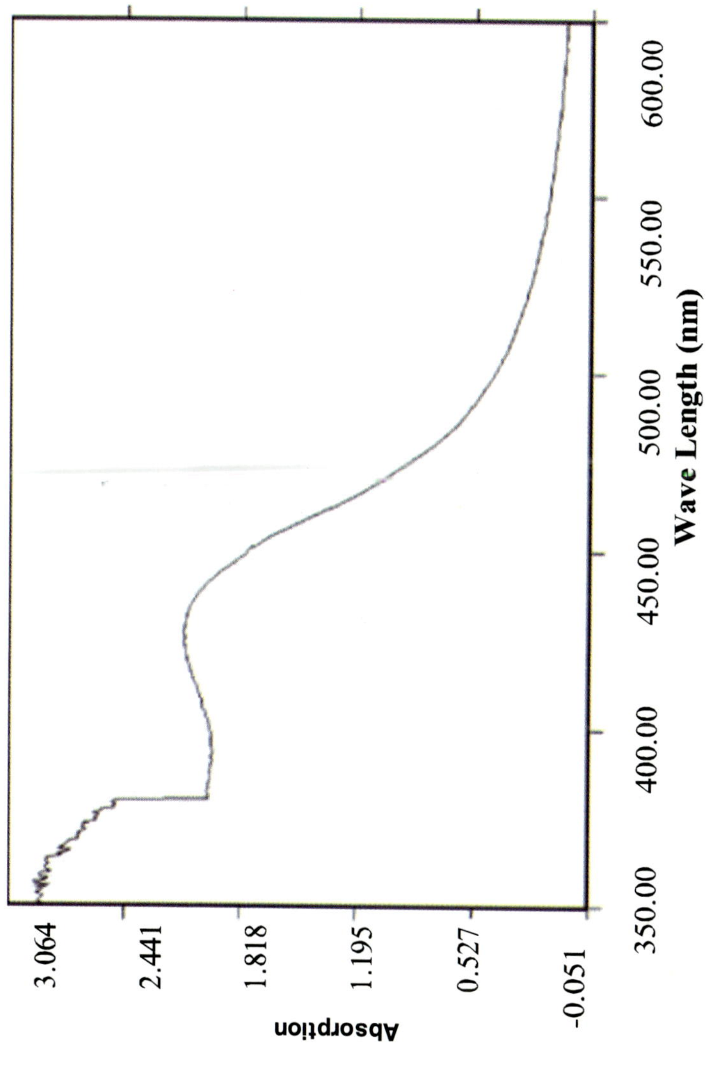

Figure 1.2: UV-Visible Absorption Spectra of Reduction of Silver Ions to Silver Nanoparticles

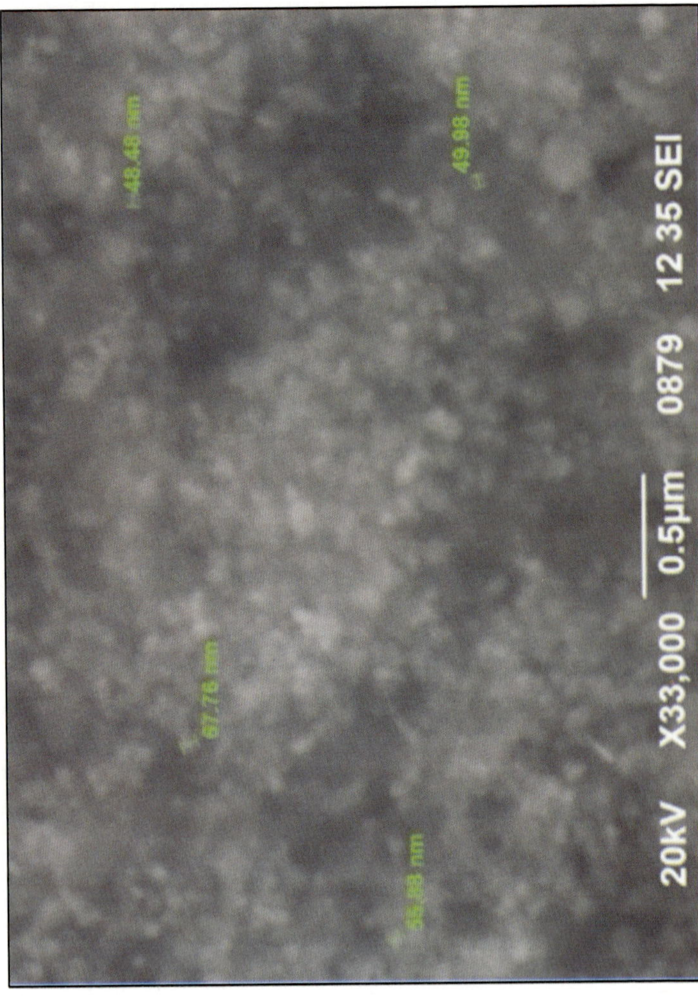

Figure 1.3: SEM Image of Ag Nanoparticles Formed by Geriforte Drug Extract

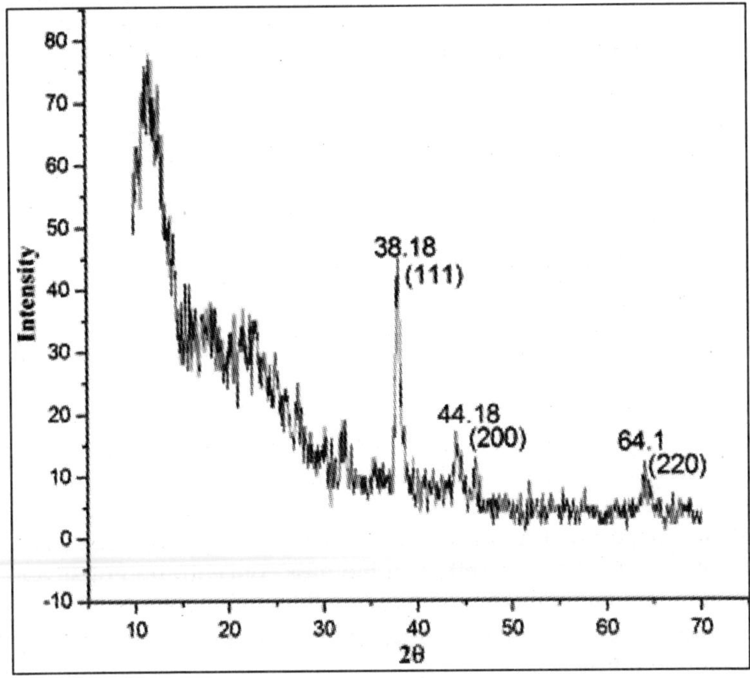

Figure 1.4: XRD Patterns Recorded for the Nanoparticles Synthesized from Geriforte Drug Extract

Antibacterial Study

The antibacterial activity of silver nanoparticles was investigated against various pathogenic bacteria of Gram positive (*Salmonella epidermidis*) and Gram negative strains (*Escherichia coli, Salmonella typhi, Pseudomonas aeroginosa, Klebsiella pneumonia*). Control is also maintained in which no zone of inhibition is observed (Figure 1.5). The diameter of inhibition zones around each well with geriforte drug extract AgNPs is represented in Table 1.1. The antibacterial activity of the bionanoparticles for *S. typhi* was maximum (40 mm) followed by *S. epidermidis* (36 mm), *Pseudomonas aeroginosa* (35 mm), *E. coli* (34 mm) and *Klebsiella pneumonia* (30 mm).

Discussion

The synthesis of nanoparticles is in lime light of modern nanotechnology. Biosynthesis of nanoparticles by plant extracts is

Table 1.1: Zone of Inhibition (mm) of Geriforte Drug Extract Silver Nanoparticles against Bacterial Strains Tested

Name of the Bacterial Species	*Zone of Inhibition (mm)*	
	Ag Nanoparticles	*Reference Drug*
Salmonella typhi	40	45
Salmonella epidermidis	36	41
Pseudomonas aeroginosa	35	40
Escherichia coli	34	39
Klebsiella pneumonia	30	35

currently under exploitation. The development of biologically inspired experimental processes for the synthesis of nanoparticles is evolved into an important branch of nanotechnology. The present study deals with the synthesis of silver nanoparticles using drug extracts of geriforte and aqueous Ag^+ ions. The experiments were carried out to study the effect of biomass concentration of silver nitrate on the rate of bioreduction of silver ions. The approach appears to be cost effective alternative to conventional methods of assembling silver nanoparticles.

Formation and stability of silver nanoparticles in aqueous colloidal solution are confirmed using UV-vis spectral analysis. It is well known that silver nanoparticles exhibit yellowish brown colour in aqueous solution due to excitation of surface plasmon vibrations in silver nanoparticles (Noginov *et al.*, 2006; Krishnaraj *et al.*, 2010). As the geriforte drug extract was mixed with aqueous solution of the silver nitrate, it started to change the colour from watery to reddish brown due to reduction of silver ion, which indicated the formation of silver nanoparticles. It is generally recognized that UV-vis spectroscopy could be used to examine size and shape-controlled nanoparticles in aqueous suspensions (Shrivastava and Dash, 2009). Absorption spectra of silver nanoparticles formed in the reaction media has absorbance peak at 440 nm and broadening of peak indicated that the particles are polydispersed.

The silver nanoparticles formed were predominantly cubical with uniform shape. It is known that the shape of metal nanoparticles considerably change their optical and electronic properties (Xu and Kall, 2002). The SEM image showed relatively spherical shape

Figure 1.5: Antibacterial Activities of Geriforte Drug Extract Silver Nanoparticles against Pathogenic Bacteria

A. *Salmonella typhi*

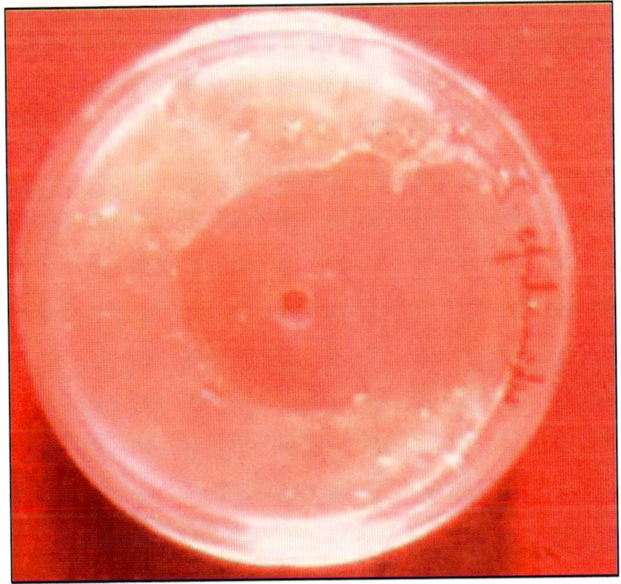

B. *Salmonella epidermidis*

Contd...

Figure 1.5–*Contd...*

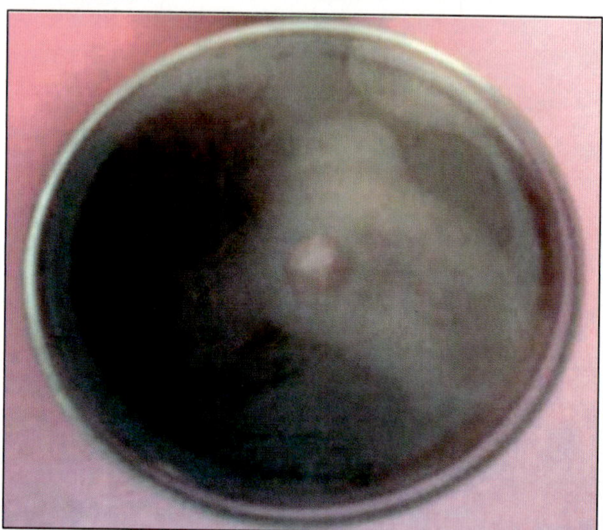

C. *Pseudomonas aeroginosa*

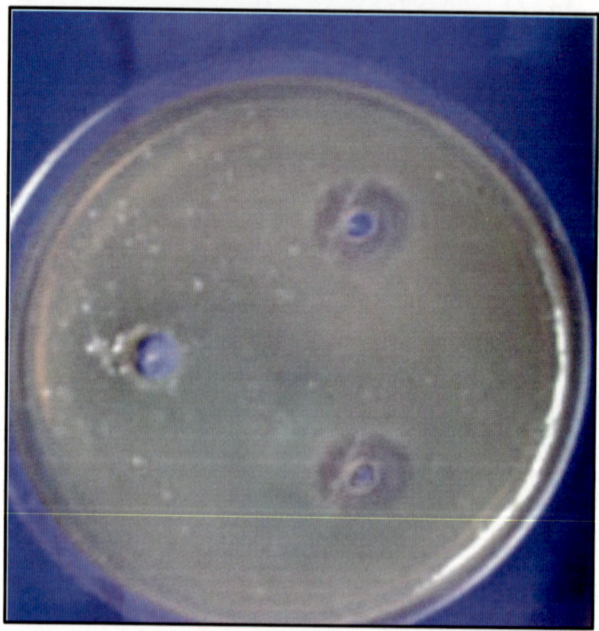

D. *Eschericia coli*

Contd...

Figure 1.5–*Contd...*

E. *Klebsiella pneumonia*

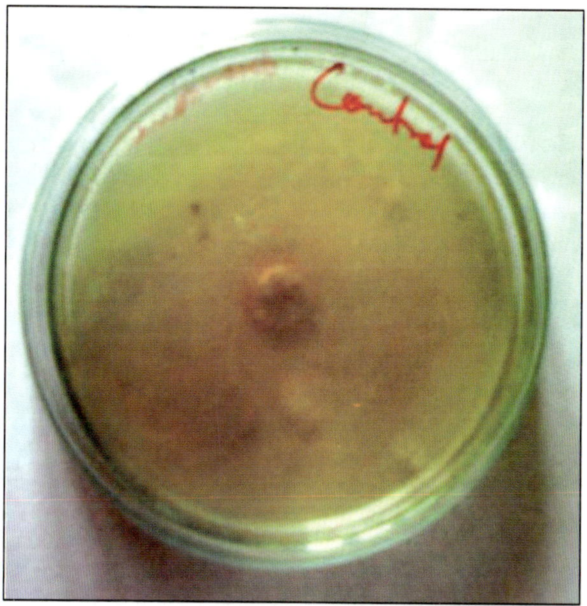

F. Control

nanoparticle formed with diameter in range 46-48 nm. Similar phenomenon was reported by Chandran et al. (2006).

Further studies using X-ray diffraction (XRD) were carried out to confirm the crystalline nature of the particles, and the XRD pattern showed numbers of Bragg's reflections that may be indexed on the basis of the face centered cubic structure of silver. A comparison of our XRD spectrum with the standard confirmed that the silver particles formed in our experiments were in the form of nanocrystals, as evidenced by the peaks at 2θ values of 38.18°, 44.18° and 64.10°, corresponding to (111), (200) and (220), respectively, for silver. The X-ray diffraction results clearly show that the silver nanoparticles formed by the reduction of Ag^+ ions by the geriforte drug extract are crystalline in nature (Huang *et al*., 2007). As mentioned in the method section, the silver nanoparticles once formed were repeatedly centrifuged and redispersed in sterile distilled water prior to XRD analysis, thus ruling out the presence of any free compound/protein that might independently crystallize and giving rise to Bragg reflections.

Silver has been widely utilized for thousands of years in human history. Its applications include jewels, utensils, currency, dental alloy, photography and explosives. Among silver's many applications, its disinfectant property is being exploited for hygienic and medicinal purposes, such as treatment of mental illness, nicotine addiction and infectious disease like syphilis and gonorrhea (Gulbrason *et al*., 2000; Huang *et al*., 2007). Silver nanoparticles have been demonstrated to exhibit antimicrobial properties against bacteria with close attachment of the nanoparticles themselves with the microbial cell and the activity being size dependent (Sondi and Salopek-Sondi, 2004).

Antibiotic resistance by the pathogenic bacteria has been observed since last decade; hence, the researchers are focusing on the development of new antibacterial agents. In the present scenario, Ag nanoparticles as antimicrobial agents have come up as a promising candidate in the medical field (Duran *et al*., 2007). The extremely small size of nanoparticles exhibits enhanced or different properties when compared with the bulk material. There are different physical and chemical methods for the synthesis of nanoparticles, but there is always a need for the development of eco-friendly route for the synthesis process (Chen et al., 2003; Ingle *et al*., 2008).

The differences between gram-positive and gram-negative bacteria essentially rest in the structure of their respective cell walls. The gram-negative bacteria have a layer of lipopolysaccharide at the exterior, followed underneath by a thin (about 7–8 nm) layer of peptidoglycan (Madigan and Martinko, 2005). Although the lipopolysaccharides are composed of covalently linked lipids and polysaccharides, they lack strength and rigidity. Negative charges on the lipopolysaccharides (Salton and Kim, 1996) are attracted towards weak positive charges available on silver nanoparticles (Sui *et al.*, 2006). On the other hand, the cell wall in gram-positive bacteria is principally composed of a thick layer (about 20–80 nm) of peptidoglycan, consisting of linear polysaccharide chains cross-linked by short peptides to form a three dimensional rigid structure (Baron, 1996). The rigidity and extended cross-linking not only endow the cell walls with fewer anchoring sites for the silver nanoparticles but also make them difficult to penetrate. The extent of inhibition of bacterial growth reported in this study was dependent on the concentration of nanoparticles in the medium. Interaction between nanoparticles and the cell wall of bacteria would be facilitated by the relative abundance of negative charges on the gram-negative bacteria, which was congenial to the fact that growth of gram-negative bacteria was more profoundly affected by the silver nanoparticles than that of the gram-positive organisms. As demonstrated by electron microscopy, interaction with nanoparticles resulted in perforations in the cell wall, contributing to the antibacterial effects of the nanoparticles. Inferences from the re-culture experiments were consistent with the entry of nanoparticles inside bacterial cells and/or strong association with bacterial cellular components. Once inside the cell, nanoparticles would interfere with the bacterial growth signalling pathway by modulating tyrosine phosphorylation of putative peptide substrates critical for cell viability and division.

Lee *et al.* (2003) investigated the antibacterial effect of nanosized silver colloidal solution against *Klebsiella pneumonia* after padding the solution on textile fabrics. Shrivastava *et al.* (2007) studied antibacterial activity against *Escherichia coli* (*E. coli*) (ampicillin resistant), *E. coli* and *Salmonella typhi* (multi-drug resistant). They reported that the effect was dose dependent and was more pronounced against gram-negative organisms than gram-positive ones. They found that the major mechanism through which silver nanoparticles manifest antibacterial properties was either by anchoring or penetrating the bacterial cell wall and modulating cellular signaling by dephosphorylating putative key peptide substrates on tyrosine residues (Shrivastava *et al.*, 2007). The antibacterial efficacy of the biogenic silver nanoparticles reported in the present study may be ascribed to the mechanism described above but it still remains to clarify the exact effect of the nanoparticles on important cellular metabolism like DNA, RNA and protein synthesis.

In the present study, the highest antimicrobial activity was observed against *S. typhi*, followed *by S. epidermidis, Pseudomonas aeroginosa, E. coli* and the least was noticed against *Klebsiella pneumonia*. Silver has been used for its well known antimicrobial properties since roman time however the advances in generating AgNPs have made possible a revival of the use of silver as a powerful bactericide (Song *et al.*, 2006). Many researchers (Feng *et al.*, 2003) used *Escherichia coli* as a model for gram-negative bacteria and proved that AgNPs may be used as an antimicrobial agent. Other workers (Yamanaka *et al.*, 2005) also opined that the AgNPs have an antimicrobial effect on *E. coli*.

We have developed a biological method to synthesize silver nanoparticles using the aqueous drug extract of geriforte, whose herbal extract is environmentally benign and renewable, act as reducing agent. Particles are mostly polydispersed in shape. Ag nanoparticles prepared in this process are quiet fast and low cost. We believe that development of eco-friendly process for the synthesis of metallic nanoparticles is an important step in the field of application of nanotechnology.

Conclusion

In conclusion we have reported the antibacterial activity of geriforte drug extract AgNPs over multidrug resistant bacteria's (both Gram-positive and Gram-negative). The zone of inhibition using high antibiotic zone scale clearly shows that all the bacterial strains which were resistant to antibiotics are highly susceptible to silver nanoparticles.

The bio-reduction of aqueous Ag^{+} ions by the herbal drug extract of the geriforte has been demonstrated. The reduction of the metal ions through leaf extracts leading to the formation of silver nanoparticles of fairly well-defined dimensions. In the present study we found that geriforte drug can be also good source for synthesis of silver nanoparticles. This green chemistry approach toward the synthesis of silver nanoparticles has many advantages such as, ease with which the process can be scaled up, economic viability, etc. Applications of such eco-friendly nanoparticles in bactericidal, wound healing and other medical and electronic applications, makes this method potentially exciting for the large-scale synthesis of other inorganic materials (nanomaterials).

References

Abu Bakar, N.H.H., Ismail, J. and Abu Bakar, M., 2007. Synthesis and characterization of silver nanoparticles in natural rubber. *Materials Chemistry and Physics*, 104: 276–283.

Bae, E., Park, H.J., Lee, J., Kim, Y., Yoon, J. and Park, K., 2010. Bacterial cytotoxicity of the silver nanoparticle related to physico-chemical metrics and agglomeration properties. *Environmental Toxicology and Chemistry*, 29(10): 2154–2160.

Baker, C., Pradhan, A., Pakstis, L., Pochan, D.J. and Shah, S.I., 2005. Synthesis and antibacterial properties of silver nanoparticles. *Journal of Nanoscience and Nanotechnology*, 5(2): 244–249.

Bansal, A., Sairam, M., Prasad, D., Sharma, S.K., Illavazhagan, G., Kumar D. and Selvamurthy W., 2001. Cytoprotective and immunomodulatory properties of geriforte, a herbomineral preparation, in lymphocytes. *Phytomedicine*, 86 (6): 438–444.

Bar, H., Bhui, D.K, Sahoo, G.P., Sarkar, P., De, S.P. and Misra, A., 2009. Green synthesis of silver nanoparticles using latex of *Jatropha curcas*. *Colloids and Surfaces A: Physico–chemical and Engineering Aspects*, 339: 134–139.

Bardhan, P., Sharma, S.K. and Garg, N.K., 1985. Effect of geriforte *in vivo* and *in vitro* on age related enzyme changes in liver and brain of rats. *Indian Journal of Experimental Biology*, 23: 323–327.

Baron, S., 1996. *Medical Microbiology*. 4th edn., Galveston: University of Texas Medical Branch.

Begum, N.A., Mondal, S., Basu, S., Laskarm R.A. and Mandal, D., 2009. Biogenic synthesis of Au and Ag nanoparticles using aqueous solutions of black tea leaf extracts. *Colloids and Surfaces B: Biointerfaces*, 71(1): 113–118.

Bhainsa, K.C. and D'Souza, S.F., 2006. Extracellular biosynthesis of silver nanoparticles using the fungus *Aspergillus fumigatus*. *Colloids and Surfaces B: Biointerfaces*, 47: 160–164.

Bhattacharya, D. and Gupta, R.K., 2005. Nanotechnology and potential of microorganisms. *Critical Review of Biotechnology*; 25: 199–204.

Chandran, S.P., Chaudhary, M., Pasricha, R., Ahmad, A. and Sastry, M., 2006. Synthesis of gold nanotriangles and silver nanoparticles using *Aloe vera* plant extract. *Biotechnology Progress*, 22: 577–583.

Chen, J.C., Lin, Z.H. and Ma, X.X., 2003. Evidence of the production of silver nanoparticles via pre–treatment of *Phoma* sp. with silver nitrate. *Letter Applied Microbiology*, 37: 105–108.

Choi, O. and Hu, Z., 2008. Size dependent and reactive oxygen species related nanosilver toxicity to nitrifying bacteria. *Environmental Science and Technology*, 42(12): 4583–4588.

Dubey, G.P., Agrawal, A., Agrawal, S., Srivastava, V.K. and Udupa, K.N., 1984. Antistress, anabolic properties of an indigenous compound and its significance in the management of psychosomatic disorders. *Asian Medical Journal*, 11: 723–729.

Duran, N., Marcato, P.D., Desouza, G.I.H., Alves, O.L. and Esposito, E., 2007. Ecosystem protection by effluent bioremediation: Silver nanoparticles impregnation in a textile fabrics process. *Journal of Biomedical Nanotechnology*, 3(2): 203–208.

Esau, S.R., Roberto, S.B., Ocotlan–Flores, J. and Saniger, J.M., 2010. Synthesis of AgNPs by sonochemical induced reduction application in SERS. *Journal of Nanoparticle Research*, 9: 77–83.

Feng, Q.L., Wu, J., Chen, G.Q., Cui, K.Z., Kim, T.M. and Kim, J.O., 2003. A mechanistic study of the antibacterial effect of silver ions on *Escherichia coli* and *Staphylococcus aureus*. *Journal of Biomedical Materials Research*, 52: 662– 668.

Fernandez, A., Soriano, E., Hernandez–Munoz, P. and Gavara, R., 2010. Migration of antimicrobial silver from composites of polylactide with silver zeolites. *Journal of Food Science*, 75(3): 186–193.

Gulbrason, S.H., Hud, J.A. and Hansen, R.C., 2000. Argyria following the use of dietary supplements containing colloid protein. *Cutis*, 66: 373–374.

Gurunathan, S., Kalishwaralal, K., Vaidyanathan, R., Venkataraman, D., Pandian, S.R., Muniyandi J., Hariharan, N. and Eom, S.H., 2009. Biosynthesis, purification and characterization of silver nanoparticles using *Escherichia coli*. *Colloids Surface B: Biointerfaces*, 74(1): 328–335.

Huang, J., Li, Q., Sun, D., Lu, Y., Su, Y., Yang, X., Wang, H., Wang, Y., Shao, W., He, N., Hong, J. and Chen, C., 2007. Biosynthesis of silver and gold nanoparticles by novel sundried *Cinnamomum camphora* leaf. *Nanotechnology*, 18: 105104–105114.

Ingle, A, Gade, A., Pierrat, S., Sonnichsen, C. and Rai, M., 2008. Mycosynthesis of silver nanoparticles using the fungus *Fusarium acuminatum* and its activity against some human pathogenic bacteria. *Current Nanoscience*, 4: 141–144.

Kishore, B., Hazra, D.K., Wahal, P.K., Khandelwal, S., Dodija, A.K. and Jain, R.K., 1983. Effect of indigenous drug formulation (Geriforte) on hormonal status. *Asian Journal of Medicine*, 11: 770–775.

Krishnaraj, C., Jagan, E.G., Rajashekar, S., Selvakumar, P., Kalaichelvan, P.T. and Mohan, N., 2010. Synthesis of silver nanoparticles using *Acalypha indica* leaf extracts and its antibacterial activity against water borne pathogens. *Colloids and Surfaces B: Biointerfaces*, 76: 50–56.

Kwakye–Awuah, B., Williams, C., Kenward, M.A. and Radecka, I., 2008. Antimicrobial action and efficiency of silver–loaded zeolite X. *Journal of Applied Microbiology*, 104(5): 1516–1524.

Lee, H.J., Yeo, S.Y. and Jeong, S.H., 2003. Antibacterial effect of nanosized silver colloidal solution on textile fabrics. *Journal of Material Science*, 38: 2199–2204.

Lobo, E., Kulkarni, R.D. and Desai, R.R., 1975. Special pharmacology of geriforte. *Probe*, 4: 266–270.

Lok, C.N., Ho, C.M., Chen, R., He, Q.Y., Yu, W.Y., Sun, H., Tam, P.K., Chiu, J.F. and Che, C.M., 2007. Silver nanoparticles: partial oxidation and antibacterial activities. *Journal of Biological Inorganic Chemistry*, 12(4): 527–534.

Madigan, M. and Martinko, J., 2005. *Brock Biology of Microorganisms*, 11th edn., Englewood Cliffs, NJ: Prentice Hall.

Mohanpuria, P., Rana, N.K. and Yadav, S.K., 2007. Biosynthesis of nanoparticles: technological concepts and future applications. *Journal of Nanoparticle Research*, 10: 507–517.

Navaladian, S., Viswanathan, B., Viswanath, R.P. and Varadarajan, T.K., 2007. Thermal decomposition as route for silver nanoparticles. *Nanoscale Research Letter*, 2: 44–48.

Nidhi, N., Santosh, K., Ghosh, T. and Dutta, P.K., 2009. Preparation of chitosan based silver nanocomposite by a facile method. *International Conference on Optics and Photonics*. Chandigiah, CSIO.

Noginov, M.A., Zhu, G., Bahoura, M., Adegoke, J., Small, C., Ritzo, B.A., Drachev, V.P. and Shalaev, V.M., 2006. The effect of gain and absorption on surface plasmon in metal nanoparticles. *Applied Physics B*, 86: 455–460.

Pal, S., Tak, Y.K. and Song, J.M., 2007. Does the antibacterial activity of silver nanoparticles depend on the shape of the nanoparticle? A study of the gram–negative bacterium *Escherichia coli*. *Applied and Environmental Microbiology*, 73: 1712–1720.

Perez. C., Paul, M. and Bazerque, P., 1990. Antibiotic assay by agar–well diffusion method. *Acta Biomedical Experiments*, 15: 113–15.

Peterson, B.R., Ayusman, S., Varun, S. and Megan, M.M., 2006. Silver bromide nanoparticle/polymer composites: dual action tunable antimicrobial materials. *Journal of American Chemical Society*, 128(30): 9798–9808.

Rajesh, R., Naik, S.J., Stringer, G.A., Sharon, E.J. and Morely, O., 2002. Actin–binding proteins sensitively mediate F–actin bundle stiffness. *Nature Materials*, 5: 748–753.

Rassaei, L., Sillanpaa, M., French, R.W., Compton, R.G. and Marken, F., 2008. Arsenite determination in the presence of phosphate at electro aggregated gold nanoparticle deposits. *Electroanalysis*, 20: 1286–1292.

Sabbani, S., Gallego–Perez, D., Nagy, A., James Waldman, W., Hansford, D. and Dutta, K.P., 2010. Synthesis of silver–zeolite films on micropatterned porous alumina and its application as an antimicrobial substrate. *Microporous and Mesoporous Materials*, 135: 131–136.

Salton, M.R.J. and Kim, K.S., 1996. *Structure Baron's Medical Microbiology*, 4th edn., Galveston: University of Texas Medical Branch.

Shankar, S.S., Rai, A., Ahmad, A. and Sastry, M., 2004. Rapid synthesis of Au, Ag, and bimetallic Au core–Ag shell nanoparticles using neem (*Azadirachta indica*) leaf broth. *Journal of Colloidal Interface Science*, 275: 496–502.

Sharma, S.P., James, T.J., Goyal, R., Singh, B., Kochar, R. and Gupta, S., 1991. Effect of geriforte, an ayurvedic drug, on the spinal cord and heart of adult wistar rats. *Indian Journal of Gerontology*, 2(2): 24–31.

Sharma, S.P., Pandey, V., James, T.J. and Singh, B., 1992. Lipofuscinolytic effect of geriforte: fluorescent microscopical study. *Indian Journal of Gerontology*, 2(4): 32–38.

Shrivastava, S., Bera, T., Roy, A., Singh, G., Ramachandrarao, P. and Dash, D., 2007. Characterization of enhanced antibacterial effect of novel silver nanoparticles. *Nanotechnology*, 18: 225103.

Singh, N., Nath, R., Mishra, N. and Kohli, R.P., 1978. An experimental evaluation of anti–stress effect of geriforte (an ayurvedic drug). *Quarterly Journal of Crude Drug Research*, 3: 125–130.

Sondi, I. and Salopek–Sondi, B., 2004. Silver nanoparticles as antimicrobial agent: a case study on *E. coli* as a model for gram–negative bacteria. *Journal of Colloid Interface Science*, 275: 177–182.

Song, H.Y., Ko, K.K., Oh, I.H. and Lee, B.T., 2006. Fabrication of silver nanoparticles and their antimicrobial mechanisms. *Journal of European Cells and Material*, 11: 46–58.

Sreeram, K.J., Nidin, M. and Nair, B.U., 2008. Microwave assisted template synthesis of silver nanoparticles. *Bulletin of Material Science*, 31(7): 937–942.

Starowicz, M., Stypula, B. and Banas, J., 2006. Electrochemical synthesis of silver nanoparticles. *Electrochemical Communication*, 8(2): 227–230.

Su, H.L., Chou, C.C., Hung, D.J., Lin, S.H., Pao, I.C., Lin, J.H., Huang, F.L., Dong, R.X. and Lin, J.J., 2009. The disruption of bacterial membrane integrity through ROS generation induced by nanohybrids of silver and clay. *Biomaterials*, 30(30): 5979–5987.

Sui, Z.M., Chen, X., Wang, L.Y., Xu, L.M., Zhuang, W.C., Chai, Y.C. and Yang, C.J., 2006. Capping effect of CTAB on positively charged Ag nanoparticles. *Physica, 33*: 308–14

Upadhyaya, L., Shukla, S.S., Agarwal, A. and Dubey, G.P., 1988. Changes in brain biogenic amines under influence of an indigenous drug. Geriforte, following immobilization stress. *Indian Journal of Experimental Biology*, 26: 911–916.

Upadhyaya, L., Tiwari, A.K., Agrawal, A. and Dubey, G.P., 1990. Role of an indigenous drug geriforte on blood levels of biogenic amines and its significance in the treatment of anxiety neurosis. *Activitas Nervosa Superior*, 32: 1–4.

Vilchis–Nestor, A.R., Sanchez, M.V., Camacho–Lopez, M.A. and Gomez, R.M., 2008. Solventless synthesis and optical properties of Au and Ag nanoparticles using *Camellia sinensis* extract. *Mater Letter*, 62: 3103–3105.

Xu, H. and Kall, M., 2002. Surface-plasmon-enhanced optical forces in silver nanoaggregates. *Physics Review Letter*, 89: 246802.

Yamanaka, M., Hura, K. and Kudo, J., 2005. Bactericidal Actions of a Silver Ion Solution on *Escherichia coli*, Studied by Energy–Filtering Transmission Electron Microscopy and Proteomic Analysis. *Applied and Environmental Microbiology*, 71: 7589–7593.

Chapter 2

Bioremediation of Heavy Metal Pollution from Industrial Waste Water by Microorganisms

Sabyasachi Chatterjee, Sikha Dutta and Narayan Chandra Chatterjee*

Department of Botany, Burdwan University, Burdwan- 713104, West Bengal, India

ABSTRACT

Contamination of water and sediments with heavy metals is a major environmental problem all over the world (Baldrian *et al.*, 2000; Gavrilescu, 2004; Malik, 2004; Srivastava and Thakur, 2006). Metal pollutants are usually produced through industrial processes such as mining, refining and electroplating and are released by effluents generated from various industries such as metal finishing industries, metallurgy, tannery, battery manufacturing industries, glass, plastic industries etc. Bioremediation technology uses microorganisms to reduce, eliminate, or to transform metal-contaminants present in soils, sediments, water and air. Bioremediation was used over 100 years ago with the opening of the first biological sewage treatment plants in Sussex, UK (Nabir Primer, 2003). However, the word 'bioremediation' is fairly new; first appearing in a

* Corresponding author: E-mail: schatterjee227@gmail.com

peer-reviewed scientific literature in 1987 (Nabir Primer, 2003). It has been known that bioremediation can be a cost efficient and reliable method for removing hazardous waste from heavily contaminated sites with organic compounds. Even more recent researches have revealed that microbial processes may well be used to cleanup radioactive and metallic contaminants.

Keywords: *Heavy metals, Bioremediation, Clean up, Effluents, Transform.*

Introduction

Contamination of water and sediments with heavy metals is a major environmental problem all over the world (Baldrian *et al.*, 2000; Gavrilescu, 2004; Malik, 2004; Srivastava and Thakur, 2006). Metal pollutants are usually produced through industrial processes such as mining, refining and electroplating and are released by effluents generated from various industries such as metal finishing industries, metallurgy, tannery, battery manufacturing industries, glass, plastic industries etc. Bioremediation technology uses microorganisms to reduce, eliminate, or to transform metal-contaminants present in soils, sediments, water and air. Bioremediation was used over 100 years ago with the opening of the first biological sewage treatment plants in Sussex, UK (Nabir Primer, 2003). However, the word 'bioremediation' is fairly new; first appearing in a peer-reviewed scientific literature in 1987 (Nabir Primer, 2003). It has been known that bioremediation can be a cost efficient and reliable method for removing hazardous waste from heavily contaminated sites with organic compounds. Even more recent researches have revealed that microbial processes may well be used to cleanup radioactive and metallic contaminants.

Heavy metals are non biodegradable pollutants that is why they can be transformed only through absorption, methylation, complexation and changes in valence state. These transformations affect the mobility and bioavailability of metals. At low concentrations, metals can serve as important components in life processes, often serving important functions in enzyme activity. However, above certain threshold concentrations, metals may certainly be toxic to the biological system. Introduction of heavy metal compounds into the environment generally induces morphological, cytological and physiological changes in the microbial communities hence exerting a selective pressure on the microorganisms (Verma *et al.*, 2001). Generally, the sites contaminated with heavy metals are the sources of metal resistant microorganisms (Gadd, 1993). Fortunately, microorganisms can affect the reactivity and mobility of metals and thus can be used to detoxify some metals and prevent further metal contamination.

Staphylococcus, Bacillus, Pseudomonas, Citrobacteria, Klebsiella, Rhodococcus, Rhodotorula and *Pichia* are the microorganisms that are commonly used in bioremediation processes which include bio-augmentation in which microbes and nutrients are added to supplement the intrinsic microbes of the site and biostimulation in which nutrients and enzymes are added to supplement the intrinsic microbes of the site. Metal contaminants are not just at superfund sites. Sites managed by the Department of Defense and the Department of energy are also contaminated with heavy metals. These metal contaminants pose adverse health effects to those who live near the polluted sites. Metal waste is commonly found in soil, sediments and water. Breathing, eating, drinking and skin contact are all possible exposure routes for metal contaminants. Metals such as mercury, lead and arsenic potentially can be toxic to the kidneys, decrease mental capabilities and cause weakness, headaches, abdominal cramps, diarrhoea and anaemia. Chronic exposure to these pollutants can cause permanent kidney and brain damage (Adebowale Adeniji, 2004). *In situ* bioremediation is a technique that can be used to reduce the spread of metal contaminants by applying biological treatment to hazardous chemicals in soil and ground water (Connor *et al.*, 1988). *In situ* bioremediation has the ability to transform contaminants to less toxic compounds, making this is a promising environmental cleanup technique. It accelerates contaminant desorption and dissolution by treating pollutants close to their source.

Toxicokinetics of Metals

We should be aware of the ways by which our body gets exposed to heavy metals, the mechanism by which they cause toxicity is discussed below:

Aluminum (Al^{+3})

Aluminum (Al^{+3}) is present in antacids, baking powder, beer, cigarette filters, dental amalgams, toothpaste etc. is causes neurofibrillary degeneration and neuro-degeneration. This can lead to diseases like Alzheimer's, anaemia, dementia, N-M disorders, and Parkinson's disease, kidney and spleen disorders (Ray and Ray, 2009).

Arsenic (As^{+3})

Arsenic (As^{+3}) is used as colouring agent in wall paper and toys, in insecticides and rat poisons and to protect timber and hide from termites. Its toxicity increases the permeability of small blood vessels, causes necrosis of intestinal mucosa and interferes with mitochondrial enzyme system. The effect of arsenic toxicity is abdominal pain, anorexia, vomiting, liver/lung/skin cancers. Napolean Bonaparte, emperor of France was rumored to have been effected by arsenic poisoning (Ray and Ray, 2009).

Cadmium (Cd^{+2})

Cadmium (Cd^{+2}) is found in batteries, cigarette smoke (1 cigarette contains 1-2 micrograms of cadmium), incineration of tyres/rubber/plastic, vending machine soft drinks. Their toxicity causes liver and kidney damage, prostate dysfunction, bone diseases and cancer (Ray and Ray, 2009).

Mercury (Hg^{+2})

Mercury (Hg^{+2}) enters the body to dental amalgams, felt, laxatives, tattooing etc. and can cause adrenal dysfunction, damage to the central nervous system, hypothyroidism and nerve fiber degeneration. Most sad-full happening of mercury poisoning is the occurrence of 'Minamata' disease in Japan (Harada M,1968).

Nickel (Ni^{+6})

Nickel (Ni^{+6}) is present in hydrogenated fats and oils, stainless cookware, tea, tobacco smoke. Nickel toxicity causes kidney dysfunction, disruption of hormone and lipid metabolism and intestinal cancer (Ray and Ray, 2009).

Lead (Pb^{+2})

Lead (Pb^{+2}) is present in auto exhaust, hair dyes, in paints, in sindoor (vermillion), in surma (collyrium), glazing of pottery and enamel ware. Lead toxicity causes co-ordinatioin and concentration lost, decreases I.Q, causes infertility, memory loss, especially long term memory, causes mood swing and sterility. Heavy metal overload in the environment causes growth of stealth pathogens and also combinations of certain metals increases toxicity while other combinations reduces it *e.g.*, lead makes mercury 100 times more toxic, while Zn and Mg reduces cadmium toxicity (Ray and Ray, 2009).

One of the major mechanisms behind heavy metal toxicity has been attributed to oxidative stress. The metals are capable of interacting with nuclear proteins and DNA causing oxidative deterioration of biological macromolecules (Adebowale Adeniji, 2004). Studies show that heavy metal possesses the ability to generate reactive radicals resulting in cellular damage to lipid bilayer and DNA. These reactive radicals' species or reactive oxygen species (ROS) include a wide variety of radicals and chelate of amino acid, peptides and proteins complexed with toxic metals (Adebowale Adeniji, 2004). The risk assessments for an essential metal such as Mn, Zn, Fe, Se presents a challenge because of the need to consider the balance between essentiality and toxicity. For each metal there are two ranges of intake associated with adverse health effects. Intakes that are too low and can lead to nutritional toxicity and intakes that are too high and can lead to genotoxicity and immunotoxicity. Gloves, using protective breathing apparatus and eating food grown organically.

Chromium (Cr^{+6})

Chromium (Cr^{+6}) with its great economic importance in industrial use, is a major metal pollutant of the environment. The discharge of effluents by a variety of industries constitutes one of the major causes of land and water pollution by chromium compounds. The effluents from these industries contain Cr(III) and Cr(VI) at concentrations ranging from tenths to hundreds of mg/l. Hexavalent chromium species are considered as the most toxic form of chromium, whereas trivalent chromium is much less toxic. Because of these differences, the discharge of Cr(VI) into surface water is regulated to below 0.05 mg/l by the U.S. EPA, while total Cr (including Cr(III), Cr(VI) and its other forms) is regulated to below 2 mg/l.In reaching for these goals in pollution standards, many methods have been used for the removal of Cr(VI) from wastewaters including ion-exchange resins, filtration and chemical treatment. However, these methods are not completely satisfactory and have the following disadvantages:(i) generation of large amount of secondary waste products due to various reagents used in a series of treatments such as reduction of Cr(VI), neutralization of acidic solution and precipitation, and (ii) the instability of ion-exchange resins due to serious oxidation by Cr(VI). Thus, there is a need for the development of new cost effective methods that are more environmentally friendly. Ever since the first report of chromium reduction by bacteria was published in the late 1970s, a number of bacterial species have been isolated and shown to be capable of reducing Cr(VI) (Ray and Ray, 2009).

Species in the genus of *Pseudomonas, Escherichia, Bacillus, Enterobacter* and sulphate-reducing bacteria (SRB) including Desulfovibrio, Desulfomicrobium and Desulfotomaculum are being studied most extensively. Although it have been proposed that biological treatment could significantly reduce the costs for chemicals and energy compared with conventional physical and chemical treatment, there are still some problem in applying this technology to the real wastewater treatment process, *i.e.*, a low biological reducing rate compared with the high generation rate of wastewaters (above thousands tones per day), a cell death by high concentration of Cr(VI), the high cost of a bioreactor system, continuous supplies of costly nutrients and difficulty in separation of cells after treatment. As an alternative, various biomaterials have been developed as biosorbents for heavy metals (Park *et al.*, 2005). Fungi are easy to grow and produce high yield of biomass and at the same time can be manipulated genetically and morphologically. Fungal organisms are widely used in a variety of large-scale industrial fermentation processes (Kapoor and Viraraghavan, 1995). For example, strains of Aspergillus are used in the production of ferrichrome, gallic acid, citric acid and enzymes such as amylases, glucose isomerase, lipases and glucanases. *Saccharomyces cerevisiae* is extensively used in the food and beverage industries. Furthermore, since dead fungal biomass is of little use and is

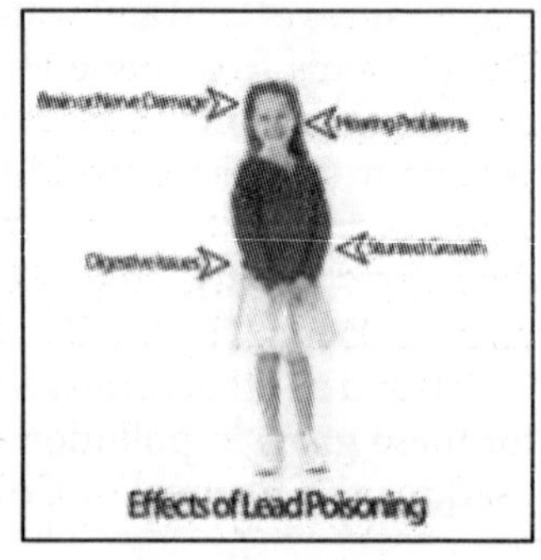
Effects of Lead Poisoning

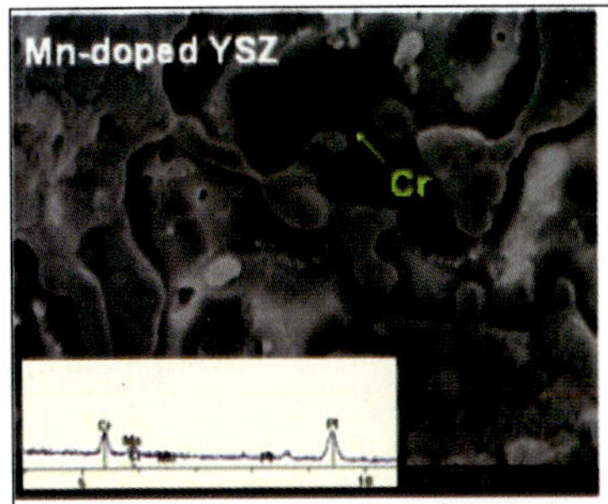

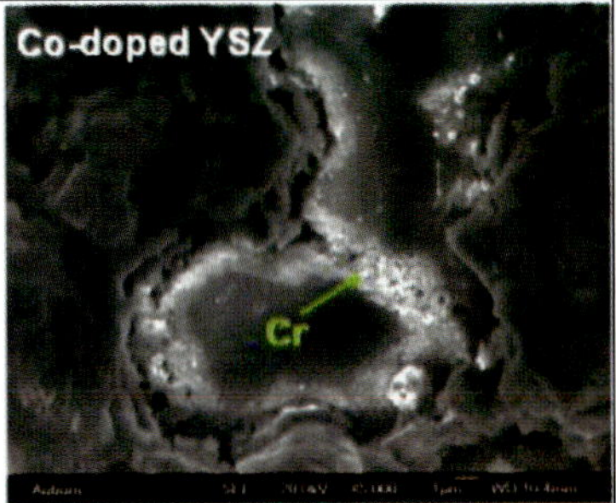

Chromium deposition on yttria-stabilzed zirconia (YSZ) with a Pt electrode after cathodic polarization in the presence of a chromia-containing atmosphere for 25 hours at 800°C.

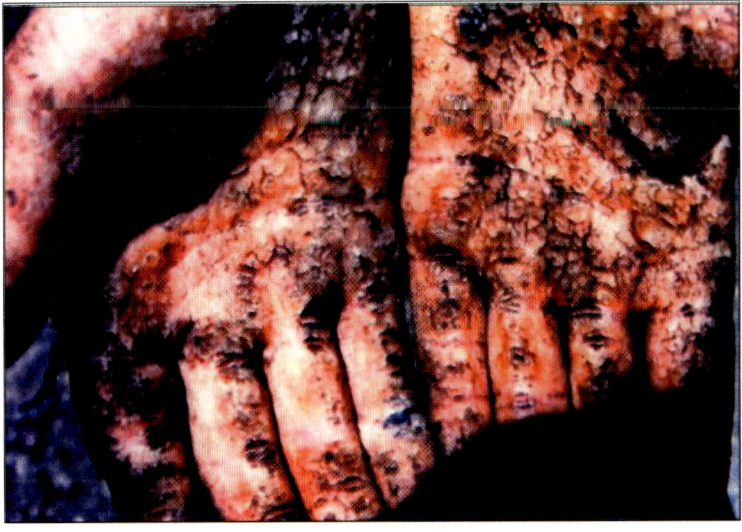

abundant, it may be good source of biomaterial for the removal of Cr(VI) from industrial wastewaters (Kapoor and Viraraghavan, 1995).

Until comparatively recently, most of studies claimed that the Cr(VI) was removed from aqueous solution through an "adsorption mechanism", that the anionic chromate ion binds to the positively charged groups such as amines of the dead fungal biomass (Kumar *et al.,* 2008). While Zhao and Duncan observed that a "partial

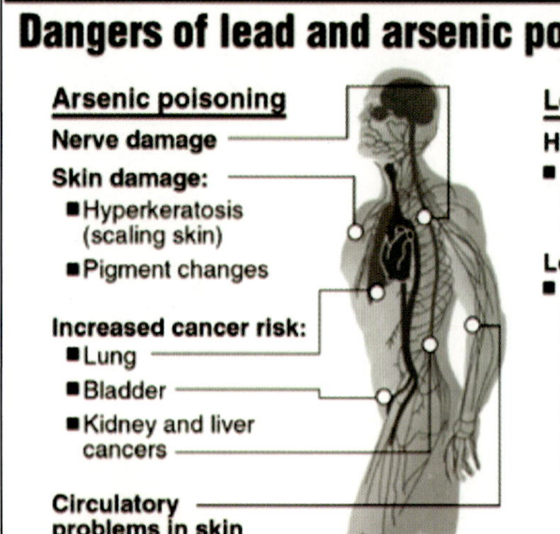

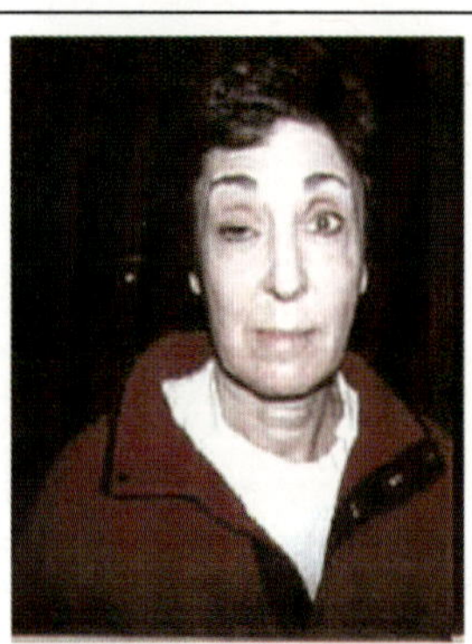

Freya Koss
March'1998
after having a tooth drilled
& filled with mercury amalgam

Freya Koss
April'2002
Four years after removal of
amalgam & detox

reduction along with anion adsorption" occurred in a column packed with formaldehyde cross-linked *S. cerevisiae*, they did not deal with the reduction phenomena in detail. Recently, scientists found that the dead fungal biomass of *A. niger* could reduce Cr(VI) to Cr(III) through a redox reaction which was not related with any enzyme activity (Park *et al.*, 2005). The objectives of this work were (i)

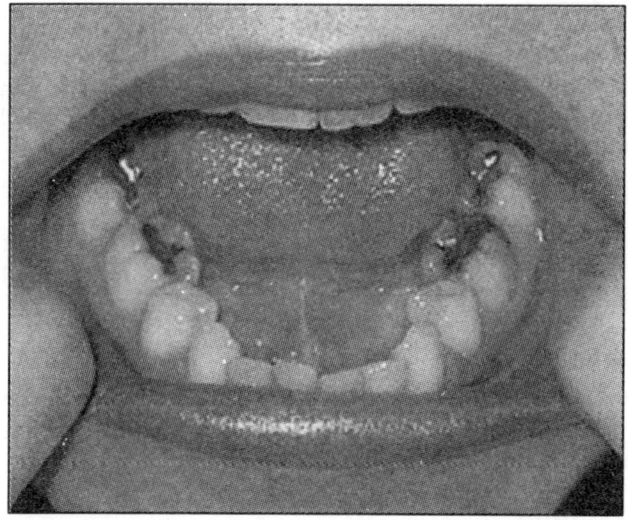

to screen the fungal biomass having high reducing capacity of Cr(VI), (ii) to investigate the effects of various parameters such as contact time, pH, initial Cr(VI) concentration and biomass concentration, and (iii) to develop a rate equation supporting our mechanism.

Types of Bioremediation

Bioremediation explores the genetic diversity and metabolic versality of microbes for the transformation of contaminants into less harmful end products which are then integrated into natural bio-geochemical cycles. The diversity of bioremediation system depends on the constituent of contaminants and it involves a multidisciplinary approach requiring knowledge from individuals with expertise in chemistry, microbiology, geology, environmental engineering, chemical engineering, soil science etc. Bioremediation techniques are of different types:

1) Bioaugmentation

Addition of bacterial cultures to a contaminated medium, frequently used in bioreactors and *ex situ* system. It is used when the proportion of degradative microbial flora to contaminant needs to be increased.

2) Biomineralization/Biocrystallization

The process whereby microbially generated ligands or microbially mediated changes in the cellular microbial environment cause precipitation of heavy metals as biomass-bound crystalline deposits.

3) Biostimulation

Stimulation of indigenous microbial population in soils and/ or ground water (*in situ* or *ex situ*). It can be applied when the addition of nutrients is necessary.

4) Bioattenuation

A method of monitoring the natural program of degradation to ensure that contaminant concentration decreases with time at relevant sampling point.

5) Bioventing

Treating contaminated soils by drawing oxygen through the soil to stimulate microbial growth and activity. It is needed when it becomes necessary to supply oxygen from air.

6) Biofilters

Use of microbial stripping columns to treat air emissions.

7) Bioreactors

A container/reactor used to treat liquid or slurries.

8) Composting

Aerobic, thermophilic treatment process in which contaminated material is mixed with a bulking agent.

9) Land Farming

Solid phase treatment system for contaminated soils, may be done *ex situ* or in a constructed soil treatment cell.

Bioremediation Mechanisms

Some of the technologies are fully developed and are being practiced as follows:

A) Biosorption

Biosorption is a property of certain types of inactive, dead microbial biomass to bind concentrate contaminants from even very dilute aqueous solutions *e.g.* heavy metals. Biomass that exhibits this property acts as a chemical substance (ion exchanger) of biological origin. The cell wall structure of certain algae, fungi, yeast and bacteria is responsible for this phenomenon, which is a passive process that requires no energy. Biosorption often results from the formation of metal-organic complexes with constituents of microbial cell walls, capsules, or extracellular polymers synthesized and excreted by the organisms. It may result from the positively charged metallic cation being retained electrostatically by negatively charged functional groups in the walls, capsules or polymers. The mechanism of sorption and the cell constituents involved in this process are usually unknown, although considerable research has been developed to trace the possible mechanisms. It involves mechanisms like ion exchange, chelation and complexation. Inorganic precipitation may occur by hydrolysis. Inorganic deposition occurs via adsorption by physical forces and ion entrapments in inner and intra-fibrillar capillaries and space of the structural polysaccharide network as a result of diffusion through cell walls and membranes (Vasudevan *et al.*, 2003). Several active groups of cell constituents like acetamido group of chitin, structural polysaccharide of fungi, amine (amino and peptidoglycosides), sulphahydral and carboxyl groups in protein, phosphodiester (teichoic acids), and phosphate, and hydroxyl in polysaccharides participate in biosorption.

B) Bioaccumulation

The active mode of accumulation of contaminants by living cells is usually referred to as bioaccumulation.This process is metabolically dependent and can be significantly affected by the contaminants especially the metals. The use of living systems certainly has application in a selected recovery system. However, the toxicity and complexity of many wastes may limit the use of living system. The mechanisms by which microorganisms actively accumulate include precipitation, intracellular accumulation and oxidation or reduction (Ahmad *et al.*, 2006). The concentration of a foreign substance within a biological system is dependent upon the rate of uptake, the duration of exposure and the rate at which it is being eliminated or reached upon by the system. The extent upto which a chemical shall be absorbed and bioaccumulated by an organism dependent on its solubility in fats or its lipophilicity.

C) Precipitation

Contaminants react with a product of microbial metabolism to yield water in soluble derivative. The removal of such precipitate constitutes remediation. Sulphides and phosphates are the common precipitates formed in microbes due to the production of H_2S from sulphates and inorganic phosphates from organic phosphate compounds. The H_2S/phosphates forms insoluble derivatives with a number of metallic ions.

D) Reductive Halogenation

This is potentially important in the detoxification of halogenated organic contaminants. The halogen atom (such as chlorine) of the contaminant molecule is replaced by a hydrogen atom due to catalytic reaction of microbes. Two electrons are added, and the contaminant becomes less toxic and susceptible to further microbial decay than the parent compounds. In most cases this reaction does not generate energy but is an incidental reaction that may benefit the cell by eliminating the toxicity of the compound.

E) Reduction

Microbes can bring about the reduction of a wide array of inorganic anions and cations. Nitrate, sulphate and carbonate are the non-metallic anions that are reduced microbiologically. The reductions convert a higher oxidation state of the element to a lower one, *e.g.*, Hg(II) to Hg(0), Fe(III) to Fe(II), Se(VI) to Se(0), Mn(IV) to Mn(II), and As(V) to As(III). The reduction changes the toxicity, water solubility and mobility of the element. In other words, one oxidation state is toxic but the reduced form has less or no toxicity at environmental conditions. An increase in water solubility and mobility can be exploited to bioremediate the insoluble form of an element in soil because the product of reduction would move out the solids into the water. A decrease in solubility of an element as a result of reduction can be used to remove it from surface to ground water. In some cases, the product of reduction is volatile, as in the reduction of nitrate to N_2 or Hg(II) to Hg(0), and the loss of the element from soil or water is (as with nitrate) or could be (as with Hg) the basis for bioremediation (Alexander, 1977). The reduction reaction of microbes are catyalyzed enzymatically, but some result from the excretion by the organism of a reducing compound that abiotically converts the ion or metabolic compound of a higher to a lower oxidation state. The enzyme–catalysed reactions serve as terminal electron acceptors permitting growth, much as O_2 functions for aerobes, simple organic molecules for fermentative microbes, sulphate for *Desulfovibrio,* nitrate for denitrifying bacteria and CO_2 for methanogenic bacteria.

F) Solubilization/Oxidation

The oxidation reaction that has practical use is associated with the oxidation of sulphides. Fe(II) and elemental sulphur. Chemoautotrophic bacteria of the genus *Thiobacillus ferroxidans* use sulphides and Fe(II) as its energy source present in the ores and produce H_2SO_4 and Fe(III).The acidification process can be used in a beneficial way under a controlled or managed process known as bioleaching (Brierly, 1989).Though several microbes have been designed for bioremediation, the physical potential of microbial population to remediate environment of relevant size, homogeneity and variability has not been adequately tested.

Metabolic Process Involved in Bioremediation

Microbes that live virtually everywhere are the vital components for bioremediation. Microbial enzymes act as catalysts in degradative (catabolic) reactions that provide energy and material for synthesis of additional microbial cells. In general the biochemical process can be divided into two groups: fermentation and respiration. (Microbial Bioremediation by Rajendran and Gunasekaran, 2006, MJP Publishers). Fermentation metabolism is characterized by production of a mixture of end products with different levels of oxidation, such as organic acids, alcohols, hydrogen and carbon dioxide. Co-metabolism (synergistic transformation): Microbes can transform contaminants even though the conversion reaction yields no benefit to the cell. The non-beneficial bioremediation is termed co-metabolism/secondary utilization. It can be defined as degradation of a compound only in the presence of other organic material that serves as the primary energy source, *e.g.* polynuclear aromatic hydrocarbons, halogenated aliphatic and aromatic hydrocarbons and pesticides. A co-metabolite does not support the growth of the organism concerned and the products of transformation are accumulated stoichiometrically. The transformation does not require energy and it is a very slow process. Co-metabolism occurs since the enzymes produced by organism for metabolic activities of their major carbon source are not substrate specific and can act on other compounds.

Microbial Interaction for Bioremediation Optimization

Microbes exhibit positive (commensalisms, synergism and mutualism) and negative (competition, amensalism, parasitism and predation) interactions that result in growth, survival and metabolism that are essential for bioremediation.

1) Commensalism

Many contaminants are co-metabolized, a process in which microbes transform the contaminants without being able to use the energy derived from the metabolism to support growth. Co-metabolizing microbes may interact with other community members commensalistically, since co-metabolites serve as substrates for other microbes. Co-metabolism is the predominant mechanism for the transformation of many substrates, *e.g.* polychlorinated biphenyls (PCBs). Analogue enrichment and cross-accumulated are the bioremedial approaches developed after understanding co-metabolism. In analogue enrichment, ecosystem may be enriched for co-metabolizing microbes by the addition of biodegradable compounds. In cross-accumulation, microbes acclimated to the biodegradation of a particular substrate can be acclimated to another contaminant that is structurally related. These approaches are advantageous since they can reduce the length of time between the first exposure to the contaminant and the onset of biodegradation.

2) Synergism

Microbes interacting syntrophically often supply the nutritional needs of each other. Syntrophism is common in anaerobic microbial communities, possibly due to the small enthalpy changes associated with most anaerobic bioconversions. The excretion of metabolic end products with high free-energy content creates a niche for other organisms to exploit and this can result in community succession. Microbial interaction can have a thermodynamic basis, as in anaerobic microbial consortia that transfer hydrogen. Interspecific hydrogen transfer forms the basis for syntrophic interactions, *e.g.* metabolism of 3-chlorobenzoate.

3) Competition

Degradative microbes face competition with other organisms for nutritional and physical resources. Due to this competition, the expected bioconversion does not occur, even though the relevant degradative microbe is found in the environment. The dominance of one species over another can be due to its faster growth rate under the imposed selection pressure or its greater affinity for a substance, nutrient or physical resource. An understanding of the community structure is important for predicting the fate of contaminants in the environment and the remedial attempts.

Strategies for the Improvement of Bioremediation Techniques

1. Addition of air or oxygen as the terminal electron acceptor for catabolic activity of microbes involved in bioremediation process.
2. Since the oxygen supplied to the *in situ* process is limited, pump and treat option improves biodegradation.
3. As the water solubility of alternative electron acceptors like nitrate, sulphate, etc. are high; they have the potential to improve electron acceptor bioactivity.

Generally microbial biomasses have evolved various measures to respond to stresses *via* processes such as transport across the cell membrane, biosorption to cell walls, entrapment in extracellular capsules, as well as precipitation and transformation of metals (Malik, 2004). Vadkertiova and Slavikova (2006) have studied the metal tolerance property of Yeast isolated from polluted environments and found that there exist some intraspecific and interspecific variations in the metal tolerance amongst the tested strains. Lead nitrate could be entrapped into the binding sites present in the exopolysaccharide of *Rhodotorula mucilaginosa* by biosorption (Das *et al.*, 2008).

It has been found that most of the natural strains were more tolerant to chromium and were able to grow in the presence of 5mM Cr (III) or 0.5mM Cr (VI), that is at concentrations which substantially inhibited the growth of laboratory strains. The cellular Cr content after treatment was similar for both strain types and ranged from 1.2–4.0 mg/g d.w. and 0.4–0.9 mg/g d.w., for Cr (III) and Cr (VI) forms, respectively; however, in one case of a natural strain it reached the value of 10 mg Cr (III)/g dry mass. Natural-source strains were grouped into four groups based on the yeasts' differential response to Cr (III) and Cr (VI). Hexavalent Cr-resistant mutants of a *P. guilliermondii* laboratory strain, which revealed markedly changed capabilities of chromium accumulation, were obtained by means of UV-induced mutagenesis. Cr (VI) treatment triggered over synthesis of riboflavin and the addition of exogenous riboflavin increased *P. guilliermondii* resistance to both Cr (III) and Cr (VI). Electrophoretic protein profiles revealed the induction and/or suppression of several proteins in response to toxic Cr (VI) levels (Ksheminska *et al.*, 2004). Organisms respond to heavy metal stress using different defence systems, such as exclusion, compartmentalization, making complexes and the synthesis of binding proteins such as metallothioneins (MTs) or phytochelatins (PCs). Organisms with unique abilities of metal absorption, accumulation or resistance can be searched for among naturally occurring organisms. Alternatively, these systems can be utilized in engineering bacteria and plants for the remediation of polluted waters and soils. The introduction and/or over expression of metal binding proteins have been widely exploited to increase the metal binding capacity, tolerance or accumulation of bacteria and plants. Modification in the biosynthesis of PCs in plants has recently been accomplished to enhance the metal accumulation whereas in bacteria, various peptides consisting of metal-binding amino acids (mainly histidine and cysteine residues) have been studied for enhanced heavy metal accumulation by bacteria.The mechanisms involved in bacterial metal resistance result from either the active efflux pumping of the toxic metal out of the cell, or the enzymatic detoxification (generally redox chemistry) converting a toxic ion into a less toxic or less available metal ion. Occasionally, bioaccumulation or sequestration of the metal in a physiologically inaccessible form is used. The genes encoding these resistance systems are most often located on plasmids or transposons. Detoxification of metals by the formation of complexes is used by most eukaryotes. MTs are low molecular weight (6–7 kDa), cysteine-rich proteins found in animals, higher plants, eukaryotic microorganisms and some prokaryotes. They are divided into three different classes on the basis of their cysteine content and structure. The Cys-Cys, Cys-X-Cys and Cys-X-X-Cys motifs (in which X denotes any amino acid) are characteristic and invariant for metallothioneins. No aromatic amino acids or histidines are found in MTs. Mammalian MTs comprise class I MTs, which are usually composed of 61 amino acids, of which 20 are cysteine residues. They contain two distinct domains that coordinate seven divalent metal ions or monovalent metal ions. MTs found in yeast, cyanobacteria and a few higher plants are also low molecular weight proteins with a high cysteine content but the cysteines are distributed differently than they are in mammalian MTs; these proteins are therefore designated class II. The biosynthesis of MTsis regulated at the transcriptional level and is induced by several factors such as hormones, cytotoxic agents and metals including Cd, zinc (Zn), mercury (Hg), copper (Cu), gold (Au), silver (Ag), cobalt (Co), nickel (Ni) and bismuth (Bi). Non-essential metals, such as Cd, Hg, lead (Pb), Bi, Ag, Au and platinum (Pt), are sequestered by MTs (Kägi, J.H.R.,1991).The only prokaryotic MT identified so far is found in a few cyanobacterial strains of the genus *Synechococcus*. This MT, which is encoded by the *smtA* gene, contains fewer cysteine residues than mammalian MTs (nine compared with 20; Ji. and Silver,1995). Possible mechanisms that govern heavy metal tolerance in plant cells are: (1) metal binding to the cell wall; (2) reduced transport across the cell membrane; (3) active efflux; (4) compartmentalization and (5) chelation. The prevailing mechanism is chelation through the induction of metal-binding peptides and the formation of metal complexes. The members of the third class of MTs, the PCs, are responsible for the formation of complexes with heavy metals in plants (Mejáre and Bülow, 2001). These peptides are enzymatically derived and synthesized on exposure of the cell to toxic metals. PCs are induced in all autotrophic plants analysed so far, as well as in some fungi. The structure of PCs is (γ-Glu-Cys)nX, in which X is Gly, γ-Ala, Ser or Glu and n = 2–11 depending on the organism, although the most common forms have 2–4 peptides. The biosynthesis of PCs is induced by many metals including Cd, Hg, Ag, Cu, Ni, Au, Pb and Zn; however, Cd is by far the strongest inducer. The metal binds to the constitutively expressed enzyme, γ-glutamylcysteinyl dipeptidyl transpeptidase (PC synthase), thereby activating it to catalyse the conversion of glutathione (GSH) to phytochelatin. Glutathione, the substrate for PC synthase, is synthesized from its constituent amino acids in two steps. The first step is catalysed by γ-glutamyl-Cys synthetase (γ-ECS) and the second by glutathione synthetase (GS). γ -ECS is feedback regulated by glutathione and is dependent on the availability of cysteine. Plant MT-like genes have been isolated from several plant species including maize, soybean, rice, wheat, tobacco and *Brassica napus* although their role in metal detoxification has yet to be established. Type I MT-like genes are expressed predominantly in the roots whereas type II MT-like genes are expressed primarily in the leaves.

Expression of Heterologous MTs in Bacteria

In the aim of enhancing the metal tolerance, sequestration or accumulation of bacteria, the high metal-binding capacity of MTs has been widely exploited. MTs from various sources have been expressed intracellularly in *Escherichia coli* including monkey MT (Murooka and Nagaoka, 1987), yeast MT (Berka *et al.*, 1988), human MT-II (Odawara *et al.*, 1995), mouse MT-I (Hou *et al.*, 1988), rainbow trout MT (Kille *et al.*, 1990), plant MT (Kille *et al.*, 1991) and pea PsMTA (Tommey *et al.*, 1991). Intracellular expression of MTs is, however, not devoid of complications and in many instances there have been problems with the stability and short half-life of the expressed heterologous proteins. The cysteine content of MTs is very high, which might interfere with cellular redox pathways in the cytosol. However, the expression of a fusion protein of human MT and are B increased the stability of MT (Romeyer *et al.*, 1988). *Escherichia coli* expressing the fusion protein accumulated five times more Cd than control cells, although no increased metal resistance was observed. Only when human MT-II was expressed intracellularly in *E. coli* was an increase in the metal resistance of the cells reported. One approach to circumvent this problem is to express the metal binders on the cell surface where they will not be able to interfere with intracellular activities. Expression of MTs on the surface of *E. coli* has been successful and in several instances an increased Cd accumulation by the cells expressing MTs has been observed, compared with control cells lacking the MT29–35. The metal-binding proteins have been anchored to the LamB and the OmpA, proteins that span the outer membrane, or to PAL, a protein that is anchored to the peptidoglycan layer. Yeast (CUP1) and mammalian (HMT1A) MTs expressed on the surface of *E. coli* as fusions to Lam B enhanced the metal binding capacity of the cells between 15 and 20-fold. The *Neurospora crassa* MT was expressed in 1–12 repeats as a fusion with the maltose-binding protein in the periplasm of *E. coli* and the Cd-binding capacity was increased tenfold with one repeat and a maximum 65-fold with nine repeats. The metal accumulation of *E. coli* cells was compared when only the α- or β-domains of a mammalian MT were expressed on the cell surface as fusions to Lam B (Kotrba *et al.*, 1999). Surface expression of the α-domain alone has shown to be more effective in metal accumulation in the cell than expression of the entire MT. A 17-fold increase in Cd accumulation by cells that express the α-domain compared with control cells was achieved, whereas cells that express the β-domain exhibited a three-fold increase in Cd accumulation. In addition, cells expressing only the α-domain accumulated more Cd than cells expressing the entire MT protein when grown in 30μM cadmium chloride ($CdCl_2$).The use of different outer membrane proteins (OMPs) as anchors for MTs at the cell surface has been investigated by Valls *et al.*, who found Lam B to be the best candidate for sequestering Cd (Valls *et al.*, 1998). Moreover, surface display is preferred if non-viable cells are to be used for bioremediation and if the biomass is to be recycled by desorption.

Expression of Metal-Binding Peptides

Novel metal binding peptides might offer a higher affinity, higher metal-binding capacity and/or specificity and selectivity, for a target metal ion than known metal-binding proteins. Peptides with unique binding properties can either be designed *de novo* or selected by screening peptide libraries. Several metal-binding peptides have been studied with the aim of increasing the Cd resistance or accumulation by *E. coli* cells. Naturally occurring Cd-binding proteins and peptides, such as MTs and PCs, are very rich in cysteine residues. In addition, histidines are known to have high affinity for transition metal ions such as Zn^{2+}, Co^{2+}, Ni^{2+} and Cu^{2+}. Because of its position in the periodic table Zn2+ bears a resemblance to Cd^{2+} (Therefore, various peptides comprising different sequences of cysteines or histidines have been exploited for binding Cd. Recently, metal-binding peptides that contain either histidines (GHHPHG)2 (HP) or cysteines (GCGCPCGCG) (CP) were engineered to LamB and expressed on the surface of *E. coli*. Surface display of CP and HP increased the bioaccumulation four-fold and two-fold, respectively. Expression of a repetitive cysteine rich metal-binding motif (CGCCG)3 as a fusion with the maltose-binding protein in *E. coli* enhanced $Cd^{2+}$and Hg^{2+} binding 10-fold. Moreover, a His6 peptide has been expressed on the surface of *E. coli* as a fusion to the

OMP LamB. This resulted in a five-fold increase in Cd accumulation when the peptide was expressed as a single copy and an 11-fold increase when expressed in tandem39. However, no increase in metal tolerance was observed. Following the screening of a peptide library for peptides with affinity for Cd, a hexapeptide (CdBP) containing no cysteines and only one histidine was selected. This hexapeptide (HSQKVF) was expressed as a fusion to the surface exposed area of the OmpA in *E. coli* and provided the cells with a higher tolerance to toxic concentrations of Cd, compared with control cells. The fusion of this peptide in quadruplicate to the N-terminal of either lactate dehydrogenase (LDH) or green fluorescent protein(GFP) resulted in increased tolerance of the cells to toxic Cd concentrations when expressed intracellularly. A hexapeptide that contained four histidines (SYHHHH) was selected for binding affinity to zinc and was expressed on the surface of *E. coli.* The expression of this peptide in *E. coli* did not render these cells more tolerant to Cd than those expressing CdBP on the cell surface, thus indicating that a peptide consisting of only one histidine in combination with other specific amino acids might be just as good as a multi-histidine peptide, as a metal chelator. A summary of Cd accumulation and tolerance obtained when expressing metal binding peptides and MTs on the surface of *E. coli.* In some cases the expression of Cd-binding peptides, or proteins in *E. coli,* has revealed a higher tolerance to Cd, and an increased accumulation of the cells in other instances. Thus, it has been argued that cell viability and bioadsorption are independent properties. It is possible that, for the intracellular expression of metal-chelating peptides, the use of peptides devoid of cysteines is less toxic to the cell compared with expression of cysteine-rich MTs. This could be the case, especially if the peptide is under the control of a constitutive promoter and not regulated by the metal ion concentration within the cell. Metal-binding peptides and MTs have mainly been exploited to increase the Cd tolerance or accumulation of *E. coli* cells. It would be of interest to evaluate Cd-binding peptides and proteins, engineered into more environmentally robust bacteria, such as *Pseudomonas,* for their potential use in bioremediation. The tolerance to Cr varied depending on the strain tested and the yeast cultures proved to be generally more sensitive to Cr(VI) (concentration range: 0.1–0.5mM) than to Cr(III) (0.25–5mM). The levels of cellular Cr content ranged from 0.29 to 11.10 mg/g d.w. and 0.21–3.3 mg/g d.w. for Cr(III) and Cr(VI), respectively. Distribution diagrams of the cell-accumulated Cr were constructed for the tested strain population, and the general uptake tendency of middle-range amounts of Cr (III), and low-range levels of Cr (VI) was revealed. The cell-accumulated Cr levels were similar at identical, non-toxic concentrations of either Cr form supplemented to the medium. Electron microscopic images proved that cytoplasm and cellular organelles were the ultimate targets for accumulation of both valences of the metal. The extreme cases of the strains revealing either the lowest or the highest Cr tolerance and uptake capabilities are discussed in terms of possible bioremediation mechanisms (Ksheminska *et al.*, 2004).

Acknowledgements

The corresponding author is grateful to Prof. Narayan Chandra Chatterjee, Dr. Sikha Dutta, Prof. Chandan Sengupta and Prof. Pranab Roy for their assistance and continuous support throughout the review process. He is also grateful to his parents, wife (Pubali Chatterjee), sister (Saheli Chatterjee) and Agniswar Sarkar (Research Scholar) for their continuous encouragement throughout the work.

References

Adeniji, Adebowale, 2004. *Bioremediation of Arsenic, Chromium, Lead and Mercury*. National Network of Environmental Management Studies Fellow prepared this paper for U.S Environment Protection Agency.

Alexander, M., 1977. *Introduction to Soil Microbiology*, 2nd edn. Wiley and Sons, New York.

Baldrian, P., Wiesche, C., Gabriel, J., Nerud, F.. and, Zadrazil, F., 2000. Influence of Cadmium and Mercury on activities of lignolytic enzymes and degradation of polycyclic aromatic hydrocarbons by *Pleurotus ostreatus* in soil. Journal of Applied and Environmental Microbiology, 66: 2471–2478.

Berka, T., Shatzman, A., Zimmerman, J., Strickler, J., Rosenberg, M. Efficient expression of the yeast metallothionein gene in *Escherichia coli*. *J. Bacteriol*, 170: 21–26.

Brierley, C.L., 1989. Bioremediation of metal contaminated surface and groundwater. *Geomicrobiology Journal*, 8: 201–223.

Connor, J.R., 1988. Case Study of Soil Venting. *Pollution Engineering*, 20(1).

Das, N., Vimala, R. and Karthika, P., 2008. Biosorption of heavy metals: An ovreview. *Indian Journal of Biotechnology*, 7: 159–169.

Gadd, G.M. and White, C., 1993. Microbial treatment of metal pollution: A working biotechnology. *Trends in Biotechnology*.

Gavrilescu, M., 2004. Removal of Heavy metals from the Environment by Biosorption. *Eng. Life Sci.*, 4(3): 219–232.

Harada, M., 1968. *Minamata Disease and the Mercury Pollution of the Globe.*

Hou, Y., 1988. Expression of the mouse metallothionein–I gene in *Escherichia coli*: increased tolerance to heavy metals. *Biochim. Biophys. Acta.*, 951: 230–234.

Ji, G. and Silver, S., 1995. Bacterial resistance mechanisms for heavy metals of environmental concern. *J. Ind. Microbiol.*, 14: 61–75.

Kägi, J.H.R., 1991.Overview of metallothioneins. *Methods Enzymol.*, 205: 613–626.

Kapoor, A. and Viraraghavan, T., 1995. Biosorption of heavy metals on *Aspergillus niger*: Effect of pretreatment. *Bioresorse Technology*, 63: 109–113.

Kille Peter, Stephens Paul, Cryer Anthony, and Kay John, 1990. The expression of a synthetic rainbow trout metallothionein gene in *E. coli*. *Biochim. Biophys. Acta*, 1048: 178–186.

Kille Peter, Stephens Paul, Cryer Anthony, and Kay John, 1991. A plant metallothionein produced in *E. coli*. *FEBS Lett.*, 295: 171–175.

Kotrba Pavel, Lucie Dolecková, Víctor de Lorenzo, and Tomas Ruml, 1999. Enhanced bioaccumulation of heavy metal ions by bacterial cells due to surface display of short metal binding peptides. *Appl. Environ. Microbiol.*, 65: 1092–1098.

Kotrba Pavel, Lucie Dolecková, Víctor de Lorenzo, and Tomas Ruml., 1999. Enhanced metallosorption of *Escherichia coli* cells due to surface display of beta- and alpha-domains of mammalian metallothionein as a fusion to LamB protein. *J. Receptor Signal Transduct. Res.*, 19: 703–715.

Ksheminska, H., Jaglarez, A., Federovych, D., Babyak, L., Yanovych, D., Kaszycki, P and Koloczek, H., 2004. Bioremediation of chromium by the yeast *Pichia guilliermondii*: Toxicity and accumulation of Cr(III) and Cr(VI) and the influence of riboflavin on Cr tolerance. *Microbiol. Res.*, 158: 59–63.

Kumar, R., Bishnoi, N.R.. and Garima, Bishnoi, K., 2008. Biosorption of chromium from aqueous solution and electroplating wastewater using fungal biomass. *Chem. Engg. J.*, 135(3): 202–208.

Liu, S. and Suflita, J.M., 1993. Ecology and evolution of microbial population for bioremediation. *Trends Biotechnol.*, 11(8): 344–352.

Malik, A., 2004. Metal bioremediation through growing cells. *Environ. Int.*, 30: 261–278.

Mejáre, M., Ljung, S. and Bülow, L., 1998. Selection of cadmium specific hexapeptides and their expression as OmpA fusion proteins in *Escherichia coli*. *Protein Eng.*, 11: 489–494.

Murooka, Y. and Nagaoka, T., 1987. Expression of cloned monkey metallothionein in *Escherichia coli*. *Appl. Environ. Microbiol.*, 53: 204–207.

Natural and Accelerated Bioremediation Research (NABIR) Program, 2003. Office of Biological and Environmental Research, Office of Science, US Department of Energy. *What is Bioremediation*? pp. 9.

Nedkovska, M. and Atanassov, A.I., 1998. Metallothionein genes and expression for heavy metal resistance. *Biotechnology*, 12: 11–16.

Odawara, F., Kurasaki, M., Suzuki-Kurasaki, M., Oikawa, S., Emoto, T., Yamasaki, F., Linde Arias, A.R. and Kojima, Y., 1995. Expression of human metallothionein–2 in *Escherichia coli*: Cadmiumtolerance of transformed cells. *J. Biochem.*, 118: 1131–1137.

Park, D., Yun, Y.S. and Park, J.M., 2005. Use of dead fungal biomass for the detoxification of hexavalent chromium: Screening and kinetics. *Process Biochem.*, 40: 2559–2565.

Rajendran, P. and Gunasekaran, P., 2006. *Microbial Bioremediation*. MJP Publishers, pp. 29–61.

Ray, S. and Ray, M. K., 2009. Bioremediation of heavy metal toxicity with special reference to chromium. *Al Ameen J. Med. Sci.*, 2(2): 57–63.

Srivastava, S. and Thakur, I.S., 2006. Biosorption Potency of *Aspergillus niger* for Removal of Chromium (VI). *Current Microbiology*, (53): 232–237.

Tommey, A.M., Shi, J., Lindsay, W.P., Urwin, P.E. and Robinson, N.J., 1991. Expression of the pea gene *PsMTA* in *E. coli*. *FEBS Lett.*, 292: 48–52.

Vadkertiova, R. and Slavikova, E., 2006. Metal tolerance of yeasts isolated from water, soil and plant environments. *J. Basic Microbiol.*, 46: 145–152.

Vasudevan, P., Padmavathy, V. and Dhingra, S.C., 2003. Kinetics of biosorption of cadmium on baker's yeast. *Bioresour. Technol.*, 89: 281–287.

Verma, T., Srinath, T., Gadpayle, R.U., Ramteke, P.W., Hans, R.K., and Garg, S.K., 2001. Chromate tolerant bacteria isolated from tannery effluent. *Bioresour. Technol.*, 78: 31–35.

Chapter 3

Time Activity Budget and Relative Frequency of Behaviour in *Diplonychus rusticus* (Insecta: Hemiptera)

*P. Venkatesan**

UGC Professor Emeritus in Zoology,
Loyola College, Chennai – 600 034, T.N., India

ABSTRACT

One of the most fascinating characteristics of aquatic insects as a group is their diverse and often intriguing behaviour. The behavioural repertoire of each insect consists of an entire collection of interacting 'ways of behaving that have been accumulated genetically through generation of selection. Such behavioural adaptation pronounced in aquatic insects may be categorized into *regulatory behaviour* that increases the control that an individual exerts over its own metabolic status, *reproductive behaviour* that is responsible for the successful continuation of life to the next generations and *foraging behaviour* that involves the gathering and processing of energy and other key resources from the environment.

The present study deals with the behavioural repertoire and calculates the time activity budget, relative frequency and

---

* E-mail: p_venkatesan@yahoo.com

all occurrences of selected behaviours in the water bug–*Diplonychus rusticus*. Results reveal that the bugs are more active during night than day hours. Males are more active than the females. The bugs spend greater portion of their time being inactive. Females have a greater tendency to cling on to plants while males prefer bottom of the container and rough corners of the tank. Next to resting, they spend considerable portion of time in respiration and searching for food. Clustering and hiding in plants are the protective strategies. The bugs are gregarious in nature and can also remain out of water for a period of time.

Keywords: *Behavioural repertoire, Time activity budget, All occurrence of selected behaviours, Water bug, Diplonychus rusticus.*

Introduction

One of the most fascinating characteristics of aquatic insects as a group is their diverse and often intriguing behaviour. The behavioural repertoire of each insect consists of an entire collection of interacting "ways of behaving" that have been accumulated genetically through generation of selection. Every individual behaviour pattern has been retained because–alone or in concert–it has improved for its bearer the chances of surviving in a dangerous environment. It is in this sense that we refer behaviour as an adaptation. The diversity of aquatic environment influences insect life. As for the apparent ingenuity of insect behaviour, this too is the product of natural selection operating in a milieu of harsh constraints and high reproductive potential.

Much of what has been written in the past about aquatic insect behaviour is anecdotal in nature. This is unfortunate because it tends to de-emphasize the fundamental importance of behavioural adaptation that helps to promulgate the myth that insect behaviour is somehow mechanical and uninteresting. On the contrary, the study of behaviour as a coherent system of adaptation is essential to a complete understanding of the ecology of aquatic insects.

Such behavioural adaptation pronounced in aquatic insects may be categorized into regulatory behaviour that increase the control that an individual exerts over its own metabolic status, reproductive behaviour that is responsible for the successful continuation of life to the next generation and foraging behaviour that involves the gathering and processing of energy and their key resources from the environment (Wiley and Kohler, 1984).

Behavioural Studies

Work has also been done on the behavioural aspects of the aquatic insects. Maier (1977) analyzed the diurnal activities, especially feeding of the water measurer *Hydrometra championa.* He also found that sizeable population of *Hydrometra championa* and *Tenagogonus quadrilineatus* co-existed in the same stream even though their diets overlapped. The co-existence was possible because of spatial resource partitioning which was reinforced by morphological adaptation. Experiments conducted by Scott and Murdoch (1983) on *Notonecta* suggest that the behaviour underlying *Notonecta*'s selective predation is inflexible.

Peckarsky (1984) reported that aquatic insects show a range of predatory behaviour and that they use visual, mechanical and chemical cues to detect the prey and said that hemipterans use a combination of cues while the use of visual cues is probably not more widespread. Bailey (1985) reported the multi-prey capture behaviour exhibited in *Ranatra dispar*, and added that the insect remains as a "Sit and Wait predator". Peckarsky and Penton (1985) conducted behavioural experiments in streams to see if predation in stonefly is affected by competition and it was found that the presence of competitor generally reduced the number of attacks and the time that predaceous stoneflies spent in eating mayflies. Most competitive or evasive behaviour by one or both stoneflies ultimately leads to avoidance of further contact.

Behaviour Responses in Belostomatids

Venkatesan and Ravisankar (1993) investigated the mode of copulation and the behavioural pattern of both sexes in nepids. Venkatesan (2011) has reported on moulting condition in *Diplonychus rusticus*. It was found that male *Laccotrephes griseus* always enjoys a free ride on the female. Copulation is frequent and random. It occurs more during the day than the night. Predatory behaviour of *Diplonychus rusticus* has been studied extensively. Cloarec (1989) described seven different categories of predatory behaviour in *Diplonychus indicus*, five include ambushing and no active search, simple capture, strike lunge and strikes following a preliminary vertical or horizontal orientation. Two include active search and dive and swimming bout.

First instar of *Diplonychus rusticus* possesses a behavioural plasticity similar to that of adults (Venkatesan, 2011). This highly versatile predator alternates frequently from foraging actively to ambushing and vice-versa (Venkatesan, 2005). Cloarec (1991) studied the similarities between the use of different predatory mode on young and old larvae by adults and found that changes per unit time from one predatory mode to the other increased with age as frequency of attempts increased with age. Adults changed predatory modes approximately seven times per hour; most changes were made after one, two or more rarely, three of one predatory attempts of the same mode. Cloarec (1995) in her observations on *Diplonychus rusticus* found that there is no relationship between locomotory activity and active foraging.

Brooding Behaviour

Belostomatid males are well known for carrying the eggs on their dorsum and extensive work has been done on their brooding behaviour. Brooding behaviour of a male water bug *Belostoma fluminium* was described by Smith (1976a). A variety of brooding behaviour includes keeping the eggs wet, frequently exposing to atmospheric air, maintaining an intermittent flow with the hind legs while below the surface. These behaviours enhance egg survival and development. Static eggs failed to develop when covered with water or left in open air. Male brooding behaviour of water bug *Abedus herberti* was also studied by Smith (1976b). He reported that probable costs of brooding to the individual male included reduced predatory efficiency. Increased exposure to predation reduced dispersal options and upper limits on reproductive potential in addition to direct caloric expenditures. These costs were buffered by the male's ability to abort brooding when conditions were sub-optimal. *Abedus herberti* execute brood pumping to move water over the eggs and aerate them while below the surface. Male brooding behaviour of *Diplonychus indicus* was described and interpreted in relation to phylogenetic significance by Venkatesan (1983). He reported that surfacing is a common brooding behaviour in males of all genera. In species of *Diplonychus*, sub-surface wandering, sub-surface resting and brood paddling were also observed. He further reported the brood-pumping in *Diplonychus* as observed by Smith (1976b) in *Abedus* but absent in *Belostoma*.

Behaviour of Normal Male and Female *Diplonychus indicus*

Normal males and females exhibit swift movements and are always found to move and cling to other individuals on their ventral sides. By these movements, they chase the prey and feed on them. They are voracious feeders, feeding on small aquatic insects such as larvae of dipterans and in the absence of prey they are cannibalistic. When the prey in the containers was changed, the insects were seen to be more active and take an average period of 3 minutes for their settlement on the substratum. In addition, on captivity, the insects attempt to escape from the light and remain in dark corners. During this period, they exhibit high rate of activity. When they are not active, they lurk amongst rushes and reeds at the edges of aquarium. In non- encumbered males, predation has been carried out by capturing prey and holding it between tarsus of the raptorial legs and hind legs vertically downwards. This prevents escape of the prey and ensures successful feeding of the prey. Also they surface the water for the acquisition of air in the form of air bubbles by protruding the respiratory straps to the surface (Venkatesan, 1978, 2005).

Aim of Present Study

Our knowledge on aquatic insects is still in the infancy. For the vast majority of species, we know only the broadest outlines of their behavioural strategies and we are entirely ignorant of most of the relevant details. Even among those few aquatic insects that have received detailed study, the information available seldom extends beyond a single facet of their behavioural repertoire, more often foraging and reproduction. Hence it has prompted the author to do work on the general behavioural strategies of *Diplonychus rusticus,* in order to prepare behavioural repertoire, and calculate the time activity budget, relative frequency and all occurrence of selected behaviours.

Materials and Methods

Collection and Maintenance of the Bug

The belostomatid bugs *Diplonychus rusticus* were collected from the edges of Cooum River, near Koyambedu, Chennai in early hours of the day. Bugs were collected using a hand net along with the vegetation such as *Hydrilla* sp. and *Eichhornia* sp. The vegetation was spread over the ground and the bugs clinging to their rootlets were hand picked. The early hours of the day was preferred for collection, since the bugs were observed in abundance near the edges of the river on the water surface.

The bugs were brought to the laboratory in a transporting medium. The transporting medium consisted water and vegetation taken in a plastic container with a few holes in the lid for aeration. In the laboratory, males and females were sorted out and transferred into separate aquaria containing aged tap water and vegetation to make the habitat as natural as possible. The bugs were regularly fed with *Culex* larvae. Water in the container was periodically changed and the dead ones were removed to avoid the fouling of the medium.

Experimental Set-Up

Two aquaria of 8" × 7" size were taken, filled with aged tap water and some vegetation. Ten male bugs were selected at random and were marked to observe them individually on their dorsal surface differently using fabric paint. Then the sampling methods were adopted to study their behavioural strategies. The experiment was then repeated with females also.

Sampling Methods

Three kinds of sampling methods were used in the study to observe various behavioural aspects.

Ad libitum Sampling

As *ad libitum* implies no restraints are employed in sampling behaviour with this method. Generally the behaviour of the chosen bugs, that were most easily observed were recorded as much as possible during the observation period and a descriptive catalogue of the behavioural repertoire of the bugs was prepared to get preliminary information.

Instantaneous Scanning

Using this method, data on the time distribution of behavioural states in an individual was obtained. The chosen bugs were scanned at predetermined points in time and their behavioural states were scored. Behavioural categories chosen for study were clearly delineated to assist in quick scoring. Percentage of time spent in various activities by the bugs was estimated (*i.e.*Time budget).

All Occurrences

Using this method, one or a limited number of behaviours was recorded and the rate of all occurrences of the selected behaviour was estimated. Observations were made for 30 minutes every day on males and females separately for 15 days. Data were taken on seven consecutive days. Observations were made during day and night hours to verify the diurnal variation, if any, in their activities. During night hours, very low intensity of light was used only at the time of observations. Care was taken to maintain the bugs in the natural condition as possible to avoid any disturbances.

Behavioural Repertoire

Different behavioural responses were observed in *D. rusticus* and their repertoire is given below:

- ☆ *Clinging to plants (CP)*–The bug during resting period was noted clinging to plants, anywhere from surface to bottom
- ☆ *Settling (SE)* -The bug after any activity settles in the bottom of the tank, any where from centre to edges.
- ☆ *Clustering (CU)*–The bug clusters in large numbers generally in the corners of the aquaria, sometimes along the edges.
- ☆ *Floating (FL)* -The bug floats aloof after active swimming, and during the time of respiration.
- ☆ *Passive rising (PR)*–The bug rises passively to the surface of water for air acquisition.
- ☆ *Surfacing (SU)*–Surfacing is the active swimming stage of the bug to the surface.
- ☆ *Bouting (BO)*–Bouting is observed when the bug searches for its prey. During bouting, it disturbs all the materials found in its vicinity.
- ☆ *Swimming (SW)*- Swimming is the common and predominant behaviour observed.
- ☆ *Mass swimming (MS)*–Sometimes the bug swims in groups.
- ☆ *Circling (CC)* -The bug at times swims in a circular pattern.
- ☆ *Crawling (CR)*–The bug crawls either on the plants or the surface of the aquarium.
- ☆ *Pausing (PU)*–After active swimming, the bug pauses to continue swimming or bouting.

- ☆ *Clinging to another (CA)*–The bug clings to each other generally ventral to ventral and they float, swim or remain static.
- ☆ *Approaching (AP)*–The bug approaches each other and either cling to each other or remain besides each other without contact.
- ☆ *Disturbing Another (DA)*–The bug as it swims or bouts dashes and disturbs the other bugs, then they either cling to a stem or leave them.
- ☆ *Being disturbed (BD)*–The bug is being disturbed by another.
- ☆ *To sit on another (ST)*–The bug rarely shows the tendency to sit on the other.
- ☆ *Being sat on by another (BS)*–Sometimes the bug is burdened by another sitting over it.
- ☆ *Avoiding (AV)*–Some bugs avoid the others when they try to cling on to them.
- ☆ *Disturbing materials (DM)*–The bugs disturb the dead leaves and other debris found in the bottom of the tank as they swim or bout.
- ☆ *Producing air bubble (PB)* -The bug produces air bubble and releases them through its respiratory siphon. When it is in the bottom of the tank it either releases the air bubble from its place, swim or rise up to surface to break the air bubble
- ☆ *Crawling and remaining out of water (OW)*–Some bugs tend to crawl out of the water along the sides of the aquaria for some time, which may last a few seconds to minutes
- ☆ *Falling back into water (FB)*–After crawling along the sides of aquaria (*i.e.* out of water) they fall back into water and they either float or swim swiftly down into the water.
- ☆ *Trying to catch (TC)*–The bug tries to catch the dead leaves and debris found in the bottom of aquaria.
- ☆ *Spurting fluid (SF)* -The bugs spurt fluid through their anal region.
- ☆ *Lingering (LI)*–Bugs linger when they remain in the same place and position for a longer period.
- ☆ *Jerking (JR)*–Some bugs show slight jerk when they are inactive for a very long period.

- *Partly Immersed (PI)*–Some bugs settle in the corners of the aquaria and they remain partly immersed in water with either anterior or posterior region immersed in water.
- *Hiding (HI)*–Most bugs show hiding behaviour; they either hide themselves inside the plants or hide behind other individuals.
- *Stirring (SR)*–Bugs move to a very short distance occasionally, when they remain inactive for a longer period.
- *Cleaning (CL)*–The bugs rub their legs against each other and on their body, thus cleaning themselves.
- *Back resting (BR)*–The bugs occasionally rest on their dorsum in the bottom of the aquaria.
- *Being inactive (IA)*–The bugs remain inactive either along the sides of the tank or in the bottom and rest. However these multivariate behavioural responses were categorized based on the possibilities of either resemblance or overlapping of one behavioural strategy with the other–crawling, swimming, producing air bubble, clustering, being out of water, being inactive, bouting, floating, cleaning, and clinging to plants.

Time Activity Budget of Individual Bugs

Various behavioural responses of *Diplonychus rusticus* were observed and the time spent for each was converted to percentage and summarized in Table 3.1.

Time Activity Budget of Behaviour in Selected 10 Male Bugs

During Day Hours

- *Crawling (CR)*–Bug number 1, 7, 9 and 10 spent very lesser portion of their time in crawling, while the others 2, 3, 4, 5, 6 and 8 never exhibited crawling.
- *Being out of water (OW)*–None of the males crawled out of water through out the observation period.
- *Clustering (CU)*–All the 10 bugs showed clustering with a high degree of inter individual variation.

Table 3.1: Time Activity Budget of Male *Diplonychus rusticus* during Day Hours

	Time Spent in Selected Behaviours (in per cent)									
Bug No	*CR*	*OW*	*CU*	*SW*	*BO*	*FL*	*IA*	*CP*	*CL*	*PB*
1	0.0793	–	3.8095	0.7142	0.2380	10.2619	56.8809	27.6190	0.3968	
2	–	–	1.656	6.2142	2.5952	9.2142	57.4206	22.8968	–	0.0317
3	–	–	6.1666	4.9285	2.2619	2.4523	51.6190	32.3809	–	0.1904
4	–	–	1.9047	4.6428	0.9285	2.1190	72.7777	17.6111	–	0.0158
5	–	–	7.7777	3.1111	0.0793	12.6428	59.4841	16.9047	–	
6	–	–	0.8730	3.9682	1.0317	3.8095	72.3809	17.9365	–	
7	0.6349	–	6.9841	3.7301	1.1111	0.5793	58.2777	28.5476	–	0.1349
8	–	–	2.3809	7.1428	3.0158	25.7142	37.1984	24.5317	–	0.0158
9	0.2380	–	2.8571	0.7142	0.7936	1.3888	23.3730	70.6349	–	–
10	0.2380	–	3.3333	2.6984	0.3968	2.6190	45.1190	45.5952	–	–

CR: Crawling; CU: Clustering; SW: Swimming; BO: Bouting; PB: Producing Bubbles; CL: Cleaning; FL: Floating; CP: Clinging to plants; HI: Hiding; DA: Disturbing Another; SE: Settling; AV: Avoiding.

- ☆ *Swimming (SW)*–All the bugs showed swimming, Bug No.2 and 8 spent their maximum time in swimming, Bug No.1 and 9 spent very lesser portion of their time in swimming. All the others spent a moderate amount of their time in swimming.
- ☆ *Cleaning (CL)*–Bug No.1 spent very lesser portion of its time in cleaning and all the others did not spend their time in cleaning.
- ☆ *Bouting (BO)*–All the bugs exhibited bouting. The inter-individual variation in the time period was considerable.
- ☆ *Floating (FL)*–All the bugs showed floating, with a high degree of inter individual variation. Bug No.8 floated for a very long period, while Bug No.7 for a very short period.
- ☆ *Producing air bubble (PB)*–Bug No.2, 3, 8, 9 spent very lesser period of time producing air bubble, while others did not spend their time in it.
- ☆ *Being inactive (IA)*–All the bugs spent considerable portion of their time being inactive. The inter-individual variation was moderate.
- ☆ *Clinging to plants (CP)*–All the bugs spent considerable portion of their time clinging to plants, here again the inter-individual variation was moderate.

Various behavioural responses of *D. rusticus* were observed and the time spent for each was converted to percentage and summarized in Table 3.2.

Time Activity Budget of Selected 10 Males during Night Hours

- ☆ *Crawling (CR)*–Bug No.2, 5, 8 and 10 spent very lesser portion of their time in crawling, while the others did not crawl.
- ☆ *Crawling out of Water (OW)*–Except for Bug No.3 and 6, others did not crawl out of water.
- ☆ *Clustering (CU)*–While Bug No. 1 and 9 remained separate, all the others clustered. But a high degree of inter-individual variation was noted.
- ☆ *Swimming (SW)*–All the bugs showed swimming, Bug No.3,

Table 3.2: Time Activity Budget of Behaviour in Male *Diplonychus rusticus* during Night Hours

	Time Spent in Selected Behaviours (in per cent)									
Bug No	*CR*	*OW*	*CU*	*SW*	*BO*	*FL*	*IA*	*CP*	*CL*	*PB*
1	–	–	–	2.5396	0.9523	20.0000	46.6507	21.7460	8.0952	0.0158
2	0.4761	–	7.6190	4.3888	–	31.7460	49.0714	4.2857	–	2.4126
3	–	1.3492	0.9920	13.6190	1.1904	17.0952	48.1587	14.1666	3.3333	0.0952
4	–	–	13.8730	21.0158	1.5079	9.4761	43.4920	10.6349	–	–
5	0.3174	–	3.0952	3.6904	0.9523	8.3333	55.9920	27.6190	–	–
6	–	0.9523	7.5238	9.6428	0.4761	29.0000	48.2301	4.1269	–	0.0476
7	–	–	12.6190	3.2857	2.5396	17.9047	35.1190	28.5317	–	–
8	0.1587	–	8.3333	2.6349	0.5555	6.2936	41.0317	40.9920	–	–
9	–	–	–	2.3809	–	40.7539	19.0476	35.9126	–	1.9047
10	0.4761	–	10.4761	2.5000	0.6984	3.8253	77.7380	4.2857	–	–

CR: Crawling; CU: Clustering; SW: Swimming; BO: Bouting; PB: Producing Bubbles; CL: Cleaning; FL: Floating; CP: Clinging to plants; HI: Hiding; DA: Disturbing Another; SE: Settling; AV: Avoiding.

4 and 6 showed swimming for a longer duration, while others for a shorter duration.

- ☆ *Bouting (BO)*–Except Bug No.2 and 9, all the others showed bouting for a more or less same duration.
- ☆ *Floating (FL)*–All the bugs showed floating and inter individual variation was moderate.
- ☆ *Being inactive (BI)*–All the bugs were inactive for a considerable period of time, Bug. No. 10 was inactive for a very long period and Bug No.9 for a short period.
- ☆ *Producing air bubble (PB)*–Bug No. 2 and 9 produced air bubble for a longer period, while Bug No. 1, 3 and 6 spent very short period in producing air bubble, others did not produce air bubble.
- ☆ *Clinging to plants (CP)*–All the bugs clanged on to plants and the inter-individual variation was moderate.
- ☆ *Cleaning (CL)*–Except Bug No. 1 and 3, others did not show cleaning behaviour. Various behavioural responses of *D. rusticus* and the time spent for each were converted to percentage and summarized in Table 3.3.

Time Activity Budget of Selected 10 Females during Day Hours

- ☆ *Crawling (CR)*–Except Bug No.3, 4 and 7, others did not show crawling, while in Bug No.3 and 4 it was observed for a very short period. In Bug No.7 it was observed for a long period.
- ☆ *Clustering (CU)*–Bug No. 4 and 9 remained separate, while in Bug No.7 it was observed for a very short period; others spent considerable portion of their time in clustering.
- ☆ *Swimming (SW)*–Except Bug No.1 all the others showed swimming for a more or less equal period of time.
- ☆ *Bouting (BO)*–Except Bug No.1, 4, 5 and 10 all the others showed bouting for a more or less equal period of time.
- ☆ *Crawling out of water (OW)*–Except Bug No.10, others did not crawl out of water.
- ☆ *Producing air bubble (PB)*–Bug No. 1, 2, 5, 6 and 9 did not produce air bubble, while Bug No.4 and 10 spent equal

Table 3.3: Time Activity Budget of Behaviour in Female *Diplonychus rusticus* during Day Dours

	Time Spent in Selected Behaviours (in per cent)									
Bug No	*CR*	*OW*	*CU*	*SW*	*BO*	*FL*	*IA*	*CP*	*CL*	*PB*
1	–	–	2.3809	–	–	11.9047	14.2857	71.4228	–	–
2	–	–	4.7619	1.5476	0.3968	26.7460	30.0793	34.5634	1.9047	–
3	0.0793	–	5.2380	2.7777	1.8888	11.9126	35.2142	42.8571	–	0.0317
4	0.2380	–	–	0.4761	–	1.8253	28.0952	55.0793	11.4285	2.8571
5	–	–	–	0.2380	–	12.3809	39.0476	48.3333	–	–
6	–	–	–	1.7460	1.4285	4.2857	40.6349	51.9047	–	–
7	1.4126	–	0.9523	1.6666	1.5079	10.2222	48.7460	35.4761	–	0.0158
8	–	–	–	1.3809	0.6349	11.4444	83.6507	2.8571	–	0.0317
9	–	–	–	1.3095	0.3174	1.2698	17.3412	79.7619	–	–
10	–	1.4285	2.8571	1.1904	–	26.984	11.3492	49.0476	4.2857	2.8571

CR: Crawling; CU: Clustering; SW: Swimming; BO: Bouting; PB: Producing Bubbles; CL: Cleaning; FL: Floating; CP: Clinging to plants; HI: Hiding; DA: Disturbing Another; SE: Settling; AV: Avoiding.

period of time in producing air bubble and others spent lesser time in producing air bubble.

- *Cleaning (CL)*–Bug No.2, 4 and 10 spent some time in cleaning while others did not spend their time in cleaning.
- *Floating (FL)*–All the bugs showed floating with a considerable inter individual variation.
- *Being inactive (IA)*–All the bugs were inactive for a considerable period of time, while Bug No.8 was inactive for a long period of time.
- *Clinging to Plants (CP)*–All the bugs clanged to plants for a considerable period of time, while Bug No.8 clanged to plants for a lesser duration.

Various behavioural responses of *Diplonychus rusticus* and the time spent for each converted to percentage and summarized in Table 3.4.

Time Activity Budget of Selected 10 Females during Night Hours

- *Crawling (CR)*–Bug No.1, 3, 4 and 10 spent more or less equal period of time in crawling while others did not spend their time in crawling.
- *Crawling out of Water (OW)*–Bug No.2, 6 and 10 crawled out of water and Bug No.6 remained out of water for a longer period. Other bugs did not crawl out of water.
- *Clustering (CU)*–Bug No. 1, 2, 3, 6, 9 and 10 remained clustered for a more or less equal period and others remained separate.
- *Bouting (BO)*–Except Bug No.3,5 and 9, others showed bouting for a equal period of time.
- *Clinging to Plants (CP)*–All the bugs clanged to plants and inter individual variation was considerable.
- *Cleaning (CL)*–Cleaning was observed in Bug No.1, 4 and 10 for a short period while in others it was absent.
- *Floating (FL)*–Bug No.1 and 9 floated for a very short period and bug No.10 for a very long period. Bug No.5 did not float and others spent more or less equal period in floating.

Table 3.4: Time Activity Budget of Female *Diplonychus rusticus* during Night Dours

	Time Spent in Selected Behaviours (in per cent)									
Bug No	*CR*	*OW*	*CU*	*SW*	*BO*	*FL*	*IA*	*CP*	*CL*	*PB*
1	0.3968	–	0.9523	1.8412	0.4761	0.5000	38.4126	55.1587	1.5476	0.7142
2	–	1.9047	0.4761	6.6507	0.9920	13.6507	57.9761	18.3492	–	–
3	0.7142	–	1.2698	3.4682	–	17.4126	33.3333	43.7698	–	0.0317
4	–	–	–	2.0634	0.9523	2.4603	24.5238	67.6190	2.3809	–
5	0.3968	–	–	0.8730	–	–	55.8730	42.8571	–	–
6	–	3.8095	0.9523	4.3650	1.1111	12.1328	32.7777	44.8412	–	–
7	–	–	–	4.4365	2.1031	7.6031	66.7460	19.0476	–	0.0634
8	–	–	–	3.5793	0.2380	0.0793	80.3571	15.7142	–	0.0317
9	–	–	1.2698	2.6190	–	12.4603	31.9841	51.6666	–	–
10	0.4761	0.7936	0.1587	5.1825	1.9841	35.9365	29.7380	21.9047	3.0952	0.7301

CR: Crawling; CU: Clustering; SW: Swimming; BO: Bouting; PB: Producing Bubbles; CL: Cleaning; FL: Floating; CP: Clinging to plants; HI: Hiding; DA: Disturbing Another; SE: Settling; AV: Avoiding.

- *Swimming (SW)*–All the bugs showed swimming and the inter-individual variation was considerable.
- *Producing air bubble (PB)*–Time spent in producing air bubble was maximum in Bug No.1 and 10 and minimum in Bug No.2, 8 and 7 while others did not produce air bubble.
- *Being inactive (LA)*–All the bugs were inactive for a considerable period of time and the inter-individual variation was high.

Time Activity Budget of *Diplonychus rusticus*

In Males during Day Hours

It is evident that males were spending their maximum time to be inactive and clinging to plants, while floating, swimming and clustering were relatively less and bouting, crawling, cleaning and production of air bubble were insignificant and none of the bugs crawled out of water (Table 3.5).

Table 3.5: Percentage of Time Spent on Various Behaviours by Male and Female *Diplonychus rusticus* during Day and Night Hours

Sl.No.	*Behaviour*	*Time Taken (in percentage)*			
		Male		*Female*	
		Day	*Night*	*Day*	*Night*
1	CR	0.1190	0.1428	0.1730	0.1984
2.	OW		0.2301	0.1428	0.6507
3.	CU	3.7714	6.4531	1.6190	0.5079
4.	SW	3.7865	6.5698	1.2333	3.5079
5.	BO	1.2452	0.8873	0.6174	0.7857
6.	FL	7.0801	18.4428	11.8976	10.2246
7.	IA	53.4531	46.4531	34.8444	45.1722
8.	CP	30.4658	19.2301	47.1309	38.0928
9.	CL	0.0396	1.1428	1.7619	0.7023
10.	PB	0.0388	0.4476	0.5793	0.1571

CR: Crawling; CU: Clustering; SW: Swimming; BO: Bouting; PB: Producing Bubbles; CL: Cleaning; FL: Floating; CP: Clinging to plants; HI: Hiding; DA: Disturbing Another; SE: Settling; AV: Avoiding.

In Males during Night Hours

It is evident that inactivity was the predominant behaviour. The bugs spent considerable time in clinging to plants and in floating in equal proportions. They spent sometime in clustering and swimming and time spent in crawling, bouting, cleaning, production of air bubble and crawling out of water were insignificant, (Table 3.5).

In Females during Day Hours

It is evident that bugs spent their maximum time clinging to plants and remaining inactive. While floating was observed to some extent, crawling, clustering, swimming, bouting, cleaning and production of air bubble and crawling out of' water was insignificant (Table 3.5).

In Females during Night Hours

It is evident that the bugs spent maximum time in being inactive and clinging to plants, while they spent sometime in floating and lesser time in swimming, other behaviours were insignificant (Table 3.5).

Relative Frequency of Behaviour in *Diplonychus rusticus*

Relative frequency of the selected behaviours observed in the bugs were calculated and summarized in Table 3.6.

In Males during Day Hours

It is evident that swimming and setting were relatively high next to which floating, bouting and clinging to plants were high. The relative frequency of clustering, avoiding, disturbing other bugs and hiding were less and frequency of crawling, cleaning and producing air bubble were insignificant (Figure 3.1).

In Males during Night Hours

It is evident that the relative frequency of swimming was very high and that of settling and floating were also high and were equal while that of clustering and clinging to plants were moderate. All the other behaviours were insignificant (Figure 3.2).

Table 3.6: Relative Frequency of Selective Behaviours Observed in Male and Female *Diplonychus rusticus* during Day and Night Hours

Sl.No.	*Behaviour*	*Frequency per hour*			
		Male		*Female*	
		Day	*Night*	*Day*	*Night*
1.	HI	12.8379	9.9473	7.5672	8.4934
2.	DA	15.5126	10.8947	9.7296	13.0670
3.	CL	1.0697	3.3156	22.7028	4.5732
4.	PB	4.2793	7.5789	10.8107	9.8003
5.	GU	18.7220	35.5263	10.8107	12.4136
6.	FL	49.7475	57.3159	49.7298	54.2287
7.	SW	109.6585	120.3161	63.7840	143.0859
8.	CR	3.7442	2.8419	7.5672	6.5335
9.	AV	15.5126	12.3157	10.8107	12.4136
10.	CP	25.6759	25.5789	51.8919	39.8548
11.	SE	77.0283	58.2632	96.2166	39.8548
12.	BO	26.2108	16.1051	18.3783	15.6805

CR: Crawling; CU: Clustering; SW: Swimming; BO: Bouting; PB: Producing Bubbles; CL: Cleaning; FL: Floating; CP: Clinging to plants; HI: Hiding; DA: Disturbing Another; SE: Settling; AV: Avoiding.

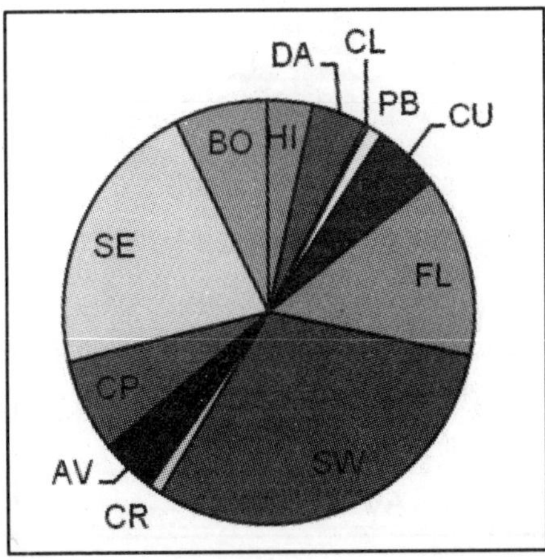

Figure 3.1: Relative Frequency of Selected Behaviours Observed in Male *D. rusticus* during Day Hours

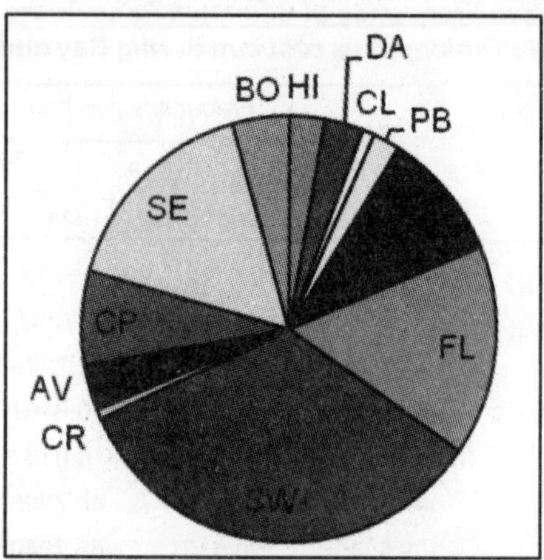

Figure 3.2: Relative Frequency of Selected Behaviours Observed in Male *D. rusticus* during Night Hours

In Females during Day Hours

It is evident that settling, swimming, clinging to plants and floating were relatively frequent while others were insignificant (Figure 3.3).

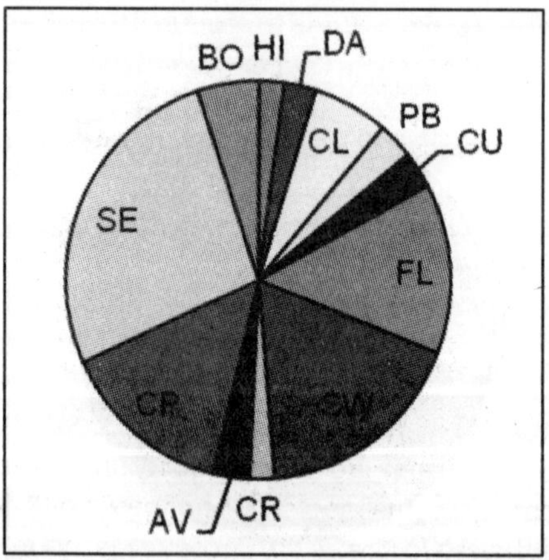

Figure 3.3: Relative Frequency of Selected Behaviours Observed in Female *D. rusticus* during Day Hours

In Females during Night Hours

It is evident that relative frequency of swimming was the highest while that of floating, clinging to plants and settling were moderate and almost equal. All the other behaviours were insignificant (Figure 3.4).

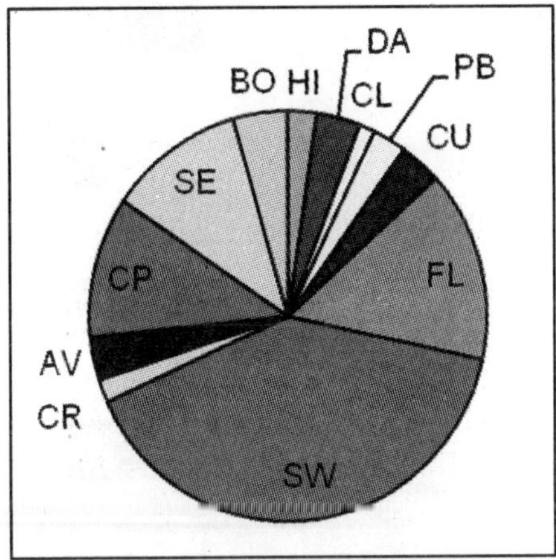

Figure 3.4: Relative Frequency of Selected Behaviours Observed in Female *D. rusticus* during Night Hours

Behaviour Patterns in *Diplonychus rusticus*

Ten behaviours were selected and their frequency of occurrence was calculated and summarized in Table 3.7.

In Males during Day Hours

It is evident that frequency of disturbing other bugs (and being disturbed) is high hence they hide frequently. Frequency of sitting on another bug (and being sat by another bug) and producing air bubble was more or less equal. While the frequency of being partly immersed, back resting and cleaning was very less and frequency of being out of water was completely absent (Figure 3.5).

In Males during Night Hours

It is evident that the frequency of disturbing others (and being disturbed) and hiding is almost equal. Frequency of producing air bubble was also high. The frequency of cleaning, crawling out of water and sitting on another bug (and being sat by another bug) was

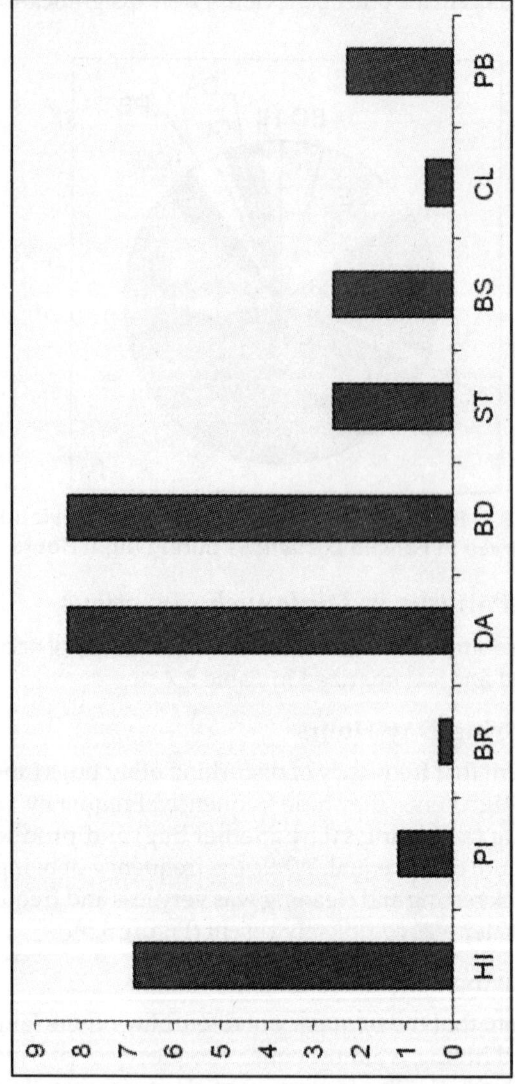

Figure 3.5: Frequency of All Occurrence Behaviour in Male *D. rusticus* during Day Hours

almost equal while the frequency of back resting and being partly immersed was in significant (Figure 3.6).

Table 3.7: Duration of Various Behavioural Patterns in *Diplonychus rusticus*

Sl.No.	*Behaviour*	*Time Taken (in percentage)*			
		Male		*Female*	
		Day	*Night*	*Day*	*Night*
1.	CR	0.1190	0.1428	0.1730	0.1984
2.	OW		0.2301	0.1428	0.6507
3.	CU	3.7714	6.4531	1.6190	0.5079
4.	SW	3.7865	6.5698	1.2333	3.5079
5.	BO	1.2452	0.8873	0.6174	0.7857
6.	FL	7.0801	18.4428	11.8976	10.2246
7.	IA	53.4531	46.4531	34.8444	45.1722
8.	CP	30.4658	19.2301	47.1309	38.0928
9.	CL	0.0396	1.1428	1.7619	0.7023
10.	PB	0.0388	0.4476	0.5793	0.1571

CR: Crawling; CU: Clustering; SW: Swimming; BO: Bouting; PB: Producing Bubbles; CL: Cleaning; FL: Floating; CP: Clinging to plants; HI: Hiding; DA: Disturbing Another; SE: Settling; AV: Avoiding.

In Females during Day Hours

It is evident that the frequency of cleaning was the highest. The frequency of disturbing other bugs (and being disturbed) and producing bubble was almost equal. Frequency of hiding was moderate while frequency of back resting, sitting on other bug (and being sat by other bug) and remaining out of water was very less. Being partly immersed was completely absent (Figure 3.7).

In Females during Night Hours

It is evident that frequency of disturbing others bugs (and being disturbed) was the highest. Producing air bubble and remaining out of water were almost equal. Hiding was also seen frequently. While the frequency of sitting on other bug (and being sat by other bug) and cleaning was very less. Back resting and being partly immersed never occurred (Figure 3.8).

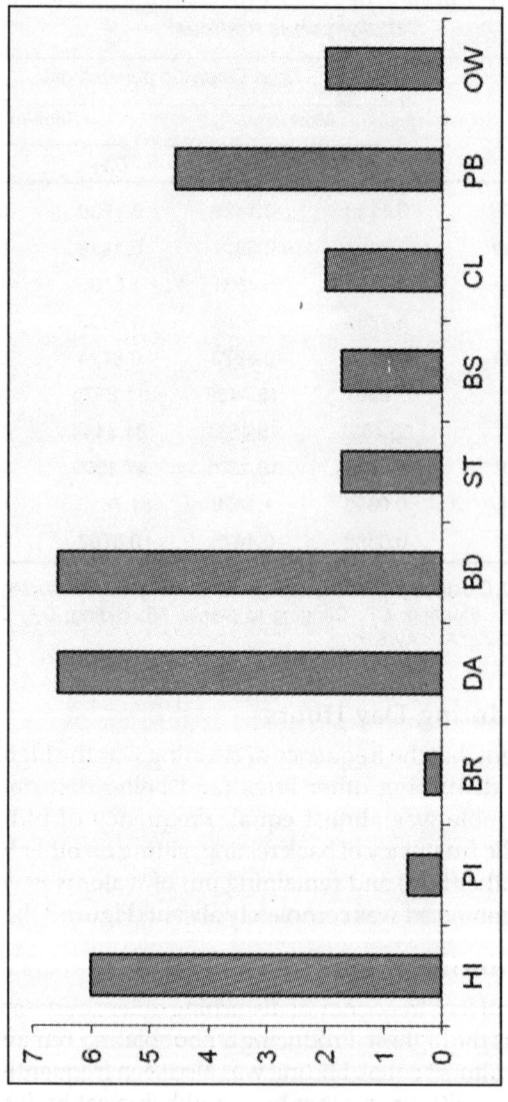

Figure 3.6: Frequency of All Occurrence Behaviour in Male *D. rusticus* during Night Hours

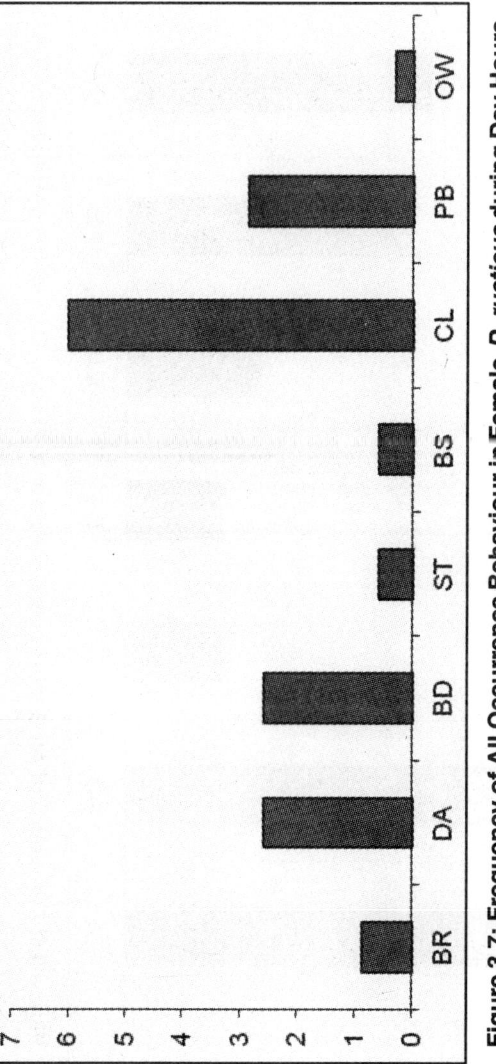

Figure 3.7: Frequency of All Occurrence Behaviour in Female *D. rusticus* during Day Hours

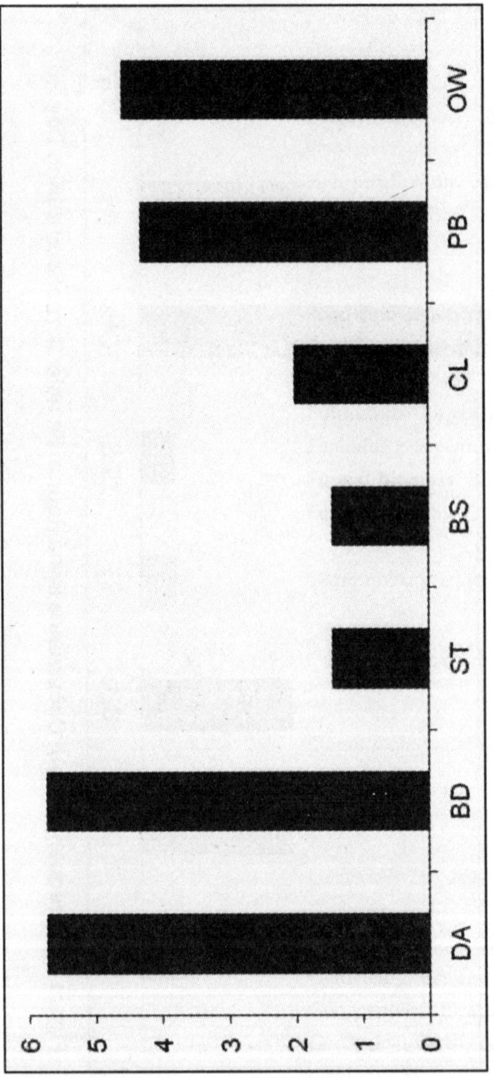

Figure 3.8: Frequency of All Occurrence Behaviour in Female *D. rusticus* during Night Hours

Table 3.8: Frequency of All Occurrence Behaviour per hour in *Diplonychus rusticus*

Sl.No.	*Behaviour*	*Time Taken (in percentage)*			
		Male		*Female*	
		Day	*Night*	*Day*	*Night*
1.	HI	6.8571	6.0000	1.9999	3.7142
2.	PI	1.1428	0.5714	–	–
3.	BR	0.2857	0.2857	0.8571	–
4.	DA	8.2857	6.5714	2.5714	5.7142
5.	BD	8.2857	6.5714	2.5714	5.7142
6.	ST	2.5714	1.7142	0.5714	1.4285
7.	BS	2.5714	1.7142	0.5714	1.4285
8.	CL	0.5714	1.9999	6.0000	1.9999
9.	PB	2.2857	4.5714	2.8571	4.2857
10.	OW	–	1.9999	0.2857	4.5714

CR: Crawling; CU: Clustering; SW: Swimming; BO: Bouting; PB: Producing Bubbles; CL: Cleaning; FL: Floating; CP: Clinging to plants; HI: Hiding; DA: Disturbing Another; SE: Settling; AV: Avoiding.

Discussion

In the present study, analysis of the selected behaviour implies that the bugs 'swim' and 'bout' to search for prey, they 'cluster' and 'tide' to secure protection, they 'float' and 'produce air bubble' in order to respire, they 'cling to plants' and 'remain inactive' in order to rest, they 'crawl out of water' because they are protoceptive. All the other behaviours were associated with these basic behaviours. It is evident from the analysis that the bugs are more active during night than day hours, and males are more active than females. This is in accordance to the report of Venkatesan and Rao (1980) that this belostomatid bug has the predatory periodicity with their maximum predation at 8 hours and 20 hours, though the peak at 20 hours is distinct than the other. Also Sankaralingam (1989) observed active bugs in the night hours. The bugs spend a great portion of their time in resting next to which respiration takes up a considerable portion of their time. This argument may hold appropriate because the bug shows a tendency to wait and attack the prey than to make unsuccessful attempts (Cloarec, 1989).During resting, males tend to settle in the bottom and corners of the aquarium while females have higher tendency for clinging to plants; hence while respiration, the males either actively swim to the surface or passively rise to the surface when they are in the bottom of the tank, or they settle on the rough corners of the tank and remain partly immersed in water helping themselves to respire and rest simultaneously.

On the other hand, females generally cling to plants found near the surface of water. When they are found deeper in the plants, they produce air bubble or crawl to the surface along plants for respiration. Hence females were not observed to remain partly immersed and production of air bubble, crawling and cleaning is higher in females than males. The frequency and time spent in clustering is higher in males than in females, this might be because of their resting places. Clustering is a protective strategy in the females because of their tendency to cling to plant, get themselves concealed within the plants and do not have the need to cluster for protection; whereas the bottom and corners of the tank, which are devoid of plants make the males conspicuous thus forcing them to cluster among themselves or to hide behind one another or to crawl into plants when necessary. Such behavioural responses indicate the males being relatively active than females. This may correlate with the paternal assurance met with in males while bearing the eggs on their dorsum. The stress would have resulted in higher respiratory activity in males (Venkatesan *et al.*, 1986).

The bug being a voracious feeder spends a considerable portion of its time in searching for prey. They do this by rapidly swimming, bouting and sometimes swimming in circles, during which they disturb the debris found in the bottom of the tank, probably looking for their prey amidst them. During these activities, they pause for a second, probably trying to locate a prey, then continue their searching and they sometimes try to catch the waste materials mistaking them for their prey. They even show mass swimming at times when they are actively searching for the prey.

Though *Diplonychus rusticus* is a true water bug, some bugs remain out of water for surprisingly a long period of time (even 30 minutes and more). This may be because of their photoreceptive nature. This may also be suggestive of their tendency to migrate to other water bodies during dry season. They crawl along the sides of the tank above the water surface for varying period of time. It is interesting to note that they do not crawl back into the water as they did while moving out, but fall back into the water. On falling back, they either float or swim to the bottom (sort of diving).

The bugs tend to disturb other individuals as they are swimming and the disturbed bugs avoid and mostly hide themselves. Sitting on another bug is rarely observed, where in fifty per cent of cases the bug- which is being sat on- avoids the encounter. Nepids and ranatrids are well known to exhibit the phenomenon of free ride (Ravisankar, 1993). Another rare behaviour observed is the spurting of fluid through the anal region. This fluid must be the uric acid which is the excretory product of the insect, which must have been excreted when the bug was on the surface of water and thus was observed. As discussed, the bugs rest for a greater portion of their time. This may reflect on their adaptability in conservation of energy, which may be profitable during predation as reported earlier (Smith, 1979).

During that long session, they exhibit some rare and insignificant strategies like jerking, lingering and stirring. Back-resting is one such behaviour, exhibited by a very few bugs that last for few seconds whereas such a behaviour is distinct in notonectids (Nishi, 1990). Another interesting behaviour observed in the bugs is their tendency to approach other bugs and remain either near them or cling to them. They sometime cling to each other on their ventral sides and remain inactive or swim about. In addition to this, mass swimming and clustering behaviour prove them to be gregarious. Such behavioural responses in *Diplonychus rusticus* clearly indicate their higher adaptability to the environmental changes and in particular for predation, reproduction and respiration. Lauck and Menke (1961) have shown that among five genera of Sub- family Belostomatinae, *Diplonychus* is the highly evolved and specialized one. Such a taxonomic inference holds good with reference to the present investigation. Further studies on the application of cluster analysis on the data obtained may substantiate the present arguments.

Acknowledgement

The Author is thankful to University Grants Commission, New Delhi for extending Emeritus Fellowship.

References

Altman, J., 1973. Observational study of behaviour sampling method. *Behaviour*, 49(3/4): 227–262.

Bailey, P.C.E., 1985. "A prey in the hand", multi prey capture behaviour in a sit and wait predator (Heteroptera: Nepidae) the water stick insect. *J. Ethol.*, 3: 105–112.

Bell, H.L. and Nebeker, A.V., 1969. Preliminary studies on the tolerance of aquatic insects to low pH. *J.Kansas. Ent. Soc.*, 42: 230–236.

Clausen, C.P., 1962. *Entomophagus Insects.* Hofner New York.

Cloarec, A., 1989. Variation of foraging factors in a water bug *Diplonychus indicus. J. Ethol.*, 7: 27–34.

Cloarec, A., 1991. Predatory versatility in water bug *Diplonychus indicus. Behavioural Processes*, 23: 231–242.

Cloarec, A., 1992. The influence of feeding on predator tactics in a water bug. *Physiol. Ent.*, 17: 25–32.

Cloarec, A., 1995. General activity and foraging factors in a water bug. *J.Ethol.*, 13: 31–39.

Curio, E., 1976. *Ethology of Predation*. Springer Verlag Berlin Germany, 250 pp.

Dufour, L., 1833. Reherches Anatomiques et physiologiques sur les A hemipteres. *Mem press Acad. Sci. Paris*, 4: 3–33.

Ellis, R.A. and Burden, J.H., 1970. Predation by *Notonecta unclulata* (Heteroptera: Notonectidae) on larvae of the yellow fever mosquito. *Ann. Ent. Soc. Amer.*, 63(4): 963–973.

Gadagker, R.S. and Joshi, N.V., 1983. Quantitative ethology of wasps. Time activity budget and caste differentiation in *Ropalidia marginata* (LEP) (Hymenopterawespidae). *Ani. Behav.*, 31: 26–31.

Hassell, M.P., 1978. The dynamics of arthoropod predator–prey systems. *Monographs in Population Biology*, 13: 1–237.

Hungerford, H.B., 1925. Notes on the giant water bugs. *Pshyche.*, 32: 88–91.

Hungerford, H.B., 1959. Hemiptera. In: *Freshwater Biology*, pp. 958–972.

Hynes H.B.N., 1984. The relationships between the taxonomy and ecology of aquatic insects. *The Ecology of Aquatic Insects*, pp. 9–21.

Jenkins, D.W., 1964. Pathogens, parasites and predators of medicaly important arthropods, annotated list and bibliography. *Bull. World. Org. Suppl.*, 30: 1–150.

Kari Vepsaiainen and Matti Nummelin 1986. Habitat selection by water strider larvae (Heteroptera: Gerridae) in relation to food and imagoes. *Oikos*, 47(2): 374–381.

Lauck, D.R. and Menke, A.S., 1961. The higher classiciation of Belostomatidae (Hemiptera). *Ann. Ent. Soc. Amer.*, 56: 644–657.

Lehner, P.N., 1979. *Handbook of Ethological Methods.* Garland series in Ethology, N.Y., p. 1–403.

Maier, C.T., 1977. The behaviour of *Hydrometra championa* (Hemiptera; Hydrometridae), and resource partitioning with *Tenagogonus quadrilineatus* (Hemiptera: Gerridae). *J. Kansas Ent. Soc.*, 50: 263–271.

Nebeker, A.V., 1972. Effect of low oxygen concentration on survival and emergence of aquatic insects. *Trans. Amer. Fish. Soc.* (101): 675–679.

Nishi Raghavan, 1990. Evaluation of predatory potential of the notonectid bug. *Anisops bovieri* Kirkaldy (Insecta: Hemiptera). *Ph.D. Thesis*, Madras University, pp. 1–109.

Peckarsky, B.L., 1980. Predator-prey interaction between stone flies and may flies behavioural observation. *Ecol.*, 61(4): 932–943.

Peckarsky, B.L., 1984. Predator-prey interaction among aquatic insects. *Ecol. of Aqu. Insects*. 196–253.

Peckarsky, B.L. and Penton, M.A., 1985. Is predaceous stonefly behaviour affected by competition? *Ecol.*, 66(6): 1718–1728.

Pimental, D., 1986. Population dynamics and the importance of evolution in successful biological control. *Fortschirtti Der Zoologie*. BD 32 *Franz (HRSG)*.

Presswala, M.J. and George, C.J., 1936. Morphology of *Sphaerodema rusticum* Fabr. *J. Univ. Bombay*, 4: 29–65.

Polhemus, T.J., 1978. Aquatic and semiaquatic hemiptera. *An Introduction to Aquatic Insects of North America*, pp. 1–441.

Pollhemus, T.J., 1982. *Hemiptera Belostomatidae*. Aq. Biota of Mexico, C.America and the Wilndies, p. 288–327.

Scott, M.A. and Murdoch, W.W., 1983. Selective predation by backswimmer *Notonecta*. *Limnol. Oceanogr.*, 28(2): 352–366.

Smith, R.L., 1976a. Brooding behaviour of a male water bug *Belostoma flumineum* (Hemiptera: Belostomatidae). *J.Kansas, Ent. Soc.*, 49(3): 333–343.

Smith, R.L., 1976b. Male brooding behaviour of the water bug *Abedus herberti* (Hemiptera: Belostomatidae). *Ann. Ent. Soc. Amer.*, 69(4): 740–747.

Smith, R.L., 1979a. Repeated copulation and sperm precedence: Paternity assurance for a male brooding water bug. *Science*, 205: 1029–1031.

Smith, R.L., 1979b. Paternity assurance and altered roles in the mating behaviour of a giant water bug. *Abedus herberti* (Heteroptera: Belostomatidae). *Anim. Behav.*, 27: 716–725.

Smith, R.L., 1980. Evolution of exclusive post copulatory paternal care in the insects. *Florida Entomol.*, 63(1): 65–78.

Spence, J.R., 1981. Experimental analysis of microhabitat selection in water striders (Heteroptera: Gerridae). *Ecol.*, 62(6): 1505–1514.

Venkatesan, P., 1978. Studies on aquatic insects. *Ph.D. Thesis*, University of Madras, p. 1–160.

Venkatesan, P. and Rao, T.K.R., 1980. Effect of nutritional status and sex on the cercadian variation in predation and haemolymph free sugars of *Diplonychus indicus* sp. (Hemiptera: Belostomatidae). *Chronobiologia*, 7: 15–20.

Venkatesan, P. and Rao, T.K.R., 1981. Description of new species and a key to Indian species of Belostomatidae. *J. Bomb. Nat. Hist. Soc.*, 77: 299–303.

Venkatesan, P., 1983. Male brooding behaviour of *Diplonychus indicus* Venk. and Rao (Hemiptera: Belostomatidae) in its predation of two species of mosquito larvae of varied size. *Entomon.*, 9(3): 191–196.

Venkatesan, P., 1985. Biological control potential of water bugs in the antimosquito work. *Proc. Sem. Entomoph. Ins.*, Calicut, pp. 78–83.

Venkatesan, P. and Jeyachandra, C.M., 1985. Estimation of mosquito predation by the water bug *Diplonychus indicus* (Venk and Rao). *Indian J. Exp. Biol.,* 23: 227–229.

Venkatesan, P., Guillareme, G.C. and Halberg, F., 1986. Modelling prey-predator cycles using hemipteran predators of mosquito larvae for reducing world-wide mosquito born disease incidence. *Chronobiologia,* 13(B51).

Venkatesan, P. and Sankaralingam, A., 1989. Larvicidal properties of water bug *Diplonychus indicus* and its use in mosquito control. *Indian J. Exp. Biol.,* 27: 174–176.

Venkatesan, P., 1990. Influence of prey size on choice by the water bug. *Diplonychus indicus. J. Ent. Res.,* 14(2): 130–138.

Venkatesan, P., 1993. Aspects of biodiversity among belostomatid bug. *Hexapoda,* 5(2): 151–156.

Venkatesan, P. and Ravisankar, S., 1993. Copulatory behaviour of *Laccotrephes griseus* and its phylogenetic relevance in Nepidae. *J. Eco. Biol.,* 5(3): 197–202.

Venkatesan P., 1995. Perspectives of Aquatic entomology. *Bicovas.,* 1–33; 66–75.

Venkatesan, P., 2005. Contribution to biocontrol of vector mosquitoes with special reference to the water bug – *Diplonychus rusticus* (= *indicus* Venk. and Rao). *D.Sc. Thesis,* University of Madras.

Venkatesan, P., 2011. Effect of moulting on prey-predator interaction of the water bug *Diplonychus rusticus* (Hemiptera). *J. Ecobiol.,* 29(1): 7–13.

Wiyey, M. and Kohler, 1984. Behavioural adaptation of aquatic insects; *The Ecology of Aquatic Insects,* p. 101–125.

Wollkind, D., Ahastings, J. and Logan, J.A., 1980. Functional response, numerical response and stability in arthropod predator-prey ecosystems involving age structure. *Res. Popul. Ecol.,* 22: 323–338.

Chapter 4

Acridoids as Bio-indicators of Contaminant Accumulation in the Industrial Areas of the State of Jharkhand (India)

Mohd. Kamil Usmani, Md. Rashid Nayeem and Md. Humayoon Akhtar*

Department of Zoology, Aligarh Muslim University, Aligarh – 202 002, U.P., India

ABSTRACT

Grasshoppers have long been studied as pests but they are important and intimate part of majority of the ecosystems. They are the primary consumers and support many other creatures that occupy niches above them. The diversity, functional importance and sensitivity of grasshoppers make them excellent indicators for using assessment of ecological change associated with land use. The state of Jharkhand (India) was surveyed during 2009-2011 that yielded quite surprising results. The concerned area is rich in biodiversity but also very rich in mineral deposits and thereby in present scenario industrialization is proceeding with a fast pace in the state. The state is still having large forest cover, mainly tropical zone dry and wet forests housing rich floral and faunal diversity. During

* Corresponding author: E-mail: usmanikamil94@gmail.com

the study it was observed that the grasshopper fauna was directly related with presence or absence of industries in the area and this study formed the basis of the manuscript prepared. Grasshoppers are useful indicators for conservation planning and habitat disturbances.

Keywords: *Acridoidea, Diversity, Indicator, Pollution, Jharkhand.*

Introduction

Superfamily Acridoidea comes under suborder Caelifera (Short-horned grasshoppers with three segmented tarsi, antenna shorter than body and a short ovipositor and tympana, when present, are on the first abdominal tergite). Members of the superfamily acridoidea are a dominant group of herbivorous insects round the globe, and their high diversity, functional importance, sensitivity to disturbance and ease of sampling makes them potentially useful bioindicators for land management. In Jharkhand, however, the dynamics of grasshopper assemblages are very poorly understood. Grasshoppers are very important from ecological point of view as these are promising members of trophic level two *i.e.*, herbivores. Encroachment of habitats and decreasing green cover are the main factors behind the reason for tagging pest status to these insects. The diversity, functional importance and sensitivity of grasshoppers make them excellent bioindicators for assessment of ecological changes associated with land use (Armstrong and Van Hensbergen 1997; Samways 1997; Samways and Sergeev 1997).

The state of Jharkhand has a good green cover along with very rich mineral deposits. The state is endowed with 72.2 billion tonnes of coal of all categories, 3700 million tonnes of Iron ore deposits, 68.1 million tonnes of bauxite deposits, 511 million tonnes of limestone deposits, 15000 tonnes of barites, 45.335 million tonnes of china-clay, 3.45 million tonnes of fire clay, 2.19 million tonnes of feldspar, 72000 tonnes of garnet, 6.39 million tonnes of graphite, 0.90 million tonnes of kyanite, 29.86 million tonnes of dolomite, 0.96 million tonnes of quartz and silica sand and 37.22 million tonnes of quartzite. It is because of this reason that the area has been thickly industrialized since the past years and the practice continues till date.

Coal mining areas are depressingly black, iron ore town are red, ochre lends a yellow colour and limestone makes everything chalky white. Fine particles of the ore spread as a uniform layer of dust on every available surface in these areas. Mine areas often have a monochromatic appearance. The pervasiveness of dust is the strongest indication that mining is not a benign activity. Unless it is carefully planned and thoughtfully carried out, it can barren the land, pollute water, denude forests, defile the air and degrade the quality of life for people who live and work in the vicinity. Modern technology has enormously magnified our ability to extract minerals. In the process, it has also gravely threatened human lives and the environment. In Jharkhand mining activity ranges from the small, completely manual stone quarries to mechanical mines. With increasing mechanization, mining equipment has grown larger and more powerful. Entire landscapes are altered in a relatively short period of time.

Mining ruins the land, water, forests and air. The loss or pollution of natural resources degrades the quality of human life in these areas. Increasingly, mineral-based production units like coal-fired power plants, steel plants and cement factories are located near the mines. A cluster of thermal power stations are planned near the major rivers of Jharkhand. Air pollution in the mining areas in Jharkhand state has caused respiratory diseases and eye ailments. The inputs here have been released by Jharkhand State Pollution Control Board on its official website. The area has been an industrial hub since past years and the count is still increasing.

Though Jharkhand State is endowed with vast and rich natural resources mainly those of minerals and forest, 80 per cent of its population residing in 32620 villages and depend mainly on agriculture and allied activities for their livelihood. This region has been far the ages and the homeland of aboriginal races such as the Mundas, Asurs, Santhals, Oraons, Ho, Kharias etc. These indigenous groups have been the worst hit by the large scale exploitation of the natural resources of the region through the development of mines, industries and commercial exploitation of forests. This degree of resource utilization is polluting the waters of rivers like Subarnarekha and Damodar that drain the area and a large majority of population suffers from water borne diseases. The distribution pattern of grasshoppers gives some indication to the extent of damage done by rapid industrialization and so they can be used as models to study the pollutant levels in the particular environment. This type of study based on correlation of acridoid fauna with industrialization has been carried out for the first time from the state.

Grasshoppers have long been studied as a pest by many workers may be because of its swarming nature. Uvarov (1942) contributed a lot to the Acridoid fauna but the study of grasshoppers in relation to ecology and habitat destruction or alterations started late in history. Tandon and Khera (1978) studied their distribution in Andhra Pradesh and also dealt with impact of anthropogenic activities on the distribution pattern and ecology of this group. Succeeding him, Julka *et al.* (1982) worked on Acridoid of Himachal Pradesh. Notable work regarding taxonomy of the group has been carried out by Usmani and Shafee (1990) and Kumar and Viraktamath (1991a) further added literature regarding ecology and behavioural patterns to the taxonomy.

Grasshoppers of Kolkata were studied by Dey and Hazra (2003). These insects that have long been treated as pests have recently been studied as bioindicators by Saha and Haldar (2009).

Human activities have made or rather induced a lot many changes to the ecosystems and as a result have reshaped the composition and functionality of the same. The consequences are very much evident on the biodiversity and extinction of species is progressing at a much faster pace than before and also a number of species have been rendered endangered.

Material and Methods

The state of Jharkhand was surveyed during 2009-2011 and 562 specimens were collected from various sites. The samples were preserved and carried back to the laboratory and further processed which may be summarized as under:

Sample Collection

Grasshoppers were collected by using aerial net and forceps. The vegetation, even the bark of tree was thoroughly inspected for infestation and the insects were killed in cyanide bottles. Sampling by nets was done by random sweeping method for half an hour at a particular site.

Sample Preparation for Morphological Studies

The samples were pinned, relaxed, stretched and later on labeled. Collection was stored in insect storage boxes. The stretched specimens are easy to diagnose for morphological characters like size, colour, texture etc.

Preparation of Genetalic Slides for Correct Identification

Permanent slides of genetalia were prepared to assist in identification. Drawings were made by using camera lucida and conventional microscopic examination was utilized to fill in the details.

Summarizing the Results

Finally the samples were sorted out, list was prepared, and the data was tabulated.

Results and Discussion

There are two synergistic components of destruction of habitat, first one is habitat loss that means overall decrease in habitat area that further leads to decline in species diversity and species abundance while the second one is habitat fragmentation in which the habitat remains fragmented and discontinuous and may affect the available species positively, negatively or in some cases may be neutral (Grez *et al.*, 2004; Yaacobi *et al.*, 2007). The distribution pattern of grasshoppers varied greatly within the area surveyed. The presence of insect concerned in different areas has been detailed as follows:

Faunal Elements from Sampled Sites

Bokaro

(Coal fields and steel industries. Major operative units are Steel Authority of India Limited, Electrosteel Castings Limited, Damodar Valley Corporation, Coal India Limited, Chandrapura Thermal Power Station, Oil and Natural Gas Commission, JayPee Cement)

Spathosternum prasiniferum prasiniferum.

Deogarh

Spathosternum prasiniferum prasiniferum, Cyrtacanthacris tatarica tatarica, Stenocatantops splendens, Diabolocantops pinguis, Trilophidia repleta, Oedipoda miniata miniata, Acorypha glaucopsis.

Dhanbad

(Coal mining, coal washing and coke making are major activities; Giants operative in the area are Tata Steel, Bharat Coking Coal Limited, Eastern Coalfields Limited, Steel Authority of India Limited).

Spathosternum prasiniferum prasiniferum, Catantops pinguis.

Dumka

Oxya hyla hyla, Spathosternum prasiniferum prasiniferum, Stenocatantops splendens, Diabolocantops pinguis, Xenocatantops karnyi, Phlaeoba infumata, Oedipoda miniata miniata.

Garhwa

Spathosternum prasiniferum prasiniferum, Acrida exaltata, Acrida gigantea, Aiolopus simulatrix, Tristria pulvinata, Oedipoda miniata miniata.

Giridih

(Coal mines in southern side of Giridih owned by Central Coalfields Limited)

Spathosternum prasiniferum prasiniferum, Aiolopus simulatrix, Acorypha glaucopsis, Schistocerca gregaria gregaria.

Godda

Oxya velox, Spathosternum prasiniferum prasiniferum, Chondacris rosea, Cyrtacanthacris tatarica tatarica, Epistaurus sinetyi.

Gumla

Acorypha glaucopsis, Hieroglyphus nigrorepletus, Tristria pulvinata, Stenocatantops splendens, Diabolocantops pinguis, Acrida exaltata, Aiolopus thalassinus thalassinus.

Hazaribagh

(Some coal mines owned by Central Coalfields Limited are operative in the region but still the area has a dense forest cover and a national park that are striving to preserve the biodiversity of the region).

Oxya hyla hyla, Spathosternum prasiniferum prasiniferum, Teratodes monticollis, Cyrtacanthacris tatarica tatarica, Stenocatantops splendens, Diabolocantops pinguis, Xenocatantops karnyi, Acrida exaltata, Trilophidia repleta, Aiolopus simulatrix, Chloebora grossa, Chloebora marshalli.

Jamshedpur

(Tata Steel, Tata Motors, Tata Power, Lafarge Cement, Telcon, Brin's Oxygen Company, Tata Technologies Limited, Praxair, Tata Consulting Engineers Limited, Tata Consultancy Services Limited, Timken India, Tinplate and many more).

Spathosternum prasiniferum prasiniferum, Diabolocatantops pinguis.

Koderma (Thermal Power Plant)

Oxya japonica japonica, Spathosternum prasiniferum prasiniferum.

Latehar

Oxya japonica japonica, Tylotropidius varicornis, Cataloipus indicus, Eyprepocnemis alacris alacris, Acrida exaltata, Acrida gigantea, Aiolopus simulatrix, Aiolopus thalassinus thalassinus, Oedaleus abruptus, Leva indica, Chorthippus Indus.

Lohardaga

Spathosternum prasiniferum prasiniferum, Acrida exaltata, Acrida gigantea, Aiolopus thalassinus tumulus, Tristria pulvinata.

Pakur

Spathosternum prasiniferum prasiniferum, Tristria pulvinata, Phlaeoba infumata, Aiolopus simulatrix, Oedipoda miniata, Oedaleus senegalensis.

Palamu

Spathosternum prasiniferum prasiniferum, Acrida exaltata, Acorypha glaucopsis, Tristria pulvinata, Oedaleus abruptus, Diabolocatantops pinguis.

Namkom

Diabolocantops pinguis, Oxya velox, Spathosternum prasiniferum prasiniferum, Tristria pulvinata, Stenocatantops splendens, Diabolocantops pinguis, Orthochtha indica, Acrida exaltata, Phlaeoba infumata, Phlaeoba panteli, Trilophidia annulata, Aiolopus simulatrix, Aiolopus thalassinus tumulus, Acrotylus insubricus insubricus, Oedaleus abruptus, Oedipoda miniata miniata, Stenohippus mundus, Dociostaurus apicalis, Aulacobothrus luteipes luteipes.

Sahibganj

Acrida exaltata, Acrida gigantean, Acorypha insignis, Spathosternum prasiniferum prasiniferum, Oedipoda miniata, Morphacris fasciata.

West Singhbhum (Major haematite mining belt)

Spathosternum prasiniferum prasiniferum, Stenocatantops splendens, Diabolocantops pinguis.

Mining activity is carried out either from surface mines or from underground mines, whatever is the case, these activities pollute the ecosystem to a great extent. Mining of ore and extraction process releases a lot of particulate matter, oxides of sulphur, nitrogen, carbon and a potent greenhouse gas named methane along with mineral dust of the particular ones present in the area. The coal dust or soot released during the mining and transport of coal causes many dreaded respiratory tract diseases as for example chronic bronchitis and asthma. The rocks or soil overlaid with coal seam that are removed before extraction are dumped on surface and therefore it requires large surface area. It is because of this purpose that forest areas are cleared. Apart from these nuisances mining also generates solid wastes, causes land degradation and water pollution.

Cement dust contains heavy metals like nickel, cobalt, lead, chromium and other pollutants hazardous to the biotic environment, with adverse impact for vegetation, human and animal health and ecosystems (Baby *et al.*, 2008). People of cement dust zone area are badly affected by respiratory problems, gastrointestinal diseases etc (Adak *et al.*, 2007). The iron ore mine and smelting of ore releases oxides of carbon, sulphur, nitrogen that are major air pollutants besides being causative agents for acid rain. These may be some of potential reasons for dwindling Acridoid fauna as observed during survey.

1. Pi- chart of listed data (Figure 4.1)
2. Major and minor diversity areas (Figure 4.2)

Acknowledgements

We extend our gratitude to Department of Science and Technology, New Delhi for providing financial assistance during the tenure of a major research project (Ref. No. SR/SO/AS 32/2008) being carried out on Biosystematics and Biodiversity of Acridoidea (Orthoptera) in North India. Thanks are also due to Prof. Irfan Ahmad, Chairman, Department of Zoology, Aligarh Muslim University, Aligarh for providing necessary facilities.

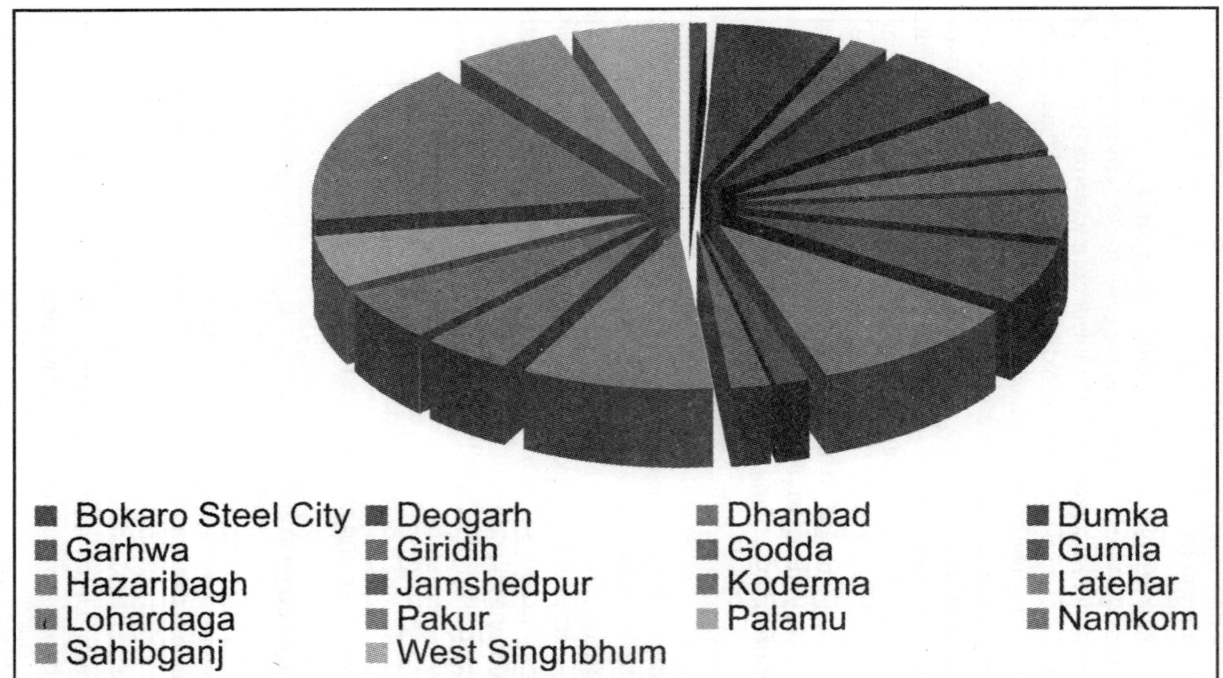

Figure 4.1: Pi-Chart Showing the Distribution of Acridoid Fauna in Various Areas of Jharkhand

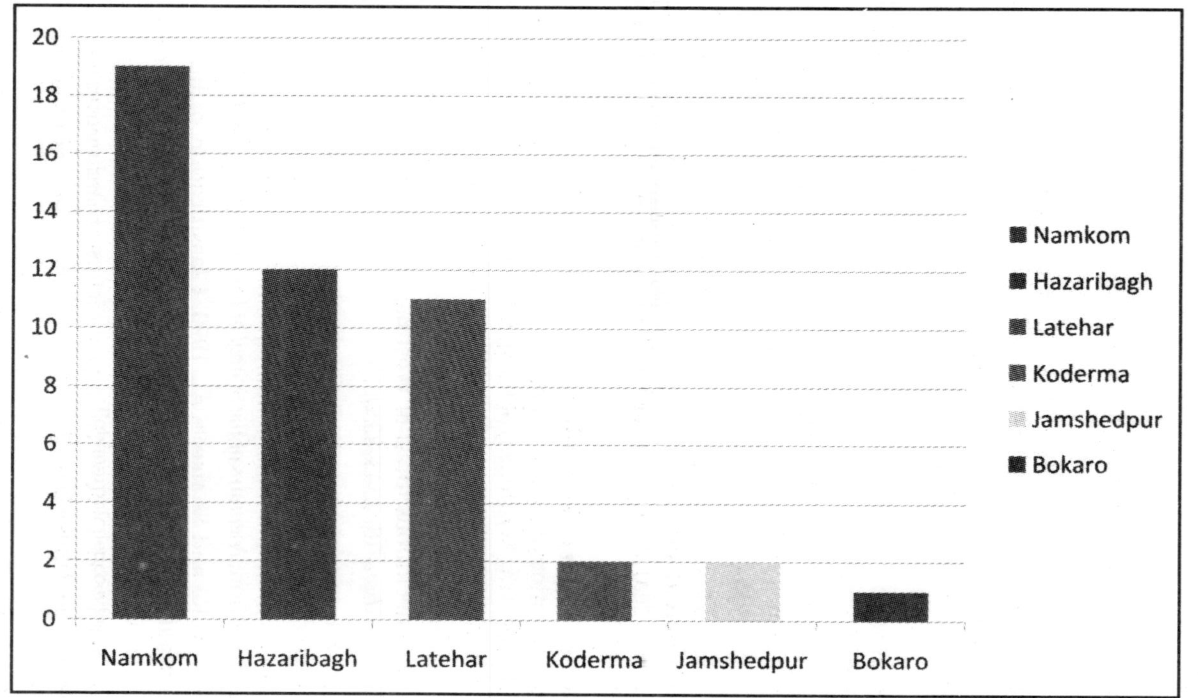

Figure 4.2: Major and Minor Diversity Areas of Jharkhand

References

Adak, M.D., Adak, S. and Purohit, K.M., 2007. Ambient air quality and health hazards near mini cement plants. *Pollution Research*, 26(3): 361–364.

Baby, S.N., Singh, A., Shrivastava, P., Nath, S.R., Kumar, S.S., Singh, D. and Vivek, K., 2008. Impact of dust emission on plant vegetation of vicinity of cement plant. *Environmental Engineering and Management Journal*, 7(1): 31–35.

Dey, A. and Hazra, A.K., 2003. Diversity and distribution of grasshopper fauna of Greater Kolkata with notes on their ecology. *Memoirs*, 19 (3): 1–118.

Grez, A., Zaviezo, T., Tischendorf, L., and Fahrig, L., 2004. A transient, positive effect of habitat fragmentation on insect population densities. *Oecologia*, 141: 444–451.

Julka, J.M., Tandon, S.K., Halder, P. and Shishodia, M.S., 1982. Ecological observation on grasshoppers (Orthoptera: Acridoidea) at Solan, Himachal Pradesh, India. *Oriental Insects*, 16: 63–75.

Kumar, P. and Viraktamath, C.A., 1991a. Illustrated keys for identification of common species of short-horned grasshoppers (Orthoptera: Acridodea) of Karnataka and short notes on their ecology and behaviour. *Hexapoda*, 3(1–2): 53–70.

Saha, H.K. and Haldar, P., 2009. Acridids as indicators of disturbance in dry deciduous forest of West Bengal in India. *Biodiversity Conservation*, 18: 2343–2350.

Tandon, S.K. and Khera, P., 1978. Ecology and distribution of grasshoppers (Orthoptera: Acridoidea) in Arunachal Pradesh, India and impact of human activities on their ecology and distribution. *Memoirs of the School of Entomology, Agra*, 6: 73–92.

Usmani, M.K. and Shafee, S.A., 1990. Classification of Indian Acrididae (Orthoptera: Acridoidea). *Indian Journal of Systematic Entomology*, 7(2): 89–102.

Uvarov, B.P., 1942. Differentiating characters of Oedipodinae and Acridinae. *Transactions of American Entomological Society*, 67: 303–361.

Yaacobi, G., Ziv, Y. and Rosenzweig, M.L., 2007. Habitat fragmentation may not matter to species diversity. *Proceedings of the Royal Society B: Biological Sciences*, 274: 2409–2412.

Chapter 5

Novel Biopesticides for the Sustainable Management of Mosquito Vectors (Insecta : Diptera : Culicidae)

K. Murugan[1]*, *P. Madhiyazhagan A. Noortheen*[1], *T. Nataraj*[1], *D. Amerasan*, *K. Kalimuthu*[1] *and Jiang-Shiou Hwang*[2]

[1]*Department of Zoology, Bharathiar University, Coimbatore – 641 046, Tamil Nadu, India*

[2]*Institute of Marine Biology, National Taiwan Ocean University, Keelung, Taiwan, 202*

ABSTRACT

Mosquitoes are vectors of several diseases like malaria, filariasis, dengue fever, yellow fever, etc., causing serious health problems to human beings. In the past, synthetic organic chemical insecticides based intervention measures for the control of insect pests and disease vectors have resulted in development of insecticide resistance in some medically important vectors of malaria, filariasis and dengue fever. In view of the above, we have investigated on the impact of biopesticides for the control of mosquito species. There was

* Corresponding author: E-mail: kmvvkg@gmail.com; kmvvk@rediffmail.com

considerable toxicity was evident after the individual treatment of Azadiracthin with other microbial insecticides (*Bacillus thuringiensis*, Spinosad, nanomaterials and copepods). Study was also undertaken to test the predatory efficiency of copepode against mosquito larvae which were co-existing dengue vector in the breeding sites (stored container). Predatory efficiency was greatly increased after the combination of Azadiracthin and microbial insecticides. Green synthesis of silver nanoparticles was also done in the *Catharanthus roseus* and study was investigated for the toxicity with microbial insecticides. The results showed significant mortality after combined treatment with nanoparticles.

Keywords: *Biopesticides, Nanomaterials, Anopheles stephensi, Aedes aegypti, Culex quinquefasciatus, Larval toxicity, Predatory efficiency.*

Introduction

The major disease burden in India is contributed by vector-borne diseases like Malaria, Filaria and Dengue fever. Mosquito breeds in waters polluted with organic debris such as rooting vegetation, household refuse, drains and ditches, soakaway pits, septic tanks, and in village pots. Mosquitoes are insects of sanitary and economical importance both due to their irritant or allergenic bites and to the parasites they transmit (Snow *et al.*, 2005). Vector control remains a primary element in the current global strategy for the control of vector-borne diseases (Zaim and Jambulingam, 2004). Mosquito transmitted disease continues to be a major source of illness and death. Dengue is a mosquito-borne infection that has become a major international public health concern. The mosquito gets the virus by biting an infected person. The first symptom of the disease appears in about 5–7 days after the infected mosquito bites a healthy person. According to the World Health Organization, 50 to 100 million dengue infections occur yearly, including 500000 Dengue Haemorrhagic Fever cases and 22000 deaths, mostly among children.

Malaria is world's largest parasitic disease, killing more people than any other communicable disease except tuberculosis. Malaria is a major public health problem in more than 100 countries, inhabited by a total of some 2.4 billion people, or close to half of the world's population. *A.stephensi* is recognized as a major vector for urban malaria in India. India contributes 77 per cent of the total malaria in

Southeast Asia (Kumar *et al.*, 2007). Malaria, caused by *Plasmodium falciparum*, is one of the leading causes of human morbidity and mortality from infectious diseases, predominantly in tropical and subtropical countries (Snow *et al.*, 2005). In India, malaria is transmitted by six vector species, in which *Anopheles stephensi* is responsible in urban areas (Senthilkumar *et al.*, 2009). The highest number of malaria, *P. falciparum* cases, and malaria-related deaths are recorded from the state of Orissa located in the eastern part of India (Sharma *et al.*, 2010).

C. quinquefasciatus is one of the most annoying and important man-biting mosquitoes. It works as the vector of lymphatic filariasis and Japanese encephalitis in India (Mourya *et al.*, 1989; Das *et al.*, 2002). Pandian *et al.* (1989) observed the repellent activity of herbal smoke on the biting activity of *C. quinquefasciatus* it is a major public health and socio-economic problem; approximately 420 million people reside in endemic areas, and 48.11 million are infected. *C. quinquefasciatus* is one of the potential vectors of *Wuchereria bancrofti*, the causative agent of human lymphatic filariasis infecting over 120 million people all over the world (Terranella *et al.*, 2006). Vector control is of serious concern in developing countries like India due to the lack of general awareness, development of resistance and socio-economic reasons.

Management of disease vector using synthetic chemicals has failed because of resistance, effect on nontarget organisms and environmental pollution. On the other hand, the recent public perception against the vector control using synthetic chemicals has shifted the research effort towards the development of environmentally sound and biodegradable agents. Plants are rich source of bioactive organic chemicals and offer an advantage over synthetic pesticides as these are less toxic, less prone to development of resistance, and easily biodegradable. Many of the herbs and shrubs are found to have promising medicinal properties, mosquito larvicidal and mosquito repellent properties. Many studies on plant extracts against mosquito larvae have been conducted around the world.

Biological control, using the natural enemies of *Aedes*, appears to be an alternative approach to the systematic failure of use of insecticides (Lardeux *et al.*, 2002). Homeowners also have used biological control methods instead of pesticides as an alternate means to treat their properties and keep arthropods away. Copepods are small aquatic crustaceans. Observations of copepod predation on first-instar mosquito larvae (Hurlbut, 1938), led to the first investigation in 1981 (Riviere *et al.*, 1981) of their potential as biological control agents. Since then, various species of predacious copepods have been tested for their potential to control mosquitoes. Several species of copepods, including *Mesocyclops aspericornis*, *M. thermocyclopoides*, *M. guangxiensis*, and *M. longisetus*, have been reported as potential biological control agents of *Ae. aegypti* (Jekow *et al.*, 1998; Locantoni *et al.*, 2006; Manrique-Saide *et al.*, 1998; Schaper, 1999; Rawlins *et al.*, 1997). It has been shown to be an effective predator of *Ae. aegypti* larvae in both laboratory experiments and field trials (Kay *et al.*, 1992; Russell *et al.*, 1996).

The discovery of the mosquitocidal strains of *B. thuringiensis* var. *israelensis* de Barjac (Bti) and *B. sphaericus* Neide (Bs) during the mid- 1970s, microbial larvicides have become available that are highly effective, yet selective (Charles and Nielsen-LeRoux, 2000), and therefore environmentally safe to non-target organisms (WHO, 1999). Bacillus products are easy to handle, cost-effective, require inexpensive equipments to apply, and can be organized locally (Becker, 1992). More importantly, the low probability of mosquito vectors to develop resistance against Bti (Charles and Nielsen-LeRoux, 2000) and its high general acceptability makes it a good candidate. The advantages of targeting the larval stages in mosquito control are that mosquitoes are killed before they disperse to human habitations. Unlike adult mosquitoes (Charlwood and Graves, 1987; Yohannes *et al.*, 2005), mosquito larvae cannot escape control activities targeted at the larval habitat (Killeen *et al.*, 2002).

Spinosad (spinosyn A and spinosyn D) are a new chemical class of insecticides that are registered by the United States Environmental Protection Agency (EPA) to control a veriety of insets. The active ingredient is derived from a naturally occurring soil dwelling bacterium called *Saccharopolyspora spinosa*. Spinosad kills susceptible species by causing rapid excitiation of the insect nervous system. Due to this unique mode of action, Spinosad is valued in resistance management programs. Spinosad must; be ingested by the insect, therefore it has little effect on sucking insects and non-target predatory insects. Spinosad is relatively fast acting. The insect dies within 1 to 2 days after ingesting the active ingredient. There appears to be 100 per cent mortality.

Nanoparticles, generally considered as particles with a size of up to 100 nm, exhibit completely new or improved properties as compared to the larger particles of the bulk material that they are composed of based on specific characteristics such as size, distribution, and morphology (Willems and van den Wildenberg, 2005). The growing environmental concerns and the strict environmental regulations in various countries and demand for developing clean, non-toxic, and eco-friendly synthetic approaches (green chemistry) in the synthesis of these nanoparticles has led to the fabrication of bio-based advanced materials. The properties of these particles in applications as diverse as catalysis, sensors and medicine depend critically on the size and composition of the nanoparticles (Haverkamp *et al.*, 2007). Production of nanoparticles can be achieved through mainly three methods such as, chemical, physical and biological methods. The synthesis of metal nanoparticles using biological systems is an expanding research area due to the potential applications in nanomedicines. Nanoparticles synthesized by chemical method are not eco-friendly. The biological synthesis of silver nanoparticles is convenient and extracellular method which is environmentally safe. In recent years, the biosynthetic method using plant extracts has received more attention than chemical and physical methods, and even than the use of microbes, for the nano-scale metal synthesis due to the absence of any requirement to maintain an aseptic environment. Nanoparticles have attracted considerable attention owing to their various applications. The silver nanoparticles are reported to possess anti-bacterial (Sathishkumar *et al.*, 2009), antiviral (Rogers *et al.*, 2008), anti-fungal activity (Panacek *et al.*, 2009).

Hence, in the present investigations an attempt has been to evaluate the impact of neem active principle, *Azadiracthin* with other microbial insecticides (*Bacillus thuringiensis*, Spinosad, nanomaterials and copepods) on the larvicidal, pupicidal toxicity effect of dengue vectors, such as *Aedes aegypti*.

Materials and Methods

Collection of Eggs and Maintenance of Larvae

The eggs of *A. stephensi*, *A. aegypti* and *C. quinquefasciatus* were collected from National Centre for Disease Control (NCDC) field station of Mettupalayam, Tamil Nadu, India, using an 'O' type brush.

These eggs were brought to the laboratory and transferred to 18×13×4 cm enamel trays containing 500 ml of water for hatching. The mosquito larvae were pedigree dog biscuits and yeast at 3:1 ratio. The feeding was continued until the larvae transformed into the pupal stage.

Maintenance of Pupae and Adults

The pupae were collected from the culture trays and transferred to plastic containers (12×12 cm) containing 500 ml of water with the help of a dipper. The plastic jars were kept in a 90×90×90 cm mosquito cage for adult emergence. Mosquito larvae were maintained at 27+2°C, 75–85 per cent RH, under a photoperiod of 14 L: 10D. A 10 per cent sugar solution was provided for a period of 3 days before blood feeding.

Blood Feeding of Adult Mosquito Vectors

The adult female mosquitoes were allowed to feed on the blood of a rabbit (a rabbit per day, exposed on the dorsal side) for 2 days, to ensure adequate blood feeding for 5 days. After blood feeding, enamel trays with water from the culture trays were placed in the cage as oviposition substrates.

Collection of Neem Seeds and Preparation of Concentration

The Neem seeds were collected from Neem trees in and around Coimbatore, Tamil Nadu, India. The seeds were ground to a fine powder (200–250 lm) using an electric blender. One-hundred gram aliquots of the powder were extracted with 300 ml of distilled water for 8 h using a Soxhlet apparatus. One ml of extract was added to 99 ml of acetone to make a 1 per cent stock solution that was used in mosquito bioassays. 50, 100, 150, 200 and 250 ppm of Neem seed Kernel Extract (NSKE) were prepared by diluting the stock solution with distilled water.

Larval Toxicity Test of NSKE

Laboratory colonies of mosquito larvae were used to test the larvicidal activity of NSKE. Twenty-five first, second, third and fourth instars were introduced into four separate 500-ml glass beakers containing 249 ml of de-chlorinated water and 1 ml of the desired concentration of NSKE. Larval food was provided to the test larvae and the experiment consisted of five replicates for each NSKE concentration. The control consisted of 1 ml of acetone in 249 ml of dechlorinated water. Mortality in the treatments was corrected for mortality in the control using Abbott's formula (Abbott, 1925). The LC_{50}, LC_{90}, regression equation and 95 per cent confidence limits were calculated from toxicity data using probit analysis (Finney, 1971).

Pupal Toxicity Test of NSKE

A laboratory colony of mosquito pupae was used to assess the pupicidal activity of NSKE. Twenty freshly emerged pupae were kept in a 500-ml glass beaker containing 249 ml of de-chlorinated water and 1 ml of the desired NSKE concentrations. The experiment consisted of five replicates. The control consisted of 1 ml of acetone in 249 ml of dechlorinated water. Mortality in the treatments was corrected for mortality in the control using Abbott's formula (Abbott, 1925). The LC_{50} and LC_{90} were calculated from toxicity data using probit analysis (Finney, 1971).

Collection and Culture of Copepods

The copepod, *M. aspericornis* was collected from a pond (Muthanamkulam, Coimbatore-641 046, India) using a 100 lm-mesh net. Culture techniques followed by methods used by Kosiyachinda *et al.* (2003). Isofemale lines were established from gravid females and maintained at the Department of Zoology, School of Life Sciences, Bharathiar University, Coimbatore, India. All the collected samples were identified and confirmed by Dr. Y. Ranga Reddy, Department of Zoology, Acharya Nagarjuna University, Andhra Pradesh, India to be *M. aspericornis.* Once identified, gravid female *M. aspericornis* from different isofemale lines were pooled and massreared in dechlorinated water (pH 7) in a 15-l fish tank at 27±2_C (mean±SD). *Paramecium* spp. prepared from boiled rice straw water extract and commercial powdered fish foods were used to support the copepod cultures. We sampled 100 copepods to apply as the initial number to each treatment that contained copepods as a control agent.

Predatory Efficiency Test

Adult copepods were used to measure predatory activity towards the four instars of mosquito larvae. One hundred mosquito larvae of each instar and ten adult copepods were introduced into five separate 500-ml glass beakers containing 250 ml of dechlorinated water. The mosquito larvae were replaced daily with the new ones. Each mosquito instarcopepod treatment was replicated five. Controls consisted of 250 ml of de-chlorinated water without copepods. The glass beakers were checked after 24, 48 and 72 h and the number of prey consumed by the predator was recorded.

Predatory efficiency of a copepod was calculated by the following formula:

$$\text{Predator Efficiency} = \frac{\dfrac{\text{No. of Prey Consumed}}{\text{No. of Predators}}}{\text{Total Number of Prey Introduced}} \times 100$$

Microbial Bioassay

B. thuringiensis subsp. israelensis was obtained from Tuticorin Alkali Chemicals and Fertilizers Limited, Chennai, India. *B. thuringiensis* subsp. var *israelensis*, 630 ITU/mg (a.i.) 5 per cent w/w; total proteins (including the active ingredient 5 per cent w/w), 10 per cent w/w; fermentation solids, 10 per cent w/w; inert ingredient, 48 per cent w/w; non-ionic surfactant, 0.2 w/w; food grade preservative, 0.3 per cent; UV protectant, 0.1 per cent; and water, 71.4 per cent were used. Total 100 per cent w/w was active specifically against mosquito larvae. The required quantity of *B. thuringiensis* subsp. Var. *israelensis* was thoroughly mixed with distilled water and prepares various concentrations, ranging from 0.6 to 0.96 μg/ml.

Spinosad was obtained from T-Stanes and Company Limited, Research and Development Centre, Coimbatore, Tamil Nadu, India. The required quantity of spinosad was thoroughly mixed with distilled water to prepare at various concentrations ranging from 0.02 to 0.10 ppm, respectively.

Larval Toxicity Test

A laboratory colony of *A. stephensi*, *A. aegypti* and *C. quinquefasciatus* larvae were used for the larvicidal activity. Hundred individuals of first, second, third, and fourth instar larvae were kept in a 500-ml glass beaker containing 249 ml of dechlorinated water and 1 ml of desired concentration of plant extracts, and *B. thuringiensis* subsp. Var. *israelensis* was added. Larval food was given for the test larvae. At each tested concentration, two to five trials were made, and each trial consisted of five replicates. The control was set up by mixing 1 ml of acetone with 249 ml of dechlorinated water. The larvae exposed to dechlorinated water without acetone served as control. The control mortalities were corrected by using Abbott's formula (Abbott, 1925). The LC_{50} and LC_{90} were calculated from toxicity data by using probit analysis (Finney, 1971).

$$\text{Percentage Mortality} = \frac{\text{Number of Dead Larvae}}{\text{Number of Larvae Introduced}} \times 100$$

Plant Collection

Fresh leaves of *Catharanthus roseus* were collected from local (in and around Bharathiar University, Coimbatore Tamil Nadu, India) and the taxonomic identification made by BSI (Botanical survey of India), Coimbatore. The voucher specimen was numbered and kept in our research laboratory for further reference. Silver nitrate was obtained from the Precision Scientific Co., Coimbatore, India.

Synthesis of Silver Nanoparticles

The fresh leaf of *Catharanthus roseus* broth solution was prepared by taking 10 g of thoroughly washed and finely cut leaves in a 300-mL Erlenmeyer flask along with 100 mL of sterilized double distilled water and then boiling the mixture for 5 min before finally decanting it. The extract was filtered through Whatman Filter Paper No. 1 and stored at "15°C and could be used within 1 week. The filtrate was treated with aqueous 1 mM $AgNO_3$ solution in an Erlenmeyer flask and incubated at room temperature. As a result in a brown-yellow solution indicating the formation of AgNPs was found that aqueous silver ions can be reduced by aqueous extract of plant parts to generate extremely stable silver nanoparticles in water.

Characterization of the Synthesized Silver Nanoparticles

Synthesis of AgNPs solution with leaves extract may be easily observed by UV–vis spectroscopy. The bio-reduction of the Ag^+ ions in solutions was monitored by periodic sampling of aliquots (1mL) of the aqueous component and measuring the UV–vis spectra of the solution. UV–vis spectra of these aliquots were monitored as a function of time of reaction on a Vasco 1301 spectrophotometer in 400–600-nm range operated at a resolution of 1 nm.

Scanning Electron Microscope (SEM) and Energy Dispersive X-ray

Spectrometer (EDX) Analysis

SEM is a type of electron microscope that images a sample by scanning it with a high-energy beam of electrons in a raster scan patterns. In this experiment after the synthesis of nanoparticles using the plants and then lyophilisation was done using VIRTIS BENCHTOP machine. SEM and EDX analysis was done using JEOL-MODEL 6390 machine. Thin films of the sample were prepared on a carbon coated copper grid by just dropping a very small amount of the sample on the grid, extra solution was removed using a blotting paper and then the film on the SEM grid were allowed to dry by putting it under a mercury lamp for 5 min.

X-ray Diffraction Analysis

The particle size and nature of the silver nanoparticle was determined using X-ray diffraction (XRD). This was carried out using Shimadzu XRD-6000/6100 model with 30kv, 30mA with Cu kα radians at 2θ angle. X-ray powder diffraction (XRD) is a rapid analytical technique primarily used for phase identification of a crystalline material and can provide information on unit cell dimensions. The analyzed material is finely ground, and average bulk composition is determined. The particle or grain size of the particles on the silver nanoparticles was determined using Debye Sherrer's equation.

$$D = 0.94\lambda / B \cos\theta$$

Statistical Analysis

The average larval mortality data were subjected to probit analysis for calculating LC_{50} and LC_{90}, and other statistics at 95 per cent fiducidal limits of upper confidence limit and lower confidence limit, and $\chi 2$ values were calculated using Finney's method (1971). SPSS software package 16.0 version was used. Results with $P<0.05$ were considered to be statistically significant.

Results

Table 5.1 provides the effect of NSKE at different concentrations on the mortality and toxicity of different larval instars of *Aedes aegypti*. Considerable mortality was evident after the treatment of NSKE.

Table 5.1: Larval and Pupal Toxicity of *Azadirachta indica* Seed Kernel Extract on Dengue Vector, *Aedes aegypti*

Larval and Pupal Stage	*Larval and Pupal Mortality (per cent) (Mean±S.D)*					*LC_{50} (LC_{90})*	*95 per cent Confidence Limit*				*Regression Equation*	*Chi Square Value*
	Concentration (ppm)						*LC_{50}*		*LC_{90}*			
	50	*100*	*150*	*200*	*250*		*LCL*	*UCL*	*LCL*	*UCL*		
I	23.6±3.91	40.2±3.50	61±4.35	86.6±3.04	98.2±1.78	115.78 (215.02)	104.91	125.79	200.38	234.58	X=+0.01291 Y=–1.49521	4.079
II	32±3.39	52±3.80	69.4±3.04	87.4±2.40	96.2±3.03	95.95 (213.57)	81.33	108.22	196.52	236.62	X=+0.01090 Y=–1.045490	0.631
III	26.8±2.86	45.6±2.60	60.6±4.21	82.2±3.89	92.6±3.84	113.36 (238.83)	99.51	125.53	219.48	265.34	X=+0.01021 Y=–1.15794	0.923
IV	19.2±3.34	27.2±2.28	55.8±2.77	66.8±3.27	81.8±2.58	149.53 (288.23)	136.46	162.57	263.07	323.92	X=+0.00924 Y=–1.38160	2.681
Pupa	10.2±3.27	22.8±1.15	41.6±1.28	61.4±1.02	79±1.20	171.73 (295.37)	150.02	184.37	271.83	328.02	X=+0.01037 Y=–1.78013	0.015

LCL: 95 per cent of Lower Confidence Limit; UCL: 95 per cent of Upper Confidence Limit.

Within a column means followed by the same letter(s) are not significantly different at 5 per cent level by DMRT.

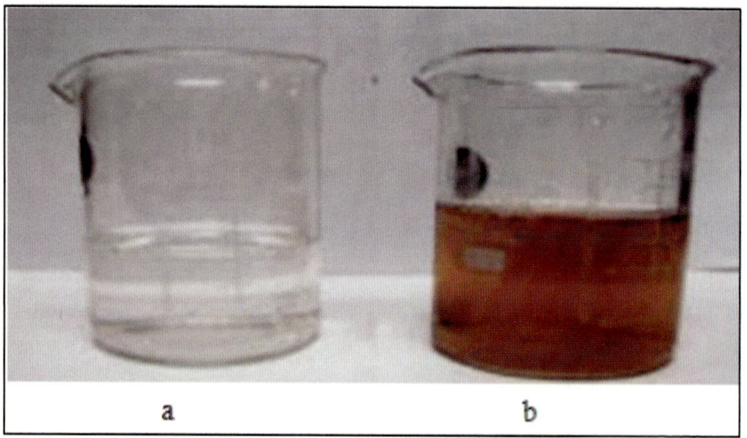

Figure 5.1: Photograph Showing Colour Changes After Adding $AgNO_3$ Before Reaction (a) and After Reaction Time 6h (b)

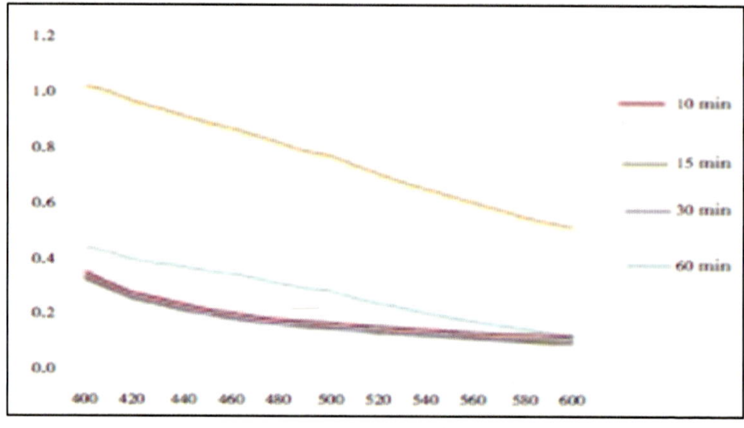

Figure 5.2: UV-vis Absorption Spectra of Aqueous Silver Nitrate with *C. roseus* Leaf Extract at Different Time Intervals

The lowest mortality recorded after treating with NSKE was 10.2 per cent at 50ppm on the pupa and the highest mortality was 98.2 per cent at 250 ppm on I instars. The LC_{50} and LC_{90} values were 115.78 and 215.02. The mortalities of II instar *Aedes aegypti* after the treatment of NSKE at 50, 100, 150, 200 and 250 ppm concentration were 32, 52, 69.4, 87.4 and 96.2 respectively. The LC_{50} and LC_{90} values were 95.95 and 213.57. Mortality of III instars 26.8, 45.6, 60.6, 82.2 and 92.6 per

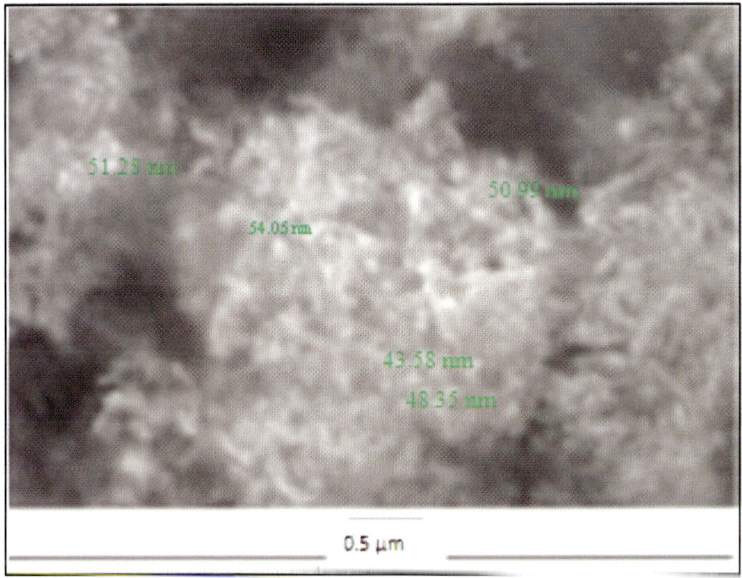

Figure 5.3: SEM Image of Silver Nanoparticles Formed by *C. roseus*

Figure 5.4: EDX Spectra Recorded from a Film, After Formation of Silver Nanoparticles with Different X-ray Emission Peaks Labeled

cent, respectively. The LC_{50} and LC_{90} values were 113.36 and 238.83. The mortality of IV instars was 19.2, 27.2, 55.8, 66.8 and 81.8 per cent, respectively. The LC_{50} and LC_{90} values were 149.53 and 288.23.

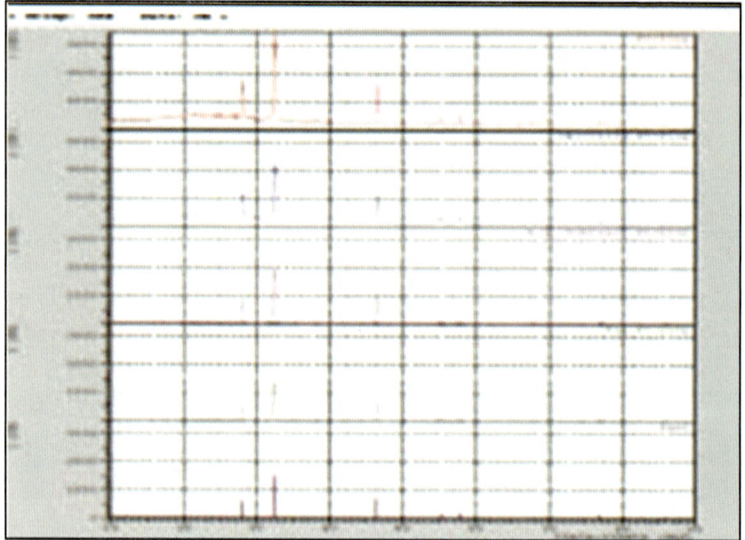

Figure 5.5: XRD Pattern of Silver Nanoparticles Formed After Reaction of Plant Extracts *C. roseus*

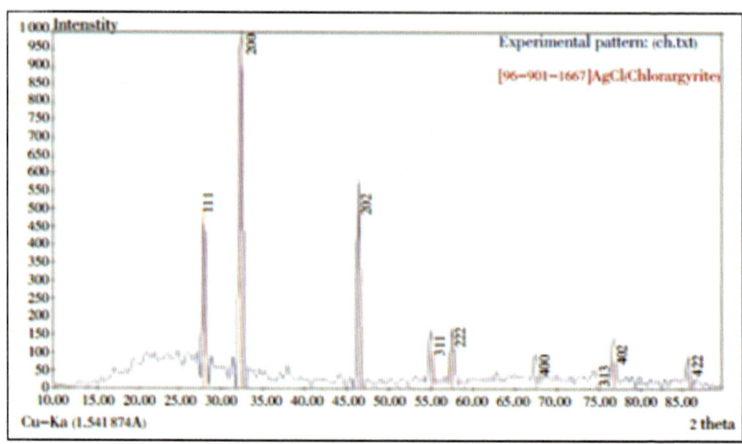

Figure 5.6: Phase Matched XRD Pattern of Silver Nanoparticles

NSKE was also effective against the pupae and the recorded pupal mortality was 10.2, 22.8, 41.6, 61.4 and 79 per cent respectively. The LC_{50} and LC_{90} values were 171.73 and 295.37.

Table 5.2: Larval and Pupal Toxicity of *Azadirachta indica* Seed Kernel Extract on Malarial Vector, *Anopheles stephensi*

Larval and Pupal Stage	*Larval and Pupal Mortality (per cent) (Mean±S.D)*					*LC_{50} (LC_{90})*	*95 per cent Confidence Limit*				*Regression Equation*	*Chi Square Value*
	Concentration (ppm)						*LC_{50}*		*LC_{90}*			
	50	*100*	*150*	*200*	*250*		*LCL*	*UCL*	*LCL*	*UCL*		
I	21.8±3.11	41±4.94	60.4±4.61	87.6±3.43	97.6±2.30	116.82 (214.27)	106.16	123.68	199.57	233.40	X=+0.01215 Y=–1.53636	3.216
II	31.4±3.91	50.6±3.04	68.4±4.66	90.2±3.96	98.2±2.48	97.24 (203.53)	84.22	108.42	188.09	244.02	X=+0.01206 Y=–1.17244	2.721
III	24.6±3.64	46.4±3.36	60.8±4.32	81.8±3.49	94±3.80	144.50 (233.82)	101.40	126.15	215.58	258.52	X=+0.01074 Y=–1.22986	1.241
IV	18.8±2.94	28.8±3.11	54.6±2.70	66.4±3.20	83.8±3.89	160.27 (277.61)	89.91	250.49	210.54	703.67	X=+0.01092 Y=–1.75067	23.41
Pupa	11.2±1.92	23.2±3.27	43±2.91	64±3.31	77.4±3.97	169.79 (297.12)	157.81	182.66	272.86	330.99	X=+0.01006 Y=–1.70893	00.42

LCL: 95 per cent of Lower Confidence Limit; UCL: 95 per cent of Upper Confidence Limit.

Within a column means followed by the same letter(s) are not significantly different at 5 per cent level by DMRT.

Table 5.3: Larval and Pupal Toxicity of *Azadirachta indica* Seed Kernel Extract on Filarial Vector, *Culex quinquefasciatus*

Larval and Pupal Stage	*Larval and Pupal Mortality (per cent) (Mean±S.D)*					*LC_{50} (LC_{90})*	*95 per cent Confidence Limit*				*Regression Equation*	*Chi Square Value*
	Concentration (ppm)						*LC_{50}*		*LC_{90}*			
	50	*100*	*150*	*200*	*250*		*LCL*	*UCL*	*LCL*	*UCL*		
I	22.8±2.28	42±3.80	60.6±3.04	86.6±3.57	97.8±3.19	115.70 (215.46)	104.76	125.75	200.42	235.12	X=+0.01285 Y=–1.48627	3.420
II	33.4±3.64	51.2±4.08	66.4±2.70	89.4±2.60	99±1.41	96.09 (204.98)	61.15	120.15	174.31	264.80	X=+0.01177 Y=–1.13097	5.432
III	22.8±3.56	45.2±3.34	59.8±3.96	80.4±4.15	91.8±3.03	144.50 (233.82)	101.40	126.15	215.58	258.52	X=+0.01074 Y=–1.22986	1.241
IV	19.8±2.86	27.2±3.70	53.6±2.96	64.6±3.36	83.4±4.27	143.75 (260.86)	104.97	180.83	213.65	386.55	X=+0.01094 Y=–1.57296	9.829
Pupa	9.8±2.38	20.2±1.28	42.6±3.43	62±2.64	76.6±3.13	174.28 (298.02)	162.53	187.05	274.23	331.06	X=+0.01036 Y=–1.80498	0.749

LCL: 95 per cent of Lower Confidence Limit; UCL: 95 per cent of Upper Confidence Limit.

Within a column means followed by the same letter(s) are not significantly different at 5 per cent level by DMRT.

The mortality and toxicity of different larval instars of *Anopheles stephensi*. Considerable mortality was evident after the treatment of NSKE (Table 5.2). The lowest mortality recorded after treating with NSKE was 11.2 per cent at 50ppm on the pupa and the highest mortality was 97.6 per cent at 250 ppm on I instars. The LC_{50} and LC_{90} values were 116.82 and 214.27. The mortalities of II instar *Anopheles stephensi* after the treatment of NSKE at 50, 100, 150, 200 and 250 ppm concentration were 31.4, 50.6, 68.4, 90.2 and 99 respectively. The LC_{50} and LC_{90} values were 96.09 and 204.98. Mortality of III instars 24.6, 46.4, 60.8, 81.8 and 94 per cent, respectively. The LC_{50} and LC_{90} values were 144.50 and 233.82. The mortality of IV instars was 18.8, 28.8, 54.6, 66.4 and 83.8 per cent, respectively. The LC_{50} and LC_{90} values were 160.27 and 277.61. NSKE was also effective against the pupae and the recorded pupal mortality was 11.2, 23.2, 43, 64 and 77.4 per cent respectively. The LC_{50} and LC_{90} values were 169.79 and 297.12.

The mortality and toxicity of different larval instars of *Culex quinquefasciatus*. Considerable mortality was evident after the treatment of NSKE (Table 5.3). The lowest mortality recorded after treating with NSKE was 9.8 per cent at 50ppm on the pupa and the highest mortality was 97.8 per cent at 250 ppm on I instars. The LC_{50} and LC_{90} values were 115.70 and 215.46. The mortalities of II instar *C. quinquefasciatus* after the treatment of NSKE at 50, 100, 150, 200 and 250 ppm concentration were 33.4, 51.2, 66.4, 89.4 and 99, respectively. The LC_{50} and LC_{90} values were 96.09 and 204.98. Mortality of III instars 22.8, 45.2, 59.8, 80.4 and 91.8 per cent, respectively. The LC_{50} and LC_{90} values were 144.50 and 233.82. The mortality of IV instars was 19.8, 27.2, 53.6, 64.6 and 83.4 per cent, respectively. The LC_{50} and LC_{90} values were 143.75 and 260.86. NSKE was also effective against the pupae and the recorded pupal mortality was 9.8, 20.2, 42.6,62 and 76.6 per cent respectively. The LC_{50} and LC_{90} values were 174.28 and 298.02.

Table 5.4 provides the effect of at *B. thuringiensis* different concentrations on the mortality the larval mortality of *Aedes aegypti* (I to Pupa) after the treatment of *B. thuringiensis* subsp. *israelensis* at different concentrations (0.06 to 0.96 µg/ml). Twenty-four per cent mortality was noted at I instar larvae by the treatment of *B. thuringiensis* subsp. *israelensis* at 0.06 µg/ml, whereas it has been increased to 98.6 per cent at 0.96 µg/ml of *B. huringiensis* subsp.

Table 5.4: Larval and Pupal Toxicity of Bacterial Pesticide *Bacillus thruingiensis* on Dengue Vector, *Aedes aegypti*

Larval and Pupal Stage	Larval and Pupal Mortality (per cent) (Mean±S.D)					LC_{50} (LC_{90})	95 per cent Confidence Limit				Regression Equation	Chi Square Value
	Concentration (ppm)						LC_{50}		LC_{90}			
	0.06	0.12	0.24	0.48	0.96		LCL	UCL	LCL	UCL		
I	28.4±2.60	34.8±3.03	55.4±3.84	78±2.91	98.6±2.19	0.229 (0.638)	0.187	0.270	0.562	0.746	X=+3.13656 Y=–0.72012	0.983
II	27.6±3.04	37.2±2.77	55±3.67	76.2±3.27	96.2±1.64	0.236 (0.713)	0.188	0.282	0.628	0.835	X=+2.68789 Y=–0.63594	2.422
III	23.4±3.57	30.8±3.42	52.4±2.79	72.6±2.40	90.2±2.86	0.297 (0.857)	0.128	0.459	0.635	1.490	X=+2.28556 Y=–0.67885	8.238
IV	18.6±3.20	29.6±2.88	48.2±3.83	64.2±3.03	89.2±3.76	0.350 (0.919)	0.217	0.498	0.707	1.428	X=+2.25235 Y=–0.79007	6.302
Pupa	10.6±3.36	28.6±2.96	36.8±3.76	53.6±3.04	73.2±3.70	0.522 (1.272)	0.292	1.022	0.866	3.362	X=+1.70895 Y=–0.89222	11.826

LCL: 95 per cent of Lower Confidence Limit; UCL: 95 per cent of Upper Confidence Limit.

Within a column means followed by the same letter(s) are not significantly different at 5 per cent level by DMRT.

Table 5.5: Larval and Pupal Toxicity of Bacterial Pesticide *Bacillus thruingiensis* on Malarial Vector, *Anopheles stephensi*

Larval and Pupal Stage	*Larval and Pupal Mortality (per cent) (Mean±S.D)*					*LC_{50} (LC_{90})*	*95 per cent Confidence Limit*				*Regression Equation*	*Chi Square Value*
	Concentration (ppm)						*LC_{50}*		*LC_{90}*			
	0.06	*0.12*	*0.24*	*0.48*	*0.96*		*LCL*	*UCL*	*LCL*	*UCL*		
I	27.8±2.86	32.8±4.14	55.8±3.83	77.2±3.27	97.6±1.14	0.239 (0.668)	0.195	0.281	0.590	0.780	X=+2.98537 Y=–0.71418	2.416
II	24.4±3.20	38.2±4.49	56.8±4.20	75±3.31	98±1.58	0.236 (0.668)	0.192	0.279	0.589	0.781	X=+2.96491 Y=–0.70091	3.496
III	25.4±4.15	29.4±2.40	50.6±3.04	71.8±4.60	92.2±5.40	0.295 (0.827)	0.244	0.346	0.730	0.964	X=+2.40895 Y=–0.71197	4.717
IV	20.8±4.54	30.2±3.96	50±4.30	65.8±4.76	87.8±3.56	0.339 (0.942)	0.184	0.503	0.709	1.562	X=+2.12394 Y=–0.72120	7.017
Pupa	9.4±2.70	26.8±2.58	37.2±3.42	55.4±4.15	76.4±4.39	0.500 (1.179)	0.281	0.926	0.816	2.859	X=+1.88702 Y=–0.94496	12.719

LCL: 95 per cent of Lower Confidence Limit; UCL: 95 per cent of Upper Confidence Limit.

Within a column means followed by the same letter(s) are not significantly different at 5 per cent level by DMRT.

Table 5.6: Larval and Pupal Toxicity of Bacterial Pesticide *Bacillus thruingiensis* on Filarial Vector, *Culex quinquefasciatus*

Larval and Pupal Stage	Larval and Pupal Mortality (per cent) (Mean±S.D)					LC_{50} (LC_{90})	95 per cent Confidence Limit				Regression Equation	Chi Square Value
	Concentration (ppm)						LC_{50}		LC_{90}			
	0.06	0.12	0.24	0.48	0.96		LCL	UCL	LCL	UCL		
I	27.4±4.66	32±3.67	51.8±3.56	76.6±4.21	98.6±1.94	0.248 (0.653)	0.208	0.289	0.577	0.761	X=+3.16901 Y=–0.78792	0.579
II	23±4.74	35.8±4.65	54.8±5.63	76.8±4.32	98.2±2.16	0.244 (0.650)	0.203	0.285	0.575	0.758	X=+3.15767 Y=–0.77295	2.768
III	26.6±5.54	27.2±5.80	49.8±4.49	72.6±4.87	94.6±3.36	0.289 (0.774)	0.242	0.337	0.685	0.898	X=+2.64423 Y=–0.76596	3.613
IV	18.2±3.19	23.8±5.40	45.8±4.81	67±4.47	91.2±4.49	0.355 (0.859)	0.239	0.488	0.673	1.270	X=+2.54068 Y=–0.90288	6.111
Pupa	8.8±2.16	18.6±3.57	42±4.06	61.8±4.76	88.4±2.96	0.423 (0.909)	0.251	0.678	0.662	1.728	X=+2.63318 Y=–1.11445	13.566

LCL: 95 per cent of Lower Confidence Limit; UCL: 95 per cent of Upper Confidence Limit.

Within a column means followed by the same letter(s) are not significantly different at 5 per cent level by DMRT.

israelensis treatment. Similar trends have been noted for all the instars of *Aedes aegypti* at different concentrations. The LC_{50} and LC_{90} values were represented as follows: the LC_{50} value of I instar was 0.229 per cent, II instar was 0.236 per cent, III instar was 0.297 per cent, IV instar was 0.350 per cent, and pupa was 0.522 per cent respectively. The LC_{90} value of I instar was 0.638 per cent, II instar was 0.713 per cent, III instar was 0.857 per cent, IV instar was 0.919 per cent, and pupa was 1.272 per cent, respectively.

Table 5.5 the larval mortality of *Anopheles stephensi* (I to Pupa) after the treatment of *B. thuringiensis* subsp. *israelensis* at different concentrations (0.06 to 0.96 µg/ml). Twenty-four per cent mortality was noted at I instar larvae by the treatment of *B. thuringiensis* subsp. *israelensis* at 0.06 µg/ml, whereas it has been increased to 97.6 per cent at 0.96 µg/ml of *B. huringiensis* subsp. israelensis treatment. Similar trends have been noted for all the instars of *C. quinquefasciatus* at different concentrations. The LC_{50} and LC_{90} values were represented as follows: the LC_{50} value of I instar was 0.239 per cent, II instar was 0.236 per cent, III instar was 0.295 per cent, IV instar was 0.339 per cent, and pupa was 0.500 per cent respectively. The LC_{90} value of I instar was 0.668 per cent, II instar was 0.668 per cent, III instar was 0.827 per cent, IV instar was 0.942 per cent, and pupa was 1.179 per cent, respectively.

Table 5.6 the larval mortality of *C. quinquefasciatus* (I to Pupa) after the treatment of *B. thuringiensis* subsp. *israelensis* at different concentrations (0.06 to 0.96 µg/ml). Twenty-four per cent mortality was noted at I instar larvae by the treatment of *B. thuringiensis* subsp. israelensis at 0.06 µg/ml, whereas it has been increased to 98.6 per cent at 0.96 µg/ml of *B. thuringiensis* subsp. *israelensis* treatment. Similar trends have been noted for all the instars of *C. quinquefasciatus* at different concentrations. The LC_{50} and LC_{90} values were represented as follows: the LC_{50} value of I instar was 0.248 per cent, II instar was 0.244 per cent, III instar was 0.289 per cent, IV instar was 0.355 per cent, and pupa was 0.423 per cent respectively. The LC_{90} value of I instar was 0.653 per cent, II instar was 0.650 per cent, III instar was 0.774 per cent, IV instar was 0.859 per cent, and pupa was 0.909 per cent respectively.

Table 5.7 provides the effect of at spinosed different concentrations on the mortality the larval mortality of *Aedes aegypti* (I to Pupa) after the treatment of spinosed at different concentrations

Table 5.7: Larval and Pupal Toxicity of Bacterial Pesticide Spinosed on Dengue Fector, *Aedes aegypti*

Larval and Pupal Stage	*Larval and Pupal Mortality (per cent) (Mean±S.D)*					*LC_{50} (LC_{90})*	*95 per cent Confidence Limit*				*Regression Equation*	*Chi Square Value*
	Concentration (ppm)						*LC_{50}*		*LC_{90}*			
	0.02	*0.04*	*0.06*	*0.08*	*0.10*		*LCL*	*UCL*	*LCL*	*UCL*		
I	42.6±4.3	58.4±4.7	70.2±4.4	85.5±3.8	100±0	0.0316 (0.0832)	0.0003	0.0457	0.0666	0.1295	X=+24.8358 Y=–0.7853	8.377
II	38.4±4.4	61.4±4.2	78.2±3.4	88.8±4.2	100±0	0.0310 (0.0756)	0.0245	0.0362	0.0694	0.0840	X=+28.7609 Y=–0.8936	4.099
III	32.2±3.96	54.8±5.0	68.2±3.8	78.8±5.4	92±4.6	0.0386 (0.0974)	0.0312	0.0445	0.0884	0.1104	X=+21.7954 Y=–0.8422	1.380
IV	21.2±3.7	42.4±4.8	58.6±4.0	79.2±5.8	91.2±5.1	0.0495 (0.0980)	0.0444	0.0541	0.0903	0.1084	X=+26.4078 Y=–1.3073	0.484
Pupa	23.6±5.1	42.4±4.6	61±4.6	72±5.2	91.8±2.0	0.0494 (0.1021)	0.0439	0.0543	0.0936	0.1141	X=+24.2883 Y=–1.2003	2.385

LCL: 95 per cent of Lower Confidence Limit; UCL: 95 per cent of Upper Confidence Limit.

Within a column means followed by the same letter(s) are not significantly different at 5 per cent level by DMRT.

(0.02 to 0.10 µg/ml). Twenty-four per cent mortality was noted at I instar larvae by the treatment of spinosed at 0.02 µg/ml, whereas it has been increased to 100 per cent at 0.96 µg/ml of spinosed treatment. Similar trends have been noted for all the instars of *Aedes aegypti* at different concentrations. The LC_{50} and LC_{90} values were represented as follows: the LC_{50} value of I instar was 0.0316 per cent, II instar was 0.0310 per cent, III instar was 0.0386 per cent, IV instar was 0.0495 per cent, and pupa was 0.0494 per cent respectively. The LC_{90} value of I instar was 0.0832 per cent, II instar was 0.0755 per cent, III instar was 0.0974 per cent, IV instar was 0.0980 per cent, and pupa was 0.1021 per cent,respectively.

Table 5.8 provides the effect of at spinosed different concentrations on the mortality the larval mortality of *Anopheles stephensi* (I to Pupa) after the treatment of spinosed at different concentrations (0.02 to 0.10 µg/ml). Twenty-four per cent mortality was noted at I instar larvae by the treatment of spinosed at 0.02 µg/ml, whereas it has been increased to 100 per cent at 0.96 µg/ml of spinosed treatment. Similar trends have been noted for all the instars of *Anopheles stephensi* at different concentrations. The LC_{50} and LC_{90} values were represented as follows: the LC_{50} value of I instar was 0.0316 per cent, II instar was 0.0286 per cent, III instar was 0.0424 per cent, IV instar was 0.0488 per cent, and pupa was 0.0506 per cent respectively. The LC_{90} value of I instar was 0.0813 per cent, II instar was 0.0739 per cent, III instar was 0.0895 per cent, IV instar was 0.0989 per cent, and pupa was 0.1031 per cent,respectively

Table 5.9 provides the effect of at spinosed different concentrations on the mortality the larval mortality of *Culex quinquefasciatus* (I to Pupa) after the treatment of spinosed at different concentrations (0.02 to 0.10 µg/ml). Twenty-four per cent mortality was noted at I instar larvae by the treatment of spinosed at 0.02 µg/ml, whereas it has been increased to 100 per cent at 0.96 µg/ml of spinosed treatment. Similar trends have been noted for all the instars of *Culex quinquefasciatus* at different concentrations. The LC_{50} and LC_{90} values were represented as follows: the LC_{50} value of I instar was 0.0324 per cent, II instar was 0.0310 per cent, III instar was 0.0372 per cent, IV instar was 0.0470 per cent, and pupa was 0.0494 per cent respectively. The LC_{90} value of I instar was 0.0835 per cent, II instar was 0.0779 per cent, III instar was 0.0886 per cent, IV instar was 0.0979 per cent, and pupa was 0.1002 per cent,respectively.

Table 5.8: Larval and Pupal Toxicity of Bacterial Pesticide Spinosed on Malarial Vector, *Anopheles stephensi*

Larval and Pupal Stage	*Larval and Pupal Mortality (per cent) (Mean±S.D)*					*LC_{50} (LC_{90})*	*95 per cent Confidence Limit*				*Regression Equation*	*Chi Square Value*
	Concentration (ppm)						*LC_{50}*		*LC_{90}*			
	0.02	*0.04*	*0.06*	*0.08*	*0.10*		*LCL*	*UCL*	*LCL*	*UCL*		
I	43.6±5.5	57±3.9	69.6±5.7	83.6±5.1	100±0	0.0316 (0.0813)	0.0018	0.0454	0.0652	0.1251	X=+25.7813 Y=–0.8152	8.471
II	41.4±5.1	64±5.0	79±3.6	90.2±4.9	100±0	0.0286 (0.0739)	0.0216	0.0340	0.0676	0.0823	X=+28.2909 Y=–0.80934	3.607
III	29.8±3.83	49.4±6.3	62.4±4.5	80.4±4.7	98.8±1.7	0.0424 (0.0895)	0.0246	0.0542	0.0741	0.1257	X=+27.1655 Y=–1.1518	7.515
IV	25.4±4.3	41±5.9	57.2±5.8	76.8±4.6	93.4±2.4	0.0488 (0.0989)	0.0436	0.0536	0.0909	0.1098	X=+25.5945 Y=–1.2506	2.222
Pupa	21±3.8	39.8±2.4	64±5.0	73.8±4.9	88.2±5.6	0.0506 (0.1031)	0.0452	0.0555	0.0945	0.1149	X=+24.4178 Y=–1.2364	1.555

LCL: 95 per cent of Lower Confidence Limit; UCL: 95 per cent of Upper Confidence Limit.

Within a column means followed by the same letter(s) are not significantly different at 5 per cent level by DMRT.

Table 5.9: Larval and Pupal Toxicity of Bacterial Pesticide Spinosad on Filarial Vector, *Culex quinquefasciatus*

Larval and Pupal Stage	*Larval and Pupal Mortality (per cent) (Mean±S.D)*					*LC_{50} (LC_{90})*	*95 per cent Confidence Limit*				*Regression Equation*	*Chi Square Value*
	Concentration (ppm)						*LC_{50}*		*LC_{90}*			
	20	*40*	*60*	*80*	*100*		*LCL*	*UCL*	*LCL*	*UCL*		
I	40.8±5.2	58.2±5.4	71.2±4.6	83.8±2.8	100±0	0.0324 (0.0835)	0.0033	0.0462	0.0671	0.1276	X=+25.1132 Y=–0.81552	8.180
II	40.4±4.7	60.6±5.5	74.4±4.5	88.4±5.8	100±0	0.0310 (0.0779)	0.0242	0.0364	0.0714	0.0869	X=+27.3257 Y=–0.8489	5.133
III	35.2±5.6	55.6±5.4	66±4.5	81.6±4.9	98.4±2.3	0.0372 (0.0886)	0.0163	0.4939	0.0730	0.1254	X=+24.9395 Y=–0.9289	6.624
IV	28±4.4	40.8±4.3	59.2±4.4	79.6±5.7	92.8±2.2	0.04707 (0.0979)	0.0416	0.0519	0.0899	0.1089	X=+25.1831 Y=–1.1853	1.737
Pupa	21.8±5.4	41.8±4.6	63±4.4	74.4±5.3	91±3.1	0.0494 (0.1002)	0.0441	0.0542	0.0921	0.1113	X=+25.2220 Y=–1.2464	1.262

LCL: 95 per cent of Lower Confidence Limit; UCL: 95 per cent of Upper Confidence Limit.

Within a column means followed by the same letter(s) are not significantly different at 5 per cent level by DMRT.

Table 5.10 provides the effect of at silver nanoparticles *Catharanthus roseus* different concentrations on the mortality the larval mortality of *Aedes aegypti* (I to Pupa) after the treatment of *Catharanthus roseus* at different concentrations (0.312 to 2.50 µg/ml). Twenty-four per cent mortality was noted at I instar larvae by the treatment of *Catharanthus roseus* at 0.312 µg/ml, whereas it has been increased to 100 per cent at 2.50 µg/ml of *Catharanthus roseus* treatment. Similar trends have been noted for all the instars of *Aedes aegypti* at different concentrations. The LC_{50} and LC_{90} values were represented as follows: the LC_{50} value of I instar was 0.220 per cent, II instar was 0.304 per cent, III instar was 0.354 per cent, IV instar was 0.436 per cent, and pupa was 0.493 per cent respectively. The LC_{90} value of I instar was 0.786 per cent, II instar was 0.983 per cent, III instar was 1.326 per cent, IV instar was 1.560 per cent, and pupa was 1.814 per cent,respectively

Table 5.11 provides the effect of at silver nanoparticles *Catharanthus roseus* different concentrations on the mortality the larval mortality of *Anopheles stephensi* (I to Pupa) after the treatment of *Catharanthus roseus* at different concentrations (0.312 to 2.50 µg/ml). Twenty-four per cent mortality was noted at I instar larvae by the treatment of *Catharanthus roseus* at 0.312 µg/ml, whereas it has been increased to 100 per cent at 2.50 µg/ml of *Catharanthus roseus* treatment. Similar trends have been noted for all the instars of *Anopheles stephensi* at different concentrations. The LC_{50} and LC_{90} values were represented as follows: the LC_{50} value of I instar was 0.253 per cent, II instar was 0.389 per cent, III instar was 0.343 per cent, IV instar was 0.475 per cent, and pupa was 0.444 per cent respectively. The LC_{90} value of I instar was 0.728 per cent, II instar was 0.983 per cent, III instar was 1.383 per cent, IV instar was 1.599 per cent, and pupa was 1.823 per cent,respectively

Table 5.12 provides the effect of at silver nanoparticles *Catharanthus roseus* different concentrations on the mortality the larval mortality of *Culex quinquefasciatus* (I to Pupa) after the treatment of *Catharanthus roseus* at different concentrations (0.312 to 2.50 µg/ml). Twenty-four per cent mortality was noted at I instar larvae by the treatment of *Catharanthus roseus* at 0.312 µg/ml, whereas it has been increased to 100 per cent at 2.50 µg/ml of *Catharanthus roseus* treatment. Similar trends have been noted for all the instars of *Culex quinquefasciatus* at different concentrations. The LC_{50} and LC_{90} values

Table 5.10: Larvicidal and Pupicidal Activity of Synthesized Silver Nanoparticles Using *Catharanthus roseus* Leaf Extract against Larvae and Pupae of *Aedes aegypti*

Larval and Pupal Stage	*Larval and Pupal Mortality (per cent) (Mean±S.D)*				*LC_{50} (LC_{90})*	*95 per cent Confidence Limit*			*Regression Equation*	*Chi Square Value*
	Concentration (ppm)					*LC_{50}*		*LC_{90}*		
	0.312	*0.625*	*1.25*	*2.50*		*LCL*	*UCL*	*LCL*	*UCL*	
I	68.3±2.36	82.4±3.6	96.9±2.2	100.0±0	0.220 (0.786)	0.146	0.282	0.655	1.017	1.884
II	52.4±2.8	78.2±2.4	92.7±3.5	100.0±0	0.304 (0.983)	0.235	0.364	0.825	1.258	1.748
III	44.9±3.5	74.2±3.2	86.5±3.0	98.1±1.2	0.354 (1.326)	0.277	0.423	1.088	1.753	1.659
IV	38.0±1.4	64.9±1.8	82.4±2.9	98.6±1.0	0.436 (1.560)	0.359	0.509	1.278	2.063	2.094
Pupa	32.3±3.4	62.7±2.4	78.6±2.7	96.0±2.1	0.493 (1.814)	0.411	0.573	1.472	2.430	1.847

LCL: 95 per cent of Lower Confidence Limit; UCL: 95 per cent of Upper Confidence Limit.

Within a column means followed by the same letter(s) are not significantly different at 5 per cent level by DMRT.

Table 5.11: Larvicidal and Pupicidal Activity of Synthesized Silver Nanoparticles Using *Catharanthus roseus* Leaf Extract against Larvae and Pupae of *Anopheles stephensi*

Larval and Pupal Stage	*Larval and Pupal Mortality (per cent) (Mean±S.D)*				*LC_{50} (LC_{90})*	*95 per cent Confidence Limit*				*Chi Square Value*
	Concentration (ppm)					*LC_{50}*		*LC_{90}*		
	0.312	*0.625*	*1.25*	*2.50*		*LCL*	*UCL*	*LCL*	*UCL*	
I	75.4±2.2	84.8±2.0	92.1±2.8	100.0±0	0.253 (0.728)	0.176	0.273	0.600	1.049	1.843
II	62.9±2.8	79.2±1.4	96.7±1.0	100.0±0	0.389 (0.983)	0.226	0.358	0.876	1.227	1.779
III	58.3±3.9	67.1±3.8	86.8±2.5	98.0±1.4	0.343 (1.383)	0.268	0.423	1.091	1.730	1.609
IV	36.6±2.9	62.1±2.7	81.3±2.9	97.8±0	0.475 (1.599)	0.396	0.571	1.275	2.089	2.045
Pupa	31.3±1.9	55.1±2.7	73.5±3.4	94.6±1.0	0.444 (1.823)	0.456	0.529	1.483	2.411	1.892

LCL: 95 per cent of Lower Confidence Limit; UCL: 95 per cent of Upper Confidence Limit.

Within a column means followed by the same letter(s) are not significantly different at 5 per cent level by DMRT.

Table 5.12: Larvicidal and Pupicidal Activity of Synthesized Silver Nanoparticles Using *Catharanthus roseus* Leaf Extract against Larvae and Pupae of *Culex quinquefasciatus*

Larval and Pupal Stage	*Larval and Pupal Mortality (per cent) (Mean±S.D)*				*LC_{50} (LC_{90})*	*95 per cent Confidence Limit*				*Chi Square Value*
	Concentration (ppm)					*LC_{50}*		*LC_{90}*		
	0.312	*0.625*	*1.25*	*2.50*		*LCL*	*UCL*	*LCL*	*UCL*	
I	78.3±2.7	84.9±3.1	95.5±1.5	100±0	0.256 (0.711)	0.152	0.236	0.646	1.063	1.863
II	66.5±3.2	80.7±2.9	92.9±1.4	100±0	0.307 (0.975)	0.244	0.396	0.879	1.299	1.797
III	52.3±1.8	74.1±3.6	86.3±2.1	100±0	0.366 (1.345)	0.269	0.475	1.052	1.772	1.652
IV	37.0±3.4	56.9±2.6	80.5±2.9	92.9±1.4	0.496 (1.588)	0.396	0.563	1.202	2.096	2.152
Pupa	28.4±3.6	42.7±2.2	62.1±1.7	86.3±3.4	0.473 (1.829)	0.428	0.586	1.486	2.450	1.878

LCL: 95 per cent of Lower Confidence Limit; UCL: 95 per cent of Upper Confidence Limit.

Within a column means followed by the same letter(s) are not significantly different at 5 per cent level by DMRT.

Table 5.13: Predatory Efficiency of Copepods, *Mesocyclops aspericornis* on Dengue Vector, *Aedes aegypti*

Larval Stage	*No. Introduced*		*Predation (Mean±SD)*			*Total Predation*	*Predatory Percentage*	*Predatory Efficiency of Single Copepod*
	Mosquito Larvae	*Copepods*	*Duration (days)*					
			1	*2*	*3*			
I	100	10	63±2.38	61±1.12	59±0.82	183	61	6.1
II	100	10	51±0.82	49±0.82	46±0.70	146	49	4.9
III	100	10	39±3.5	33±3.5	27±4.06	99	33	3.3
IV	100	10	10±0.9	9±0.7	7±1.6	26	8.66	0.86

Within a column means followed by the same letter(s) are not significantly different at 5 per cent level by DMRT.

were represented as follows: the LC_{50} value of I instar was 0.256 per cent, II instar was 0.307 per cent, III instar was 0.366 per cent, IV instar was 0.496 per cent, and pupa was 0.473 per cent respectively. The LC_{90} value of I instar was 0.711 per cent, II instar was 0.975 per cent, III instar was 1.345 per cent, IV instar was 1.588 per cent, and pupa was 1.829 per cent,respectively

The copepod *Mesocyclops aspericornis* showed effective predation against *Ae. aegypti* larval instars. Table 5.13 provides the percentage of predatory efficiency of copepod adults towards the different instars of *Ae. aegypti*. The percentage predation was reduced as the mosquito larvae grew older. The early instars were susceptible and much preferred by the copepods. Very low predation was observed on the IV instars of *Ae. aegypti*. The predatory efficiency of a Single adult copepod was 6.1, 4.9, 3.3 and 0.86 larvae/day on I, II, III and IV instars, respectively.

Discussion

Mosquitoes cause more human suffering than any other organism. More than one million people die from mosquito-borne diseases every year. Not only can mosquitoes carry diseases that afflict humans, they also transmit several diseases and parasites of dogs and horses. Dengue is a serious arborvirus disease of the Americas, Asia and Africa. Many approaches have been developed to control mosquito menace. One such approach to prevent mosquito borne disease is by killing mosquito at larval stage.

The current mosquito control approach is based on synthetic insecticides. At present, the main threat to effective mosquito control is resistance to insecticide in the mosquito (Chandra *et al.*, 1998). Even though they are effective they created many problems like insecticide resistance (Liu *et al.*, 2005), pollution, toxic side effect on human beings (Lixin, 2006). Today, the environmental safety of an insecticide is considered to be of paramount importance. Botanical insecticides provide an alternative to synthetic insecticides because they are generally considered safe, are biodegradable, and can often be obtained from local sources (Prabhakar and Jabanesan, 2004). In addition, the use of medicinal plants for mosquito control is likely to generate local employment, reduce dependence on expensive imported products, and stimulate efforts to enhance public health (Bowers *et al.*, 1995).

In the present study, the larvicidal activity against *A. stephensi*, *A. aegypti* and *C. quinquefasciatus* was more pronounced with the neem, Bacillus thuringuinessis, Spinosad and nanomaterial. These results suggest that neem acts as best interpreter with the microbial pesticides for the production of greater toxicity rate in the immature stages of mosquitoes. Sukumar *et al.* (1991) has pointed out the most promising botanical mosquito control agents are in the families' Astraceae, Cladophoraceae, Lamiaceae, Meliaceae, Oocystaceae and Rutaceae.

Azadirachtin is the predominant insecticidal active ingredient, having antefeedent, ovipositional deterrence repellency, growth disruption, sterility and larvicidal action against insects (Schmutterer, 1990). Secondary metabolites from Neem tree have insecticidal, antifeedant activity on a variety of pests of agricultural and medical importance. The active principles in the neem are azadirachtin and nimbin which have insecticidal property. Azadirachtin, a complex tetranortriterpenoid limonoid from the neem seeds, is the main component responsible for the toxic effects in insects. considering the importance of botanicals in the management of insect vectors of diseases. In this study, we sought to determine whether an ethanol extract from *Azadirachta indica* could be used for mosquito control. Different concentrations of NSKE showed appreciable mortality against the different larval instars. An emulsion of neem oil in water was found to be effective in controlling breeding of *Cx. quinquefasciatus, An. stephensi* and *Ae. aegypti* in pools, tanks and coolers up to 2 to 3 weeks (Batra *et al.*, 1998). An application of neem cake powder resulted in a drastic reduction in the late-instar larvae and pupae of culicine mosquitoes in paddy fields (Rao *et al.*, 1992). In the previous study Babu and Murugan, (2000) investigated that the larvicidal effect of resinous exudates from the tender leaves of *Azadirachta indica.* The potential toxicity leading to high larval mortality exhibited by the extracts could be attributed to the group of toxic bimolecules possessing insecticidal properties in the plant extracts. (L.) (Vineetha and Murugan 2009) were investigated. Nirmal Sharma *et al.* (1998) reported that larvicidal activity of *Gliricidia sepium* crude ethanol extracts of dried leaves, fresh leaves, dried petioles and stem bark were tested for their activities against third instar larvae of *A. stephensi*, *A. aegypti* and *C. quinquefasciatus*. Murugan and Jeyabalan (1999) studied the effect of some indigenous plants on the larvicide and ovipositional properties on *A. stephensi.* Babu and Murugan (2000) investigated that the larvicidal effect of resinous exudate from the tender leaves of *Azadirachta indica*. *A. paniculata* is a traditional medicinal plant that has been used for pest control (Kuppusamy and Murugan, 2006). Among the three solvents, maximum efficacy was observed in methanol (Govindarajan, 2010).

The addition of *B. thuringiensis* var. *israelensis* with the plant extracts had an adverse effect upon the larval mortality. The addition of *B. thuringiensis* var. *israelensis* with plant extracts caused a significant mortality due to the avoidance of treated diet and may be due to increased toxicity (Gould *et al.*, 1991). In the present study, it can therefore be concluded that *B. thuringiensis* var. *israelensis* and plant compounds caused swelling of the gut epithelial cells (Nasiruddin and Mordue Luntz, 1993). At naturally occurring concentrations, allelochemicals produce midgut lesions, reduce feeding and growth and increase mortality (Lindroth *et al.*, 1988). Larvae of *C. quinquefasciatus* treated with *B. thuringiensis* var. israelensis exhibited reduced weight gain and delayed development compared with the control and resulted in a delayed (or) inhibition of molting (Sieber and Rembold, 1983). Mortality of *C. quinquefasciatus* after the treatment of *B. thuringiensis* var. *israelensis* at different dose levels showed considerable effect on the various larval instars, and dose-dependent mortality was evident at *B. thuringiensis* toxin treatment. Larval feeding behaviour of these two strains was different when they were exposed to potato foliage treated with Cry IIIA (Alter *et al.*, 1996).

Spinosad is a natural fermentation product produced by an actinomycete, *Saccharopolyspora spinosa*. This compound is a mixture of spinosyns A and D. It has shown activity against Lepidoptera, Thysanoptera, and other insect orders such as Diptera. Spinosad proved to be highly toxic to larvae of both species of mosquitoes in laboratory assays. Laboratory assays on *C. capitata* indicated LC_{50} values of 0.013 ppm. For neonate larvae continuously exposed in diet (Ada'n *et al.*, 1996), spinosad is slower acting than many conventional chemical insecticides, such that observations of mortality shortly after exposure may not accurately reflect the proportion of the population that have acquired a lethal dose of the compound (Williams *et al.*, 2003). In demographic studies on

Daphnia pulex, continuous exposure to spinosad resulted in population extinction at >0.01 ppm. However, spinosad was at least five times less toxic than the organophosphate diazinon during continuous exposure studies.

This naturally derived insecticide has been reported to have no adverse effects on predatory insects such as ladybirds, lacewings, big-eyed bugs, or minute pirate bugs (Kirst *et al.*, 1992; DeAmicis *et al.*, 1997; Copping and Menn, 2001; Williams *et al.*, 2003). Spinosad acts as a stomach and contact poison and degrades rapidly in the environment (Cisneros *et al.*, 2002). An immediate effect of ingestion is the cessation of feeding, followed 24 h later, by paralysis and death. This compound is a neurotoxin with a novel mode of action involving the nicotinic acetylcholine receptor and GABA receptors (Watson, 2001). Spinosad has little toxicity to birds and mammals (Bret *et al.*, 1997; Breslin *et al.*, 2000). There is no cross-resistance to synthetic and traditional biological insecticides (Salgado, 1997; 1998). Moreover, spinosad is classified by the US Environmental Protection Agency as an environmentally and toxicologically reduced risk material (Saunders and Bret, 1997).

In the present study copepods are very much booming as predators with first and second instar larvae but are not effective against the late instars. Biological control, using the natural enemies of *Aedes*, appears to be an alternative approach to the systematic failure of use of insecticides (Lardeux *et al.*, 2002). Employing the copepods for the mosquito control is another option and the acquired result shows that copepods and other predators can be the best alternative to synthetic insecticides. The development of resistance to botanical insecticides may also be compromised by the predatory copepods. Mesocyclops has been studied as an antagonist of mosquito larvae and its effectiveness has been demonstrated in different countries, including the United States (Marten, 1990), Honduras (Marten *et al.*, 1994), Vietnam (Nam *et al.*, 1998), and the French Polynesia (Lardeux *et al.*, 2002). This work demonstrates that the predatory efficacy of *M. aspericornis* is substantial against the different larval instars of *Ae. aegypti.* The predator *M. aspericornis* consumed first and second instars in greater numbers than third and fourth instars. The active movements and large size of the older larval instars may have reduced the predation rate of the copepods. Though there was little consumption of the late instars, punctures and injures to late instars of mosquitoes lead to constrained development and death. As a support reports from earlier work states that, *Mesocyclops thermocyclopoides* is a very common species in Costa Rica (Collado *et al.*, 1984, Herna´ndez- Chavarri´a and Schaper, 2000) and was evaluated as a biological control agent against *Aedes.* This copepod feeds on the first and second instars of the mosquito larvae, fatally wounding about seven individuals per Copepods are effective predators of first and second instars of mosquitoes but are not effective against the late instars day (Schaper, 1999).

Human beings have been exposed to airborne nanosized particles throughout their evolutionary stages, and such exposures have increased dramatically over the last century. The rapidly developing field of nanotechnology will result in new sources of this exposure, through inhalation, ingestion, and injection. Although nonmaterial is currently being widely used in modern technology, there is a serious lack of information concerning the human health and environmental implications of manufactured nonomaterials. Since these are relatively new particles, it is necessary to investigative their toxicological behaviour. The objective of this review was to trace the cellular response to nanosized particle exposure. Therapeutic application of selected nanoparticles together with their range of toxic doses was also reviewed. Effect of therapeutically used nanoparticles on cell membrance, mitochondrial function, prooxidant/antioxidant status, enzyme leakage, DNA, and other biochemical endpoints was elucidated. This chapter highlights the need for caution during the use and disposal of such manufactured nanomaterials to prevent unintended environmental impacts (Ansary and Daihan, 2009). In the present study it is also evident that the percentage of mortality of *Aedes aegypti* against gold nanoparticles was increased as the concentration was increased and the mortality was high in early instar when compared with the later ones. This is the first time nanoparticles have been used for the control of mosquito vectors and it showed a good result.

The finding of the present of investigation revealed that *Azadirachta indica* good larvicidal and pupicidal activity against *Aedes aegypti* at different breeding sites. Their mode of action and effect of non target organism are presently under the investigations. To evaluate the impact of neem active principle, Azadiracthin with other microbial insecticides (*Bacillus thuringiensis*, spinosad, nanomaterials and copepods) on the larvicidal, pupicidal toxicity effect of dengue vectors, such as *Aedes aegypti*. It is an obligate aerobe bacterium showed that has good effective mosquito control properties and also can act as an eco-friendly, biopesticide for further vector control programs.

References

Ada´N, A., Del Estal, P., Budia, F., Gonza´lez, M. and Vin˜uela, E., 1996. Laboratory evaluation of the novel naturally derived compound spinosad against *Ceratitis capitata*. *Pesticide Science*, 48: 261–268.

Alder, H.L. and Rossler, E.B., 1977. Introduction to probability and statistics. Freeman, San Francisco, p 246

Babu, R. and Murugan, K., 2000. Larvicidal effect of resinous exudate from the tender leaves of *Azadirachta indica*. *Neem Newsletter*, 17(3)

Babu, R. and Murugan, K., 1998. Interactive effect of neem seed kernel and neem gum extracts on the control of *Culex quinquefasciatus* say. *Neem Newsletter*, 15 (2): 9–11.

Badam, L., Deolanker R. P., Kulkami, M., Wagsampgi, M. and Wagh U. V., 1987. *In vitro* antimalaria activity of neem leaf and seed extracts. *Indian J. Malariology*, 24: 111–117.

Batra, C. P., P. K. Mittal, T. Adak and Sharma, V. P., 1998. Efficacy of neem–water emulsion against mosquito immatures. *Indian Journal of Malariol* 35: 15–21.

Bowers, W.S., Sener, B., Evans, P.H., Bingol, F. and Endoganl, G.H., 1995. Activity of Turkish medicinal plants against mosquitoes *Aedes aegypti* and *Anopheles gambiae*. *Insect Sci Appl.*, 16: 339–342.

Breslin, W.J., Marty, M.S., Vedula, U., Liberacki, A.B. and Yano, B.L., 2000. Developmental toxicity of spinosad administered by gavage to CD rats and New Zealand white rabbits. *Food Chem Toxicol.*, 38: 1103–1112.

Bret, B.L., Larson, L.L., Schoonover, J.R., Sparks, T.C. and Thompson, G.D., 1997. Biological properties of Spinosad. *Down to Earth*, 52: 6–13.

Chandra, F., Darriet, F., Darder, M., and Cuany, A., 1998. Pyrethroid resistance in *Culex quinquefasciatus* from West Africa. *Med. Vet. Entomol.*, 12: 359–366.

Charles, J.F. and Nielsen–LeRoux, C., 2000. Mosquitocidal bacterial toxins: diversity, mode of action and resistance phenomena. *Mem Inst Oswaldo Cruz.*, 95: 201–206.

Charlwood, J.D. and Graves, P.M., 1987. The effect of permethrin impregnated bednets on a population of *Anopheles farauti* in coastal Papua New Guinea. *Med Vet Entomol.*, 1: 319–327.

Cisneros, J., Goulson, D., Derwent, L.C., Penagos, D.I., Hernández, O. and Williams, T., 2002. Toxic effects of spinosad on predatory insects. *Biol Control.*, 23: 156–163.

Collado, C. D., D. Defaye, B. H. Dussart and Fernando, C. H., 1984. The freshwater Copepoda of Costa Rica with notes on some species. *Hydrobiologia,* 119: 89–99.

Das, P.K., Pani, S.P. and Krishnamoorthy, K., 2002. Prospects of elimination of lymphatic filariasis in India. *ICMR Bull.*, 32: 41–54

DeAmicis, C.V., Dripps, J.E., Hatton, C.J. and Karr, L. L., 1997. Physical and biological properties of the spinosyns: novel macrolide pestcontrol agents from fermentation. In: Hedin PA, Hollingworth RM, Masler EP, Miyamoto J, Thompson DG (eds) Phytochemicals for pest control, Symposium Series 658. American Chemical Society, Washington, pp 144–154

El–Ansary, A. and Al–Daihan, S., 2009. On the toxicity of therapeutically used nanoparticles: an overview. *J Toxicol.*, 754–810.

Gould, F., Anderson, A., Landis, D. and Van Mellaert, J., 1991. Feeding behaviour and growth of *Heliothis virescens* larvae on diets containing *Bacillus thuringiensis* formulations or endotoxins. *Entomol Exp Appl.*, 58: 199–210.

Govindarajan, M., 2010. Larvicidal efficacy of Ficus benghalensis L. plant leaf extracts against *Culex quinquefasciatus Say, Aedes aegypti* L. and *Anopheles stephensi* L. (Diptera: Culicidae). *Eur Rev Med Pharmacol Sci.*, 14: 107–111.

Haverkamp, R.G., 2007. Pick your carats: nanoparticles of gold–silver–copper alloy produced *in vivo*. *J Nanoparticle Research,* 9: 697–700.

Herna´ndez–Chavarri´a, F. and Schaper, S., 2000. *Mesocyclops thermocyclopoides* (Copepoda: Cyclopoidea): a scanning electron microscopy study. *Review Latino Am Microbiology,* 42: 53–56.

Hurlbut, H. S., 1938. Copepod observed preying on first instar larva of *Anopheles quadrimaculatus. J Parasitology,* 24: 281.

Jekow, P., S. Schaper, D. Gu¨nther, P. Tavares and Hinrichs, W., 1998. Crystallization and preliminary X–ray crystallographic studies of the 13–fold symmetric portal protein of bacteriophage SPP1. *Acta Crystallographica Section D Biological Crystallography,* 54(Pt 5): 1008–1011.

Kay, B. H., C. P. Cabral, A. C. Sleigh, M. D. Brown, Z. M. Ribeiro and Vasconcelos, A. W., 1992. Laboratory evaluation of Brazilian *Mesocyclops* (Copepoda: Cyclopidae) for mosquito control. *J Med Entomology,* 29: 599–602.

Killeen, G., Fillinger, U., Kiche, I., Gouagna, L., and Knols, B., 2002. Eradication of *Anopheles gambiae* from Brazil: lessons for malaria control in Africa. *Lancet Infect Dis.,* 2: 618–627.

Kirst, H.A., Michel, K.H., Mynderse, J.S., Chio, E.H., Yao, R.C., Nakatsukasa, W.M., Boech, L.D., Occlowitz, J.L., Paschal, J.W., Deeter, J.B. and Thompson, G.D., 1992. Discovery, isolation, and structure elucidation of a family of structurally unique fermentation–derived tetracyclic macrolides. In: Baker DR, Fenyes JG, Steffans JJ (eds) Synthesis and chemistry of agrochemicals III. American Chemical Society, Washington, pp 214–225.

Kosiyachinda, P., A. Bhumiratana and Kittayapong, P., 2003. Enhancement of the efficacy of a combination of *Mesocyclops aspericornis* and *Bacillus thuringiensis* var. *israelensis* by community–based products in controlling *Aedes aegypti* larvae in Thailand. *Am J Trop Medi Hygiene,* 69: 206–212.

Kumar, A., Valecha, N., Jain, T., Aditya, and Dash, P., 2007. Burden of malaria in India: Retrospective and prospective view. *Am. J. Trop. Med. Hyg.,* 77(6): 69–78.

Kuppusamy, C. and Murugan, K., 2006. Mosquitocidal effect of ethanolic extracts of *Andrographis paniculata* Nees on filarial vector *Culex quinquefasciatus* Say (Diptera: Culicidae). In: International Conference on Biodiversity of Insects: Challenging

Issues in Management and Conservation, 30 January–3 February 2006, Tamil Nadu, India, p 194.

Lardeux, F., F. Riviere, Y. Sechan and Loncke, S., 2002. Control of the Aedes vectors of the dengue viruses and *Wuchereria bancrofti*: the French Polynesian experience. *Annals of Tropical Medicine and Parasitology*, 96: S105–S116.

Lindroth, R.L., Scriber, J.M. and Hsia M.T.S., 1988. Chemical ecology of the tiger swallowtail: mediation of host use by phenolic glycosides. *Ecology*, 69: 814–822.

Liu, H., Xu, Q., Zhang, L. and Liu, N., 2005. Chlorpyrifos resistance in Mosquito *Culex aegypti* and *Culex pipiens* pallens. *J Med Entomology*, 42(5): 815 – 820.

Lixin, S., Huiquin, D., Chongxia, G., Jin, Q., Jing, S., Lei, M. and Chang liang, Z., 2006. Larvicidal activity of extracts of *Ginko biloba* Exocarp for three different strains of *Culex pipiens pallens*. *J Med Entomology*, 43(2): 258–261.

Locantoni, L., F. Guioti, M. Crictofaro, L. Pasqualini, F. Esposito, P. Lupetti and Habluetzel, A., 2006. Effect of neem extract on blood feeding oviposition and oocyte ultra structure in *Anopheles stephensi* Liston (Diptera: Culicidae). *Tissue Cell*, 38: 361–371.

Manrique–Saide, P., S. Ibanez–Bernal, H. Delfin–Gonzalez and Parra Tabla, V., 1998. *Mesocyclops longisetus* effects on survivorship of *Aedes aegypti* immature stages in car tyres. *Med Vet Entomology*,12: 386–390.

Marten, G. G., 1990. Elimination of *Aedes albopictus* from tire piles by introducing *Macrocyclops albidus* (Copepoda, Cyclopidae). *J Am Mosquito Control Association*, 6: 689–693.

Marten, G. G., G. Borjas, M. Cush, E. Ferna´ndez and Reid, J. W., 1994. Control of larval *Aedes aegypti* in peridomestic breeding containers. *J Med Entomology*, 31: 36–44.

Mourya, D.T., Iikal, M.A., Mishra, A.C., Jacob, P.G., Pant, U., Ramanujam, S., Mavale, M.S, Bhat, H.R. and Dhanda, V., 1989. Isolation of Japanese encephalitis virus from mosquitoes collected in Karnataka state, India from 1985 to 1987. *Trans R Soc Trop Med Hyg.*, 83: 550–552.

Murugan, K. and Jeyabalan, D., 1999. Effect of certain plant extracts against the mosquito, *Anopheles stephensi* Liston. *Curr Sci.*, 76: 631–633.

Nam, V. S., N. T. Yen, B. H. Kay, G. G. Marten and Reid, J. W., 1998. Eradication of *Aedes aegypti* from a village in Vietnam, using copepods and community participation. *Am J Trop Medicine and Hygiene*, 59: 657–660.

Nasiruddin, M. and Mordue Luntz, A.J., 1993. The effect of azadirachtin on the midgut histology of the locusts *Schistocerca gregaria* and *Locusta migratoria*. *Tissue Cell*, 25: 875–884.

Nirmal Sharma, J.S., Qadry, B., Subramanium, T., Verghese, S.J., Rahman, S.K., Sharma, and Jalees, S., 1998. Larvicidal activity of *Gliricidia sepium* against mosquito larvae of *Anopheles stephensi*, *Aedes aegypti* and *Culex quinquefasciatus*. *Pharmaceutical Biol.*, 36 (1): 3–7.

Panacek, A., Kolar, M., Vecerova, R., Prucek, R., Soukupova, J., Krystof, V., Hamal, P., Zboril, R. and Kvitek, L., 2009. Antifungal activity of silver nanoparticles against Candida spp. *Biomaterials*, 30, 6333–6340.

Pandian, R.S., Dwarkanath, S.K. and Martin, P., 1989) Repellent activity of herbal smoke on the biting activity of mosquitoes. *J Ecobiol.*, 1: 287–289.

Prabhakar, K. and Jabanesan, A., 2004. Mosquitocidal effect of bitter gourd leaf extracts. *Bioresour Technol.*, 95: 113–114.

Rao, D. R., R. Reuben, M. S. Venugopal, B. A. Nagasampgi and Schmutterer, H., 1992. Evaluation of neem–*Azadirachta indica* with and without water management for the control of culicine mosquito larvae in rice field. *Med Vet. Entomology*, 6: 318–324.

Rawlins, S. C., R. Martinez, S. Wiltshire, D. Clarke, P. Prabhakar and Spinks, M., 1997. valuation of Caribbean strains of *Macrocyclops* and *Mesocyclops* (Cyclopoida: Cyclopidae) as biological control tools for the dengue vector *Aedes aegypti*. *J Am Mosquito Control Association*, 13: 18–23.

Riviere, F. and Thirel, R., 1981. La predation du copepods *Mesocyclops leuckarti* pilosa sur les larves de *Aedes* (*Stegomyia*) *aegypti* et Ae. St. polynesiensis essais preliminaries d'utilization comme de lutte biologique. *J Entomophaga*, 26: 427–439.

Rogers, J.V., Parkinson, C.V., Choi, Y.W., Speshock, J.L. and Hussain, S.M., 2008. A preliminary assessment of silver nanoparticle inhibition of monkey pox virus plaque formation. *Nanoscale. Res. Lett.*, 3, 129–133.

Russell, B. M., L. E. Muir, P. Weinstein and Kay, B. H., 1996. Surveillance of the mosquito *Aedes aegypti* and its biocontrol with the copepod *Mesocyclops aspericornis* in Australian wells and gold mines. *Medical and Veterinary Entomology*, 10: 155–160.

Salgado, V.L., 1997. The mode of action of spinosad and other insect control products. *Down to Earth*, 52: 35–44.

Salgado, V.L., 1998. Studies on the mode of action of spinosad: insect symptoms and physiology correlates. *Pest Biochem Physiol.*, 60: 91–102.

Sathishkumar, M., Sneha, K., Won, S.W., Cho, C.W.S., and Kim Yun, Y.S., 2009. *Cinnamon zeylanicum* bark extract and powder mediated green synthesis of nano–crystalline silver particles and its bactericidal activity. Colloids Surf. B: *Biointerfaces*, 73: 332–338.

Saunders, D.G. and Bret, B.L., 1997. Fate of spinosad in the environment. *Down to Earth*, 52: 14–20.

Schaper, S., 1999. Evaluation of Costa Rican copepods (Crustacea: Eudecapoda) for larval *Aedes aegypti* control with special reference to *Mesocyclops thermocyclopoides*. *J Am Mosquito Control Association*, 15: 510–519.

Schmutterer, H., 1990. Properties and potential of natural pesticides from the neem tree, *Azadirachta indica*. *Ann. Rev. Ent.*, 35, 271–297.

Senthilkumar, N., Verma, P. and Gurusubamanian, G., 2009) Larvicidal and adulticidal activities of some medicinal plants against the malarial vector *Anopheles stephensi* (Liston). *Parasitol Res.*, 104: 237–244.

Sharma, S.K., Upadhyay, A.K., Haque, M.A., Tyagi, P.K., Raghavendra, K. and Dash, A.P., 2010. Wash–resistance and field evaluation of alpha cypermethrin treated long–lasting insecticidal net (Interceptor) against malaria vectors *Anopheles culicifacies* and *Anopheles fluviatilis* in a tribal area of Orissa, India. *Acta Trop.*, 116(1): 24–30.

Sieber, K.P. and Rembold, H., 1983. The effects of azadirachtin on the endocrine control of moulting in *Locusta migratoria*. *J Insect Physiol.*, 29: 523–527.

Snow, R.W, Guerra, C.A, Noor, A.M., Myint, H.Y. and Hay, S.I., 2005. the global distribution of clinical episodes of *Plasmodium falciparum* malaria. *Nature*, 434: 214–217.

Sukumar, K., Perich, M.J. and Boobar, L.R., 1991. Botanical derivatives in mosquito control: A review. *J Am. Mosquito Control Association*, 7: 210–237.

Terranella, A., Eigiege, A., Gontor, I., Dagwa, P., Damishi, S., Miri, E., Blackburn, B, McFarland, D., Zingeser, J, Jinadu, M.Y. and Richards, F.O., 2006. Urban lymphatic filariasis in central Nigeria. *Ann Trop Med Parasitol.*, 100(2): 163–172.

Vineetha, A. and Murugan, K., 2009. Larvicidal and smoke repellency effect of *Toddalia asiatica* and *Aegle marmelos* against the dengue vector, *Aedes aegypti* (Insecta: Diptera: Culicidae). *Entomol Res.*, 39: 61–65.

Watson, G.B., 2001. Actions of insecticidal spinosyns on caminobutyric acid responses from small–diameter cockroach neurons. *Pest Biochem Physiol.*, 71: 20–28.

WHO (1999. International programme on chemical safety (IPCS): microbial pest control agent *Bacillus thuringiensis Environ Health Criteria*, I217: 1–105.

Willems, van den Wildenberg, 2005. Roadmap Report on Nanoparticles. W&W Espana s.l., Barcelona, Spain.

Williams, T., Valle, J. and Vin˜uela, E., 2003. Is the naturally–derived insecticide Spinosad1 compatible with insect natural enemies? Bio Sci Technology, 13, 459–475.

Yohannes, M., Haile, M., Ghebreyesus, T.A., Witten, K.H., Getachew, A., Byass, P. and Lindsay, S.W., 2005. Can source reduction of mosquito larval habitat reduce malaria transmission in Tigray, Ethiopia? *Trop Med Int Health*, 10: 1274–1285.

Zaim, M., and Jambulingam, P., 2004. Global insecticide use for vector-borne disease control. In: World Health Organization Communicable Disease Control PaE (ed). WHO Pesticide Evaluation Scheme (WHOPES).

Chapter 6

Use of Biological Insecticides for the Control of Dengue Vector, *Aedes aegypti* (L.) (Insecta : Diptera : Culicidae)

K. Murugan[1]*, *A. Noortheen*[1], *T. Nataraj*[1], *D. Amerasan*[1], *P. Madhiyazhagan*[1], *K. Kalimuthu*[1] *and* *Jiang-Shiou Hwang*[2]

[1]*Department of Zoology, Bharathiar University, Coimbatore – 641 046, Tamil Nadu, India*

[2]*Institute of Marine Biology, National Taiwan Ocean University, Keelung, Taiwan, 202*

ABSTRACT

Combined effect of *Canarium strictum* (CSME) with neem based biopesticide (Econeem), microbial (*Bacillus thuringiensis* var. *israelensis*–Bti) and synthetic insecticides (chlorpyriphos) against *Aedes aegypti* (Diptera: Culicidae) was studied. The growth and development of mosquito had affected considerably after the above said treatment. The EI_{50} and EI_{90} (Emergence Inhibition) values were 8.473 and 13.646 per cent, respectively. The larval and pupal mortality were tested by individual and combined treatments. The higher mortality was

* Corresponding author: E-mail: kmvvkg@gmail.com; kmvvk@rediffmail.com

evident at CSME with Bti combined treatment. LC_{50} values were 1.00, 0.84, 1.46, 1.68 and 1.85 per cent of I, II, III, IV instars and pupa, respectively, and LC_{90} values were 2.76, 2.56, 3.12, 3.25 and 3.27 per cent of I, II, III, IV instars and pupa, respectively. In the present study CSME with Bti treatment had higher mortality to the immature larvae at the laboratory. Smoke toxicity test against the adult mosquito vectors showed interesting results and smoke exposed adult had mortality, and treated mosquito laid minimum number of eggs and further progeny production was affected during subsequent generations. The Repellent activity conducted by 8hrs between 18.00hrs-2.00hrs, 2ppm concentration of CSME showed higher mosquito repellency (79.09 per cent). The smoke toxicity effect of plant against biting activity and ensured population of *Ae. Aegypti* showed higher result of unfed over control I was 49 per cent and reduction in population over control I was 53.7 per cent using control I and II was Negative (blank without plant material) and positive control (mortein coil). The percentage of larval reduction after the treatment (24hrs, 48hrs and 72hrs) at the breeding sites of *Aedes aeygpti* showed considerable efficacy and hence, these combination may be used for the integrated vector management programme.

Keywords: *Aedes aeygpti, Canarium strictum, Bacillus thuringiensis, Pesticides, Larvicidal, Smoke toxicity, Control.*

Introduction

Dengue is a vector–borne disease of tropical and subtropical human populations. The world Health Organization estimates that 50 to 100 million cases of dengue infection occur each year (WHO, 1996; Monath, 1994; Gubler, 1998). More than three lakh cases of dengue haemorrhagic fever are diagnosed each year. Dengue cased a fatality of 24,000 deaths per year. The Dengue fever is caused by a positive single stranded RNA virus of the flaviviridae family; this virus infection is transmitted from the bite of *Aedes aegypti* and *Aedes albopictus* mosquitoes (Gubler and Meltzer, 1999; Platt *et al.*, 1997).

Canarium strictum showed a pesticidal activity against insects, and its seeds contain a large percentage of resin acids. *Azadirachta indica* is known to contain diverse array of biologically active principles of which Azadirachtin (tetranortriterpenoid) is one of the best known derivatives (Broughton, 1986). Chlorpyriphos is a chlorinated organophosphate insecticide used to control mosquitoes, fire ants, household insects, turf and ornamental plant insects and for other purposes (Berg, 1983). Recently, the workers have shifted their focus from synthetic insecticides to botanicals because the plant materials are non-toxic to non-target animals, have no phytotoxic properties and leave no residues in the environment. Scientists, therefore, have embarked on a mission to survey the flora extensively to discover more and more potential plants having insecticidal properties.

The biopesticides *Bacillus thuringiensis* is a cosmopolitan, gram positive, spore forming bacterium. It is a pathogen that has been isolated from insects, soil and leaves in most terrestrial environment. This causes the cell to swell and lyse which paralyzes the gut and eventually kills the insect host. *Bacillus thuringiensis* var. *israelensis* is undoubtedly the most useful pathogen as it can be easily mass produced is toxicologically safe to humans and wild life and is more or less specific in killing mosquito larvae. It is formulated as a powder that is sprayed on the water surface and kills mosquito larvae when it is ingested (Davidson, 1982).

The interference of the phytochemicals in combination with the microbial insecticide *Bacillus thuringiensis* on the growth, development and reproduction mechanism in relation to biochemical changes of insect is very important (Murugan,1998). The plant extracts are active only at high concentrations, which made them uneconomical for field use (Kalyanasundaram and Das, 1985; Babu and Murugan, 2000). In view of above fact, an attempt has been made to evaluate the combined effect of *Canarium strictum*, Microbial insecticide, *Bacillus thuringiensis*, Neem based botanical insecticide, Econeem and synthetic organophosphorus insecticide, chlorpyriphos on the larvicidal, pupicidal and smoke toxic effect on the dengue vector, *Aedes aegypti*.

Materials and Methods

Collection of Eggs and Mosquitoes

The eggs of *Aedes aegypti* were collected from National Institute for Communicable Diseases (NICD), Mettupalayam, Coimbatore, Tamil Nadu, and India without exposure to any insecticide, the eggs were then brought to the laboratory and transferred to 18 X 13 X 4 cm size enamel trays containing 500 ml water and kept for larval hatching.

Maintenance of Larvae, Pupae and Adult

The mosquito larval was maintained in our laboratory at 27+2°C, 75-85 per cent RH, under14L: 10D photoperiod cycles. The mosquito larvae were fed with dog biscuits and yeast at 3:1 ratio. The feeding were continued till the larvae are transformed into the pupa stage. The pupae were collected from the culture trays and will be new transferred to plastic containers (12 X12 cm) containing 500 ml of water with the help of a dipper. The plastic jars will be kept in 90 X 90X 90 cm size mosquito cage for adult emergence. The cage is made up of wooden frames and covered with polythene sheets on four sides (two laterals, one back and other one upper) and the front part as covered with a muslin cloth bottom of the cage is fitted with 10 per cent sugar solution for a period of three days before they will be provided with animal for blood feeding.

Blood Feeding of Adult *Aedes aegypti*

The adult female mosquitoes were allowed to feed on the blood of a rabbit (exposed on the dorsal side) for two days, to ensure adequate blood feeding for 5 days. After blood feeding enamel trays with water from the culture trays were placed in the cage for the adults to lay eggs.

Collection of Plant Materials

The *Canarium strictum* (Araucariaceae) seeds were collected from the Kolli Hills, Eastern Ghats, Namakkal, Tamil Nadu, India. The plants were authentified at BSI (Botanical Survey of India), Coimbatore, India and the specimens were deposited at Zoology Department, Bharathiar University, Coimbatore, India.

Preparation of Plant Extracts

The plant materials of *Canarium strictum* seeds were washed with tap water, shade dried at room temperature and powdered by an electrical blender. From each sample, 100g of the plant material was extracted with 300 ml of methanol solvents for 8hrs in a Soxhlet apparatus (Vogel, 1978). The crude plant extracts were evaporated to dryness in rotary vaccum evaporator.

Larval/Pupal Toxicity Test

Laboratory colonies of mosquito larvae/pupae were used for the larvicidal/pupicidal activity. Twenty-five numbers of I to IV instar larvae and pupae were introduced into 500-ml glass beaker containing 249 ml of dechlorinated water and 1 ml of desired concentrations of plant extract, and spinosad were added. Larval food was given for the test larvae. At each tested concentration, two to five trials were made, and each trial consisted of three replicates. The control was set up by mixing 1ml of acetone with 249ml of dechlorinated water. The larvae and pupae which were exposed to dechlorinated water without acetone served as control. The control mortalities were corrected by using Abbott's formula (Abbot, 1925). The LC_{50} and LC_{90} were calculated from toxicity data by using probit analysis (Finney, 1971).

$$\text{Corrected mortality} = \frac{\text{Observed mortality in treatment} - \text{Observed mortality in control}}{100 - \text{Control mortality}} \times 100$$

$$\text{Percentage mortality} = \frac{\text{Number of dead larvae/pupae}}{\text{Number of larvae/pupae introduced}} \times 100$$

Repellent Activity of *Canarium strictum*

Repellent activity of plant compounds tested with human volunteers. For the repellent activity of *Canarium strictum* percentage protection in relation to dose method was adopted. Three to four days old blood starved female of adult mosquitoes (100) were kept in a net cage. The arms of the test person cleaned with isopropanol. After air-drying the arm only 25 cm^2 of the dorsal side of the skin on each arm will be exposed, the remaining area being covered by rubber gloves.

The *Canarium strictum* was dissolved in isopropanol and this alcohol served as control. The *Canarium strictum* at 0.5, 1.0 and 2.0 ppm concentrations were applied. The control and treated arms introduced simultaneously into the cage. The number of bites counted over 5 min every 60 min, from 20:00 to 6.00 hrs. The experiment was conducted five times. The percentage protection was calculated by using the following formula.

$$\%\ \text{Protection} = \frac{\text{Number of bites received by control arm} - \text{Number of bites received by treated arm}}{\text{Number of bites received by control arm}} \times 100$$

Smoke Toxicity Test

Canarium strictum seed kernel was used for smoke toxicity assay. The mosquito coils were prepared following method of Saini *et al.* (1986) with minor modifications by using 4 grams from each plant powdered sample considered active ingredient two grams of saw-dust as binding material and two grams of coconut shell charcoal powder as burning material. All the three were thoroughly mixed with distilled water to form a semisolid past. Mosquito coils (0.6 cm thickness) were prepared manually from the semisolid paste and were shade dried. The control coils were prepared by without plant ingredient.

The experiments were conducted in glass chamber measuring 140 × 120 × 60 cm. A window measuring 60 × 30 cm was situated at mid bottom of one side of the chamber. Hundred of three to four days old blood starved adult female, fed with sucrose solution and were released into the chamber. A belly shaven pigeon was kept tied inside the cage in immobilized condition. The coil was kept on a stand inside the cage and was allowed to burn out. The experimental chamber was tightly closed. The experiment was conducted at night for eight hours from 10.00 pm to 6 am. Each experiment was repeated five times on five separate days including control using mosquitoes of same age groups. The data were pooled and average values were subsequently used for calculations. Control was maintained in two sets. One set was run with coil lacking the active ingredient of plant powder (control I) another one is mortein coil which was used for positive control to compare the effectiveness of plant coils.

After the experiment was over the fed, unfed (active and dead) mosquitoes were counted. The protection given by the smoke from plant samples against the biting of *Aedes aegypti* was calculated in terms of percentage of unfed mosquitoes due to treatment.

$$\text{Repellent activity (\%)} = \frac{\text{Number of unfed mosquitoes in treatment} - \text{Number of unfed mosquito in control I}}{\text{Number of mosquitoes treated}} \times 100$$

The alive mosquitoes fed with blood meal were reared in mosquito cage measuring 30 × 30 × 15 cm. The top and bottom of the cage were fit with glass and all other sides were covered with muslin cloth. Water soaked raisins and 5 per cent sucrose solution soaked in cotton balls were kept as food inside the cage. Water containing powdered yeast and dog biscuits were also kept inside the cage in a glass bowl to collect egg rafts. The egg rafts from the cage were collected daily till all the mosquitoes died. Five or six egg rafts were allowed to hatch in each plastic tray measuring 30 × 25 × 6 cm, containing about 2.5 liters of unchlorinated tap water. The larvae hatched from the egg rafts were fed with a mixture of dog biscuits and yeast powder in the ratio of 2:1 and water in the tray was changed daily. The number of larvae hatched was counted at 1st instar stage. The reduction in the population form the smoke treated mosquitoes was calculated using the following formula:

$$\text{Populatio reduction (\%)} = \frac{\text{Number of larvae hatched in control I} - \text{Number of larvae hatched in treated}}{\text{Number of larvae hatched in control I}} \times 100$$

Field Trials

CSME and *Bti* were used for the field trail the quantity of pesticide required (Based on laboratory LC_{50} value) for the combined treatment was determined by calculating the total surface area of the water in each habitat. The required quantities of CSME and *Bti* were mixed thoroughly with water in a bucket with constant agitation. Teepol was used as emulsifying agent (0.05 per cent). Field applications of CSME and *Bti* were done with the help of a knapsack sprayer uniformly on the surface of the water in each habitat. Dipper sampling and counting of larvae monitored the larval density before 24 hrs, 48 hrs and 72 hrs after the treatment. A separate sample was taken to determine the species composition of each larval habitat.

Five trails were conducted for CSME and *Bti* combined treatment. The percentage of reduction was calculated by the following formula:

$$\% \text{ of Reduction} = \frac{C-T}{C} \times 100$$

where,

C: The total number of mosquitoes in control

T: The total number of mosquitoes in treatment.

Statistical Analysis

All data were subjected to analysis of variance; the means were separated using Duncan's multiple range tests by Alder and Rossler (1977). The average larval mortality data were subjected to probit analysis, for calculating LC_{50} and LC_{90} values were calculated by using the Finney (1971) method. SPSS software package 9.0 versions were used. Results with $P<0.05$ were considered to be statistically significant.

Results and Discussion

Table 6.1 provides the effect of methanolic crude extract of *Canarium strictum* seed extract on the growth and development of *Aedes aegypti.* The plant extract concentrations were ranged from 2 to 10 per cent. In control, total days of development were 11.5±0.4, and the percentage of total emergence was 95 per cent. At the concentration 2 and 4 per cent the total number of developmental days were 13.0±2.0 and 13.0±1.4 and per cent age of total emergence was 95 and 88 per cent, the total numbers of developmental days were 16±1.0 and 20±1.2 and percentage of total emergence was 68 and 58 per cent, respectively. The EI_{50} and EI_{90} (Emergence Inhibition) values were 8.473 and 13.646 per cent. When concentration increased the total number of developmental days also increased. Among the concentration 10 per cent was more effective than the other concentrations in the total emergence of adults. Total emergence of adult was markedly affected at 10 per cent than other concentrations.

The active fractions of the plant extracts disrupted moulting and metamorphosis induced malformation, extended the larval duration and inhibited adult emergence (Muthukrishnan *et al.*, 1997). The CSME suggests the multiple effect of phytochemical active compounds (Oleo-gum-resins are similar to abietic acid, a know contact allergen) present in CSME. In contrast to the earliest work in

Table 6.1: The Effect of Methanolic Crude Extract of *Canarium strictum* Seed Extract at Various Concentrations on the Growth and Development of Dengue Vector, *Aedes ageypti*

Concentration of CSME (per cent)	*Mean Duration in Each Instars (Days)*[a]					*Total Number of Days*	*Total Mortality (per cent)*	*Total Emergency (per cent)*	*EI*$_{50}$ [EI$_{90}$] (per cent)*
	$L_1 - L_2$[b]	*$L_2 - L_3$*[b]	*$L_3 - L_4$*[b]	*Pupae*	*Adult*				
2.00	1.0 ± 1.0^{d}	3.0 ± 2.1^{d}	5.0 ± 1.4^{c}	2.0 ± 1.2^{a}	1.5 ± 1.4[illegible]	13.0 ± 2.0^{c}	5^{e}	95^{a}	8.47391
4.00	2.0 ± 1.0^{c}	3.0 ± 1.4^{d}	5.0 ± 0.8^{c}	2.0 ± 0.4^{a}	1.5 ± 1.8[illegible]	13.0 ± 1.4^{c}	12^{d}	88^{b}	[13.64654]
6.00	3.0 ± 0.8^{b}	4.0 ± 1.2^{b}	5.0 ± 1.0^{c}	2.0 ± 0.8^{c}	2.0 ± 1[illegible]	16.0 ± 1.0^{b}	32^{c}	68^{c}	
8.00	4.0 ± 0.6^{a}	6.0 ± 1.2^{b}	6.0 ± 0.8^{b}	2.0 ± 1.0^{a}	2.0 ± 1[illegible]	20.0 ± 1.2^{b}	42^{b}	58^{d}	
10.00	4.0 ± 1.2^{a}	7.0 ± 2.8^{a}	7.0 ± 1.6^{a}	2.0 ± 1.0^{a}	2.0 ± 1[illegible]	21.0 ± 1.6^{a}	65^{a}	35^{e}	
Control	1.0 ± 0.4^{d}	3.0 ± 0.6^{d}	4.0 ± 1.2^{d}	2.0 ± 1.2^{a}	1.5 ± 0.6[illegible]	11.5 ± 0.4^{d}	5^{e}	95^{a}	

Within column means followed by the same letter(s) are not significantly different at 5 per cent level by DMRT.

a: X±SE

b: L_1, L_2, L_3, L_4 Larval stages

EI*: Emergency Inhibition

the present study the CSME showed significant toxicity and growth regulatory effects on *A. aegypti.* The part of the plant, the solvents used for extraction, the geographical location where the plants were grown and the application methods (Sukumar *et al.*, 1991).

Senthil Nathan *et al.* (2005) tested the methanolic extracts of leaves from the Indian white cedar *Dysoxylum malabaricum* Bedd. (Meliaceae) against mature and immature *Anopheles stephensi* Liston (Diptera), mosquitoes under laboratory condition. The extract showed strong larvicidal, pupicidal, adulticidal and antiovipositional activity. Therefore, the present study the plant seed contain methanolic extract and showed higher toxicities.

Table 6.2 shows the larval and puapl mortality of *A. aegypti* after the treatment of CSME. Considerable larval and pupal mortality was evident after the treatment of CSME for all instars and pupae. Mortality was increased as concentration increased, for example, 42 per cent mortality was noted at I instar larvae by the treatment of CSME at 2 per cent whereas; it has been increased to 95 per cent at 10 per cent of PAME treatment. The above concentration 2 per cent -10 per cent (2, 4, 6, 8, 10 per cent) and their mortality results were consist slope value. The slope value of I instar was 0.153. Similarly, the same trend has been noted for all larval stages and pupae of *A. aegypti* at different concentrations of CSME treatment. The LC_{50} and LC_{90} values represented as follows: LC_{50} value of I instar was 3.09 per cent, II instar was 4.21 per cent, III instar was 5.02 per cent, IV instar was 5.86 per cent and pupa was 6.79 per cent, respectively. LC_{90} value of I instar was 9.17 per cent, II instar was 10.62 per cent, III instar was 11.44 per cent, IV instar was 12.82 per cent and pupa was 14.02 per cent, respectively.

In fact, many researchers have reported on the combined treatment was more effective against mosquito larvae control. Kovendan *et al.* (2011) have reported that the leaves extract of *Carica papaya* and bacterial insecticide, spinosad is promising as good larvicidal and pupicidal properties of against chikungunya vector, *A. aegypti.* In the present study, Econeem, a neem based insecticide showed larvicidal and pupicidal activity on *A. aegypti,* similar result were obtained on the pupicidal effect against malarial vector, *A. stephensi* (Murugan and Jeyabalan, 1998), *Bacilus thuringiensis* is one of the most actively researched entomopathogens and has found commercial application in several cases. *Bacillus thuringiensis* var.

Table 6.2: Larvicidal and Pupucidal of CSME at Various Concentrations against the Dengue Vector *Ae. aegypti*

Parameter			LC_{50} (ppm)		LC_{90} (ppm)	
Larval Instars	*Slope (SE)*	X^2	*Estimate*	*(95 per cent fiducial limit)*	*Estimate*	*(95 per cent fiducial limit)*
I	0.153(0.02)	1.653	3.09	2.18-3.77	9.17	8.28-10.47
II	0.144(0.02255)	1.816	4.21	3.46-4.83	10.62	9.56-12.20
III	0.138(0.02213)	2.711	5.02	4.36-5.60	11.44	10.30-13.16
IV	0.142(0.02162)	3.853	5.86	5.21-6.50	12.82	11.42-15.00
Pupa	0.148(0.02143)	2.627	6.79	6.14-7.52	14.02	12.39-16.59

Table 6.3: Combined Effect of CSME and Botanical Insecticide Econeem at Various Concentration on Larvicidal and Pupicidal Activity of 0. (0.02162) the Dengue Vector *Ae. aegypti*

Parameter			LC_{50} (ppm)		LC_{90} (ppm)	
Larval Instars	*Slope (SE)*	X^2	*Estimate*	*(95 per cent fiducial limit)*	*Estimate*	*(95 per cent fiducial limit)*
I	0.034(0.09428)	0.992	1.05	0.87-1.21	2.66	2.32-3.22
II	0.034(0.09393)	0.518	1.27	1.09-1.45	2.99	2.58-3.70
III	0.035(0.09673)	1.144	1.61	1.44-1.83	3.19	2.76-3.91
IV	0.038(0.09929)	1.685	1.83	1.64-2.12	3.46	2.96-4.33
Pupa	0.044(0.1013)	0.295	2.09	1.84-2.50	3.87	3.24-5.05

israelensis (Bti) has been shown in laboratory and field studies to be a potent larvicide of mosquitoes (Mulligan *et al.*, 1980; Molloy and Jamnback, 1981).

Tables 6.3–6.5 show the combined effect of CSME and Econeem, *Bacillus thuringiensis* or Chlorpyriphos on larvicidal and pupicidal activity at various concentration against the dengue vector *A. aegypti* at all larval and pupal stages, after the treatment of CSME with Econeem result as showed, Among the different larval and pupal stages, the I instar larvae were more susceptible than the other larval and pupal stages. The LC_{50} and LC_{90} values of I instar was 1.05 per cent and 2.66 per cent, respectively (Table 6.3). The Table 6.4 provides the result as among the different larval and pupal stages, the II instar larvae were more susceptible than the other larval and pupal stages. LC_{50} and LC_{90} values of II instar was 0.84 per cent and 2.56 per cent, respectively, and combined treatment of CSME with chlorpyriphos result showed. Among the different larval pupal stages, the I instar larvae were more susceptible than the other larval and pupal stages. LC_{50} and LC_{90} values of I instar was 1.24 per cent and 3.02 per cent, respectively (Table 6.5).

The higher larval and pupal mortality was evident after the combined treatment of CSME with Bti treatment (The LC_{50} values were 1.00, 0.84, 1.46, 1.68 and 1.85 per cent of I, II, III, IV instars and pupa, respectively, and LC_{90} values were 2.76, 2.56, 3.12, 3.25 and 3.27 per cent of I, II, III, IV instars and pupa respectively, Table 6.4) and it due to the toxic effect of *Bt* toxins on the gut of mosquito vector. Since, the Bt toxins are activated by midgut proteases and the activated toxins interact with larval midgut epithelium causing a disruption in membrane integrity and ultimately leading to insect death.

Combined effect of CSME with *Bt* toxins not only affected the larva but also inhibited the growth of pupa and the treated mosquitoes did not emerge from the pupa as adults. It suggests that the plant compounds also had an impact on growth pattern of mosquito.

The repellent activity of CSME extract on *Ae. aegypti* is given in the Table 6.6. The repellent activity was carried out in the evening from 18.00-02.00 hrs. The repellency effect was very low (58.02 per cent protection) at 0.5 ppm. The percentage protection was 76.54 per cent and 79.09 per cent at 1.0ppm and 2.0ppm, respectively.

Table 6.4: Combined Effect of CSME and Microbial Insecticide *Eti* at Various Concentrations on Larvicidal and Pupicidal Activity of the Dengue Vector *Ae. aegypti*

Parameter			LC_{50} (ppm)		LC_{90} (ppm)	
Larval Instars	*Slope (SE)*	X^2	*Estimate*	*(95 per cent fiducial limit)*	*Estimate*	*(95 per cent fiducial limit)*
I	0.034 (0.0937)	3.882	1.00	0.84-1.16	2.76	2.38-3.36
II	0.035 (0.09461)	0.840	0.84	0.67-1.00	2.56	2.22-3.10
III	0.035 (0.09493)	0.665	1.46	1.30-1.66	3.12	2.70-3.78
IV	0.035 (0.09815)	1.048	1.68	1.51-1.91	3.25	2.83-3.92
Pupa	0.036 (0.10491)	1.790	1.85	1.68-2.09	3.27	2.87-3.89

Table 6.5: Combined Effect of CSME and Synthetic Insecticide Chlorpyriphos at Various Concentrations on Larvicidal and Pupicidal Activity of the Dengue Vector *Ae. aegypti*

Parameter			LC_{50} (ppm)		LC_{90} (ppm)	
Larval Instars	*Slope (SE)*	X^2	*Estimate*	*(95 per cent fiducial limit)*	*Estimate*	*(95 per cent fiducial limit)*
I	0.036 (0.09331)	1.642	1.24	1.08-1.42	3.02	2.60-3.70
II	0.036 (0.09542)	2.365	1.60	1.43-1.84	3.34	2.87-4.11
III	0.038 (0.09843)	0.814	1.79	1.60-2.06	3.45	2.97-4.23
IV	0.040 (0.1024)	0.885	1.98	1.77-2.30	3.62	3.10-4.47
Pupa	0.046 (0.11051)	2.257	2.21	1.96-2.60	3.78	3.23-4.72

Table 6.6: Repellent Activity of *Canarium strictum* Methanolic Leaf Extract against Dengu Vector, *Ae. aegypti*

Repellent Activity Observation Time (hr)	*No. of Mosquitoes Fed*			
	C. strictum Methanolic Leaf Extract Concentration (ppm)			
	0.5	*1.0*	*2.0*	*Control*
18.00-19.00	0	0	0	0
19.00-20.00	5	3	2	12
20.00-21.00	4	2	2	14
21.00-22.00	3	1	4	13
22.00-23.00	5	3	3	12
23.00-24.00	4	5	3	13
24.00-01.00	6	3	2	8
01.00-02.00	7	2	1	9
Fed mosquito	34	19	17	81
Unfed mosquito	66	81	83	19
Percentage of mosquito repellency	58.02	76.54	79.09	–

Regarding the anti oviposition activity of insect repellents against mosquitoes, Bar-Zeev and Den- Tamar, (1968) reported on the ovipositional repellency of DEET and other repellents to *Ae. aegypti*. Zebitz (1984) and Mohsen *et al*. (1995) reported antiovipositional and larvicidal activity of plant extracts against mosquitoes. Xue *et al*. (2001) showed that the skin repellents DEET and 2 experimental compounds (A13-37220 and AI3- 35765) were effective oviposition deterrents against *Ae. albopictus* in the laboratory as well as in a variety of field situations. The same 3 repellents also were potent larvicides of *Ae. albopictus* and *Anopheles albimanus* Wied. in the laboratory or field, and provided sustained reduction of larvae for several weeks after treatment (Xue *et al.*, 2001). However, the present study revealed that the volatile extract obtained from *Canarium strictum* seed exhibited repellency against mosquito *Ae. aegypti* at a very lower concentration than those of the plants studied earlier, but the repellent effect was only for 2 hrs.

Canarium strictum seed powder demonstrated smoke toxicity on adult mosquitoes. After the treatment of *C. strictum* seed was 21 fed and 49 unfed mosquitoes. Mortality was significantly higher in the positive control (Table 6.7).

Table 6.8 shows the population reduction effect and egg laying of *Ae. aegypti* after exposure to *C. strictum* seed smoke. After treatment of *C. strictum* seed smoke, 25 mosquitoes oviposited a total, of 955 eggs of which only 440 eggs hatched. The percentage of reduction was 53.7. In the negative control, 25 mosquitoes laid 1357 eggs and only 950 eggs hatched. The percentage of reduction was 82.6 of the plant *C. strictum* seed smoke had a higher efficacy of percentage reduction on the selected population of *Ae. aegyupti*.

Plant derived smoke contains an array of chemicals with different modes of action, which kill mosquitoes. Subramania Thangam and Kathiresan (1992) reported that smoke from burning various dry materials has been used since early times to deter insects especially mosquitoes. The roots and stems of *A. marmelos* are used as antipyretic agents and they contain alkaloids, coumarine, flavonoids and sterois (Waterman and Grundon, 1983). Vineetha and Murugan (2009) have reported the mosquito adult exposed to smoke oviposited fewer eggs : possible the vapor of plant active compound affected fecundity. In the present toxic volatile ingredients of oleo-gum-resins from *C. strictum* seed used into mosquito coil. The

Table 6.7: Smoke Toxicity Effect of Plants against Biting Activity of *Aedes aegypti*

Plant Used	*No. of Mosquitoes Tested*	*Fed Mosquitoes*	*Unfed Mosquitoes*		*Total*	*Per cent Unfed over Control I*
			Alive	*Dead*		
Canarium seed	100	21[b]	49[c]	30[b]	79[bc]	49[ab]
Control I*	100	70[d]	30[a]	0[e]	30[d]	0[c]
Control II*	100	18[a]	40[b]	42[a]	82[a]	–

Table 6.8: Smoke Toxicity Effect of Plants Ensured Population of *Aedes aegypti*

Plant used	*No. of Mosquitoes Tested*	*Total No. of Eggs*	*Total No. of Larvae Hatched from the Eggs*	*Per cent of reduction in Population Over Control I*
Canarium seed	25	955[c]	440[c]	53.7[c]
Control I*	25	1357[a]	950[a]	82.6[ab]
Control II*	25	380[e]	165[d]	–

Within coloum means followed by the same letter(s) are not significantly different at 5 per cent level by DMRT

Control I*: Negative control – blank without plant material

Control II*: Positive control – mortein coil

smoke emerged from the coil from the plant part of *C. strictum* seed not only showed toxin against the mosquito and had chemosterilant effect. The smoke exposed adult mosquito does not feed well, as well as the fecundity and progeny production was also greatly affected and it revealed that smoke had impact on the reproductive organ of egg development.

Table 6.9 provides after the application of combined effect of CSME and microbial insecticide Bti at waste cement tank, which was one of the breading sites of *Ae. aegypti.* The larval mortality was assessed at the interval of 24, 48 and 72 hrs duration. After the treatment of 24hrs the percentage reduction of larval density was 90.7 per cent likewise 48 hrs and 72 hrs percentage reductions were 97.5 and 100 per cent.

Table 6.9: Field Trail by Using CSME with Microbial Insecticide, *Bti* against Dengue Vector, *Ae. aegypti*

Place: Vadavalli panchayat, Coimbatore

Habitat: Waste Cement Tank

Size: 2.5 × 2.5 m

Depth: 1 cm

Species: *Aedes aegypti*

Stage: Larvae

Calculation: 2.5 × 2.5 = 6.25 × 1.46 × 10 = 91.25

Required Concentration: 91.25 percentage

Sl.No.	*Larval Density*			
	Before Treatment	*After Treatment*		
		24 hrs	*48 hrs*	*72 hrs*
1.	178	18	7	—
2.	142	14	4	—
3.	114	10	3	—
4.	163	15	2	—
5.	143	13	5	—
6.	131	11	1	—
Total	871	81	22	—
Average	145.17	13.5	3.67	—
Reduction	—	90.70%	97.50%	100%

In the present study, The CSME with Bti applied on breeding place spread on the water surface and from a thin layer which cuts of oxygen supply to mosquito larvae come up to the water surface to breathe the oil penetrates the larva through the respiratory tube and kills it either by suffocation or by poisoning (Murugan *et al.*, 2003). The extract showed higher field trial activity against *Ae. aegypti* dissolves into the water, methanolic extract and combined also killed the mosquito population. The methanolic extract of *Plectranthus amboinicus* leaf had promising larvicidal effect on *Ae. aegypti* and this was done by Amudhadevi, *et al.* (1999) using pongamia oil against *Culex quiquefasciatus.*

Conclusion

The percentage reduction was considerably increased in combined treatment that the individual treatment. Hence, *Canarium strictum* plants are also employing for the mosquito control programmes. Since, lower dose with combined plant products may facilitate the slow release active compound.

References

Abbot, W.S., 1925. A method of computing the effectiveness of insecticides. *J.Econ.Entmol.*, 18: 265–26.

Alder HL. and Rossler, E.B., 1977. Introduction to probability and statistics. Freeman, San Francisco, p 246.

Amudhadevi, P., Murugan, K. and Jeyabalan, D., 1999. Evaluation of certain plant oils on the control filarial vector, *Culex* spp. Proceedings of the National Symposium on Biological Control of insects in Agriculture, forestry, medicine and veterinary science. pp. 38–40.

Babu, R. and Murugan, K., 2000. Larvicidal effect of resinous exudates from the tender leaves of *Azadirachta indic,. Neem Nsletter*, 17(1).

Bar-Zeev, M. and Den-Tamar, D., 1968. The effectiveness of repellents on cloth as determined by oviposition of *Aedes aegypti* L. *Mosq News*, 28: 396–403.

Berg, G.L., 1983. Farm Chemicals Handbook Eds., Meister Publishing Company, Willoughby, Ohio. pp. C55–56.

Broughton, H. B., Ley, S. V., Slawin, A. M. Z., Williams, D. J. and Moragan, E.D., 1986. X–ray crystallographic structure determination of detigloyldihydro azadirachtin and reassignment of the structure of the limonoids insect antifeedant azadirachtin. *J. Chem. Sec. Chem. Commun.*, pp. 46–47.

Davidson, G., 1982. Developments in malaria vector control. *Br. Med. Bull.*, 38(2): *201–206.*

Finney, D.J., 1971. Probit Analysis. Cambridge University, London. pp. 68–78.

Gubler, D.J., 1998. Climate change: implications for human health. *Hlth. Env. Digest.*, 12: 54–55.

Gubler, D.J. and Meltzer, M., 1999. Impact of Dengue and Dengue Haemorrhagic Fever on the developing world. *Adv. Virus Research.*, 53: 35–70.

Kalimuthu Kovendan, Kadarkarai Murugan, Arjunan Naresh Kumar, Savariar Vincent and Jiang–Shiou Hwang, 2011. Bioefficacy of larvicdial and pupicidal properties of *Carica papaya* (Caricaceae) leaf extract and bacterial insecticide, spinosad, against chikungunya vector, *Aedes aegypti* (Diptera: Culicidae). *Parasitol Res.*, DOI 10.1007/s00436–011–2540–z

Kalyanasundaram, M. and Das., 1985. Larvicidal synergistic activity of plant extracts for mosquito control. *Indian J. Med. Res.*, 82: 19–23.

Mohsen, Z.H., Jawad, A.M., Al–Saadi, M., Al–Naib, A., 1995. Anti–oviposition and insecticidalactivity of *Imperata cylindrica* (Gramineae). *Med Vet Entomol.*, 9: 441–442.

Molloy, D. and Jamnback, H., 1981. Field evaluation of *Bacillus thuringiensis* var. *israelensis* as a black fly biocontrol agent and its effect on nontarget stream insects. *J. Econ. Entomol.*, 74: 314–318.

Monath, T.P., 1994. Dengue: The risk to developed and developing countries. *Proctl. Natl. Acad. Sci.*, USA; 91: 2395–400.

Mulligan, F.S. III, Schaefer, C.H. and Wilder, W.H., 1980. Effectiveness of the bacterial pathogen *Bacillus thuringiensis* var. *israelensis* against mosquito larvae. *Proc. Pop. Annu. Conf. Calif. Mosq. Control. Assoc.*, 48: 25–27.

Murugan K Vahitha R Baruah I and Das S. C., 2003. Integration of botanicals and microbial pesticides for the control of filarial vector, *Culex quinquefasciatus. Annals of Medical Entomology,* 12 (1&2); 11–23.

Murugan, K and Jeyabalan, D., 1998. Neem enhance the activity of microbial pesticides, *Insect Environ.,* 4(10): 3–4.

Murugan, K., Raja, NS., Jeyabalan, D., Kumar, NS. and Sivaramakrishnan, S., 1998. Evaluation of certain tropical plant extracts for their antifeedant and toxic properties against *Spodoptera litura* (Fab.). *Journal of Insect Science,* 11, 186–187.

Muthukrishnan, J., Pushpalatha, E. and Kasthuribhai, A., 1997. Biological effect of four plant extract on *Culex quinquefasciatus* Say. Larval stages. *Insect Sci. Appli.,* 17: 389–394.

Platt, K.B., Linthicum, K.J., Myint, K.S., Innis, B.L., Lerdthusnee, K. and Vaughn, D.W., 1997. Impact of dengue virus infection on feeding behaviour of *Aedes aegypti. Am J Trop Med Hyg.,* 57: 119–25.

Saini, H.K., Sharma, R.M., Bami, H.L. and Sidhu, K.S., 1986. Preliminary study on constituents of mosquito coil smoke. *Pesticides,* 20: 15–18.

Senthil Nathan, S., Kalaivani, K. and Murugan, K., 2005. Effects of neem limonoids on the malarial vector *Anopheles stephensi* Liston (Diptera: Culicidae). *Acta Tropica.,* 96: 47–55.

Subramonia thangam, T. and Kathiresan, K., 1992. Smoke repellency and killing effect of marine plants against *Culex quinquefasciatus. Trop. Biomed.,* 9: 35–38.

Sukumar, K., Perich, M.J. and Boobar, L.R., 1991. Botanical derivatives in mosquito control: A review. *J. Am. Mosq. Control Assoc.,* 7: 210–237.

Vineetha, A., Murugan, K., 2009. Larvicidal and smoke repellency effect of *Todaalia asiatica* and *Aegle marmelos* against the dengue vector, *Aedes aegypti* (Insecta: Diptera: Culicidae). *Entomol Res.,* 39: 61–25.

Waterman, PG. and Grundon, MF., 1983. *Chemistry and Chemical Taxonomy of the Rutales.* Academic ress, London.

WHO, 1996. Report of the WHO in formal consultation on the evaluation on the and testing of insecticides CTD/WHO PES/ IC/96(1): 69.

Xue, R.D., Barnard, D.R. and Ali, A., 2001. Laboratory an field evaluation of insect repellent a *oviposition deterents* against the mosquito *Aedes albopictus. Med vector. Entomol.,* 15: 126–131.

Zebitz, C.P.W., 1984. Effect of some crude and Azadarachtin enriched neem *Azadirachita indica* seed kernel extracts on larvae of *Aedes aegypti. Ent. Exp. Appl.,* 35: 11–14.

Chapter 7

Larvicidal Effect of Spinosad and Azadirachtin Against Dengue Vector, *Aedes aegypti* (Insecta : Diptera : Culicidae) and *Chironomus kiiensis* (Chironomidae)

P. Madhiyazhagan[1*], *K. Murugan*[1], *A. Nareshkumar*[1], *T. Nataraj*[1], *D. Amerasan*[1], *K. Kalimuthu*[1] *and Jiang-Shiou Hwang*[2]

[1]*Department of Zoology, Bharathiar University, Coimbatore – 641 046, Tamil Nadu, India*

[2]*Institute of Marine Biology, National Taiwan Ocean University, Keelung, Taiwan, 202*

ABSTRACT

Aedes aegypti is historically the primary vector for the viruses that cause human dengue and yellow fever. In Asia, this species is also considered the principal vector of chikungunya virus. The family Chironomidae of the order Diptera contains many medically important species that produce serious nuisance and allergy problems upon their outbreaks. spinosad was purchased from Kalpatharu pesticide Limited,

* Corresponding author: E-mail: madhiphd@yahoo.com

Coimbatore, Tamil Nadu, India. Required quantity of Spinosad was thoroughly mixed with distilled water to prepare various concentrations, ranging from 0.002 to 0.010ppm. Azadirachtin was obtained from neem extracts kindly provided with a sample (min. 96 per cent AI) from Central Research Laboratories, Taiyo Kagaku Co Ltd., Japan. Azadirachtin and spinosad gave significant efficacy against different larval instars of dengue vector and chironomid species. Field trail have also conducted to test the efficacy of neem compound, Azadirachtin and bacterial insecticide, spinosad at the breeding sites of mosquito vectors and it showed considerable larval after the treatments. The mortality of larvae in laboratory and field suggested that the neem and spinosad had affected the neurometabolic activities of test insects. The combined treatment with spinosad and azadirachtin showed higher toxicity than on the individual treatment. This study concluded that the spinosad and azadirachtin can be used for the potential larvicides against *Aedes aegypti* and *Chironomus kiiensis*.

Keywords. *Aedes aegypti, Chironomus kiiensis, Spinosad, Azadirachtin, Larvicidal, Pupicidal.*

Introduction

Dengue is transmitted by *Aedes* sp. mosquitoes that breed in container habitats. The main vector, *Aedes aegypti*, is a cosmotropical species that proliferates in water containers in and around houses. *Ae. aegypti* breeds in many types of household containers, such as water storage jars, drums, tanks and plant or flower containers. They do not fly far, dispersing probably no more than 100 m beyond the emergence location (Harrington *et al.*, 2005; Honorio *et al.*, 2003; Muir and Kay, 1998; Reiter *et al.*, 1995) and are highly anthropophilic, rarely feeding on non-human hosts. Dengue fever (DF) and the potentially lethal version of the disease, dengue haemorrhagic fever, which has now been detected in almost all countries where DF is prevalent, has become a major cause of hospitalization and death among children and the old. Approximately 2500 million people, two fifths of the world's population, are now at risk from DF. The WHO currently estimates there may be 50 million cases of DF infection worldwide every year (WHO, 2011).

The family Chironomidae of the order Diptera contains many medically important species that produce serious nuisance and allergy problems upon their outbreaks (Resh and Grodhaus, 1983; Kondo *et al.*, 1992). Among the *Chironomidae* (Diptera), a small number of species live in the soil, which is considered an unusual habitat for the family (Hudson, 1987). These species occur in a large diversity of soils and habitats such as tundra (Sendstad *et al.*, 1977), forests (Healey and Russel-Smith, 1971; Seddon, 1986; Hövemeyer, 1992), meadows and agricultural fields (Delettre and Lagerlöf, 1992; Frouz, 1994; Delettre and Morvan, 2000) and heathlands (Delettre, 1985; 1994).

Chironomid midges have been recorded as pests of rice in many temperate rice growing countries (Surakarn and Yano, 1995). The larvae, commonly known as bloodworms, either attack the seed itself, or feed on the roots and/or shoots of young seedlings. When high larval densities are present in rice fields secondary damage can arise through their tunneling activity in the sediment, which can destabilize the root systems of young plants and increase water turbidity, reducing photosynthesis and slowing the growth of submerged seedlings. Determination of the molecular structure and properties of allergens that elicit severe immediate-type hypersensitivity diseases in humans and knowledge of the structure of their antibody-binding sites should provide new insight into the pathogenetic mechanisms of allergic diseases. Monomeric and homodimeric haemoglobins (CTT I to X) have been identified as potent allergenic components of Chironomidae, a family of Diptera (Baur *et al.*, 1986).

Chironomids are widely and abundantly distributed in the vicinity of standing waters. Larvae of *Chironomus* and some other genera are known to contain haemoglobins, which have been described as a major allergen, and the adults that have no haemoglobins also have been reported to contain allergens. In this study, we tried to establish the role of chironomid allergy and characterize the allergen of *Chironomus kiiensis* adults. Skin tests using *C. kiiensis* adult extracts were performed on patients with allergic symptoms. A cDNA library of *C. kiiensis* adults was screened with *C. kiiensis* immune mouse sera to identify allergens, and results were confirmed using skin test-positive human sera. (Jeong *et al.*, 2004).

Management of this disease vector using synthetic chemicals has failed because of insecticide resistance, vector resurgence and environmental pollution. Consequently, an intensive effort has been made to find alternative methods of control (Service, 1983). Plant-derived materials are usually safer and more ecologically acceptable. They must be tested, however, to judge their efficacy against the target hosts.

Neem contains several biologically active chemicals called limonoids such as azadirachtin, nimbin, salannin, azadirachtol, nimbidin, gedunin, etc. These compounds are responsible for diverse activities such as insect antifeedant, insect growth disrupting, insecticidal, nematicidal, fungicidal, bactericidal, etc. (Kavathekar, 2003). Azadirachtin, whose chemical structure is shown in is the most potent and the most abundant (0.2–0.6 per cent (w/w)) chemical found in seed kernels of neem. The bitter extracts from neem kernels obtained by alcoholic extraction contain various bioactive limonoids and are widely used in biopesticidal formulations (Govindachari *et al.*, 1990).

Azadirachtin is a colourless crystalline compound, found in the seeds of the neem tree, *Azadirachta indica* A. Juss. 1830. The compound is of great interest because it is a powerful feeding deterrent for a number of phytophagous insects, and has the power, at parts per million concentration, to inhibit growth and development of all insects ingesting it. The effects of neem extracts or purified azadirachtin on the blood sucking stages of hematophagous insects have been investigated in only a few species, such as the triatomine bug *Rhodnius prolixus* (Nogueira *et al.*, 1997; Gonzalez *et al.*, 1999) and *Triatoma infestans* (Kollien and Schaub, 1999), and mosquito species belonging to the genera *Culex*, *Aedes* and *Anopheles* (Mulla and Su, 1999).

Spinosad is a natural product derived from the bacterium *Saccharopolyspora spinosa*. This bacterium is normally responsible for the decomposition of organic material in soil. This organism was first isolated in certain soil samples in 1988 (Thompson *et al.*, 1997). The most active metabolites from spinosad fermentation were identified as Spinosyn A and D (Sparks *et al.*, 1997). Spinosad exhibits stomach and contact poisoning properties and affects specifically the function of g-aminobutyric acid (GABA) receptors and nicotinic acetylcholine receptors of the target insects (Salgado, 1997; 1998). This product has been widely tested against injurious insects in a variety of crops, such as cotton (Banerjee *et al.*, 2000), wheat (Fang *et al.*, 2002) and tobacco (Blanc *et al.*, 2004).

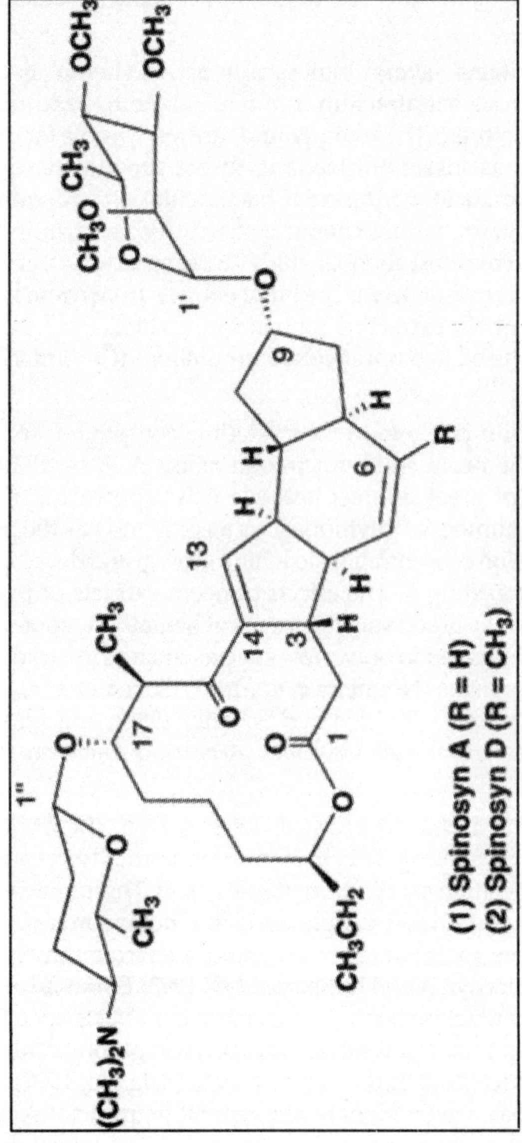

(1) Spinosyn A (R = H)
(2) Spinosyn D (R = CH_3)

Figure 7.1: Structure of Spinosad

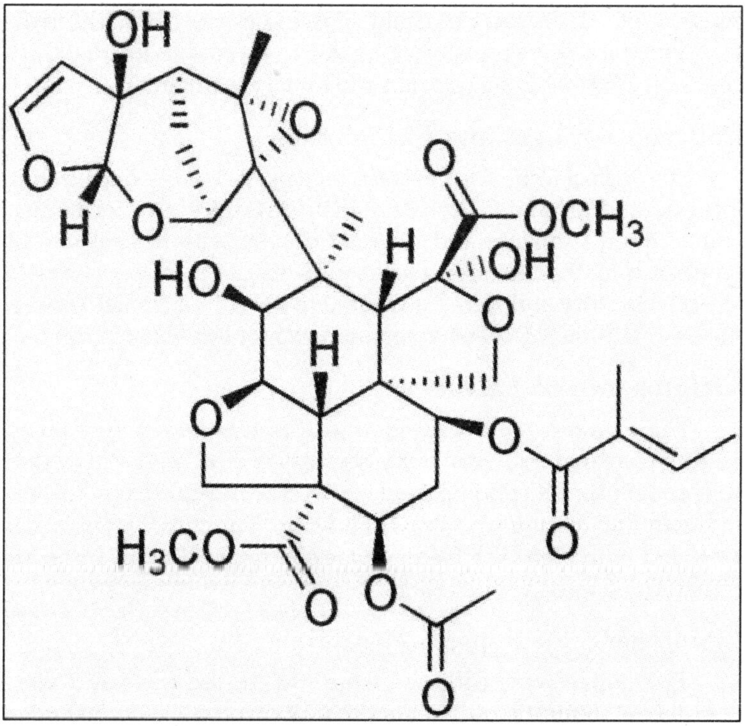

Figure 7.2: Structure of Azadirachtin

Spinosad is a mixture of tetracyclic macrolide neurotoxins, spinosyn A and D, produced during the fermentation of the soil actinomycete *Saccharopolyspora spinosa*. As such, it may be considered as a bioinsecticide (Copping and Menn, 2000). Spinosad is highly toxic to Lepidoptera, Diptera and some Coleoptera and has a unique mode of action involving the postsynaptic nicotinic acetylcholine and GABA receptors (Salgado, 1998; Watson, 2001). The objectives of the study (laboratory and field) is to evaluate the individual and combined effect of microbial insecticide, spinosad and azadirachtin against dengue vector, *Aedes aegypti* (Insecta: Diptera: Culicidae) and *Chironomus kiiensis* (Chirono1midae).

Materials and Methods

Collection of Eggs and Mosquitoes

The eggs of *Aedes aegypti*, were collected from local in and around Coimbatore district water stored container with the help of 'O' type brush, for the laboratory bioassay. These eggs were brought to the laboratory and were transferred to 18 × 13 × 4 cm size enamel trays containing 500 ml of water and keep for larval hatching.

Collection of Eggs and Chironomus

The eggs of *Chironomus kiiensis* (egg mass) were collected from local (in and around Coimbatore districts) dirking water bodies, water stored container and stagnant ditches with the help of 'O' type brush, for the laboratory bioassay. These eggs mass were brought to the laboratory and were transferred to 18 × 13 × 4 cm size enamel trays containing 500 ml of water and keep for larval hatching.

Maintenance of Larvae

The mosquito and Chironomid (*Chironomus kiiensis*) larval culture were maintained in our laboratory at 27+2°C, 75-85 per cent RH, under14L: 10D photoperiod cycles. The mosquito larvae were fed with dog biscuits and yeast at 3:1ratio. The chironomid larvae were fed with alga. The feeding was continued till the larvae are transformed into the pupa stage.

Maintenance of Pupae and Adult

The pupae were collected from the culture trays and were transferred to plastic containers (12 × 12 cm) containing 500 ml of water with the help of a dipper. The plastic jars were kept in 90 × 90 × 90 cm size mosquito cage for adult emergence. The cage is made up of wooden frames and covered with polythene sheets on four sides (two laterals, one back and other one upper) and the front part as covered with a muslin cloth bottom of the cage is fitted with 10 per cent sugar solution for a period of three days before they were provided with animal for blood feeding.

Blood Feeding of Adult Mosquito Species

The adult female mosquitoes were allowed to feed on the blood of a rabbit (showed on the dorsal side) for two days, to ensure adequate blood feeding for 5 days. After blood feeding enamel trays with water from the culture trays were placed in the cage for the adults to lay eggs. Both females and males provided with 10 per cent of glucose solution on cotton wicks.

Preparation of Spinosad

Success of spinosad was purchased from Kalpatharu pesticide Limited, Coimbatore, Tamil Nadu, India. Required quantity of Spinosad was thoroughly mixed with distilled water to prepare various concentrations, ranging from 0.002 to 0.010ppm.

Preparation of Azadirachtin

Azadirachtin was obtained from neem extracts kindly provided with a sample (min. 96 per cent AI) from Central Research Laboratories, Taiyo Kagaku Co Ltd., Japan. The powdery preparation was dissolved in pure ethanol, and the emulsions were added to the artificial diet, which contained 1ppm azadirachtin.

Larvicidal and Pupicidal Bioassay

A laboratory colony of *Aedes aegypti* and *Chironomus kiiensis* larvae and pupae were used for the larvicidal and pupicidal activity. Hundred numbers of 1st, 2nd, 3rd, 4th instar larvae and pupae were kept in 500 ml glass beaker containing 249 ml of dechlorinated water and 1 ml of desired concentration of selected microbial insecticide spinosad and azadirachtin were added. Larval food was given for the tested larvae. For each tested concentration, 2 to 5 trials were made and each trial consists of three replicates. The control was set up by mixing 1ml of acetone with 249ml of dechlorinated water. The larvae/pupae exposed to dechlorinated water without acetone was served as control. Mortality was corrected by using Abbott's formula (Abbott, 1925).

$$\text{Corrected mortality} = \frac{\text{Observed mortality in treatment} - \text{Observed mortality in control}}{100 - \text{Control mortality}} \times 100$$

$$\text{Percentage mortality} = \frac{\text{Number of dead larvae/pupae}}{\text{Number of larvae/pupae introduced}} \times 100$$

The values of LC_{50}, LC_{90} and their 95 per cent confidence limit of upper confidence limit (UCL) and lower confidence limit (LCL) and chi-square $(\chi)^2$ values were calculated by using Probit analysis (Finney, 1971).

Field Trials

Field trials were conducted in mosquito breeding sites at the endemic Districts of Tamil Nadu. For the field trial the quantity of spinosad and azadirachtin required (based on laboratory LC_{50} values) for each treatment was determined by calculating the total surface area of the water in each habitat. The required quantities of formulation were mixed thoroughly with water in a bucket with constant agitation. Water pH, water temperature and ambient temperature were taken daily. Teepol was used as emulsifying agent (0.05 per cent). Field applications of the biformulations were done with the help of a knapsack sprayer. Uniformly spraying the spinosad and azadirachtin on the surface of the water in each habitat. Dipper sampling and counting of larvae densities 24 hrs, 48 hrs and 72 hrs after the treatment. A separate sample was taken to determine the species composition of each larval habitat. Samples of same field without treatment served as controls. From all habitats and brought to the laboratory for recording and species identification to find any other species was present. The percentage of larval reduction was calculated by the following formulae:

$$\%\text{ of Reduction} = \frac{C - T}{C} \times 100$$

where,

C: The total number of mosquitoes in control

T: The total number of mosquitoes in treatment.

Statistical Analysis

The LC_{50}, LC_{90} and regression equation were calculated from toxicity data using SPSS (version 16.0) software package. The values were expressed as mean±standard deviation of five replicates. Differences between the treatments were determined by Duncan's multiple range tests.

Results

Larvicidal and pupicidal activity of spinosad at various concentration against dengue vector, *Aedes aegypti* is given in the Figure 7.3. Considerable mortality was evident after the treatment of spinosad for all larval instars and pupae. Mortality was increased as the concentration increased, for example, 52 per cent mortality

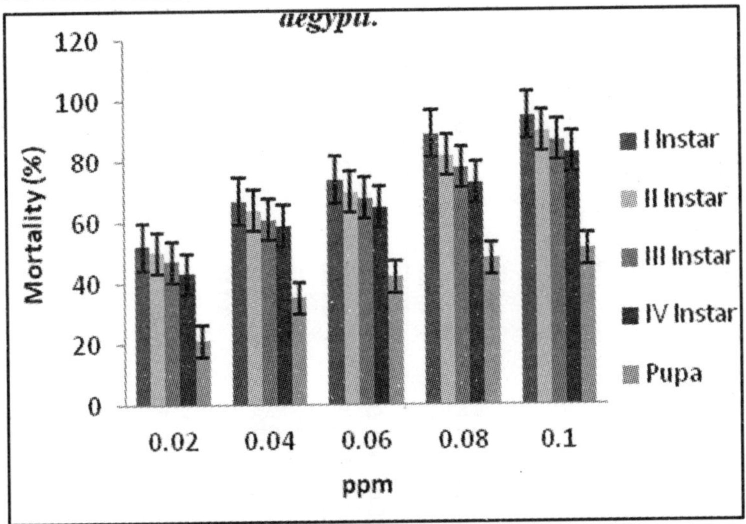

Figure 7.3: Larvicidal and Pupicidal Effect of Spinosad against Dengue Vector, *Aedes aegypti*

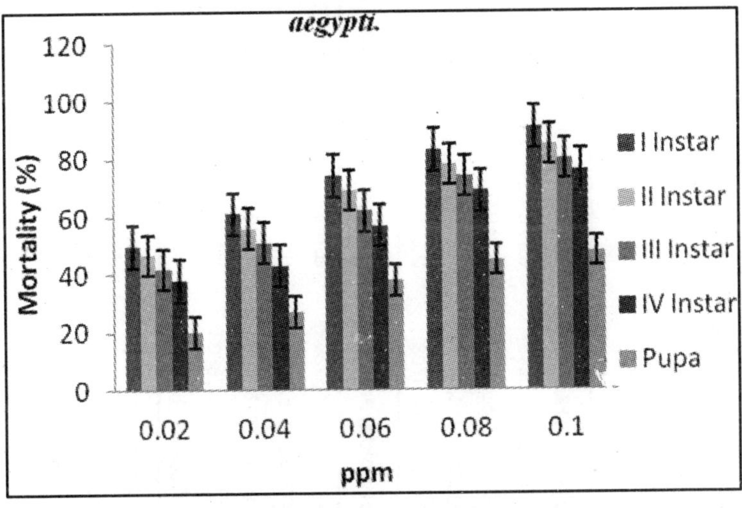

Figure 7.4: Larvicidal and Pupicidal Effect of Azadirachtin against Dengue Vector, *Aedes aegypti*

was noted at I instar larvae by the treatment of spinosad at 0.02ppm whereas it has been increased to 95 per cent at 0.1ppm of spinosad treatment. Similar trend has been noted for all the larval instars and

Table 7.1: Larvicidal and Pupicidal Effect of Spinosad against Dengue Vector, *Aedes aegypti*

Parameter			LC_{50} (ppm)		LC_{90} (ppm)	
Larval Instars and Pupa	*Slope (SD)*	*χ2*	*Estimate*	*(95 per cent fiducial limit)*	*Estimate*	*(95 per cent fiducial limit)*
I	0.00181 (17.2423)	1.596*	0.019	0.006–0.028	0.086	0.076–0.099
II	0.00201 (15.5948)	0.726	0.020	0.003–0.031	0.103	0.090–0.125
III	0.00204 (15.41752)	0.410	0.024	0.008–0.035	0.113	0.098–0.140
IV	0.002072 (15.05988)	0.860	0.030	0.013–0.040	0.127	0.108–0.161
Pupa	0.00254 (11.9708)	2.325	0.089	0.076–0.113	0.220	0.173–0.333
Larvicidal and Pupicidal Effect of Azadirachtin against Dengue Vector, *Aedes aegypti*						
I	0.001907 (16.51363)	0.145	0.021	0.007–0.031	0.098	0.087–0.117
II	0.002021 (15.57241)	0.157	0.027	0.011–0.037	0.117	0.101–0.144
III	0.002002 (15.7226)	0.242	0.036	0.023–0.045	0.131	0.112–0.165
IV	0.001922 (16.28803)	0.785	0.047	0.036–0.055	0.142	0.121–0.179
Pupa	0.002619 (11.88697)	1.073	0.098	0.084–0.125	0.223	0.176–0.332

Significant at $p<0.001$ (heterogeneity factor used in calculation of confidence limits).

SD: Standard deviation; LC: Lethal concentration.

pupae of *Aedes aegypti* at different concentrations of spinosad treatment. The LC_{50} and LC_{90} values were represented as follows: LC_{50} value of I instar was 0.019 per cent, II instar was 0.020 per cent, III instar was 0.024 per cent, IV instar was 0.030 per cent, and pupa was 0.089 per cent, respectively. LC_{90} value of I instar was 0.086 per cent, II instar was 0.103 per cent, III instar was 0.113 per cent, IV instar was 0.127 per cent, and pupa was 0.220 per cent, respectively. The slope value has been given in Table 7.1. The values were represented as follows: for I instar (0.00181, $p<0.0001$), II instar (0.00201, $p<0.0001$), III instar (0.00204, $p<0.0001$), IV instar was (0.002072, $p<0.0001$) and pupa (0.00254, $p<0.0001$), respectively. Similarly, 50 per cent mortality was noted at I instar larvae by the treatment of azadirachtin at 0.02ppm whereas it has been increased to 95 per cent at 0.1ppm of azadirachtin treatment (Figure 7.4). Similar trend has been noted for all the larval instars and pupae of *Aedes aegypti* at different concentrations of azadirachtin treatment. The LC_{50} and LC_{90} values were represented as follows: LC_{50} value of I instar was 0.021 per cent, II instar was 0.027 per cent, III instar was 0.036 per cent, IV instar was 0.047 per cent, and pupa was 0.098 per cent, respectively. LC_{90} value of I instar was 0.098 per cent, II instar was 0.103 per cent, III instar was 0.113 per cent, IV instar was 0.142 per cent, and pupa was 0.223 per cent, respectively. The slope value has been given in Table 7.1. The values were represented as follows: for I instar (0.001907, $p<0.0001$), II instar (0.002021, $p<0.0001$), III instar (0.002002, $p<0.0001$), IV instar was (0.001922, $p<0.0001$) and pupa (0.002619, $p<0.0001$), respectively.

Larvicidal and pupicidal activity of spinosad various concentration against Chironomid, *Chironomus kiiensis* is given in the Figure 7.5. Considerable mortality was evident after the treatment of spinosad for all larval instars and pupae. Mortality was increased as the concentration increased, for example, 51 per cent mortality was noted at I instar larvae by the treatment of spinosad at 0.02ppm whereas it has been increased to 94 per cent at 0.1ppm of spinosad treatment. Similar trend has been noted for all the larval instars and pupae of Chironomid, *Chironomus kiiensis* at different concentrations of spinosad treatment. The LC_{50} and LC_{90} values were represented as follows: LC_{50} value of I instar was 0.020 per cent, II instar was 0.031 per cent, III instar was 0.039 per cent, IV instar was 0.044 per cent, and pupa was 0.093 per cent, respectively. LC_{90} value of I instar was 0.094 per cent, II instar was 0.118 per cent, III instar was 0.129

Table 7.2: Larvicidal and Pupicidal Effect of Spinosad against Chironomid, *Chironomus kiiensis*

	Parameter		LC_{50} (ppm)		LC_{90} (ppm)	
Larval Instars and Pupa	*Slope (SD)*	$\chi 2$	*Estimate*	*(95 per cent fiducial limit)*	*Estimate*	*(95 per cent fiducial limit)*
I	0.001907 (16.51363)	1.498	0.020	0.005–0.029	0.094	0.083–0.111
II	0.001952 (16.16478)	1.118	0.031	0.017–0.040	0.118	0.102–0.145
III	0.001921 (16.45296)	0.308	0.039	0.027–0.048	0.129	0.112–0.160
IV	0.001958 (16.14001)	0.263	0.044	0.033–0.053	0.138	0.118–0.174
Pupa	0.002683 (11.58447)	1.088	0.093	0.079–0.120	0.225	0.176–0.343
Larvicidal and Pupicidal Effect of Azadirachtin against Chironomid, *Chironomus kiiensis*						
I	0.00197 (16.00937)	0.525	0.020	0.004–0.030	0.098	0.086–0.117
II	0.001828 (17.10848)	0.809	0.031	0.018–0.040	0.114	0.100–0.138
III	0.002097 (15.04992)	0.410	0.041	0.028–0.050	0.142	0.120–0.183
IV	0.002321 (13.61249)	0.095	0.051	0.039–0.062	0.167	0.137–0.228
Pupa	0.002767 (11.24722)	1.123	0.106	0.090–0.140	0.235	0.183–0.358

Significant at $p<0.001$ (heterogeneity factor used in calculation of confidence limits).

SD: Standard deviation; LC: Lethal concentration.

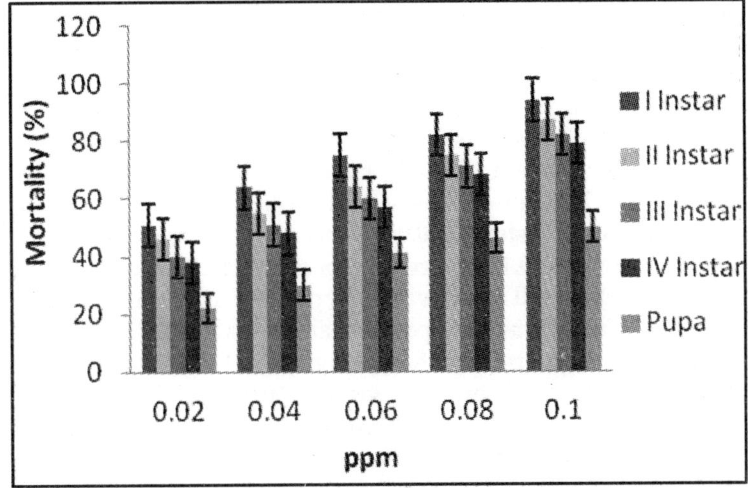

Figure 7.5: Larvicidal and Pupicidal Effect of Spinosad against Chironomid, *Chironomus kiiensis*

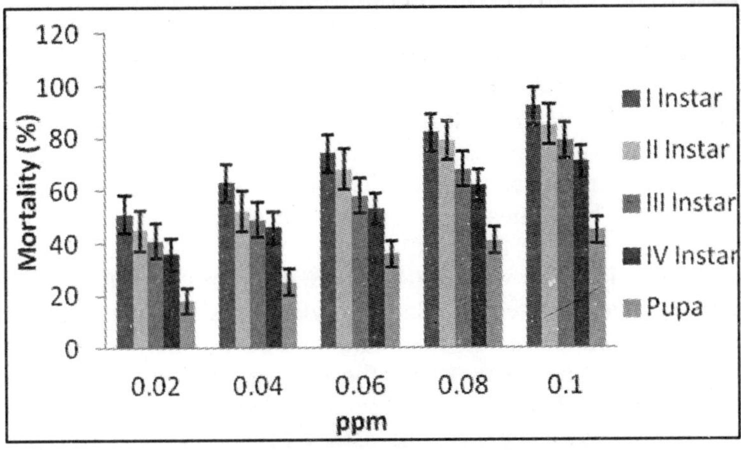

Figure 7.6: Larvicidal and Pupicidal Effect of Azadirachtin against Chironomid, *Chironomus kiiensis*

per cent, IV instar was 0.138 per cent, and pupa was 0.225 per cent, respectively. The slope value has been given in Table 7.2. The values were represented as follows: for I instar (0.001907, $p<0.0001$), II instar (0.001952, $p<0.0001$), III instar (0.001921, $p<0.0001$), IV instar was (0.001958, $p<0.0001$) and pupa (0.002683, $p<0.0001$), respectively.

Similarly, 51 per cent mortality was noted at I instar larvae by the treatment of azadirachtin at 0.02ppm whereas it has been increased to 91 per cent at 0.1ppm of azadirachtin treatment (Figure 7.6). Similar trend has been noted for all the larval instars and pupae of Chironomid, *Chironomus kiiensis* at different concentrations of azadirachtin treatment. The LC_{50} and LC_{90} values were represented as follows: LC_{50} value of I instar was 0.020 per cent, II instar was 0.031 per cent, III instar was 0.041 per cent, IV instar was 0.051 per cent, and pupa was 0.106 per cent, respectively. LC_{90} value of I instar was 0.098 per cent, II instar was 0.114 per cent, III instar was 0.142 per cent, IV instar was 0.167 per cent, and pupa was 0.235 per cent, respectively. The slope value has been given in Table 7.2. The values were represented as follows: for I instar (0.00197, $p<0.0001$), II instar (0.001828, $p<0.0001$), III instar (0.002097, $p<0.0001$), IV instar was (0.002321, $p<0.0001$) and pupa (0.002767, $p<0.0001$), respectively.

Table 7.3 shows Larvicidal and pupicidal activity of combined effect of spinosad and Azadirachtin against Dengue vector, *Aedes aegypti* the mortality is given in the Figure 7.5. The LC_{50} and LC_{90} values were represented as follows: LC_{50} value of I instar was 0.031 per cent, II instar was 0.036 per cent, III instar was 0.045 per cent, IV instar was 0.055 per cent, and pupa was 0.150 per cent, respectively. LC_{90} value of I instar was 0.167 per cent, II instar was 0.190 per cent, III instar was 0.218 per cent, IV instar was 0.251 per cent, and pupa was 0.396 per cent, respectively. The slope value has been given in Table 7.5. The values were represented as follows: for I instar (0.003875, $p<0.0001$), II instar (0.003979, $p<0.0001$), III instar (0.00407, $p<0.0001$), IV instar was (0.004186, $p<0.0001$) and pupa (0.004805, $p<0.0001$), respectively. Similarly, Larvicidal and pupicidal activity of combined effect of spinosad and Azadirachtin against Chironomid, *Chironomus kiiensis* the mortality is given in the Figure 7.6. The LC_{50} and LC_{90} values were represented as follows: LC_{50} value of I instar was 0.033 per cent, II instar was 0.040 per cent, III instar was 0.051 per cent, IV instar was 0.076 per cent, and pupa was 0.163 per cent, respectively. LC_{90} value of I instar was 0.168 per cent, II instar was 0.207 per cent, III instar was 0.240 per cent, IV instar was 0.279 per cent, and pupa was 0.409 per cent, respectively. The slope value has been given in Table 7.3. The values were represented as follows: for I instar (0.003819, $p<0.0001$), II instar (0.004063, $p<0.0001$), III instar (0.004211, $p<0.0001$), IV instar was (0.004221, $p<0.0001$) and pupa (0.004898, $p<0.0001$), respectively.

Table 7.3: The Combined Effect of Spinosad and Azadirachtin against Dengue Vector, *Aedes aegypti*

Parameter			LC_{50} (ppm)		LC_{90} (ppm)	
Larval Instars and Pupa	*Slope (SD)*	$\chi 2$	*Estimate*	*(95 per cent fiducial limit)*	*Estimate*	*(95 per cent fiducial limit)*
I	0.003875 (16.22344)	4.586	0.031	0.002–0.051	0.167	0.149–0.195
II	0.003979 (15.85244)	2.050	0.036	0.005–0.057	0.190	0.168–0.227
III	0.00407 (15.43697)	1.723	0.045	0.011–0.065	0.218	0.189–0.266
IV	0.004186 (14.82565)	2.353	0.055	0.020–0.077	0.251	0.214–0.320
Pupa	0.004805 (12.90349)	1.398	0.150	0.129–0.182	0.396	0.318–0.570
The Combined Effect of Spinosad and Azadirachtin against Chironomid, *Chironomus kiiensis*						
I	0.003819 (16.50152)	0.981	0.033	0.004–0.052	0.168	0.150–0.196
II	0.004063 (15.53061)	0.273	0.040	0.007–0.061	0.207	0.181–0.251
III	0.004211 (14.93988)	0.082	0.051	0.016–0.073	0.240	0.206–0.302
IV	0.004221 (14.92314)	0.460	0.076	0.048–0.096	0.279	0.236–0.360
Pupa	0.004898 (12.70039)	1.280	0.163	0.140–0.199	0.409	0.327–0.591

Significant at $p<0.001$ (heterogeneity factor used in calculation of confidence limits).

SD: Standard deviation; LC: Lethal concentration.

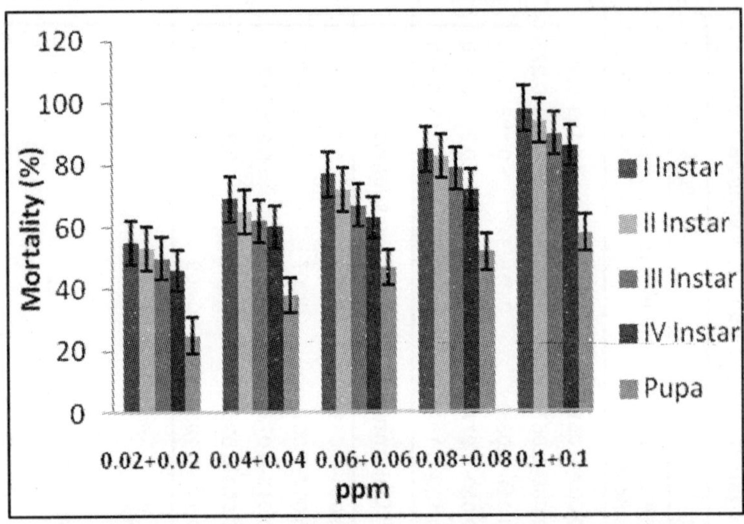

Figure 7.7: The Combined Effect of Spinosad and Azadirachtin against Dengue Vector, *Aedes aegypti*

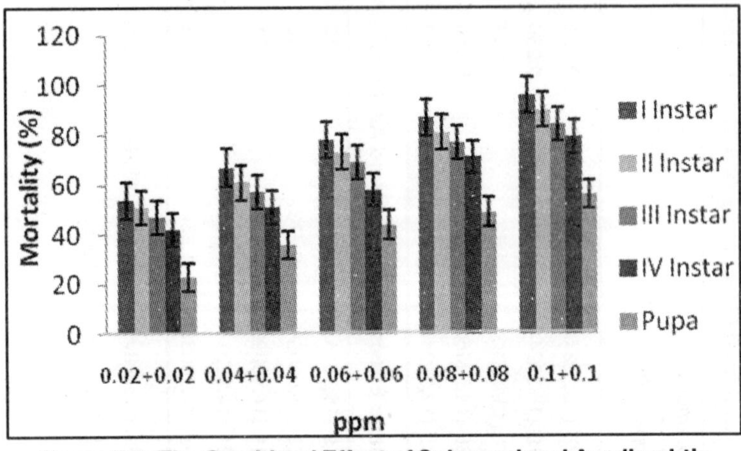

Figure 7.8: The Combined Effect of Spinosad and Azadirachtin against Chironomid, *Chironomus kiiensis*

Tables 7.4 and 7.5 provides the field trial conducted at Bommanampalayam and Kalveerampalayam in Coimbatore district, India after application of spinosad and azadirachtin at the breeding sites such as Water stored Cement drum and waste cement tank. The surface area of their breeding site is 2m X 2m and 2m X2m. It

Table 7.4: Field Trials after the Treatment of Spinosad on *Aedes aegypti* in Different time Duration

Place: Bommanampalayam, Coimbatore, India

Habitat: Water stored Cement drum

Size: 2m × 2m

Depth: 1m

Species: *Aedes aegypti*

Stage: Larvae

Calculation: 4 × 0.024 × 6 = 0.576 per cent

Required no. of concentration: 1 ppm

Sl.No.	*Larval Density*			
	Before Treatment	*After Treatment*		
		24 hrs	*48 hrs*	*72 hrs*
1.	118	45	31	26
2.	129	40	23	20
3.	142	37	17	15
4.	125	30	12	10
5.	150	47	30	20
6.	114	21	4	3
Total	778	220	117	94
Average	130	37	19	16
Reduction	–	71.72%	84.96%	87.91

was observed that *Aedes aegypti* larvae were found, and there was no mixed population. Before treatment, the number of total larval count was enumerated by dipping methods. The total numbers of larvae were 778 and 906 for six dipping in plotted area. The required concentrations of spinosad and azadirachtin were prepared according to the surface area of breeding habitat and were spread over the breeding site. Based on the laboratory IIIrd instar larval lethal concentration value was taken in which larvae less susceptible and the resulting dose was applied on the surface area. The spinosad and azadirachtin was sprayed with help of knapsack sprayer. Larval mortality was assessed at the interval of 24 hrs, 48 hrs and 72 hrs duration. After the treatment of 24 hrs, 48 hrs and 72hrs, the

Table 7.5: Field Trials after the Treatment of Azadiractin on *Aedes aegypti* in Different Time Duration

Place: Kalveerampalayam, Coimbatore, India

Habitat: Waste cement tank

Size: 2m × 2m

Depth: 1m

Species: *Aedes aegypti*

Stage: Larvae

Calculation: 4 × 0.036 × 6 = 8.64 per cent

Required no. of concentration: 10ppm

Sl.No.	*Larval Density*			
	Before Treatment	*After Treatment*		
		24 hrs	*48 hrs*	*72 hrs*
1.	126	56	49	41
2.	140	43	38	31
3.	162	37	30	23
4.	180	58	42	34
5.	158	29	27	21
6.	140	21	17	15
Total	906	244	203	165
Average	151	41	34	27
Reduction	–	73.06	77.59	81.78

percentage reduction of larval density in Bommanampalayam was 71.72 per cent, 84.96 per cent and 87.91 per cent whereas in Kalveerampalayam, the percentage reduction of larval density was 73.06 per cent, 77.59 per cent and 81.78 per cent, respectively.

Discussion

From this study *Aedes aegypti* and *Chironomus kiiensis* considerable mortality was evident after the treatment of spinosad and azadirachtin. In this study showed for all larval instars and pupae. Mortality was increased as the concentration also increased. This study recommended that the low amount of spinosad and azadirachtin showed more toxic effect on the mosquito. The toxic effect may be due to the active compound spinosyns A and D in Spinosad and azadirachtin, azadirachtols, triterpenoid. The larvae of a number of mosquito species (*Aedes* spp., *Anopheles* spp.) are sensitive to neem products and show antifeedant and growth regulating effects (Zebitz, 1987; Murugan *et al.*, 1996). Earlier studies reported (Vargas *et al.*, 2001) that that the spinosad is known to be highly active against Dipterans. Recent work (Alouani *et al.*, 2009) reported that Azadirachtin showed promising larvicidal agent against *Culex pipiens* and naturally occurring biopesticide could be an alternative for chemical pesticides.

In present study, the larvicidal and pupicidal activity was more pronounced with the spinosad against *Aedes aegypti* and chironomid, *Chironomus kiiensis*. The potential lethal activity against *Aedes aegypti* and chironomid, *Chironomus kiiensis* was observed with spinosad, indicated that the presence of chemical active compound such as presence of spinosyns A and D. Bret *et al.*, 1997, Thompson *et al.*, 2000 reported that Spinosad is a mixture of two neurotoxic macrolide compounds: spinosyn A and spinosyn D that are active mainly by ingestion. Spinosyns are produced by fermentation of the actinomycete *Saccharopolyspora spinosa* Mertz and Yao isolated from a Caribbean soil sample. Spinosyns A and D are highly toxic to Diptera, Lepidoptera, Thysanoptera, and some species of Coleoptera, but they have extremely low toxicity for mammals; therefore, spinosad is classified by the U.S. Environmental Protection Agency as a reduced-risk material. Vargas *et al.* (2001) reported that Spinosad is known to be highly active against Diptera and is registered for control of leaf mining dipteran pests of crops in many countries. Due to the favourable United States EPA classification, spinosad is also used in a bait formulation over very large areas for control of the Mediterranean fruit fly, *Ceratitis capitata*, in Central America.

Azadirachtin, a biologically active compound has been promoted as a new insecticide that is considered more eco- friendly than synthetic insecticides. The pesticidal efficacy, environmental safety and public acceptability of neem and its products for control of crop pests has led to its adoption into various mosquito control programmes (Su and Mulla, 1998a; Su and Mulla, 1998b).

A neem oil formulation containing 32 per cent neem seed oil (an equivalent of 0.03 per cent azadirachtin), an emulsifier

(5 per cent) and 63 per cent iso propanol (solvent) was investigated for its larvicidal activities against *An. gambiae* (Okumu *et al.*, 2007). It was toxic to mosquito larvae with LC50 value of 11 ppm and also reported to possess insect growth regulators. (Gianotti *et al.*, 2008). similarly, this study also well effect on growth and development like, the LC_{50} value of azadirachtin against *Aedes aegypti*, 0.021, 0.027, 0.036, 0.047 and 0.098.

In this study azadirachtin gave 91 per cent mortality of *Aedes aegypti* at 0.1ppm, 92 per cent of chironomid, *Chironomus kiiensis* at 0.1 level the toxic effect may be presence of active compounds like, azadirachtols, triterpenoid. Nakanishi (1974) and Zanno, *et al.* (1975) repoted that Azadirachtin is a tetranotriterpenoid. Rembold (1989) and Ley *et al.* (1993) reported that the biosynthesis of azadirachtin starts with a steroid precursor (*e.g.*, tirucallol) azadirone, azadiradione) and C-ring opening (*e.g.*, nimbin, salannin), after which further and proseeds via two mais levels of structural complexity: furan ring formation (*e.g.*, modifications yield azadirachtin. Some studies evaluated that the exact mode of action of azadirachtin is unknown. It has been hypothesized that azadirachtin suppresses normal synthesis of ecdysteroid or eclosion hormone (Schluter *et al.*, 1985; Barnby and Klocke, 1987; Rembold *et al.*, 1982; Sieber and Rembold, 1983). Azadirachtin-treated insects (Schluter *et al.*, 1985) or reverse its effects on their development. Alternatively, direct cytotoxic effects of azadirachtin on the epidermal cells (Schluter, 1985; Schluter, 1985), imaginal discs (Schulz, 1985; Schluter, 1985), or endocrine glands themselves (Schulz, 1984) have been cited as the primary lesions preventing molting.

In addition to having generalized anti-feedant properties that inhibit normal larval growth, azadirachtin can disrupt the normal sequence of physiological events associated with molting such as apolysis and ecdysis, culminating in death of pharate larvae within the exuvium. Consequently, some authors have attributed azadirachtin-induced anomalies to an anti-ecdysteroid or anti-eclosion hormone-like mode of action. Whereas (Kauser and Koolman, 1984) proposed that the compound binds directly to ecdysteroid receptors in target tissues, others have concluded that it suppresses release of prothoracicotropic hormone (Barnby and, Klocke, 1987; Rembold, 1987; Barnby and Klocke, 1987), interferes with the normal route of metabolism of ecdysteroids (Pener, 1987), or inhibits the release of eclosion hormone (Redfern *et al.*, 1982; Sieber and Rembold, 1983). Azadirachtin, a complex tetranotriterpenoid luminoid from the neem seeds, is the main component responsible for the toxic effects in insects (Modue and Nisbet, 2000).

In this present study the *Aedes aegypti* showed higher mortality when compare to chironomid, *Chironomus kiiensis.* The ppm level was prepared same but the novel pesticide spinosad give more effect when compare the azadirachtin. spinosad more active than the active compound of azadirachtin because its directly affect the nervous system. Salgado (1997;1998) reported that Spinosad exhibits stomach and contact poisoning properties and affects specifically the function of g-aminobutyric acid (GABA) receptors and nicotinic acetylcholine receptors of the target insects. This product has been widely tested against injurious insects in a variety of crops, such as cotton (Banerjee *et al.*, 2000), wheat (Fang *et al.*, 2002). Previous studies reported that the spinosad was most effective at the lowest concentrations (0.024 to 0.025 ppm) and Spinosad effectively prevented breeding of Culex (Diptera: Culicidae) mosquitoes and chironomids (Diptera: Chironomidae) (Bond *et al.*, 2004). The results of a recent work showed that spinosad was very effective in the control of *Spodoptera littoralis* (Aydin and Gurkan, 2006).

The combine study of spinosad and azadirachtin was more effective on when compare to individuals. *Aedes aegypti* and chironomid, *Chironomus kiiensis.* Because it may be the both compound mixed and give more mortality to compare the individuals. Previous studies reported that Spinosad and azadirachtin–low risk insecticides are now popular components of IPM and organic plant protection programme of brassica crops in many countries (Hines and Hutchison, 2001; Grišakova *et al.*, 2006; Guerena, 2006; Burkness and Hutchison, 2008). This was an effect of many experiments conducted to prove their efficacy in pest control and also on their environmental side effect. For example, some field observation indicated that spinosad caused no disruption to beneficial species in commercial orchards (Arthurs *et al.*, 2007). However, some laboratory experiments show that spinosad and azadirachtin are toxic to beneficial insects. It found that *Hyposoter didimator* (Tunberg) (Ichneumonidae) a parasitoid of early larval instars of lepidopteran pests is very susceptible to both spinosad and azadirachtin (Schneider *et al.*, 2003).

The physiological effects of azadirachtin are much more consistent than the antifeedant effects, and result from interference with growth and moulting, interference with reproduction and interference with cellular processes. In all species tested dose response effects can be seen as reduced growth, increased mortalities, abnormal moults and delayed moults. These effects are related to disruption of endocrine system controlling growth and moulting. The moulting effects are due to a disruption in the synthesis and release of ecdysteroids (moulting hormone) and other classes of hormones and this can be demonstrated by accurately timed injections of azadirachtin into the haemoulymph of vth instar nymphs of *L. migratoria* (Mordue (Luntz) *et al.*, 1986).

Spinosad was shown to be highly toxic to *Ae. aegypti* and *Anopheles albimanus* Weidemann in the laboratory, and it completely suppressed the development of *Ae. aegypti, Culex* spp., and chironomid larvae in seminatural field conditions for periods of 8 to 22 wk, depending on concentration (Bond *et al.*, 2004). Spinosad (active against larvae) has been found to have low toxicity for humans and other non-target fauna (Williams, 2003) and has the potential to be used against mosquitoes as it does not show cross-resistance with conventional insecticides (Darriet *et al.*, 2005). This study concluded that the spinosad more effect than the azadirachtin because it is directly affected by nevous system, the azadirachtin also one of the novel bio-pesticide the compounds of azadirachtin may be control to the moulting, growth and regulatory system. The novel pesticide spinosad and azadirachtin safer to human and more potential larvicides against *Aedes aegypti* and *Chironomus kiiensi.*

References

Alouani, A., Rehimi, N. and Soltani, N., 2009. Larvicidal activity of a neem tree extract Azadirachtin) against mosquito larvae in the Republic of Algeria. *Jordan J Biol Sci.*, 2(1): 15–22.

Arthurs, S.P., Lacey, L.A. and Miliczky, E.R., 2007. Evaluation of the codling moth granulovirus and spinosad for codling moth control and impact on non–target species in pear orchards. *Biological Control*, 41(1): 99–109.

Aydin, H. and Gurkan, M.O., 2006. The efficacy of spinosad on different strains of *Spodoptera littoralis* (Boisduval) (Lepidoptera: Noctuidae). *Turk J Biol.*, 30: 5–9.

Banerjee S. K., Turkar, K. S. and Wanjari, R. R., 2000. Evaluation of newer insecticides for the control of bollworms in cotton. *Pestology*, 24: 14–6.

Banerjee, D., Teriaca, L., Doyle, J. G. and Lemaire, P., 2000. *Sol. Phys.*, 194, 43

Barnby, M. A. and Klocke, J. A., 1987. Effects of azadirachtin on nutrition and development of larvae of the tobacco budworm, *Heliothis virescens* (Fabr.). In: *Proceedings of the International Conference on Endocrinological Frontiers in Physiological Insect Ecology*. Zabza A, Sehnal F, Denlinger DL, eds. Technical University of Wroclaw, Poland.

Barnby, M. A. and Klocke, J. A., 1987. Effects of azadirachtin on the nutrition and development of the tobacco budworm, *Heliothis virescens* (Fabr.) (Noctuidae). *J Insect Physiol.*, 33: 69.

Baur, X., Aschauer, H., Mazur, G., Dewair, M., Prelicz, H. and Steigemann, W., 1986. Structure, antigenic determinants of some clinically important insect allergens: *Chironomid haemoglobins*. *Science*, 233 (4761): 351–354.

Blanc, M. P., Panighini, C., Gadani, F. and Rossi, L., 2004. Activity of spinosad on stored–tobacco insects and persistence on cured tobacco strips. *Pest Manag.*, 60: 1091–8.

Bond, J. G., Marina, C. F. and Williams, T., 2004. The naturally derived insecticide spinosad is highly toxic to *Aedes* and *Anopheles* mosquito larvae. *Med Vet Entomol.*, 18: 50–56.

Bret, B. L., Larson, L. L., Schoonover, J. R., Sparks, T. C. and Thompson, G. D., 1997. Biological properties of spinosad. *Down to Earth*, 52: 6–13.

Burkness, E.C. and Hutchison, W.D., 2008. Implementing reduced–risk integrated pest management in fresh–market cabbage: improved net returns via scouting and timing of effective control. *J. Econ. Ent.*, 101, (2): 461–471.

Copping, L.G. and Menn, J.J., 2000. Biopesticides: a review of their action, applications and efficacy. *Pest Management Science*, 56: 651–676.

Darriet, F., Duchon, S. and Hougard, J. M., 2005. Spinosad: a new larvicide against insecticide resistant mosquito larvae. *J Am Mosq Control Assoc.*, 21: 495–496.

Delettre, Y. R., 1985. Importance des processus de colonisation dans la cinétique de *Limnophyes minimus* (Mg.), Diptère Chironomide à larves édaphiques des landes armoricaines. *Bulletin de la Société zoologique de France*, 110: 403–407.

Delettre, Y. R., 1994. Fire disturbance of a chironomid (Diptera) community on heathlands. *Journal of Applied Ecology*, 31: 560–570.

Delettre, Y. R. and Lagerlöf, J., 1992. Abundance and life–history of terrestrial Chironomidae (Diptera) in four Swedish agricultural cropping systems. *Pedobiologia*, 36: 69–78.

Delettre, Y.R. and Morvan, N., 2000. Dispersal of adult aquatic Chironomidae (Diptera) in agricultural landscapes. *Freshwater Biology*, 44: 399–411.

Fang, J., Feng, Q., Ketel, C.S., Wang, H., Cao, R. Xia, L., Erdjument–Bromage, H., Tempst, P., Simon, J.A. and Zhang, Y.(2002. Purification and functional characterization of SET8, a nucleosomal histone H4–lysine 20–specific methyltransferase. *Curr Biol.*, 12(13): 1086–1099.

Fang, L., Subramanyam, B. and Arthur, F. H., 2002. Effectiveness of spinosad on four classes of wheat against five stored–product insects. *J Econ Entomol.*, 95: 640–50.

Frouz, J., 1994. Changes in the terrestrial chironomid community (Diptera: Chironomidae) during secondary succession in old fields. *Pedobiologia*, 38: 334–343.

Gianotti, R. L., Bomblies, A., Dafalla, M., Issa–Arzika, I., Duchemin, J.B. and Eltahir, E.A.B., 2008. Efficacy of local neem extracts for sustainable malaria vector control in an African village. *Malar J.*, 7: 138.

Gonzalez, M.S., Nogueira, N.F.S., Mello, C.B., De Souza,W., Schaub, G.A., Azambuja, P. and Garcia, E.S., 1999. Influence of brain and azadirachtin on *Trypanosoma cruzi* development in the vector, *Rhodnius Prolixus. Exp. Parasitol.*, 92: 100–108.

Govindachari, T.R., Sandhya, G. and Ganeshraj, S.P., 1990) Simple method for the isolation of azadirachtin by preparative high–performance liquid chromatography, *J. Chromatography*, 513: 389–391.

Grišakova, M., Metspalu, L., Jõgar, K., Hiiesaar, K., Kuusik, A. and Põldma, P., 2006. Effects of biopesticide Neem EC on the large white butterfly, *Pieris brassicae* L. (Lepidoptera, Pieridae). *Agronomy Research*, 4 (Special issue): 181–186.

Guerena, M., 2006. Cole crops and other brassicas: organic production. A publication of ATTRA–National Sustainable Agriculture Information Service.1–20. available at: http: // www.attra.ncat.org/attra-pub/PDF/cole.pdf

Harrington, L.C., Scott, T.W., Lerdthusnee, K., Coleman, R. C., Costero, A., Clark, G. G., Jones, J. J., Kitthawee, S., Kittayapong, P., Sithiprasasna, R. and Edman, J. D., 2005. Dispersal of the dengue vector *Aedes aegypti* within and between rural communities. *Am J Trop Med Hyg.*, 72: 209–220.

Healey, I.N. and Russell–Smith, A., 1971. Abundance and feeding preferences of fly larvae in two woodland soils. Annales de Zoologie et d'Ecologie Animale 3, numéro hors–série, IV. Proc. 5th Colloquium Pedobiologiae, Dijon, France, pp. 177–192.

Hines R.L. and Hutchison W.D., 2001. Evaluation of Action Thresholds and Spinosad for Lepidopteran Pest Management in Minnesota Cabbage. *J.Econ. Ent.*, 94(1): 190– 196.

Honório N. A., W Da Costa Silva., Leite P. J., Gonçalves J. M., Lounibos L. P. and Lourenço–De–Oliveira R., 2003. Dispersal of *Aedes aegypti* and *Aedes albopictus* (Diptera: Culicidae) in an urban endemic dengue area in the state of Rio de Janeiro, Brazil. *Memórias do Instituto Oswaldo Cruz.*, 98: 191–198.

Hövemeyer, K., 1992. Die Dipterengemeinschaft eines Kalkbuchenwaldes: eine siebenjährige Untersuchung. *Zoologische Jahrbücher für Systematik*, 119: 225–260.

Hudson, P.L., 1987. Unusual larval habitats and life history of Chironomid (Diptera) genera. *Entomologica Scandinavica Supplements*, 29: 369–373.

Jeong, K. Y., Yum, H. Y., Lee, I. Y., Ree, H. I., Hong, C. S., Kim, D. S. and Yong, T. S., 2004. Molecular Cloning and Characterization of Tropomyosin, a Major Allergen of *Chironomus kiiensis*, a Dominant Species of Nonbiting Midges in Korea. *Clin Vaccine Immunol.*, 11(2): 320–324.

Kauser, G. and Koolman, J., 1984. Ecdysteroid receptors in tissues of the blowfly, *Culliphoru vicina*. In: Advances in Invertebrate Reproduction 3. Engels W, ed. Elsevier North–Holland, New York, p. 602.

Kavathekar, K.Y., 2003. Neem in India, NISCAIR, New Delhi, pp. 21–23.

Kollien, A.H. and Schaub, G.A., 1999. The effect of azadirachtin on *Blastocrithidia triatomae* and *Trypanosoma cruzi* in *Triatoma infestans* (Insecta, Hemiptera). *Int. J. Parasitol.*, 29: 403–414.

Ley, S.V., Denhom, A.A. and Wood. A., 1993. The chemistry of azadirachtin. *Nat. Prod. Rep.*, 109–157.

Modue, L. and Nisbet, A.J., 2000. Azadirachtin from the neem tree *Azadirachta indica*: its actions against insects. *Annual Entomological Society Brasil*, 29: 615–632.

Mordue (Luntz), A.J., K.A. Evans and M. Charlet., 1986. Azadirachtin, ecdysteroids and ecdysis in *Locusta migratoria*. *Comp. Biochem. Physiol.*, 85c: 297–301.

Muir, L. E. and Kay, B. H., 1998. *Aedes aegypti* survival and dispersal estimated by mark–release–recapture in northern Australia. *Am J Trop Med Hyg.*, 58: 277–82.

Mulla, M.S. and Su, T.Y., 1999. Activity and biological effects of neem products against arthropods of medical and veterinary importance. *J. Am. Mosq. Control Assoc.*, 15(2): 133–152.

Murugan, K., Jahanmohani, P. and Babu, R., 1996. Effect of neen kernal extract and neem oil on nutritive and reporactive physioliogy of *Helianthus armigera* Hub. Neem and Environment. Delhi: Oxford and IBH Co., Pvt. Ltd., p. 321–334.

Nakanishi, K., 1974. Structure of the insect antifeedant azadirachtin. *Recent Adv Phytochem.*, 9: 293.

Nogueira, N.F.S., Gonzales, M., Garcia, E.M. and de Souza, W., 1997. Effect of azadirachtin A on the fine structure of the midgut of *Rhodnius prolixus*. *J. Invertebr. Pathol.*, 69: 58–63.

Okumu, F. O., Knols, B. G. J. and Fillinger, U., 2007. Larvicidal effects of a neem (*Azadirachta indica*) oil formulation on the malaria vector *Anopheles gambiae*. *Malar J.*, 6: 63.

Pener, M., Rountree, D. B., Bishoff, S. and Gilbert, L. I., 1987. Azadirachtin maintains active prothoracic glands but reduces ecdysone titres in *Manduca sextu* pupae: *In vivo* and *in vitro* studies. In: Proceedings of the International Conference on Endocrinological Frontiers in Physiological Insect Ecology. Zabza A, Sehnal F, Denlinger DL, eds. Technical University of Wroclaw, Poland.

Redfern, R. E., Kelly, T. J., Borkovec, A. B. and Hayes, D. K., 1982. Ecdysteroid titers and moulting aberrations in last stage *Oncopelfus* nymphs treated with insect growth regulators. *Pestic Biochem Physiol.,* 28: 351.

Reiter, P., Amador, M. A., Anderson, R. A. and Clark. G. G., 1995. Short report: dispersal of *Aedes aegypti* in an urban area after blood feeding as demonstrated by rubidium marked eggs. *Am. J. Trop. Med. Hyg.,* 52: 177–179.

Rembold, H., 1984. Secondary plant products in insect control with special reference to the azadirachtins. In: Advances in Invertebrate Reproduction 3. Engels W eds. *Elsevier* North–Holland, New York, pp 481–491.

Rembold, H., 1989. Azadirachtins: Their structure and mode of action, p.150–163. In J.T.Arnason, B.J.R. Philogène and P. Morand (eds.), Insecticides of plant origin. ACS Symp. Ser. 387 *American Chemical Society,* Washington, DC.

Rembold, H., Sharma, G. K., Czoppelt, C. H. and Schmutterer, H., 1982. Azadirachtin: A potent insect growth regulator of plant origin. *Z Ang Entomol.,* 93: 12.

Salgado, V. L., 1997. The mode of action of spinosad and other insect control products. *Down to Earth,* 52: 35–44.

Salgado, V. L., 1998. Studies on the mode of action of spinosad: insect symptoms and physiological correlates. *Pestic Biochem Physiol.,* 60: 91–102.

Schluter, U., 1985. Die Wirkung von Azadirachtin auf Gewebe von Insekten: Epidermis und Imaginalanlagen von *Epilachna varivestis.* Muls. (Col., Coccinellidae). *Mitt Dtsch Ges Allg Angew Ent.,* 4: 197.

Schluter, U., 1985. Disturbance of epidermal and fat body tissue after feeding azadirachtin and its consequence on larval moulting in the Mexican bean beetle *Epilachna varivestis* (Coleoptera: Coccinellidae). *Entomol Gen.*, 20: 97.

Schluter, V., Bidmon, H. J. and Grewe, S., 1985. Azadirachtin affects growth and endocrine events in larvae of the tobacco hornworm, *Munducu sexfu. J Insect Physio.*, 131: 773.

Schneider, M.I., Smagghe, G., Gobbi, A. and Viñuela, E., 2003. Toxicity and pharmacokinetics of insect growth regulators and other novel insecticides on pupae of *Hyposoter didymator* (Hymenoptera: Ichneumonidae), a parasitoid of early larval instars of lepidopteran pests. *J. Econ. Entomol.*, 96(4): 1054–1065.

Schulz, V. W. D., 1984. Ultrastrukturelle Veranderungen des Corpus cardiacum–allatum–Komplexes von adulten Weibchen des Mexikanischen Bohnen Kafers *Epilachna varivesfris* Muls. nach Behandlung mit Extrakten des Neem–Baums *Azadirachta indica.* A Juss. *Verh Dtsch Zool Ges.*, 77: 330.

Schulz, V. W. D., 1985. Histologische Untersuchungen zur Wirkung von Extrakten aus Samen des Neem–Baumes *Azadirachfa indica* auf weibliche Imagines von *Epilachna varinesfis* (Coleoptera, Coccinellidae) I. Schadigungen von Ovarentwicklung und Oogenese. *Zool Jb Anat.*, 223: 193.

Seddon, A.M., 1986. Abundance and life–history of four species of terrestrial Chironomidae (Dipt.) from deciduous woodland soil in South East England. *Entomologist's Monthly Magazine,* 122: 219–228.

Sendstad, E., Solem, J.O. and Aagaard, K., 1977. Studies of terrestrial chironomids (Diptera) from Spitzbergen. *Norway Journal of Entomology,* 24: 91–98.

Service, M.W., 1983. Management of vector. In: Youdeowei, A., Service, N. (Eds.), Pest and Vector Management in the Tropics. Longman Group Ltd., England, pp. 7–20.

Sieber, K. P. and Rembold, H., 1983. The effects of azadirachtin on the endocrine control of moulting in *Locusfa migratoria. J Insect Physiol.*, 29: 523.

Sparks, T. C., Thompson, G.D. and Kirst, H. A., 1997. Fermentation–derived insect control agents –the spinosyns. In: Hall F, Menn JJ, eds. Biopesticides: use and delivery. Totowa, New Jersy: Humana Press, 171–88.

Su, T. and Mulla, M. S., 1998a). Antifeedancy of neem products containing Azadirachtin against *Culex tarsalis* and *Culex quinquefasciatus* (Diptera: Culicidae). *J Vector Ecol.*, 23: 114–122.

Su, T. and Mulla, M. S., 1998b). Ovicidal activity of neem products (azadirachtin) against *Culex tarsalis* and *Culex quinquefasciatus* (Diptera; Culicidae). *J Am Mosq Control Assoc.*, 14: 204–209.

Surakarn, R. and Yano, K., 1995. Chironomidae (Diptera) recorded from paddy fields of the world: a review. *Makunagi/Acta Dipterologica*, 18, 1–20.

Thompson G. D., Michel, K. H. and Yao, R. C., 1997. The discovery of *Saccharopolyspora spinosa* and a new class of insect control products. *Down to Earth*, 52: 1–5.

Thompson, G. D., Dutton, R. and Sparks, T. C., 2000. Spinosad–a case study: an example from a natural products discovery programme. *Pest Manag. Sci.*, 56: 696–702.

Vargas, R.I., Peck, S.L., McQuate, G.T., Jackson, C.G., Stark, J.D. and Armstrong, J.W., 2001) Potential for area wide integrated management of Mediterranean Fruit Fly (Diptera: Tephritidae) with a braconid parasitoid and a novel bait spray. *Journal of Economic Entomology*, 94, 817–825.

Watson, G.B., 2001. Actions of insecticidal spinosyns on gaminobutyric acid receptors from small–diameter cockroach neurones. *Pesticide Biochemistry and Physiology*, 71: 20–28.

Williams, T., Valle, J. and Vinuela E (2003. Is the naturally–derived insecticide spinosad® compatible with insect natural enemies?. *Bio Sci Tech.*, 13: 459–475.

World Health Organization., 2011. Dengue and dengue haemorrhaghic fever. *Factsheet*, 117.

Zanno, P. R. I., Miura, K., Nakanishi, K. and Elder, D. L., 1975. Structure of the insect phagodeterrent azadirachtin. Application of PRFTICWD carbon 13–NMR. *J Am Chem* Soc., 97.

Zebitz, C. P. W., 1987. Potential of neem seed kernel extracts in mosquito control. In: Schmutterer H, Ascher KRS (eds). Proceedings of the third international neem conference (Nairobi, Kenya). Eschborn: German Technical Cooperation GTZ, p. 555–573.

Chapter 8

Effect of Some Abiotic Factors on the Post Embryonic Development of *Spathosternum prasiniferum* Walker (Orthoptera : Acridoidea : Hemiacridini)

Uzma Rafi, Mohd. Kamil Usmani and Zeb Shamael*

Department of Entomology, Department of Zoology, Aligarh Muslim University, Aligarh – 202 002, U.P., India

ABSTRACT

The present study was conducted to investigate the effect of temperature and humidity on the growth and development of *Spathosternum prassiniferum* under laboratory conditions. The grasshoppers were reared at two different levels of temperature and humidity, 37°C±1°C and 70 per cent -80 per cent relative humidity, and 27°C±1°C and 50 per cent -60 per cent relative humidity. This work has been carried out to achieve knowledge of biology, post-embryonic development and morphometrics of *Spathosternum prassiniferum,* which is an economically

* Corresponding author: E-mail: rafiuzma@gmail.com

important insect species. It is a polyphagous pest found in small grasses and bushes, and usually feeds on crops such as rice, maize, millet, etc.

Keywords: *Temperature, Relative humidity, Spathosternum prassiniferum.*

Introduction

Orthoptera is one of the largest and most diverse groups of insects, sometimes called as a short horned grasshoppers belonging to suborder Caelifera which is divided into five super families *viz.* Acridoidea, Pyrgomorphoidea, Tetrigoidea, Tridactyloidea and Eumastacoidea. Although most species are found in warmer areas of the world, a large number occur in temperate climates, where usually they overwinter in the egg stage. Most Acridoids inhabit grassland and others low vegetation, but a number live in trees. They are predominantly diurnally active insects. They are phytophagus, eating mainly living plants. They feed on wide range of plants and are voracious feeders, consume approximately one half of their body weight. About 57 per cent of the population of India is engaged in agriculture directly or indirectly, therefore the problem of locust and grasshopper seem to be major now-a-days.

Previously Criddle (1918) observed the egg-laying habits of some of the Acrididae. Hussain and Ahmad (1936) studied the biology of *Schistocerca gregaria* with special relation to temperature. Hussain and Mathur (1944) further observed the influence of temperature on the weight and size of *Schistocerca gregaria*. Hafez and Ibrahim (1964) studied the behaviour of *Sphingonotus carinatus* Saussure towards humidity and temperature. Greathead (1966) studied the effect of biotic factors on population of the desert locust *Schistocerca gregaria*. Grewal and Atwal (1968) studied the affect of different levels of temperature and humidity on the development of *Chrotogonus trachypterus* Blanchard. Khan and Aziz (1973) studied the seasonal variations in populations of hoppers and adults of *Oedaleus abruptus* Thunberg and also observed the influence of different levels of temperature and humidity on the development and hatching of eggs of the *Oedaleus abruptus* Thunberg. Iqbal and Aziz (1973) studied the effect of different levels of temperature and humidity on the development of *Spathosternum prasiniferum* Walker. In 1974 they studied the life-history of *Spathosternum prasiniferum* Walker while in 1975 studied the food preference of *Spathosternum prasiniferum* Walker (Orthoptera: Acridoidea). Sanger (1973), Khan and Aziz (1974a, b, c), Parihar and Pal (1978) and Majeed (1990) studied the impact of temperature and humidity on growth, development, incubation period etc. of different grasshopper species.

The species under study (*Spathosternum prasiniferum*) belongs to the sub order Caelifera, super family Acridoidea and family Catantopidae and subfamily Hemiacridini. The present work has been carried out to achieve complete knowledge of biology, post embryonic development of this economically important species of grasshoppers. This species is polyphagous, usually found in short grasses and also feed on many crop like rice, maize, millet and pulses etc.

The present study is mainly based on experiments carried out in the laboratory under controlled conditions. The study deals with the life history and the successive stages of development under the influence of different environmental conditions and their impact was observed during the course of study.

Material and Methods

The present study was carried out under laboratory condition in the Department of Zoology, AMU Aligarh. Field observation and collections were made from adjacent areas of Aligarh for a period of 6 months during October 2010- March 2011.

The Grasshoppers used in the experiments were obtained from the field collection as well as from the stock maintained in the laboratory at 37°C and at 27°C. In order to maintain stock, adult males and females of *Spathosternum prasiniferum* were collected and kept in insect breeding cages.

Collection and Maintenance of Stock

Mature and immature stages of *Spathosternum prasiniferum* were collected from the fields of rice, sugarcane, wheat, grasses and bushes by using sweeping nets. They were reared in wooden cages. Three sides of the cage were made up of wood with glass panels on the front side. Either of the two sides of wooden panels was provided with small windows covered with wire-mesh wire gauge and the third side with an outlet for the purpose of feeding, cleaning, placing the insects, watering of egg-laying tubes and harvesting of egg-pods after egg laying. The central portion of the cage floor was provided with wire mesh for proper ventilation, while sides of the floor were made of wood. One side of the cage floor was provided with holes for egg-laying tubes. The wooden roof also had a lid. An electric bulb was fitted in one corner of the roof to provide light and regulate temperature. The legs of cages were placed in glass bowls filled with mixture of water and kerosene oil or phenol to keep away predators such as ants and lizards (Figure 8.1).

The adult male and female grasshoppers were fed *Cynodon* grass. The egg-laying tubes were checked regularly for eggs. The tubes were replaced by fresh ones. The tubes containing egg pods were labelled (species name, date of egg-laying) and placed in jars covered with muslin cloth and transferred to BOD at 37°C±1°C. When the second egg laying took place, it was kept at 27°C±1°C. The egg pods were moistened regularly.

On hatching, the hoppers were transferred to separate jars and two different sets of experiments were prepared, one at 37°C±1°C and the other at 27°C±1°C, and both were fed *Cynodon dactylon* grass. The moulting period of each instar varies at both temperatures. The instar duration was recorded after each moulting (Figure 8.2).

Morphometry

In the present study morphometrics was carried out for each stage for different experimental condition with the help of Vernier calliper.

Soil Preference

To check the soil preference for egg laying separate experimental set up was prepared.

Two sets of three jars were prepared, each of them contains about three centimeter thick bed of white sand, red sand and garden soil respectively. One set was kept at 37°C±1°C in BOD and the another at 27°C±1°C. Two males and two females were released in each jar to analyze the result.

Results

Spathosternum prasiniferum was reared in different experimental setup to find out the impact of different environmental factors (*i.e.*, temperature, humidity, soil type) on egg laying and developmental

Figure 8.1: Culture of *S. prasiniferrum* in Cage

Figure 8.2: Culture of *S. prasiniferrum* in BOD at 37°C and 70 per cent–80 per cent RH and 27°C &50 per cent–60 per cent RH

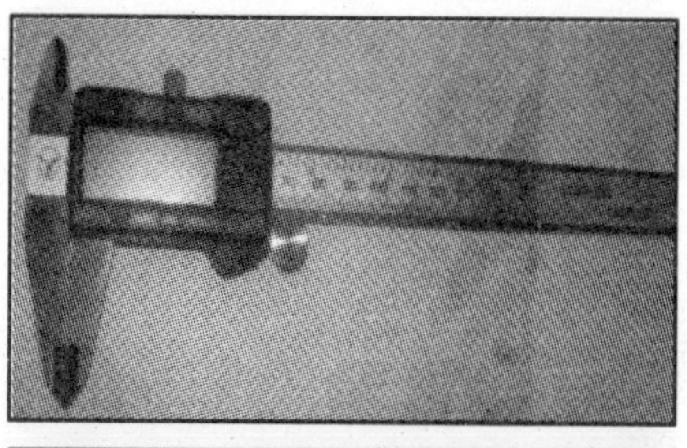

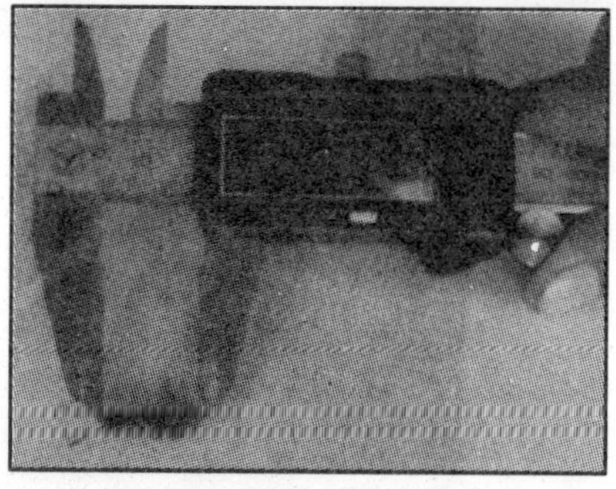

Figure 8.3: Digital Vernier Calliper Used for the Measurement

stages of these grasshopper species. Temperature has marked influence on the development of hoppers. Rate of development increased at higher temperature with least hopper duration and at lower temperature, development decreased with increased hopper duration.

The life span of an adult *Spathosternum prasiniferum* is about 60-90 days at 37°C, 70 per cent -80 per cent RH and 30-50 days at 27°C, RH 50 per cent -60 per cent. The newly emerged nymphs take about 30-40 days to become adult at 37°C and 65-75 days at room temperature, because the duration varies according to temperature. A female lays about 15-20 eggs each time in an egg pod. The incubation period of these eggs are 15- 20 days and 25-30 days at both temperatures respectively. The larval duration varies in different instar. The newly hatched nymphs are called as vermiform larva. I instar moult after 5-7 days just after hatching and transform into II instar and the average duration of second instar is about 8-9 days and the III instar 7-8 days while the duration of IV instar is about 6 days and the final instar *i.e.*, V instar takes 5-10 days to become an adult (Table 8.1, Figure 8.5).

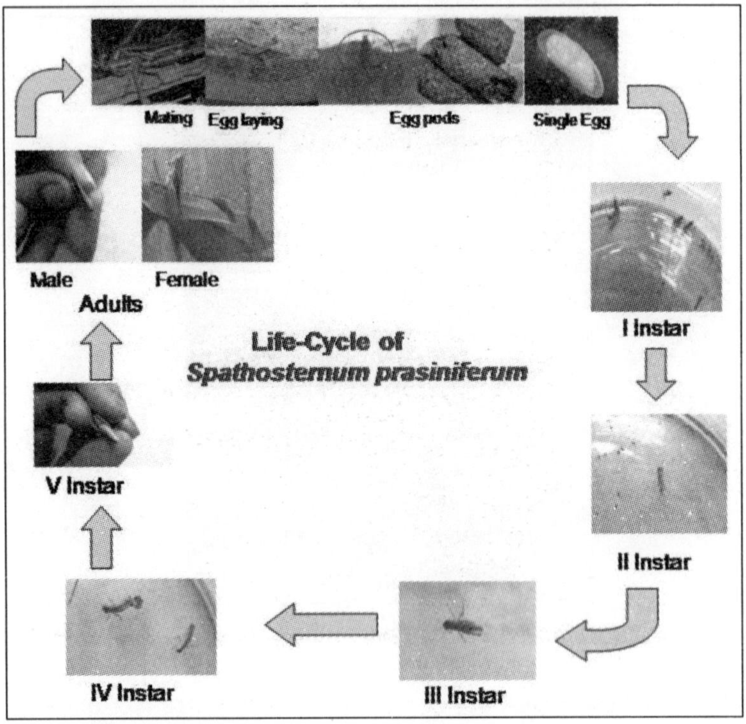

Mating Egg laying Egg pods Single Egg

Male Female

Adults

Life-Cycle of *Spathosternum prasiniferum*

I Instar

II Instar

III Instar

IV Instar

V Instar

Figure 8.4: Life Cycle of *Spathosternum prasiniferum*

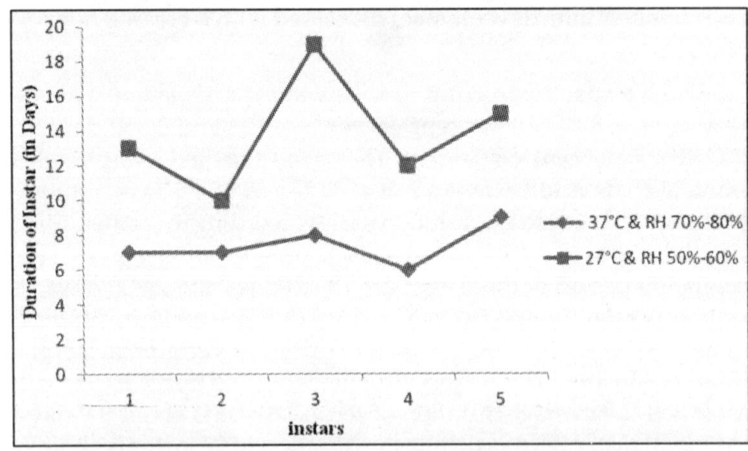

Figure 8.5: Variation in Duration (in days) of Nymphal Instars in Different Conditions of *Spathosternum prasiniferum*

Table 8.1: Development of *Spathosternum prasiniferum* (in days) at 37°C±1°C and 27°C±1°C.

Temperature and RH	*Instars*						
	I	*II*	*III*	*IV*	*V*	*Total No. of Days for Males*	*Total No. of Days for Females*
At 37°C±1°C and RH 70%–80%	7	7	8	6	9	28	37
27°C±1°C and RH50%–60	13	10	19	12	15	53	68

Morphometry

The morphometrics of hoppers was carried out on certain body parts for each experimental set up to obtain the rate of increase in various instars. Linear measurements of various body parts of male and female grasshoppers were made separately with the help of Vernier calliper (Table 8.2 and 8.3)

MORPHOMETRICAL DATA

Table 8.2: Morphometry of *Spathosternum prasiniferum* at 37°C

Sl.No.		I Instar	II Instar	III Instar	IV Instar	V Instar	Adult Female	Adult Male
1.	Body Length	5.30	7.40	8.75	11.18	13.22	22.11	14.05
2.	Max. width of Head	0.87	1.05	1.15	1.55	1.6	2.49	1.75
3.	Length of Antenna	0.90	1.35	1.95	2.7	3.35	3.88	3.50
4.	Width of vertex b/w eyes	0.35	0.22	0.32	0.40	0.60	0.90	0.86
5.	Vertical dia. of eyes	0.22	0.32	0.35	1.35	1.5	1.74	1.61
6.	Horizontal dia. of eyes	0.50	0.60	0.85	0.9	1.05	1.32	1.04
7.	Max. width of Pronotum	0.77	1.00	1.35	1.75	2.05	2.59	1.96
8.	Min. width of Pronotum	0.72	0.90	1.1	1.5	1.6	2.21	1.55
9.	Length of Pronotum	0.90	1.10	2.0	2.75	2.8	3.84	2.70
10.	Height of Pronotum	1.25	1.37	1.7	2.2	2.5	2.78	2.25
11.	Length of Ant. Femur	0.80	0.95	1.1	1.7	1.9	2.72	2.15
12.	Length of Mid. Femur	0.85	1.05	1.45	2.45	2.72	3.56	2.82
13.	Length of Hind Femur	2.40	2.75	4.85	6.2	6.9	9.28	7.25
14.	Length of Hind Tibia	2.15	2.70	3.75	5.45	5.7	7.6	6.05
15.	Length of Elytron	–	–	–	–	–	15.2	10.6

Soil Preference

It was observed that, at both temperature regimes, maximum egg laying was observed in white sand while least in garden soil. (Tables 8.4 and 8.5)

Discussion

The results obtained during the study were found to be significant. The effects of different environmental factors such as

Table 8.3: Morphometry of *Spathosternum prasiniferum* at 27°C

Sl.No.		*I Instar*	*II Instar*	*III Instar*	*IV Instar*	*V Instar*	*Adult Female*	*Adult Male*
1.	Body Length	4.2	6.07	6.8	8.96	10.5	21.05	13.83
2.	Max. width of Head	0.73	0.77	0.83	1.24	1.25	2.24	1.43
3.	Length of Antenna	0.67	0.82	1.03	1.9	3.05	3.45	3.26
4.	Width of vertex b/w eyes	0.18	0.21	0.31	0.38	0.55	0.86	0.84
5.	Vertical dia. of eyes	0.76	0.92	0.88	1.17	1.2	1.46	1.43
6.	Horizontal dia. of eyes	0.40	0.43	0.47	0.77	0.8	1.05	0.87
7.	Max. width of Pronotum	0.70	0.92	0.92	1.59	1.9	2.39	1.78
8.	Min. width of Pronotum	0.42	0.80	0.84	1.20	1.4	1.94	1.43
9.	Length of Pronotum	0.7	0.8	1.09	1.61	2.45	3.46	2.47
10.	Height of Pronotum	0.84	1.02	1.05	1.67	2.4	2.56	1.96
11.	Length of Ant. Femur	0.5	0.58	0.72	1.27	1.65	2.67	1.89
12.	Length of Mid. Femur	0.7	0.87	1.07	1.28	2.6	3.34	2.65
13.	Length of Hind Femur	2.12	2.4	3.04	4.63	6.25	9.04	7.14
14.	Length of Hind Tibia	1.86	2.38	2.59	4.05	5.05	7.35	5.83
15.	Length of Elytron	–	–	–	–	–	14.3	9.75

temperature, humidity, soil type on egg-laying, incubation period and biology were studied. The egg-laying, hatching, nymphal development and adult survival were greatly affected by the abiotic factors.

Temperature and humidity had a marked influence on egg laying, incubation period and rate of development of instars. Hatching of eggs was quick in case of higher temperature (37°C) than low temperature (27°C). The rate of development of nymphs was faster at 37°C (RH-70 per cent–80 per cent) and the hopper duration was less at higher temperature. Parker (1930) in his extensive experiments with *Melanoplus mexicanus mexicanus* Saussure and *Camnula pellucida* Scudder clearly indicated a shortening of hopper period and an accelerated rate of development with rising in temperature. Shamshad (1982) also described the same conditions when applied on *Acrida exaltata*, while Hamilton (1936, 1950) studied the relation of temperature and humidity to the development of

Table 8.4: Soil Preference for Egg Laying at 37°C

Soil type	*No. of Females in Each Jar*	*No. of Eggpods Laid by Females*	*Average Nos. of Eggs/Pods*	*No. of Hatchings*	*No. of instars Hatched*	*Incubation Period (in days)*
Garden Soil	2	5	6	3	12	26
White Sand	2	15	8	15	100	20
Red Sand	2	15	8	14	88	22

Table 8.5: Soil Preference for Egg Laying at 27°C

Soil type	*No. of Females in Each Jar*	*No. of Eggpods Laid by Females*	*Average Nos. of Eggs/Pods*	*No. of Hatchings*	*No. of instars Hatched*	*Incubation Period (in days)*
Garden Soil	2	4	5	2	8	60
White Sand	2	10	8	9	50	40
Red Sand	2	8	8	5	25	43

Locusta migratoria migratorioides, Schistocerca gregaria and *Nomadacris septemfasciata*.

The morphometrical studies were also done and were found that the instar size and the adult size were larger at 37°C±1°C than 27°C±1°C. The growth rate is highly affected according to the temperature which in turn affects the total life cycle of the grasshoppers.

The different soil types were used in the study and a great diversity of egg laying preference was observed such as the grasshopper preferred to lay eggs in white sand than in other types of sand and soil (red sand and garden soil) it may be because of the particle size of the sand.

Acknowledgements

We extend our gratitude to the Council of Science and Technology, U.P., for providing financial assistance during the tenure of a major research project carried out on "Ecology and Distribution of Acridoid Pests (Orthoptera: Acridoidea) with observations on their natural enemies in Uttar Pradesh". Thanks are also due to Prof. Irfan Ahmad, Chairman, Department of Zoology, Aligarh Muslim University, Aligarh, for providing the necessary facilities.

References

Criddle, N., 1918. The egg-laying habits of some of the Acrididae (Orthoptera). *Canadian Entomology*, 50: 145–51.

Greathead, D.J., 1966. A brief study of the effect of biotic factors on population of the desert locust. *Journal of Applied Ecology*, 3: 239–250.

Grewal, G.S. and Atwal, A.S., 1968. Development of *Chrotogonus trachypterus* Blanch. (Orthoptera: Pyrgomorphidae) in relation to different levels of temperature and humidity. *Indian Journal of Entomology*, 30(1): 1–7.

Hafez, M. and Ibrahim, M.M., 1964. Studies on the behaviour of the desert grasshopper, *Sphingonotus carinatus* Sauss., toward humidity and temperature. *Bulletin de la Société Entomologique d'Egypte (Le Caire)*, XLVIII: 229–243.

Hamilton, A.G., 1936. The relation of humidity and temperature to the development of three species of African locusts, *Locusta migratoria migratorioides* (R. and F.), *Schistocerca gregaria* (Forsk.) and *Nomadacris septemfasciata* (Serv). *Transactions of the Royal Entomological Society of London*, 85: 1–60.

Hamilton, A.G., 1950. Further studies on the relation of humidity and temperature to the development of two species of African locusts, *Locusta migratoria migratorioides* (R. and F.) and *Schistocerca gregaria* (Forsk.). *Transactions of the Royal Entomological Society of London*, 101(1): 1–58.

Hussain, M.A. and Ahmad, T., 1936. Studies on *Schistocerca gregaria* Forsk. II. The biology of the desert locust with special relation to temperature. *Indian Journal of Agricultural Sciences*, 6: 188–262.

Hussain, M.A. and Mathur, C.B., 1944. Studies on *Schistocerca gregaria* Forsk. XI. The influence of temperature on the growth in weight and size of the hoppers. *Indian Journal of Entomology*, 5: 107–115.

Iqbal, M. and Aziz, S.A., 1973. Effect of different levels of temperature and humidity on the development of *Spathosternum prasiniferum* Walker (Orthoptera: Acridoidea). *Indian Journal of Entomology*, 35(3): 211–218.

Iqbal, M. and Aziz, S.A., 1974. Life-history of *Spathosternum prasiniferum* Walker 1871, *Indian Journal of Zoology*, 2(1): 37–43.

Iqbal, M. and Aziz, S.K., 1975. Food preference of *Spathosternum prasiniferum* Walker (Orthoptera: Acridoidea). *Indian Journal of Entomology*. 37(1): 51–56.

Khan, H.R. and Aziz, S.A., 1974a. Development of hoppers of *Oedaleus abruptus* Thunberg (Orthoptera: Acrididae) under different levels of temperature and humidity. *Indian Journal of Entomology*, 36(3): 175–178.

Khan, H.R. and Aziz, S.A., 1974b. Effect of temperature and humidity on the incubation and hopper development periods of *Eyprepocnemis alacris* Serville (Orthoptera: Acrididae). *Acrida*, 3: 47–53.

Khan, H.R. and Aziz, S.A., 1974c. Effect of crowding on the hopper developmental period of *Oedaleus abruptus* Thunberg and *Eyprepocnemis alacris* Serville (Orthoptera: Acrididae) under controlled ecological conditions. *Indian Journal of Entomology*, 36(2): 145–148.

Khan, M. and Aziz, S.A., 1973a. Observations on seasonal variations in populations of hoppers and adults of *Oedaleus abruptus* Thunberg (Orthoptera: Acrididae). *Indian Journal of Entomology*, 35(4): 300–305.

Khan, M. and Aziz, S.A., 1973b. Influence of different levels of temperature and moisture on the development and hatching of eggs of *Oedaleus abruptus* Thunberg (Orthoptera: Acrididae). *Indian Journal of Entomology*, 35(1): 29–31.

Majeed, Q., 1990. Development of various stages of *Acrotylus humbertianus* Saussure (Orthoptera: Acrididae) under varying levels of temperature and relative humidity. *Journal of Entomological Research*, 14(1): 78–86.

Parihar, D.R. and Pal, S.K., 1978. Effects of temperature on development of eggs and hoppers of surface grasshoppers *Chrotogonus trachypterus* (Blanch.) (Acridoidea: Pyrgomorphidae). *Zeitschrift für angewandte Zoologie*, 65(2): 205–212.

Parker, J.R., 1930. Some effects of temperature and moisture upon *Melanoplus mexicanus mexicanus* Sauss. and *Camnula pellucida* Scudd. (Orthoptera). *Montana Agricultural Experimental Station, Bulletin*, 223: 1–132.

Sanger, K., 1973. Consumption by some field grasshoppers (Orthoptera: Acrididae) depending on different humidity rates. *Verhandlungen der zoologisch–botanischen Gesellschaft in Wien*, 113: 81–92.

Shamshad, A., 1982. Effect of temperature and humidity on the development and fertility and fecundity of *Acrida exaltata* Walk. *Proceedings of Indian Academy of Sciences* (*Animal Sciences*), 91(3): 267–273.

Chapter 9

Comparative Studies on Growth, Dietary Utilization and Survivality of Tiger Prawn (*Penaeus monodon*) Fed with Two Different Commercially Available Feeds

Basudev Mandal* and Tarani Sen Mahata

Department of Aquaculture Management and Technology, Vidyasagar University, Midnapore – 721 102, West Bengal, India

ABSTRACT

The present investigation focused on the growth, dietary utilization and survivality of *Penaeus monodon* with two easily available cheap commercial feeds having high ADG and FCR value. The studies are mainly emphasizes with the analysis of growth pattern with their FCR values of experimental prawn species (*Penaeus monodon*) in respect of two different artificial feeds, the estimation of per unit production efficiency and finally the economic analysis from the trial. The two feeds namely C.P. Prawn feed and Godrej Super Tiger feed were selected for comparative experimental trial in different ponds (1, 4, A-5, 5). These two feeds were given to the tiger prawn for their feeding

* Corresponding author: E-mail: bmandalamtvu@gmail.com, bdmandal@yahoo.co.in

at an equal dose and frequency. During the experimental trial it has been found that, by the application of Godrej Super Tiger feeds water quality and colour has become enormously changed but water quality and also colour remain acceptable for prawn culture when C.P. Prawn feed was applied to the experimental pond. From this investigation it has been observed that ADG (Average Daily Growth) and FCR (Feed Conversion Ratio) values were found very satisfied for C.P. Prawn feed as compared to the Godrej Super Tiger feeds. Through this experimental trial we conclude that C.P. prawn feeds were much more beneficial than Godrej Super Tiger feeds for intensive culture of tiger prawn in commercial scale.

Keywords: *Tiger prawn, Growth, Survivality, Dietary utilization, ADG, FCR value.*

Introduction

In traditional farming system the ponds are stocked with fry either collected from the wild or concentrated through tidal water entering the pond, but the production is concentrated in about 15 nations in Asia and Latin America. In 1992, about 12 countries have contributed 95 per cent farmed shrimp production. In India scientific shrimp farming started during 1980, after hatching seed production technology evolved in Indian condition by the" Central Marine Fisheries Research Institute" at Cochin. Shrimp production varies from year to year due to the dependence on seasonal supply of fry from the world. Rosenberg in 2001 has reported based on FAO data, that global production of farmed shrimp reached 11,30,000 metric tons 1999. Farmed production could be increased by responsively intensifying culture methods, without having to develop large areas. Traditional fish farming was a part of rural life in certain areas. The present day aquaculture has a much greater significance in socio-economic development and natural resource management. Despite some temporary increases in a few areas, the total world capture fishery production appears to around 70-75 million tons (Phillips, 2002). Now-a-days Black tiger shrimp particularly *Penaeus monodon* are among the economically important species in Asia and the World also. In 1995 Asian member countries produced about 558,000 tons of *Penaeus monodon,* about 78 per cent of the total world production of shrimp. The shrimp seed production technique and the culture system were developed by M. Fudinaga and his colleagues in the early 1960 in Japan after many ground-brackish studies over the preceding 30 years. In India shrimp farming has practiced traditionally in the states of West Bengal and Kerala through a simple method of trapping brackish water farms during high tides and culturing the species thus trapped much involvement of any scientific practice. Since today's shrimp farming practices are based on the intensive culture system, the danger of major losses due to disease also increases.

The soil condition of pond bottom plays a significant role in the production of *Penaeus monodon*. The role of soil and water condition in relation to fish production and some manure as producers and also artificial fish feeds as fish growers have been studied by a number of workers (Banerjee, 1967; Govind *et al.*, 1978; Saha *et al.*, 1979; Singit *et al.*, 1980; Nandeesha, 1982; Jhingran and Pullin 1985; Gupta *et al.*, 1990; Laha and Mitra, 1987; Patra and Ray, 1988; Patra, 1993, 1993b; Mandal *et al.*, 2008). The productivity of a pond is the resultant effect of the interplay of many factors. Saha (1969), Banerjee (1967), Ghosh (1978), and Saha *et al.* (1979) reported that initial physiological condition is required to make the pond productive. Different types of fertilizers are used for this production in Africa, Europe, Asia, and the East countries (Gooch, 1967; Prowse, 1967, 1968; Vender Lingen, 1967; Wolny, 1967; Mortimer and Hickling, 1954; Schaperclause,1959).

Tiger shrimp tolerate wide variation in salinity and is the fastest growing species among marine shrimp. It attains the marketable size of 30-40g in about 4-5 month's period under commercial culture condition. Minimum dissolve oxygen level to be maintained for this species 3ppm.Though moderate production is achieved under salinity range of 0-40 ppt, intensive production system necessitates the salinity range of 10-35ppt. Generally, it is resistant to water quality change, but sensitive to disease causing germs. Prakash and Gupta (1986) showed that comparative growth rates of *Penaeus monodon* of the Govindagarh lake are studied by means of scales. The relationship between shrimp length and scale length and time of formation of annuli are estimated. Instantaneous and absolute growth rates of fishes have been calculated separately for each age group. Bandyopadhyay *et al.* (1991) reported that, an attempt to study the growth, survival and production of shrimps by rearing in the plastic cages. Fish were reared in 3 circular cages (volume of each cage 3 cu.m) made up of hollow aluminum frame enclosed with nylon net (6 mm. mesh) installed in a two hectare water body. Shrimps fingerlings was stocked at 33, 50 and 67/cu. m. in different cages.

Keeping above point into consideration; the present study was therefore undertaken on "Comparative Study of Growth, Dietary utilization and survivality of *Penaeus monodon* with two different commercial feeds" at Narghat area of Purba Medinipur. The study includes growth, different nutritional indices, survivality, FCR values and finally quality of Black Tiger Shrimp (*Penaeus monodon*).This study may serve a clear documentation of Semi intensive *Penaeus*

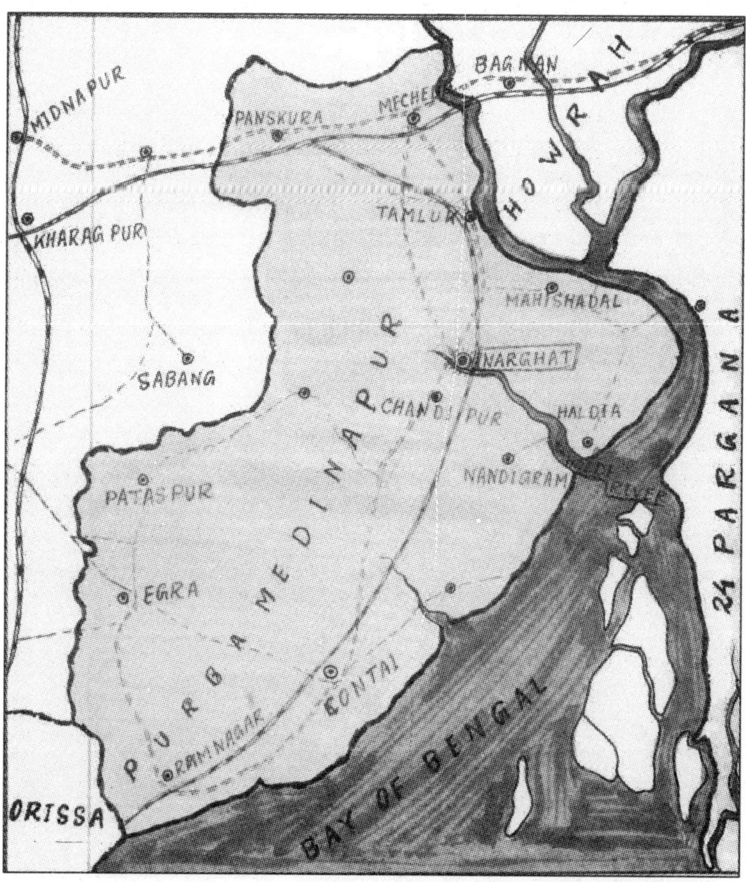

Figure 9.1: Map of Purba Medinipur District of West Bengal

monodon culture in Eastern part of West Bengal and can ensure with high economic return.

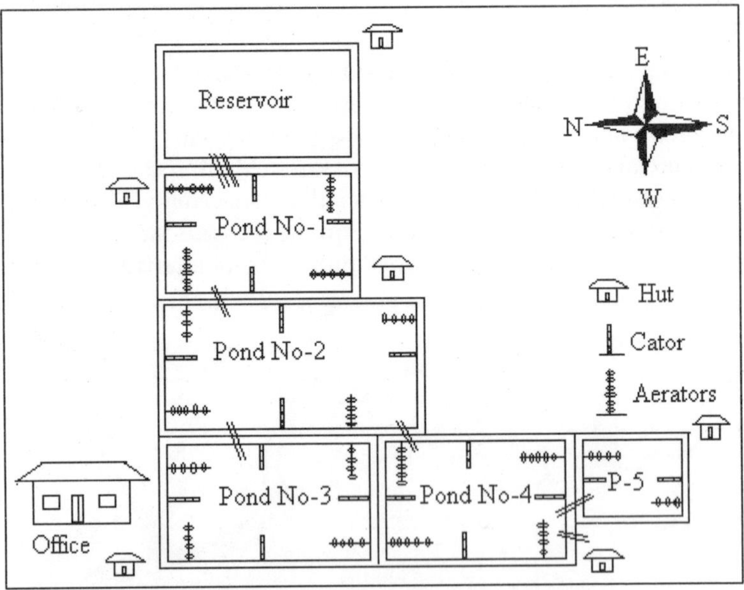

Figure 9.2: Experimental Farm Layout

Figure 9.3: Experimental Fish Species

Materials and Methods

Site Selection

The experimental work has been carried out in the different farms located at around Narghat of Nandigram Block, Purba Medinipur, West Bengal which are well connected with Haldi river for tidal inflow.

The details of the culture sheets are as follows:

- *Farm Name*: Shri Bishnu Fishery
- *Farm Location*: Kolagechia, Narghat, Purba Midnapur, West Bengal
- *Total Farm area*: 2.23ha.
- *Total Pond No.*: 5
- *Type of Culture*: Semi intensive
- *Date of Stocking*: 4th April, 2009
- *Post Larvae (PL)*: 15
- *Source of seed*: Vaisali
- *Water Source*: Haldi River
- *Experimental Sp.*: *Penaeus monodon*

Taxonomy of Experimental Species

Penaeid shrimp belongs to the largest phylum in the animal kingdom, *i.e.*, Arthropoda. This group of animals is characterized by the presence of paired appendages and protective cuticle or exoskeleton that covers the whole animals. The sub-phylum Crustacea is made up of 42,000, predominantly aquatic, species, that belongs to 10 classes. Within the class Malcostraca, shrimp, together with Cray fish, lobsters and crabs, and belong to the order Decapoda.

Pond Construction, Design and Outlet System

Before stocking for a new crop, the excessive wastes which accumulate in the pond during the previous crop must be removed and the soil and water conditioned. Growing of shrimp in an improperly prepared pond may lead to difficulty in pond management during the culture period which could result in a decrease in production capacity of the pond. A well-designed pond will facilitate the management of water exchange, harvesting of the product, waste collection and elimination, and feeding. The shapes of pond that are found to be effective for shrimp culture are rectangular, square and circular. A well-designed pond is one that would allow circulation of the water such that wastes will be accumulated at the center of the pond. Some farmers improve the water movement in the square and rectangular ponds by making the corners of the pond rounded through addition of soil. Smaller ponds are easier to manage but the construction and operation can be costly. Ponds of 0.5-1.0 ha. are commonly used in intensive culture and 1-2 ha. for semi-intensive culture. Earthen dikes, with or without lining, are found to be the most economical. Dikes are designed to impound higher than 1 m. depth of water and must be high enough to prevent flooding during the rainy seasons and the highest high tide. The slope of the dike depends on the nature of the soil. A slope of not less than 1:1.5 is normally used in the sandy soil area to avoid erosion and 1:1 is used for clay soils. Shrimp ponds are designed according to the characteristics of the selected site and the culture system. There is no unique design, but optimum and functional farm layout plan and design should be based on the physical and economic conditions prevailing in the locality.

A shrimp pond is filled with water mostly by pumping. The pumps were installed at locations where they can obtain water from the middle of the water column with least sedimentation and pollution. The pumps and inlet canal should be large enough to allow the ponds or the reservoir to be filled within 4-6 hours. A screen has been installed at the inlet canal prior to the pumps to prevent clogging at the inlets. An intensive shrimp farm have a water supply canal to convey the water from the reservoir to the ponds by gravity or pumping. The size of the supply canal will depend on the size of the culture pond, the efficiency of the pump and the required water exchange rate. The size of the gate is dependent on the size of the pond, but must allow the pond to be filled or drained within 4-6 hours. Gates of 0.5-1.0 m. wide are usually constructed, since gates wider than 1 m. will cause difficulty in screening and will allow strong currents which will cause erosion of the soil. The position of the outlet should be at the lowest point of the pond with a gradual slope of 1:200 from the inlet to allow total drainage of the pond during harvesting.

Liming

Once the pond is cleaned, it is then filled with water and left overnight before flushing out to remove debris and elevate the pH. These processes were repeated until the pH of the water remains above 7.0, and only then the lime is applied. The type of lime to be used depends on the water pH. It is recommended that agricultural lime ($CaCO_3$) or dolomite [Ca $Mg(CO_3)_2$] should be used in a pond with water pH near neutral and the hydrated lime [$Ca(OH)_2$] should be used in a pond with water pH below 5. The amount of lime to be used should be carefully calculated to avoid inducing an excessively high water pH which may increase ammonia toxicity and result in the mortality of the shrimps. During the application, lime should be spread throughout the pond bottom and up to the top of the dike.

Recommended Lime Application Rate in Pond

Soil pH	*Agriculture Lime (MT/ha)*	*Hydrated Lime (MT/ha)*
0.5-1.0	>6	1-2
5-6	2-3	1.0-1.5
<5	3-5	1.5-2.5

When the pond is properly limed and filled with water, the average water pH should be between 7.5-8.5 with daily fluctuation of less than 0.5. Agricultural lime, dolomite or hydrated lime at 100 kg./ha./day should be added to maintain the required pH.

Removal of Predators

After liming, the pond was filled to the maximum depth through a screen with fine mesh (24 per square inch) to prevent the predators and competitors from entering the pond. These animals, including fish, crustaceans and some invertebrates, may compete for food, prey on the shrimp or carry diseases and parasites. They may establish themselves in the pond that is not effectively screened effectively or is left for a long period of time. Some chemicals should be used to eradicate these animals in the pond before stocking. Fish can be killed by the application of tea seed powder or by MOC at different doses. After the application of tea seed, the pond should be left for 3 days before the post larvae can be stocked. Tea seed may also be used when the shrimp has reached a weight of more than 2 grams.

However, it must be remembered that tea seed is more toxic at high salinity and temperature, but less toxic at high pH. Application of tea seed in the evening may reduce pH and result in plankton die-off.

Snails can be eliminated by the application of quick lime (CaO) and sun dried. Then the ponds were thoroughly cleaned, filled with water and the other pests eradicated. Hypochlorite, either calcium or sodium salt, is currently used at 15-20 per cent (60 per cent active ingredient) to eliminate both vertebrates and invertebrates. The pond must be cleaned prior to the application of hypochlorite since hypochlorite may react with the organic matters and produces the toxic organochlorine compounds. Hypochlorite would apply after the pond is filled to the maximum height and left for 3 days to allow the hatching of planktonic organisms. It should be remembered that hypochlorite should be used prior to the liming since the effectiveness of hypochlorite will be lowered in high pH conditions.

Application of Fertilizers

The ponds are fertilized with organic or inorganic fertilizer to stimulate the plankton bloom in order to provide shade to the pond bottom and utilize the nitrogenous and phosphate wastes within the pond. The shade will also prevent the growth of harmful benthic algae. The manure must be soaked in water for 24 hours before it is spread over the surface of the water. Inorganic fertilizers, such as urea and compound fertilizers like, ammonium phosphate or those with N:P:K combination of (16:16:16) can also be used. The fertilizer is dissolved in water before it is spread over the water surface to avoid precipitation of the fertilizer onto the pond bottom which will enrich the soil and accelerate the growth of benthic algae. After fertilization, the planktons were growing within a few days and the colour of the water becomes slightly green. If the plankton has not bloomed within a few days, additional fertilizer must not be applied, but plankton rich water or green water from the reservoir should be added.

Selection of Fish Species for Stocking and their Ratio

The fry of PL15-20 (*Penaeus monodon*) was selected for stocking in the experimental ponds. The PL15-20 has indicated by the appearance of 4-6 spines on the rostrum. The healthy fry have the muscle-to-gut ratio in the sixth abdominal segment of about 4:1 or the thickness of the gut should be about the thickness of the muscle. Practically, fry from the first and second spawning of a brood stock with uniform size can be used. The Fry with the presence of pigment cells in the uropods may be used since this indicates the stage of development. The optimum stocking density of fry in a pond were determined in accordance with the production capacity of the farm and the culture system, which include the soil and water quality, food availability, seasonal variations, target production, and farmer's experience. It is suggested that a new crop with a low stocking density to access the production capacity of the pond. If production is successful, then the stocking density could be increased for subsequent crops. Overstocking should be avoided since it may result in management problems and loss of entire production. In this experiment the stocking density was in between 20-25 fry/m^2 and this was usually practiced in a semi-intensive culture. In an intensive culture, a well-managed pond with consistent good water quality can stock up to 30-35 fry/m^2 at 1.5 m water depth and up to 40-50 fry/m^2 at 1.7 m water depth or deeper. However, it must be emphasized that intensive cultures involve high densities and can only be sustained in well managed farms.

Application of Artificial Food and Rate of Feeding

Cost of feed constitutes a major part of the production cost and accounts for 50 per cent to 70 per cent of the total variable cost. The use of feeds will improve shrimp production and increase profits. The availability of nutrients from feeds depends on the type and quality of the raw material used, the formulation, the feed processing, feed storage conditions and the feeding management. Therefore, feed and feeding practices for semi-intensive or intensive shrimp farming require a basic understanding of nutrition and feed requirements. Approximately 40 essential nutrients are required by shrimp. These nutrients are provided in various amounts by natural food and supplemental feeds. According to recent data, an FCR as low as 1.2 has been achieved, but many farmers are still obtaining FCR of higher than 2.2. Therefore, besides the feeding management, the FCR is also closely related to the quality of feed. A high FCR or high amount of feed required to produce unit weight gain indicates overfeeding, and consequently, a poor FCR is usually associated with poor growth rate, low weight gain, stressed shrimp, mediocre water quality and adverse pond bottom conditions. Therefore, the proper amount of feed is the most critical factor of feeding management. Since many factors are involved in shrimp feed consumption, careful and frequent observation of shrimp is the most reliable approach for determining the optimal feeding amount. There are many major key factors for successful intensive shrimp culture. Use of good quality feed with better feeding management by low feed conversion ratios and improved farm management are the important goals to farmers, not only for gaining greater profit, but also for minimizing the pollution of shrimp farming area.

Management of Water Quality

Water quality, like, pH, DO, free CO_2 and salinity was monitored on regular basis as method prescribed by APHA, 2005. The level of dissolved oxygen can be controlled in 3 ways. Firstly, by increasing the water surface area by means of placing paddle wheels in the right position. This is not only causes proper water circulation, but also adds oxygen to the pond water. Secondly, by controlling plankton density to an optimum level and thirdly, by minimizing excess organic substances, such as uneaten food. To control pH level the application of lime is as follows:

1. At the beginning of a crop cycle, when water pH ranges between 7.4-7.7 about 5.7-8.7 kg/ha of dolomite should be used every 3-4 days.
2. When pH is in the range of 7.5-7.8 and there is less than 0.5 unit difference between the pH in the morning and the pH in the afternoon, 5.0-7.0 kg/ha of dolomite should be used every 2-3 days.
3. If the pH in the morning is less than 7.5, 4.5 kg/ha dolomite should be used every day until the morning pH is increased to above 7.5.
4. If the pH in the morning is higher than 8.0 and the pH in the afternoon is higher than 9.0, 4.6-8.2 kg/ha of dolomite should be used every day until the day's pH difference is less than 0.4.
5. In the second half of a crop cycle, 7.5- 8.0 kg/ha of dolomite should be used every day or at least every 2 days, depending on water colour.
6. In every time before exchanging water, 5.0-9.0 kg./ha. of dolomite should be used.

Compare among Two Commercial Feed

The different parameters of two experimentally used feed are indicated below:

Parameters	*C.P Prawn Feed*	*Godrej Super Tiger Feed*
Feed Code	9002, 9003 and 9003P	9004, 9004S and 9005
Crude Protein (per cent)	38	40
Fat (per cent)	5	6
Moisture (per cent)	12	11
Fibre (per cent)	4	3
Outlook view 		

Culture Data Sheet for C.P. Prawn Feed

- *Pond No.*: 01
- *Pond Size*: 3000m²
- *Initial Stocking*: 75,000 pcs.
- *Stocking Density*: 25pcs/m²
- *Type of Culture*: Semi intensive
- *Date of Stocking*: 4th April, 2009
- *Post Larvae* (PL): 15

Before starting the experimental work, pond was prepared by the following way:

Date	*Item*	*Quantity*
16.02.09	Dry pond operation (Soil cutting)	6" pond bed cutting
23.02.09	LSP	75 kg
12.03.09	Water loading	6'
16.03.09	Bleaching	100kg

Date	Item	Quantity
19.03.09	Dolomite	60 kg
20.03.09	Dolomite, chain and aeration	45 kg
21.03.09	Juice, chain and aeration	
23.03.09	Bio Card	3.500 kg
25.03.09	Dolomite, chain and aeration	50 kg
26.03.09	Juice, chain and aeration	
28.03.09	Dolomite, chain and aeration	15 kg
30.03.09	Juice	
01.04.09	Dolomite, chain and aeration	30 kg
02.03.09	Juice, chain and aeration	
03.04.09	Super Biotic	1.550 kg
04.04.09	Seed stocking	75,000

After stocking the tiger prawn feeds were provided as per following schedule:

Date	DOC	Feed Code	Feed Amount in Tray (gm/kg)	Feed Used/Day (gm)	Total Feed to Date (kg)
05/04/09	1	1		1500	1.5
06/04/09	2	1		1650	3.15
07/04/09	3	1		1800	4.95
08/04/09	4	1		1950	6.9
09/04/09	5	1		1950	8.85
10/04/09	6	1		2100	10.95
11/04/09	7	1		2250	13.2
12/04/09	8	1		2400	15.6
13/04/09	9	1		2550	18.15
14/04/09	10	1		2700	20.85
15/04/09	11	1+2		2924	23.774
16/04/09	12	1+2		3148	26.922
17/04/09	13	1+2		3000	29.922
18/04/09	14	1+2		3400	33.322
19/04/09	15	1+2		3600	36.922

Date	*DOC*	*Feed Code*	*Feed Amount in Tray (gm/kg)*	*Feed Used/Day (gm)*	*Total Feed to Date (kg)*
20/04/09	16	1+2		3820	40.74
21/04/09	17	1+2		3640	44.38
22/04/09	18	2		4280	48.66
23/04/09	19	2		4500	53.16
24/04/09	20	2		4800	57.96
25/04/09	21	2		6375	64.335
26/04/09	22	2		6750	71.08
27/04/09	23	2+3		7250	78.33
28/04/09	24	2+3		7625	85.955
29/04/09	25	2+3		8000	93.955
30/04/09	26	2+3		7500	101.455
01/05/09	27	2+3		7000	108.455
02/05/09	28	2+3		7000	115.455
03/05/09	29	2+3		8000	123.455
04/05/09	30	2+3	10gm/kg	7000	130.455
05/05/09	31	3	10gm/kg	6000	136.455
06/05/09	32	3	10gm/kg	8300	144.755
07/05/09	33	3	10gm/kg	9000	153.755
08/05/09	34	3	10gm/kg	10100	163.855
09/05/09	35	3	10gm/kg	12000	175.855
10/05/09	36	3	10gm/kg	13000	188.855
11/05/09	37	3	12gm/kg	12400	201.255
12/05/09	38	3	12gm/kg	12800	214.055
13/05/09	39	3	12gm/kg	12400	226.455
14/05/09	40	3	12gm/kg	12500	238.955
15/05/09	41	3	12gm/kg	12800	251.755
16/05/09	42	3	12gm/kg	14600	266.355
17/05/09	43	3	12gm/kg	14200	280.555
18/05/09	44	3+4S	12gm/kg	15200	295.755
19/05/09	45	3+4S	12gm/kg	15500	311.255
20/05/09	46	3+4S	12gm/kg	14700	325.955
21/05/09	47	3+4S	12gm/kg	16000	341.955

Date	DOC	Feed Code	Feed Amount in Tray (gm/kg)	Feed Used/Day (gm)	Total Feed to Date (kg)
22/05/09	48	3+4S	12gm/kg	17500	359.455
23/05/09	49	3+4S	12gm/kg	19000	378.455
24/05/09	50	3+4S	12gm/kg	18500	396.955
25/05/09	51	3+4S	12gm/kg	17500	414.455
26/05/09	52	3+4S	12gm/kg	19000	423.455
27/05/09	53	3+4S	13gm/kg	18000	441.455
28/05/09	54	3+4S	13gm/kg	19000	460.455
29/05/09	55	3+4S	13gm/kg	21000	481.455
30/05/09	56	3+4S	13gm/kg	21500	502.955
31/05/09	57	3+4S	13gm/kg	19500	522.455
01/06/09	58	3+4S	13gm/kg	20500	542.955
02/06/09	59	3+4S	13gm/kg	24500	567.455
03/06/09	60	3+4S	13gm/kg	25500	592.955
04/06/09	61	3+4S	13gm/kg	25500	618.455
05/06/09	62	3+4S	13gm/kg	24000	642.455
06/06/09	63	3+4S	14gm/kg	25000	667.455
07/06/09	64	3+4S	14gm/kg	25000	692.455
08/06/09	65	3+4S	14gm/kg	26000	718.455
09/06/09	66	3+4S	14gm/kg	27500	745.955
10/06/09	67	3+4S	16gm/kg	25000	770.955
11/06/09	68	4S	16gm/kg	24000	794.955
12/06/09	69	4S	16gm/kg	26000	820.955
13/06/09	70	4S	16gm/kg	27000	847.955
14/06/09	71	4S	16gm/kg	29000	876.955
15/06/09	72	4S	16gm/kg	29500	906.455
16/06/09	73	4S	16gm/kg	29300	935.755
17/06/09	74	4S	16gm/kg	29300	965.055
18/06/09	75	4S	16gm/kg	26700	991.755
19/06/09	76	4S	16gm/kg	27800	1019.555
20/06/09	77	4S	16gm/kg	29300	1048.855
21/06/09	78	4S	16gm/kg	27000	1075.855
22/06/09	79	4S	16gm/kg	29000	1104.855

Date	DOC	Feed Code	Feed Amount in Tray (gm/kg)	Feed Used/Day (gm)	Total Feed to Date (kg)
23/06/09	80	4S	16gm/kg	29300	1134.155
24/06/09	81	4S	17gm/kg	30800	1164.955
25/06/09	82	4S	17gm/kg	31000	1195.955
26/06/09	83	4S	17gm/kg	32500	1228.455
27/06/09	84	4S	17gm/kg	33500	1261.955
28/06/09	85	4S	17gm/kg	33500	1295.455
29/06/09	86	4S	17gm/kg	35300	1330.755
30/06/09	87	4S	17gm/kg	35000	1365.755
01/07/09	88	4S	17gm/kg	34500	1400.255
02/07/09	89	4S	17gm/kg	35500	1435.755
03/07/09	90	4S	17gm/kg	35800	1471.555
04/07/09	91	4S	17gm/kg	36300	1507.855
05/07/09	92	4S	17gm/kg	36000	1543.855
06/07/09	93	4S	17gm/kg	33500	1577.355
07/07/09	94	4S	17gm/kg	35000	1612.355
08/07/09	95	4S	17gm/kg	36000	1648.355
09/07/09	96	4S	17gm/kg	37000	1685.355
10/07/09	97	4S	17gm/kg	36000	1721.355
11/07/09	98	4S	17gm/kg	36500	1757.855
12/07/09	99	4S	18gm/kg	38000	1795.855
13/07/09	100	4S	18gm/kg	36500	1832.355
14/07/09	101	4S	18gm/kg	37000	1869.355
15/07/09	102	4S	18gm/kg	38000	1907.355

Few probiotics has been used as per following schedule:

Date	Doc	Probiotics	
		Water and Soil	Gut
01.05.09	27		Super Biotic(10gm/kg feed)
04.05.09	30		Super Biotic(10gm/kg feed)
05.05.09	31		Super Biotic(10gm/kg feed)
13.05.09	39	Environ-AC (3500gm)	Aqualact (10gm/kg fed)

Date	Doc	Probiotics	
		Water and Soil	Gut
14.05.09	40		Aqualact (10gm/kg fed)
15.05.09	41		Aqualact (10gm/kg fed)
16.05.09	42		Aqualact (10gm/kg fed)
17.05.09	43		Aqualact (10gm/kg fed)
18.05.09	44		SuperBiotic(10gm/kg feed)
22.05.09	48		Zymetin (10gm/kg feed)
23.05.09	49		Zymetin (10gm/kg feed)
24.05.09	50		Zymetin (10gm/kg feed)
25.05.09	51		Zymetin (10gm/kg feed)
26.05.09	52		Zymetin (10gm/kg feed)
01.06.09	58	Super-PS (3lit)	
07.06.09	64		Zymetin (10gm/kg feed)
08.06.09	65		Zymetin (10gm/kg feed)
09.06.09	66		Zymetin (10gm/kg feed)
11.06.09	68		Super Biotic(10gm/kg feed)
15.06.09	72		Zymetin (10gm/kg feed)
16.06.09	73		Zymetin (10gm/kg feed)
17.06.09	74		Zymetin (10gm/kg feed)
20.06.09	77	Super-PS (3lit)	Zymetin (10gm/kg feed)
05.07.09	92	Super-PS (5lit)	
06.7.09	93		Zymetin (10gm/kg feed)
08.07.09	95		Zymetin (10gm/kg feed)
09.07.09	96		Zymetin (10gm/kg feed)

The following vitamin mixtures were used during the culture operation:

Date	Doc	Vitamin
06.05.09	32	Herbastim (5ml/kg feed)
07.05.09	33	Herbastim (5ml/kg feed)
08.05.09	34	Herbastim (5ml/kg feed)
09.05.09	35	Herbastim (5ml/kg feed)

Date	Doc	Vitamin
10.5.09	36	Herbastim (5ml/kg feed)
22.05.09	48	Mutagen (10gm/kgfeed)
23.05.09	49	Mutagen (10gm/kgfeed)
24.05.09	50	Mutagen (10gm/kgfeed)
29.05.09	55	Mutagen (10gm/kgfeed)
30.05.09	56	Mutagen (10gm/kgfeed)
31.05.09	57	Mutagen (10gm/kgfeed)
01.06.09	58	Mutagen (10gm/kgfeed)
02.06.09	59	Immunex (10m/kg feed)
03.06.09	60	Immunex (10m/kg feed)
04.06.09	61	Immunex (10m/kg feed)
18.06.09	75	Regular AVM (10gm/kg feed)
19.06.09	76	Regular AVM (10gm/kg feed)
20.06.09	77	Regular AVM (10gm/kg feed)
21.06.09	78	Regular AVM (10gm/kg feed)
22.06.09	79	Regular AVM (10gm/kg feed)
23.06.09	80	Immunex (10m/kg feed)
25.06.09	82	Regular AVM (10gm/kg feed)
26.06.09.	83	Mutagen (10gm/kgfeed)
27.06.09	84	Mutagen (10gm/kgfeed)
28.06.09	85	Mutagen (10gm/kgfeed)
29.06.09	86	Mutagen (10gm/kgfeed)
02.07.09	89	Wockcee (2.0gm/kg feed)
03.07.09	90	Wockcee (2.0gm/kg feed)
04.07.09	91	Wockcee (2.0gm/kg feed)
11.07.09	98	Mutagen (10gm/kgfeed)
12.07.09	99	Mutagen (10gm/kgfeed)

1. Culture Data Sheet for Godrej Super Feed

- ☆ *Pond No.*: A-5
- ☆ *Pond Size*: 5100m^2
- ☆ *Initial Stocking*: 1,25,000 pcs.
- ☆ *Stocking Density*: 25pcs/m^2

☆ *Type of Culture*: Semi intensive

☆ *Date of Stocking*: 8th April, 2009

☆ *Post Larvae (PL)*: 15

The experimental ponds were prepared using following items:

Date	*Item*	*Quantity*
20.02.09	Dry pond operation (Soil cutting)	6" pond bed cutting
27.02.09	LSP	75 kg
16.03.09	Water loading	6'
20.03.09	Bleaching	100 kg
23.03.09	Dolomite	60 kg
24.03.09	Dolomite, chain and aeration	45 kg
25.03.09	Juice, chain and aeration	
27.03.09	Bio Card	3.500 kg
29.03.09	Dolomite, chain and aeration	50 kg
30.03.09	Juice, chain and aeration	
02.04.09	Dolomite, chain and aeration	15 kg
04.04.09	Juice	
05.04.09	Dolomite, chain and aeration	30 kg
06.04.09	Juice, chain and aeration	
07.04.09	Super Biotic	1.550 kg
08.04.09	Seed stocking	75,000

Feeds were provided to tiger prawn as following schedule:

Date	*DOC*	*Feed Code*	*Feed Amount in Tray (gm/kg)*	*Feed Used/Day (gm)*	*Total Feed to Date (kg)*
08/04/2009	1	1		2.5	
09/04/2009	2	1		2.75	
10/04/2009	3	1		3	
11/04/2009	4	1		3.25	
12/04/2009	5	1		3.25	
13/04/2009	6	1		3.5	

Date	DOC	Feed Code	Feed Amount in Tray (gm/kg)	Feed Used/Day (gm)	Total Feed to Date (kg)
14/04/2009	7	1		3.75	
15/04/2009	8	1		4	
16/04/2009	9	1		4.25	
17/04/2009	10	1+2		4.5	
18/04/2009	11	1+2		4.876	
19/04/2009	12	1+2		5.248	
20/04/2009	13	2		5.7	
21/04/2009	14	2		5.2	
22/04/2009	15	2		6	
23/04/2009	16	2		6.46	
24/04/2009	17	2		6.84	
25/04/2009	18	2		7.2	
26/04/2009	19	2		7.59	
27/04/2009	20	2		8.09	
28/04/2009	21	2+3		10.625	
29/04/2009	22	2+3		9	
30/04/2009	23	2+3		9.6	
01/05/2009	24	2+3		10.1	
02/05/2009	25	2+3		10.6	
03/05/2009	26	2+3		9.2	
04/05/2009	27	2+3		10.3	
05/05/2009	28	2+3		10.3	
06/05/2009	29	2+3		13.25	
07/05/2009	30	2+3		11.6	
08/05/2009	31	2+3		8	
09/05/2009	32	2+3		8.05	
10/05/2009	33	2+3		8.5	
11/05/2009	34	2+3		9.3	
12/05/2009	35	2+3		11.4	
13/05/2009	36	2+3		12.4	
14/05/2009	37	3		11.8	
15/05/2009	38	3+3S		11.9	

Date	DOC	Feed Code	Feed Amount in Tray (gm/kg)	Feed Used/Day (gm)	Total Feed to Date (kg)
16/05/2009	39	3+3S		12.3	333.179
17/05/2009	40	3S		13.3	
18/05/2009	41	3S		14.7	
19/05/2009	42	3S		16.2	
20/05/2009	43	3S		16.4	
21/05/2009	44	3S		17.9	
22/05/2009	45	3S		17.4	
23/05/2009	46	3S		16.7	445.779
24/05/2009	47	3S		18.7	
25/05/2009	48	3S		20.7	
26/05/2009	49	3S		22.2	
27/05/2009	50	3S		21	
28/05/2009	51	3S		20.5	
29/05/2009	52	3S	6gm/kg	21.5	
30/05/2009	53	3S	6gm/kg	20.5	590.879
31/05/2009	54	3S	6gm/kg	22	
01/06/2009	55	3S	6gm/kg	25.5	
02/06/2009	56	3S	6gm/kg	28	
03/06/2009	57	3S	6gm/kg	29	
04/06/2009	58	3S	6gm/kg	28.5	
05/06/2009	59	3S	6gm/kg	29.5	
06/06/2009	60	4S	8gm/kg	31.5	784.879
07/06/2009	61	4S	8gm/kg	34	
08/06/2009	62	4S	8gm/kg	34	
09/06/2009	63	4S	8gm/kg	35	
10/06/2009	64	4S	8gm/kg	34.5	
11/06/2009	65	4S	8gm/kg	33.5	
12/06/2009	66	4S	8gm/kg	35.5	
13/06/2009	67	4S	8gm/kg	36	1027.379
14/06/2009	68	4S	8gm/kg	35	
15/06/2009	69	4S	8gm/kg	37.5	
16/06/2009	70	4S	8gm/kg	38	

Date	DOC	Feed Code	Feed Amount in Tray (gm/kg)	Feed Used/Day (gm)	Total Feed to Date (kg)
17/06/2009	71	4S	8gm/kg	38.5	
18/06/2009	72	4S	8gm/kg	33.5	
19/06/2009	73	4S	8gm/kg	33.5	
20/06/2009	74	4S	8gm/kg	33.3	1276.679
21/06/2009	75	4S	8gm/kg	34.5	
22/06/2009	76	4S	8gm/kg	35.8	
23/06/2009	77	4S	8gm/kg	39.5	
24/06/2009	78	4S	8gm/kg	38.8	
25/06/2009	79	4S	8gm/kg	40	
26/06/2009	80	4S	8gm/kg	40	
27/06/2009	81	4S	9gm/kg	41	1546.279
28/06/2009	82	4S	9gm/kg	42.5	
[illegible]	[illegible]	[illegible]	9gm/kg	45	
30/06/2009	84	4S	9gm/kg	45	
01/07/2009	85	4S	9gm/kg	44.5	
02/07/2009	86	4S	9gm/kg	44.5	
03/07/2009	87	4S	9gm/kg	45	
04/07/2009	88	4S	9gm/kg	47	1859.779
05/07/2009	89	4	9gm/kg	48	
06/07/2009	90	4	9gm/kg	48.8	
07/07/2009	91	4	9gm/kg	44.8	
08/07/2009	92	4	9gm/kg	41.8	
09/07/2009	93	4	9gm/kg	40.5	
10/07/2009	94	4	9gm/kg	42	
11/07/2009	95	4	10gm/kg	45	2170.679
12/07/2009	96	4	10gm/kg	46.5	
13/07/2009	97	4	10gm/kg	48	
14/07/2009	98	4	10gm/kg	50	
15/07/2009	99	4	10gm/kg	50.5	
16/07/2009	100	4	10gm/kg	52.5	
17/07/2009	101	4	10gm/kg	52	
18/07/2009	102	4	10gm/kg	45	2516.179

Following probiotics are used with feed for better growth of fishes:

Date	*Doc*	*Probiotics*	
		Water and Soil	*Gut*
05/05/2009	27		SuperBiotic(10gm/kg feed)
06/05/2009	30		SuperBiotic(10gm/kg feed)
07/05/2009	31		SuperBiotic(10gm/kg feed)
08/05/2009	39	Environ-AC (3500gm)	Aqualact (10gm/kg fed)
09/05/2009	40		Aqualact (10gm/kg fed)
10/05/2009	41		Aqualact (10gm/kg fed)
11/05/2009	42		Aqualact (10gm/kg fed)
12/05/2009	43		Aqualact (10gm/kg fed)
13/05/2009	44		SuperBiotic(10gm/kg feed)
14/05/2009	48		Zymetin (10gm/kg feed)
15/05/2009	49		Zymetin (10gm/kg feed)
16/05/2009	50		Zymetin (10gm/kg feed)
17/05/2009	51		Zymetin (10gm/kg feed)
18/05/2009	52		Zymetin (10gm/kg feed)
19/05/2009	58	Super-PS (3lit)	
20/05/2009	64		Zymetin (10gm/kg feed)
21/05/2009	65		Zymetin (10gm/kg feed)
22/05/2009	66		Zymetin (10gm/kg feed)
23/05/2009	68		SuperBiotic(10gm/kg feed)
24/05/2009	72		Zymetin (10gm/kg feed)
25/05/2009	73		Zymetin (10gm/kg feed)
26/05/2009	74		Zymetin (10gm/kg feed)
27/05/2009	77	Super-PS (3lit)	Zymetin (10gm/kg feed)
28/05/2009	92	Super-PS (5lit)	
05/05/2009	93		Zymetin (10gm/kg feed)
06/05/2009	95		Zymetin (10gm/kg feed)
07/05/2009	96		Zymetin (10gm/kg feed)

Vitamin mixture is also used with the feed as per following schedule:

Date	*Doc*	*Vitamin*
05/05/2009	32	Herbastim (5ml/kg feed)
06/05/2009	33	Herbastim (5ml/kg feed)
07/05/2009	34	Herbastim (5ml/kg feed)
08/05/2009	35	Herbastim (5ml/kg feed)
09/05/2009	36	Herbastim (5ml/kg feed)
10/05/2009	48	Mutagen (10gm/kg feed)
11/05/2009	49	Mutagen (10gm/kg feed)
12/05/2009	50	Mutagen (10gm/kg feed)
13/05/2009	55	Mutagen (10gm/kg feed)
14/05/2009	56	Mutagen (10gm/kg feed)
15/05/2009	57	Mutagen (10gm/kg feed)
10/05/2009	58	Mutagen (10gm/kg feed)
17/05/2009	59	Immunex (10m/kg feed)
18/05/2009	60	Immunex (10m/kg feed)
19/05/2009	61	Immunex (10m/kg feed)
20/05/2009	75	Regualr AVM (10gm/kg feed)
21/05/2009	76	Regualr AVM (10gm/kg feed)
22/05/2009	77	Regualr AVM (10gm/kg feed)
23/05/2009	78	Regualr AVM (10gm/kg feed)
24/05/2009	79	Regualr AVM (10gm/kg feed)
25/05/2009	80	Immunex (10m/kg feed)
26/05/2009	82	Regualr AVM (10gm/kg feed)
27/05/2009	83	Mutagen (10gm/kg feed)
28/05/2009	84	Mutagen (10gm/kg feed)
05/05/2009	85	Mutagen (10gm/kg feed)
06/05/2009	86	Mutagen (10gm/kg feed)
07/05/2009	89	Wockcee (2.0gm/kg feed)
08/05/2009	90	Wockcee (2.0gm/kg feed)
09/05/2009	91	Wockcee (2.0gm/kg feed)
10/05/2009	98	Mutagen (10gm/kg feed)
12.07.09	99	Mutagen (10gm/kg feed)

Besides these, the periodic water exchange has been done to maintain acceptable water condition of both (Pond-1 and A5) pond water. Though four ponds were used during the culture operation, only two culture data sheet were mentioned detailed.

Results and Discussion

The physico-chemical properties of water namely pH, temperature, dissolved oxygen, and salinity have monitored on regular basis *i.e.*, every 7 days interval and their results does not varied drastically. Therefore, some of the important parameters of experimental results has been analyzed and recorded under Tables 9.1 and 9.2

Table 9.1: Average Weight Gain (gm) and FCR Values of Pond-1 (C.P Prawn Feed)

Sampling Date	*DOC*	*ABW (gm)*	*ADG (gm)*	*FCR*	*Survival (per cent)*
13.05.09	39	4.02	0.1	0.4	90
20.05.09	46	5.0	0.14	0.57	87
27.05.09	53	7.4	0.34	0.68	85
03.06.09	60	8.59	0.17	0.75	85
10.06.09	67	10.84	0.32	0.83	83
17.06.09	74	12.72	0.26	0.87	82
24.06.09	81	14.11	0.19	0.92	81
01.07.09	88	16.06	0.27	0.97	80
08.07.09	95	17.35	0.18	1.06	78
15.07.09	102	19.39	0.29	1.17	78

From the tabulated experimental results, it has been revealed that pond no. 1 shows great deviation from pond A-5 in terms of fish growth and FCR values due to the effect of C.P prawn feed, whereas growth and FCR values in pond A-5 does not satisfied as it was applied by Godrej Super feed. From the Tables 9.1 and 9.2 it has been observed that, average daily body weight gain (gm), and FCR values (per cent) of tiger prawn were higher in Pond no. 1 as compared to the Pond No. A-5 because of the application of C.P Prawn Feed and Figures 9.1 and 9.2 also indicate the same results. Figures 9.3 and 9.4 shows the FCR value and survival rate during the experimental

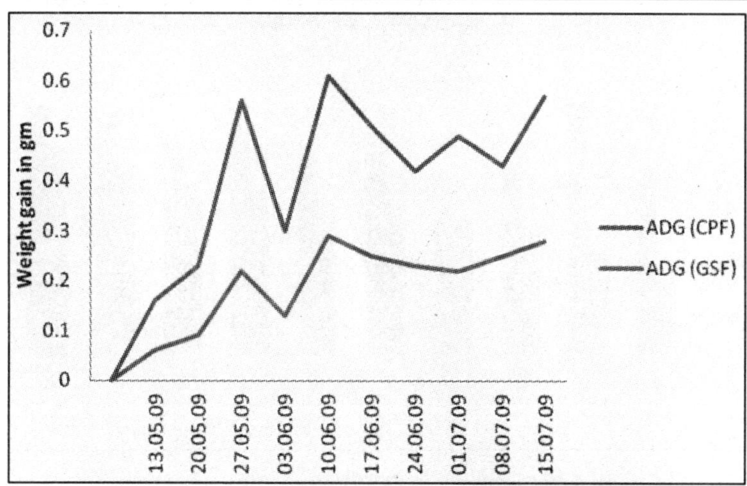

Figure 9.1: Compare of Average Daily Weight Gain Using Two Commercially Available Food (CP Food (CPF) and Godrej Super Feed (GSF))

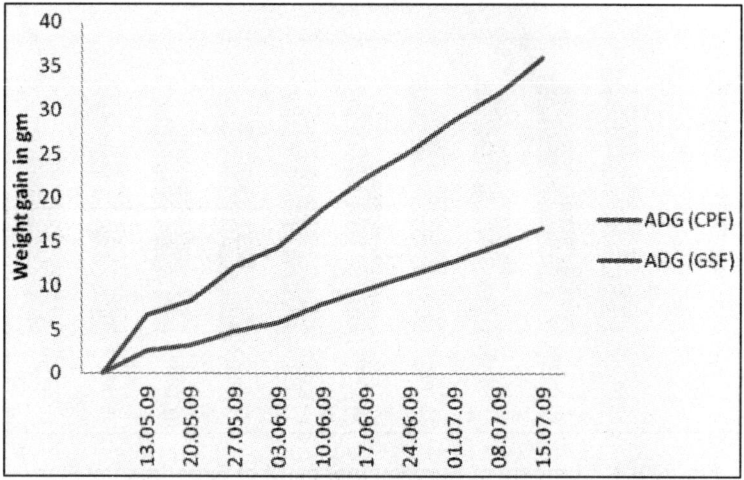

Figure 9.2: Compare of Average Body Weight gain Using Two Commercially Available Feed (CP Feed (CPF) and Godrej Super Feed (GSF))

trial. It has been observed that FCR values and survival rate were also higher in case of Pond. No. 1 where C.P Prawn feed was given. The use of supplementary feed is an important factor for the shrimp

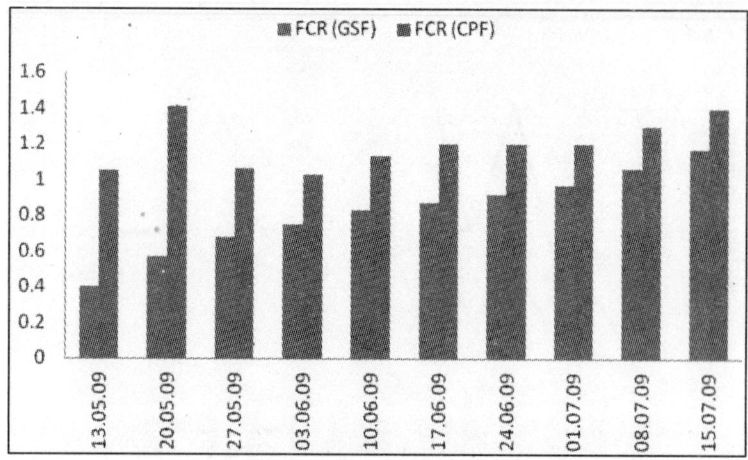

Figure 9.3: Compare of FCR Values (per cent) Using Two Commercially Available Feed (CP Feed (CPF) and Godrej Super Feed (GSF))

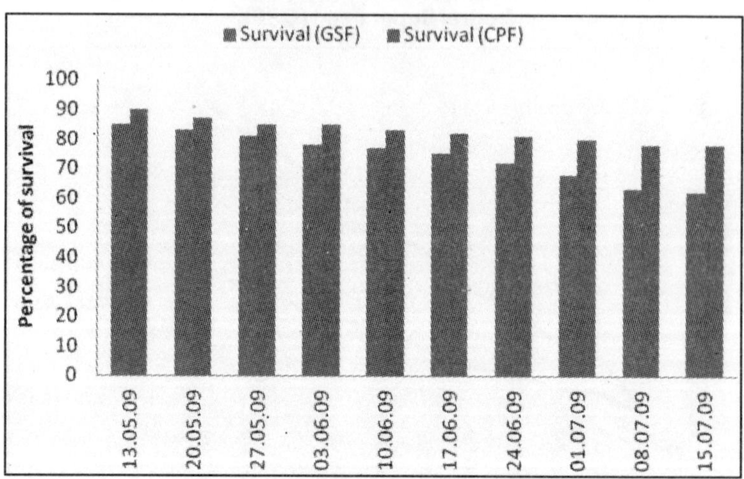

Figure 9.4: Compare of Survival (per cent) of Experimental Fish Species Using Two Commercially Available Feed (CP Feed (CPF) and Godrej Super Feed (GSF))

production in the tropical countries as suggested by Gupta *et al.* (1990) and Mandal *et al.*(2010). The nutritional value of the manure and feed remnants is preserved because losses of nitrogen and energy due to natural wastage, fermentation, evaporation and non-

Table 9.2: Average Weight Gain (gm) and FCR Values of Pond A-5 (Godrej Super Prawn Feed)

Sampling Date	*DOC*	*ABW (gm)*	*ADG (gm)*	*FCR*	*Survival (per cent)*
13.05.09	38	2.60	0.06	1.05	85
20.05.09	45	3.27	0.09	1.41	83
27.05.09	52	4.84	0.22	1.06	81
03.06.09	59	5.81	0.13	1.03	78
10.06.09	66	7.85	0.29	1.13	77
17.06.09	73	9.62	0.25	1.20	75
24.06.09	80	11.23	0.23	1.2	72
01.07.09	87	12.77	0.22	1.2	68
08.07.09	94	14.58	0.25	1.3	63
15.07.09	101	16.56	0.28	1.4	62

reversible coagulation are eliminated (Barash *et al.*, 1982; Wohlfarth and Schroeder, 1979; Plavnik *et al.*, 1983). The operating costs of shrimp production are reduced by the on farm supply of fertilizer (manure) and feed (uneaten feed residues) inputs (ADCP, 1979; Sin, 1980; FAO, 1983; Shang and Costa-Pierce, 1983; Rajbanshi and Shrestha, 1980; Vincke, 1985). Tiger shrimp production can also be increased by the proper maintaining of bio-security (Mandal *et al.*, 2009).

Through this observation it has been revealed that, different supplementary feeds are being used for culture operation at different composition and water stability condition. From this investigation C.P prawn feed show more water stability than that of Godrej Super feed. As we know water stability of feed is directly proportional to feeding duration and inversely proportional to nitrogenous waste pollution. Godrej feeds are low water stable and higher protein content that has negatively affect to the pond environment that leads to low survivality and growth rate of experimental fish species. From the above study we may conclude that, in West Bengal, especially in Purba Medinipur district is bestowed by glorious natural resources for shrimp farming because of the availability of plenty brackish water in the 'Haldi River'. Therefore, the Tiger prawn culture is increasing day by day due to the high market price and as well as consumer demand in the coastal area of South West Bengal.

References

APHA [American Public Health Association], 2005. Standard methods for the examination of water and waste water. Washington DC.

ADCP (Aquaculture Development and Coordination Programme), 1979. Aquaculture development in China. Report of a FAO/ UNDP Aquaculture study Tour to the People's Republic of China, led by T.V.R. Pillay, Aquaculture Development and Coordination Programme, FAO, Rome, Italy, 2 May–1 June 1978. Rome, UNDP/FAO, ADCP/REP/79/10: 65 p.

Bandyopadhyay, M.K., Singh, S.K. and Sarkar, B., 1991. Growth and production of *Ctenopharyngodon idella* reared in the plastic cages. National Symposium on New Horizons in Freshwater Aquaculture, 23–25, January, 1991. Proceedings.: 35–37.

Banerjee, S.M., 1967. Water Quality and soil condition of fishponds in some states of India in relation to fish production. *Indian J. Fish.*, 14 (1 and 2): 115–144.

Barash, H., Plavnik, I. and Moav, R., 1982. Integration of duck and fish farming: Experimental results. *Aquaculture*, 27: 129–140.

FAO., 1983. Freshwater aquaculture development in China. Report of the FAO/UNDP study tour organized for French–speaking African countries. 22 April–20 May, 1980. *FAO Fish. Tech. Pap.*, (215): 125 p.

Ghosh, S.R., 1978. Role of soil and water quality study in fishpond. Summer institute on Inland Aquaculture, CIFRI (ICAR), Barrackpore: 6 (Abstract).

Gooch, B.C., 1967. Appraisal of North American fish culture fertilization studies. *FAO. Fish Rep.*, 3 (44): 13–26.

Govind, B.V., Rajgopal, K.V. and Singh, G.S., 1978. Studies on the comparative efficiency of organic manures as fish producers. *J. Inland Fish Soc. India*, 10: 101–106.

Gupta, V.K., Sharma, J.P. and Srivastava, J.B., 1990. Polyculture of Indian and exotic carps using cattle manure with and without supplementary feed. In recent trends in Limnology, (Agarwal, V.P., Das, P. eds.) Muzafffarnagar, India. Society of Biosciences: 439–446.

Jhingran, V.G. and R.S.V. Pullin, (1985. A hatchery manual for the common, Chinese and Indian major carps. Manila, Philippines, ADB/ICLARM Publ., ICLARM Studies and Reviews–11, 191 p.

Laha, U.K. and Mitra, B., 1987. Effectiveness of some organic manure on the growth of three Indian major carps. *Sci. and Cult.*, 53: 84–86.

Mandal, B., Patra, B.C. and Chand, B.K., 2008. Possibilities of sustainable aquaculture practice in abandoned China clay mines of West Bengal, India. *Research Journal of Fisheries and Hydrobiology* (INSINET Publication), 3 (2): 36–40.

Mandal, B., Ganguli, D. and Chand B. K., 2009. Probiotic farming and biosecurity management for sustainable brackish water shrimp farming. *Proceedings of the International Seminar* (ISBN 978–81–89339–40–1. Published by PG Deptt. of Zoology, Raja N.L. Khan Womens' College, Midnapore, West Bengal. pp. 62 65.

Mandal, B. Mukherjee, S. and Chand, B.K., 2010. Use of Supplementary Feed for Sustainable Inland Aqua Farming: A Case Study. Perspective in Animal Ecology and Reproduction. M/S Daya Publishing House, New Delhi; vol.7 pp.146–172.

Mortimer, C.H. and Hickling, F.C., 1954. Fertilizers in fishponds. *Fish Publ. Lond.*, (5).

Nandeesha, M.C., 1982. Effect of three different organic manures on the growth of cultivable carps M.F.Sc. Thesis (unpublished) University of Agricultural Science, Bangalore.

Patra, B.C., 1993a). A report on the effect of organic manures and supplementary feeding on the production of Indian major carps in rural areas of West Bengal. *J. Freshwat. Biol.*, 5(4): 347–357.

Patra, B.C., 1993b). Satiation time, appetite and daily pattern of feed intake and faeces release by an air–breathing fish, *Anabas testudineus* (Bloch). *J. Aqua. Tropics*, 8: 41–46.

Patra, B.C. and Roy, A.K., 1988. Observations on the influence of three organic manures on the growth and yield of Indian major carps under pond culture. *J. Inland.Fish. Soc. India*, 20 (2): 61–63.

Phillips, M.J., 2002. Fresh water aquaculture in the Lower Mekong Basin. MRC Technical Paper No. 7, Mekong River Commission, Phnom Penh. 62 pp.

Plavnik, I., H. Barash and G. Schroeder, (1983. Utilization of ducks droppings in fish farming–Nutr. *Rep. Int.*, 28(3): 635–641.

Prakash, S. and Gupta, R.A., 1986. Studies on the comparative growth rates on three major carps of the Govindgarh lake. *Indian J. Fish*, 33(1): 45–53.

Prowse, G.A., 1967. A review of the methods of fertilizing warm water fishponds in Asia and far best. *FAO Fish Rep.*, 3 (44): 7 – 12.

Prowse, G.A., 1968. The latin square fertilizer trials. Trop. *Fish cult. Res. Inst. Wkg. Pab.*, (1): 28.

Rajbanshi, K.G. and M.B. Shrestha, (1980. A case study on the Economics of Integrated Farming Systems: Agriculture–Aquaculture and Animal Husbandry in Nepal. pp. 195–208. In Pullin, R.S.V. and Z.H. Shehadeh, (Editors), Integrated agriculture–aquaculture farming systems. ICLARM Conference Proceedings 4, 258p. International Center for Living Aquatic Resources Management, Manila and the Southeast Asian Center For Graduate Study and Research in Agriculture College, Los Banos, Laguna, Philippines.

Rosenberg, (2001. World Agriculture: Towards 2015/2030. An FAO Perspectives.

Saha, G.N., 1969. Relative efficiency of different nitrogenous fertilizers in fish ponds soil. *J. Indian Soc. Soil. Sci.*, 17(1): 51–57.

Saha, G.N., Ghosh, S.R. and Chatterjee, D.K., 1979. Role of soil and water in pond productivity. In Commemoration of the ICAR: Golden Jubilee year, CIFRI, Barrackpore. Part – I: 82–85.

Schaperclaus, W. Die Karpfenteichswirt Schaft, (1959. In Der Deutsche Demo Kratischen Pepublik. Sita Ber. Bd. VII. Hrt. 7 Deutsche Akad. Landwirtschaftwiss. Berlin.

Shang, Y.C. and B.A. Costa–Pierce, (1983) Integrated Aquaculture–Agriculture farming system some economic aspects. *J. World Maricul. Soc.*, 14: 523–530.

Sin, Alfred Wai–Ching, (1980. Integrated Animal–Fish Husbandry Systems in Hongkong with Case Studies on Duck–Fish and Goose–Fish Systems, pp. 113–123. In Pullin, R.S.V. and Z.H. Shehadeh, (Editors), *Integrated Agriculture–Aquaculture Farming Systems*. ICLARM Conference Proceedings 4, 258 p.

Singit, G.S., Keshavanath, P., Varghese, T.J., Konda Reddy, P. and Venkatesh prasad, G.S., 1980. Observation on fish production through composite fish culture in a farmer's pond. *Mysore J. Agric. Sci.*, 14: 236–240.

Vander Lingen, M.I., 1967. Fertilization of fishponds. *Rep. Dep. Nat. Censer. Cape. Prov.*, 16: 65–68.

Vincke, M., 1985. Developing productive systems under village conditions. Paper presented at the Commonwealth Consultative Workshop on Village Level Aquaculture Development in Africa, Sierra Leone, 14–20 February, 1985, 23 p.

Wolny, P., 1967. Fertilization of warm water fishponds in Europe. *FAO Fish Rep.*, 3 (44). 64–81.

Wohlfarth, G.W. and Schroeder, G.L., 1979. Use of manure in fish farming–A review *Agric. Wastes*, 1: 279–299.

Chapter 10

Nutritional Status of Three Species of Molluscs in Nepal

Kul Prasad Limbu[a,*], *Bharat Raj Subba*[a] *and Basanta Kumar Rai*[b]

[a]*Department of Zoology, Post Graduate Campus, Tribhuvan University, Biratnagar, Nepal*

[b]*Central Campus of Technology, Tribhuvan University, Hattisar, Dharan, Nepal*

ABSTRACT

Three species of molluscs (*Bellamya bengalensis, Lamellidens marginalis* and *Pila globosa*) were assessed for their proximate and mineral compositions. The mean nutrient levels (wet basis, standard deviation in parenthesis) of the three species were: moisture, 72 (1.87) per cent; crude protein, 15.49 (1.89) per cent; crude fat, 4.22 (0.85) per cent; total ash, 5.62 (1.76) per cent; calcium, 1668 (364.79) mg/100 g; potassium, 1.66 (1.43) mg/100 g; iron, 44.43 (41.61) mg/100 g and sodium, 9.4 (5.23) mg/100 g. The nutrient levels of the species differed significantly (P=0.05) and based on statistical analysis, *Lamellidens* and *Pila* can be considered superior to *Bellamya*. As these molluscan species are rich in minerals, it is highly recommended for regular consumption.

Keywords: *Molluscs, Nutrition, Proximate, Minerals.*

* Corresponding author: E-mail: limbukp@gmail.com

Introduction

Meat and meat products are concentrated sources of high quality protein and their amino acid composition usually compensates for shortcomings in the staple diet (Bender, 1992). The major sources of meat protein for the Nepalese populace come mainly from livestock in the form of poultry, mutton and pork. However, prices of animal products are beyond the reach of most of the population (Salawu *et al.*, 1994). Exploding population and rising standard of living have placed great pressure on the existing sources of animal protein. This situation has forced nutritionists to explore non-conventional or hitherto neglected protein resources (Hamzat *et al.*, 2002). Official database on the use of alternative or non-conventional meat protein resources in Nepal is relatively scarce. Among non-conventional sources of animal protein, freshwater molluscs–both gastropods and bivalves–have great prospect in Nepal. It is estimated that over 30 communities of the Tarai consume some 20 species of fresh water molluscs (Subba and Ghosh, 2000; Subba, 2004). However, there is general lack of nutritional data on these species. Food Composition Table released by Department of Food Technology and Quality Control (DFTQC), Nepal has two types of snails in its record. This fact notwithstanding, Subba and Ghosh (2000) and Subba (2004) have reported that some 20 species of aquatic molluscs (out of 40 species found in Nepal) are used as a delicacy by the ethnic groups of the Terai (Nepal). With scientific study on their food value, biochemical components and safety aspects, it might be possible to add molluscs to Nepalese food resource list. The nutritional value of molluscs of other regions of the world has been mentioned by different workers (Meric *et al.*, 1991; Misra *et al.*, 2002; Ponis *et al.*, 2006; Hamzat *et al.*, 2002; Gebhardt and Thomas, 2002; Fagbuaro *et al.*, 2006). This study is aimed at determining the nutritional value of 3 commonest species of molluscs (*Bellamya bengalensis, Pila globosa* and *Lamellidens marginalis*) in Nepal.

Material and Methods

Fresh edible molluscs (*Bellamya bengalensis, Pila globosa* and *Lamellidens marginalis*) were purchased from the market Duhabi, Itahari, Biratnagar, Kersiya and Damak. The shells were carefully removed so that the edible parts could be removed and dried. The powders were subjected to different processing methods of consumption. Moisture was determined on 10-12 g of the shredded meat by hot-air oven method (AOAC, 2005). Fat content was determined by soxhlet extraction (petroleum ether as the solvent, b. pt. 60-80°C) by running the apparatus for at least 40 cycles (250 ml cap extraction vessel). Protein was determined by boric acid modification of Kjeldahl method (Ranganna, 1986). Total ash was determined by dry ashing method (Ranganna, 1986). The ash was used for calcium, iron, potassium, and sodium determination. Calcium was determined by complexometric titration (Ranganna, 1986), iron by colourimetric method (Metertek, Model SP-870, China), and sodium and potassium by flame photometric method (FOTOFLAME®, single-channel, India) (KC and Rai, 2007). In the determination of sodium and potassium, attempt was made to quench the interference of iron, calcium and chlorides by supplementing the common standard (of sodium and potassium) with above ions. The data on nutritional value of the above three species of molluscs were statistically analyzed using GenStat Discovery Edition 3 (Buysse *et al.*, 2007), and MS-Excel for two-way (no blocking) ANOVA and difference between the factors (replication and mollusc type). Two variables were not analyzed in replicates, *viz.*, (i) amount of edible portion, and (ii) average weight per unit (of mollusc type) because of sample size constraint.

Results and Discussion

Data on the proximate composition of the flesh of three species of molluscs is shown in Table 10.1 reveals that mollusc meat is rich in protein, fat and ash. The table also provides average values of compositions for triplicate analyses. The superiority or inferiority of the molluscs with respect to nutrient composition has been analyzed by using least significant difference (LSD) at 5 per cent level of significance. Values of moisture, protein, fat and ash and vary according to sample species. The moisture content of the three species ranged between 74.17±1.05 in *Bellamya bengalensis* and 71.61±1.89 in *Pila globosa.* The highest composition of 17.27±0.71 of protein was recorded in *Bellamya bengalensis* while the least percentage value of 13.39±1.09 was recorded in *Lamellidens marginalis*. The mollusc samples are good sources of protein (13.39±1.09)~(17.27±0.71) compared with 8.00 per cent ~12.00 per cent in cereals (Bender, 1992). The protein values of *Bellamya bengalensis* was slightly lower than the protein values of chicken (20.2 per cent), pork (18.7 per cent), mutton (21.4 per cent) and buff (19.4 per cent) (DFTQC 1986).

Molluscs constitute the cheapest sources of protein in Nepal (Subba, 2002; Karna and Shrestha, 2006). Fat of *Pila globosa* was maximum followed by *Lamellidens marginalis* and *Bellamya bengalensis*. The least percentage value of 3.69±0.14 of ash was recorded in *Bellamya bengalensis*. The proximate composition (except protein content) of molluscs reported by various authors (Hamzat *et al.*, 2002; Ponis *et al.*, 2006; Fagbuaro, 2006; Karna and Shrestha, 2006), unfortunately, do not compare well with that of ours. This could be due to the difference in species, food habit of the mollusc, the habitat, or even the methods used for the analyses.

Table 10.1: Proximate Composition of the Flesh of Three Species of Snails

Nutrient Composition	*Mollusc Type*			*LSD (5 per cent)*
	Bellamya bengalensis	*Lamellidens marginalis*	*Pila globosa*	
Moisture, per cent (wb)	74.17^{a}±1.05	72.57^{a}±2.07	71.61^{a}±1.89	4.502
Protein, per cent (wbf)	17.27^{a}±0.71	13.39^{b}±1.09	15.83^{ab}±1.03	2.526
Fat, per cent (wbf)	3.13^{a}±0.13	4.71^{b}±0.36	4.80^{b}±0.32	0.728
Ash, per cent (wbf)	3.69^{a}±0.14	7.68^{b}±0.56	5.49^{c}±0.37	0.959

Figures preceding and following the (±) sign are arithmetic means and standard deviations of the triplicate analyses, respectively. Figures in the rows having similar superscripts are not significantly different at 5 per cent level of significance. Other notations have following explanations: wb = wet basis; wbf = wet basis, taking only the flesh portion (*i.e.*, edible portion); LSD = least significant difference.

Table 10.2 shows mineral composition of molluscs which indicates that all the three species of molluscs are good sources of minerals. Calcium was higher in *Lamellidens marginalis*. Molluscs are generally considered good sources of calcium (KC and Rai, 2007). The present finding also supports this fact. When compared with food composition data released by DFTQC (1986), the calcium content in mollusc averages those of crabs, dried vegetables and skim milk. Calcium is the most abundant nutritionally important mineral. Ca constitutes a large proportion of the bone, human blood and extracellular fluid; it is necessary for the normal functioning of cardiac muscles, blood coagulation and milk clotting, and the regulation of cell permeability. *Lamellidens marginalis* have good iron of 8.56±0.35 than the other two species. Iron level is much higher than in most meat and fish (when compared with food composition data released by DFTQC (1986). The iron from meat does not only enhance the absorption of iron from other sources such as cereal but increases con-siderably the level of iron absorption in the blood. Potassium and sodium was higher in *Pila globosa*. Na and K are essential for ionic balance of the human body and maintain tissue excitability. This finding shows that consumption of mollusks will definitely increase the level of these major elements in the body. The grand means and standard deviations of the nutrient level of three edible molluscs are shown in Table 10.3.

Table 10.2: Mineral Composition of the Flesh of Three Species of Snails (mg/100g)

Nutrient Composition	*Mollusc Type*			*LSD (5 per cent)*
	Bellamya bengalensis	*Lamellidens marginalis*	*Pila globosa*	
Ca, mg/100g (wbf)	1505.09^{a}± 61.22	2130.54^{b}± 160.45	1369.42^{a}± 91.32	285.0
Fe, mg/100g (wbf)	8.56^{a}±0.35	98.75^{b}±7.44	25.98^{c}±1.73	10.23
K, mg/100g (wbf)	1.14^{a}±0.05	0.33^{b}±0.03	3.50^{c}±0.23	0.3380
Na, mg/100g (wbf)	12.32^{a}±0.50	2.49^{b}±0.19	13.38^{a}±0.89	1.621

Figures preceding and following the (±) sign are arithmetic means and standard deviations of the triplicate analyses, respectively. Figures in the rows having similar superscripts are not significantly different at 5 per cent level of significance. Other notation has following explanations: wbf = wet basis, taking only the flesh portion (*i.e.*, edible portion); LSD = least significant difference.

Comparison of the mollusc samples for food value gives following results.

- ☆ *Protein content*: [*Bellamya*] ≥ [*Pila*] ≥ [*Lamellidens*]; [*Bellamya*] > [*Lamellidens*]
- ☆ *Moisture content*: [*Bellamya*] = [*Lamellidens*] = [*Pila*]; *i.e.*, no significant difference

Table 10.3: Mean±Standard Deviation of the Nutrient Levels in the Three Molluscs

Nutrient Composition	*Mean±Standard Deviation*
Moisture, per cent (wb)	72.78±1.87
Protein, per cent (wbf)	15.49±1.89
Fat, per cent (wbf)	4.22±0.85
Ash, per cent (wbf)	5.62±1.76
Ca, mg/100g (wbf)	1668.35±364.79
Fe, mg/100g (wbf)	44.43±41.61
K, mg/100g (wbf)	1.66±1.43
Na, mg/100g (wbf)	9.4±5.23

- ✰ *Fat content*: [*Lamellidens / Pila*] > [*Bellamya*]
- ✰ *Total ash*: [*Lamellidens*] > [*Pila*] > [*Bellamya*]
- ✰ *Calcium content*: [*Lamellidens*] > [*Bellamya / Pila*]
- ✰ *Iron content*: [*Lamellidens*] > [*Pila*] > [*Bellamya*]
- ✰ *Potassium content*: [*Pila*] > [*Bellamya*] > [*Lamellidens*]
- ✰ *Sodium*: [*Pila / Bellamya*] > [*Lamellidens*]

If the frequency of occurrence in the first position (for superiority in terms of a particular nutritional property) is to be taken as the criterion for superiority, the order (based on the above results) would be [*Pila/Lamellidens*] > [*Bellamya*]. There is a tie between *Pila* and *Lamellidens*.

Consumers of molluscs belonged exclusively to ethnic communities of Tarai in the past. These days, however, non-ethnic peoples also have gradually begun to appreciate the taste of mollusc protein, thereby increasing the demand for this item. Most of the consumers of molluscs belonging to ethnic people are still poor and cannot afford conventional animal protein sources (mutton, chicken, buff, pork and fish). Molluscs have played a major role in meeting a significant part of the dietary protein requirement, though there is no record of it. Consumption of these three species of mollusc is therefore highly recommended for both old and young as it provides essential elements to the body.

Acknowledgements

We acknowledge our thanks to National Academy of Science and Technology (NAST) for providing the research fund for this project. Our sincere thanks go to Dr. Ganga P. Kharel for kindly providing us the laboratory facility at Central Campus of Technology, Hattisar, Dharan.

References

AOAC., 2005. *Association of Official Analytical Chemists*. Horwitz, W. and Latimer, G. W. (eds). AOAC International, Maryland, USA.

Bender, A., 1992. Meat and meat products in human nutrition in developing countries. *Food and Nutrition Paper 53*. FAO of the United Nations.

Buysse, W.R., Coe Stern, R. and Crispin, C., 2007. GenStat Discovery Edition 3 for everyday use. ICARF Nairobi, Kenya. p. 117.

DFTQC, 1986. *Nutrient Content of Nepalese Foods*. Department of Food Technology and Quality Control, Babarmahal, Kathmandu.

Fagbuaro, O., Oso, J.A, Edward, J.B. and Ogunleye, R.F., 2006. Nutritional status of four species of giant land snails in Nigeria. *J. Zhejiang Univ.*, 7(9): 686–689.

Gebhardt, S.E. and Thomas, R.G., 2002. Nutritive value of foods. United States Department of Agriculture, Agriculture Research Service, Home and Garden Bulletin No. 72.

Hamzat, R.A., Faiyelo, C.O. and Longe, O.G., 2002. Nutritional qualities of snails (*Archachatina marginata*) fed solely with fresh kola. *Nutri. Food Sci.*, 32(4): 134–136.

Karna, R.R. and Shrestha, J., 2006. Molluscs as a cheap source of animal protein in Sarlahi district. *Nepal J. Sci. Technol.*, 7: 45–48.

KC, J.B. and Rai, B.K., 2007. *Essentials of Food Chemistry*. Anamnagar, Kathmandu, Nepal, pp. 61–63, 81.

Meric, E., Sakinc, M., Ozdogan, M. and Ackurt, F., 1991. Mollllusc shells found at the Yarimburgaz cave. *J. Islamic Acad. Sci.*, 4(1): 6–9.

Misra, K.K., Shkrob, I., Rakshit, S. and Dembitsky, V.M., 2002. Variability in fatty acids and fatty aldehydes in different organs of two prosobranch gastropod mollusc. *Biochemical Syst. and Ecol.*, 30: 749–761.

Ponis, E., Probert, I., Veron, B., Le Coz, J.R., Mathieu, M. and Robert, R., 2006. Nutritional value of six Pavlovophyceae for *Crassostrea gigas* and *Pecten maximus* larvae. *Aquaculture,* 254(1–4): 544–553.

Ranganna, S., 1986. *Handbook of Analysis and Quality Control for Fruit and Vegetable Products,* 2nd edn. Tata McGraw-Hill Pub. Co. India, pp. 21–24, 126, 127.

Salawu, M.B., Adedeji, S.K. and Hassan, W.H., 1994. Performance of broilers and rabbits given diets containing full-fat neem (*Azadirachta indica*), seed meal. *Animal Production,* 58: 285–289.

Subba, B.R., 2004. Edible mollusc. *Zoologue,* 2: 2.

Subba, B. R. and Ghosh, T.K., 2000. Some freshwater molluscs from eastern and central Nepal. *J. Bombay Nat. His. Soc.,* 97(3): 452–455.

Chapter 11

Effects of Xylene, Benzene and Gear Oil-WSF Exposures on Certain Biochemical Parameters of *Gafrarium divaricatum* (Gmelin)

B.G. Kulkarni[1], *Shirley Agwuocha*[2] *and A.K. Pandey*[3]*

[1]*Department of Zoology, Institute of Science, 15 Madam Cama Road, Mumbai – 400 032, Mh., India*

[2]*Department of Zoology, Thakur College of Science and Commerce, Kandivali (East), Mumbai – 400 101, Mh., India*

[3]*National Bureau of Fish Genetic Resources, Canal Ring Road, Lucknow – 226 002, U.P., India*

ABSTRACT

Gafrarium divaricatum were exposed to sublethal concentration of xylene (4.25 and 8.50 mg/l), benzene (4.35) and 8.70 mg/l) and gear oil-WSF (1 and 2 per cent) for 30 days under the laboratory conditions. There were decline in condition index of the treated clams while moisture content recorded an increase on day 30. Glycogen content of hepatopancreas of the clams treated with the pollutants registered an initial increase on day 4 whereas the level declined in gill and adductor muscle during this period. Glycogen content of the hepatopancreas, gill and adductor muscle of the treated

* Corresponding author: E-mail: akpandey_cifa@yahoo.co.in

clams registered drastic decline by day 30. Lactic acid, pyruvic acid and protein content in the three tissues increased significantly on day 4 and 30 of the treatments. It appears that the clam resort to anaerobic mode of respiration under the pollution stress induced due to prolong chronic exposure of xylene, benzene and gear oil-WSF.

Keywords: *Biochemical parameters, Xylene, Benzene, Gear oil-WSF, Gafrarium divaricatum.*

Introduction

It is well known that intertidal organisms are adaptive to environmental variations like temperature, salinity and pH by physiological and biochemical mechanisms. Among the biochemical mechanisms, tissue metabolites play an important role during such conditions. Consequently, considerable amount of data is generated regarding the effects of salinity, temperature and pH variations on tissue metabolites like glycogen, lactic acid, pyruvic acid, proteins, fats and water content in the intertidal organisms (Mane, 1974; Schoffeniels, 1976; Nagabhushanam and Mane, 1978; Rodriguez, 1981; Akabarali and Truman, 1985). Therefore, the measurement of a specific biochemical metabolite in the tissues of intertidal organisms exposed to environmental stressors provides a sensitive method for predicting the effects of such stress on metabolism of these organisms.

In toxicological investigations also, the change in the tissue metabolites proved to be a sensitive indicator of stress caused by toxicants. Therefore, many investigators have resorted to the assessment of tissue metabolites like glycogen, proteins, fats, lactic acid and pyruvic acid in fishes exposed to various pollutants under field and laboratory investigations (Venkatrama and Radhakrishniah, 1987; Al-Akel *et al.*, 1988; Ferrando and Andreu-Moliner, 1991; Durairaj and Selvarajan, 1992). These studies have proved the utility of assessment of such metabolites in organisms to know the toxic mode of action of pollutants on fishes. Such methods have also been applied in assessment of the effects of various pollutants on metabolism of invertebrates like crustaceans and molluscs (Mane *et al.*, 1986; Biswas, 1986; Krishna *et al.*, 1987; Shankara *et al.*, 1989; Shreedevi *et al.*, 1992; Reddy and Bhagyalakshmi, 1994; Roy, 1994). Bayne *et al.* (1982) have shown that xenobiotics induced stress in intertidal bivalve, *Mytilus edulis* could be detected by assessing a number of metabolic parameters such as blood sugar levels, disruption of energy balance represented by alterations in glycogen, proteins and lipid content of tissues. Similarly, many investigators have proved the utility of this concept while studying the effects of oil and related hydrocarbons on aquatic organisms (Sabo and Stegeman, 1977; Augenfeld *et al*, 1980; Dange and Masurekar, 1982; Kulkarni, 1983; Deshmukh, 1983; Axiak and George, 1987; Neff *et al.*, 1987; Moles *et al.*, 1987; Kostlev, 1988; Khan and Nag, 1993).

In India, considerable importance is given to the studies on ecophysiological aspects of marine bivalves (Deshmukh, 1972; Mane, 1975). Some of the investigators have also used intertidal bivalves as test organisms in toxicological research (Mane and Nagabhushanam, 1975; Eapen, 1987; Kulkarni and Kulkarni, 1987; Eapen and Patel, 1989; Kulkarni, 1990a,b; Roy, 1994). However, the literature on biochemical responses with respect to tissue metabolites in intertidal clams exposed to petroleum products in very scanty. Therefore, the present investigation was conducted to assess the effects of sublethal concentration of oil and petroleum hydrocarbons on certain metabolites such as moisture content, glycogen, lactic acid, pyruvic acid and soluble proteins in the tissues of clam, *Gafrarium divaricatum*.

Materials and Methods

G. divaricatum (Gmelin) were collected during low tide period from Nariman Point area of Bombay coast and after cleaning with sea water were brought to the laboratory as single stock. The clams were acclimatized for 24 hours in a medium sized aquarium (60x30x30) containing sea water of salinity 30-32 ppt, temperature 27-29°C and pH 7.7-8.0. The same conditions were maintained throughout the bioassay tests. No special food was given to the clams as these animals thrived well under laboratory conditions. The sea water used during the acclimatization and experimentation was brought from Nariman Point coastal area. Only those sets of stock animals which did not show any mortality during the acclimation period were used for the acute toxicity experiments. The active clams of more or less uniform size (30-32 mm) were selected for the bioassay experiments. The clams with protruding siphon and foot were considered as active clams.

Standard solutions of the hydrocarbons were prepared in acetone and added to experimental aquaria to obtain desired concentrations. Water soluble fraction (WSF) was prepared by adding 500 ml of test oil to 500 ml of sea water and stirring with an electric stirrer in a closed container for 24 h at 2000 rpm. The oil and water layers were allowed to separate for 6 h. The water phase at the bottom (considered 100 per cent WSF) was siphoned off and utilized immediately for experimentation. LC_{50} values of xylene, benzene and water soluble fractions (WSFs) of gear and mobile oil for *G. divaricatum* were determined (Agwuocha *et al.*, 2009) and the clams were exposed to sublethal concentrations of xylene (4.25 and 8.5 mg/l), benzene (4.35 and 8.7 mg/l) and gear oil-WSF (1 and 2 per cent) for the test period of 4 and 30 days. At the end of each period, the control and exposed clams were sacrificed and hepatopancreas, gills and adductor muscles were removed for the analysis of glycogen, lactic acid, pyruvic acid, proteins and moisture content. A brief account of the methods followed to assess the various metabolites is given below. Optical density (OD) of various solutions and extracts were measured on Spectronic 20.

Condition Index

At the end of the exposure period of 30 days, the control and exposed clams were removed from aquaria and their shell length was measured. Then they were sacrificed to remove the whole soft tissues inside their shell. They were dried in an oven at 60°C and weighed accurately. The condition index was calculated from the following formula:

$$\text{Condition index} = \frac{\text{Dry weight}}{\text{Shell length}} \times 100$$

Moisture Content

Moisture content was determined according to the method described by AOAC (1975). The whole of soft tissue from the clam was removed and dried in an oven at 60°C. The dried tissue was weighed to a constant weight. Per cent moisture content was then calculated with the help of following formula:

$$\%\ \text{moisture content} = \frac{\text{Wet weight} - \text{Dry weight}}{\text{Wet weight of tissue}} \times 100$$

Glycogen

Tissue glycogen content was estimated by anthrone method of Seifter (1950). Freshly excised tissues (hepatopancreas, gills and adductor muscles) were taken in test tubes and digested in 3 ml 30 per cent KOH kept in water bath at 70°C. Each KOH digest was diluted to 100 ml with distilled water. To 5 ml aliquot of this solution, 10 ml of freshly prepared anthrone reagent (0.2 per cent in 95 per cent concentrated H_2SO_4) was added gradually. Similarly, blank and standard tubes were also prepared using distilled water and glucose solution, respectively. The mixture was then boiled for 10 minutes in boiling water bath. After cooling, optical density of the mixture was read at 625 nm. Standard calibration curve was also prepared for calculating glucose values. The glucose values thus obtained were converted into glycogen using the Morrisons factor 1.11.

Lactic Acid

Lactic acid levels were determined according to the methods of Barker and Summerson (1941). 2 ml aliquot of the deproteinized supernatant was treated with 1 ml 20 per cent $CuSO_4$ solution and then diluted to 10 ml with distilled water. 1 gm of $Ca(OH)_2$ powder was added and the mixture was shaken vigorously and allowed to stand for 30 minutes. The mixture was then centrifuged at 5,000 rpm for 5 minutes. To 1 ml of supernatant, 0.05 ml 4 per cent $CuSO_4$ solution was added followed by 6 ml concentrated H_2SO_4. The solution was heated in a boiling water bath for 5 minutes and cooled at room temperature. After cooling, 0.1 ml p-hydroxydiphenyl reagent (1.5 gm p-hydroxydiphenyl dissolved in 10 ml of 5 per cent NaOH and diluted to 100 ml with distilled water) was added and the solution incubated at 30°C for 30 minutes. The test tubes were then placed in a vigorously boiling water bath for 90 seconds and then allowed to cool down to room temperature. The optical density was read at 560 nm and the amount of lactic acid was calculated using a standard graph.

Pyruvic Acid

Pyruvic acid content of the tissues from the control as well as experimental clams was determined by the method of Friedmann and Haugen (1943) as modified by Chaykin (1966). Tissue of known weight was homogenized in 3 ml of 30 per cent trichloroacetic acid.

The mixture was centrifuged at 5,000 rpm for 5 minutes and clear protein-free supernatant was used for pyruvic acid assay. An appropriate aliquot of the deproteinized supernatant was diluted to 2 ml with distilled water and then treated with 0.5 ml 2,4-dinitrophenyl hydrazine reagent (0.1 per cent in 2 N HCl) followed by 3 ml 2.5 N NaOH. The reaction mixture was allowed to stand for 10 minutes and OD was read at 540 nm. The amount of pyruvic acid was calculated using standard graph.

Soluble Proteins

Soluble proteins were assayed by the method of Lowry *et al.* (1951). Freshly removed hepatopancreas, gills and adductor muscles were weighed and homogenized with chilled glass distilled water. The homogenates were centrifuged at 10,000 rpm for 10-15 minutes and the clear supernatants were used for estimation of soluble proteins. 0.2 ml supernatant was diluted to 1.0 ml by glass distilled water. 5 ml alkaline copper solution (prepared freshly by mixing 50 ml 2 per cent $NaHCO_3$ in 0.1N NaOH and 1 ml of 0.5 per cent $CuSO_4$ in 1 per cent Na-K tartrate) was added. After incubation for 10 minutes at room temperature, 0.5 ml folin-phenol reagent was added rapidly and the solution was thoroughly mixed. After 30 minutes, the optical density was read at 500 nm. A standard graph was also prepared using bovine serum albumen and protein content from different tissues was calculated from this graph.

Results and Discussion

Condition Index

The condition index of clam showed significant reduction after their exposure to xylene, benzene and gear oil WSF concentration after 30 days test period (Table 11.1).

Moisture Content

A significant elevation in total moisture content of G. *divarcatum* was noticed at all test concentrations after 30 days exposure (Table 11.2).

Glycogen

Except a slight elevation in hepatopancreas, glycogen content of clam tissues exposed to all test pollutants for 96 hours depleted. After 30 days of test period, glycogen content in all test clams reduced

Table 11.1: Condition Index in *G. divaricatum* Exposed to Xylene, Benzene and Gear Oil-WSF

Exposure Period (Day)	*Control*	*Xylene (mg/l)*		*Benzene (mg/l)*		*Gear Oil-WSF (mg/l)*	
		4.25	*8.50*	*4.35*	*8.70*	*1*	*2*
30	8.03±0.32	5.89±0.51*	2.86±0.36*	5.99±0.34*	3.91±0.44*	6.73±0.46*	4.68±0.24*

Values are mean±S D of 5 specimens. Significant response: *P <0.01.

Table 11.2: Moisture Content (gm per cent) in *G. divaricatum* Exposed to Xylene, Benzene and Gear Oil-WSF

Exposure Period (Day)	*Control*	*Xylene (mg/l)*		*Benzene (mg/l)*		*Gear Oil-WSF (mg/l)*	
		4.25	*8.50*	*4.35*	*8.70*	*1*	*2*
30	76.58±2.38	80.33±0.27**	83.13±1.30**	80.85±1.04**	81.35±0.88**	79.26±0.70*	79.85±0.73*

Values are mean±S D of 5 specimens. Significant response: *P<0.05, **P <0.01.

Table 11.3: Glycogen Content (mg/gm wet weight) in Tissues of *G. divaricatum* Exposed to Xylene

Exposure Period (Day)	*Tissues*	*Control*	*Xylene (mg/ml)*	
			4.25	*8.50*
4	HP	26.60±2.63	28.08±1.49	30.59±2.90
	G	29.45±2.40	18.48±2.37*	14.40±1.20*
	AM	29.49±1.66	18.73±1.87	13.41±0.93*
30	HP	24.63±2.84	13.78±0.51*	12.13±0.25*
	G	27.88±2.84	10.50±0.86*	9.15±0.44*
	AM	30.75±3.24	20.74±1.51*	18.17±2.55*

Values are mean±S D of 5 specimens. Significant response: *P <0.01. Abbreviations: AM, adductor muscle; G, gill; HP, hepatopancreas.

Table 11.4: Glycogen Content (mg/gm wet weight) in Tissues of *G. divaricatum* Exposed to Benzene

Exposure Period (Day)	*Tissues*	*Control*	*Benzene (mg/ml)*	
			4.3	*8.7*
4	HP	23.47±3.16	25.33±3.52	28.43±1.69*
	G	27.97±2.61	22.44±3.22*	15.41±1.32**
	AM	30.49±2.98	20.51±2.15**	16.151±1.52**
30	HP	24.57±3.97	14.38±0.48**	13.18±0.63**
	G	25.72±2.33	10.60±0.82**	9.81±0.36**
	AM	29.53±2.69	13.90±0.62**	17.23±2.51**

Values are mean±S D of 5 specimens. Significant response: *P<0.05, **P <0.01.

significantly. The reduction in gills and adductor muscles was more pronounced at 30 days exposure (Tables 11.3–11.5).

Lactic Acid

Lactic acid levels in all tissues of exposed clams found elevated after short- (96 hours) and long-term (30 days) exposure to xylene, benzene and WSF of gear oil (Tables 11.6–11.8).

Pyruvic Acid

Pyruvic acid content in clam tissues exposed to all test concentrations was found elevated after 96 hours of exposure

Table 11.5: Glycogen Content (mg/gm wet weight) in Tissues of *G. divaricatum* Exposed to Gear Oil-WSF

Exposure Period (Day)	*Tissues*	*Control*	*Gear Oil-WSF (mg/ml)*	
			1	*2*
4	HP	25.58±4.82	26.47±3.43	28.36±1.55
	G	28.44±2.39	24.26±3.58	17.48±2.50*
	AM	30.81±3.16	20.70±2.24*	18.541±2.46*
30	HP	23.38±1.44	12.94±2.16*	12.38±1.78*
	G	20.46±2.75	8.53±0.97*	7.44±0.72*
	AM	23.42±0.75	9.10±0.67*	15.26±2.36*

Values are mean±S D of 5 specimens. Significant response: *P <0.01.

Table 11.6: Lactic Acid Content (mg/gm wet weight) in Tissues of *G. divaricatum* Exposed to Xylene

Exposure Period (Day)	*Tissues*	*Control*	*Xylene (mg/ml)*	
			4.25	*8.50*
4	HP	4.53±1.90	5.41±1.73	6.39±2.03
	G	7.40±2.05	8.47±1.75	11.62±3.88
	AM	8.65±1.47	10.53±3.46	11.46±2.46
30	HP	3.32±1.24	5.76±1.53*	17.28±2.06*
	G	6.25±1.88	7.34±2.17	9.15±0.44**
	AM	7.45±2.10	8.47±1.53	11.43±2.56*

Values are mean±S D of 5 specimens. Significant response: *P<0.05, **P<0.01.

whereas an extended exposure of clams to xylene, benzene and gear oil WSF resulted into marked depletion in pyruvic acid content (Tables 11.9–11.11).

Proteins

There was an elevation in protein content in all tissues of clams exposed to both the concentrations of all the three pollutants for 96 hours whereas in case of clams exposed for 30 days to these pollutants the protein content in all the tissues was found depleted significantly (Tables 11.12–11.14).

Table 11.7: Lactic Acid Content (mg/gm wet weight) in Tissues of *G. divaricatum* Exposed to Benzene

Exposure Period (Day)	*Tissues*	*Control*	*Benzene (mg/ml)*	
			4.35	*8.7*
4	HP	5.34±1.70	6.58±2.12	6.36±1.89
	G	7.43±1.87	7.69±2.04	9.41±3.36
	AM	11.34±2.40	12.47±2.43	14.66±1.72*
30	HP	5.56±1.80	6.42±1.97	8.35±1.56*
	G	6.62±1.74	6.50±2.06	9.62±3.28
	AM	8.61±1.61	9.39±3.06	11.58±2.30*

Values are mean±S D of 5 specimens. Significant response: *P<0.05.

Table 11.8: Lactic Acid Content (mg/gm wet weight) in Tissues of *G. divaricatum* Exposed to Gear Oil-WSF

Exposure Period (Day)	*Tissues*	*Control*	*Gear Oil-WSF (mg/ml)*	
			1	*2*
4	HP	4.42±2.12	4.43±2.17	5.45±1.65
	G	6.52±1.95	7.51±2.03	8.56±1.54
	AM	7.58±1.88	8.44±1.60	10.35±3.59
30	HP	3.31±1.23	4.42±2.07	6.60±1.92
	G	7.22±2.03	8.23±1.58	12.56±2.26*
	AM	5.37±1.55	6.27±1.79	7.47±2.12

Values are mean±S D of 5 determinations. Significant response: *P <0.01.

The clams exposed to different concentrations of petroleum hydrocarbons and water soluble fraction (WSF) of oil showed signs of stress, probably as a result of negative energy balance (Tables 11.3–11.14). The results on alterations in biochemical metabolites of *Gafrarium divaricatum* in response to oil and petroleum hydrocarbon exposures are in agreement with the findings of other workers (Bayne *et al.*, 1982; Deshmukh, 1983; Axiak and George, 1987; Neff *et al.*, 1987; Moles *et al.*, 1987; Kostylev, 1988; Patel and Eapen, 1989). Rate of growth is a fundamental measure of physiological fitness and therefore growth has often been used as measure of environmental quality and pollution effects. However, in adult

Table 11.9: Pyruvic Acid Content (mg/gm wet weight) in Tissues of *G. divaricatum* Exposed to Xylene

Exposure Period (Day)	*Tissues*	*Control*	*Xylene (mg/ml)*	
			4.25	*8.50*
4	HP	8.41±2.46	11.47±1.37*	12.36±1.34*
	G	5.36±1.11	7.37±1.59*	9.35±2.54*
	AM	4.40±1.24	5.43±1.11	7.50±1.43**
30	HP	9.57±2.64	7.34±1.61	4.39±1.40**
	G	4.50±1.51	3.38±1.01	2.70±0.94
	AM	5.26±1.21	4.30±1.42	3.26±1.17*

Values are mean±S D of 5 specimens. Significant response: *P<0.05, **P<0.01.

Table 11.10: Pyruvic Acid Content (mg/gm wet weight) in Tissues of *G. divaricatum* Exposed to Benzene

Exposure Period (Day)	*Tissues*	*Control*	*Benzene (mg/ml)*	
			4.35	*8.7*
	HP	7.34±1.62	9.46±2.59	10.61±1.89*
	G	5.29±1.34	7.38±1.56	8.37±2.68
	AM	6.19±1.77	7.45±1.57	8.54±2.58
30	HP	8.37±2.69	7.41±1.75	4.39±1.39*
	G	5.21±1.28	4.30±1.53	4.29±1.29
	AM	6.31±1.73	5.32±1.15	4.07±1.42

Values are mean±S D of 5 specimens. Significant response: *P<0.05.

bivalves, it is often difficult to quantify and interpret the growth, especially in relation to pollution. Therefore, the condition index in bivalves can be utilized as an indication of whole body response of bivalve to the environment. Subsequently, there have been many laboratory studies in which condition index has been used to assess the toxic effects of various pollutants (Augenfeld *et al.*, 1980; Axiak and George, 1987; Eapen, 1987). The observed decline of condition index in the experimental clams may be due to the use of energy reserves demanded by stressful conditions resulting into the decrease in dry weight. Stekoll *et al.* (1980) have noted the decrease in condition index of *Maconma balthica* exposed to higher concentrations of oil.

Table 11.11: Pyruvic Acid Content (mg/gm wet weight) in Tissues of *G. divaricatum* Exposed to Gear Oil-WSF

Exposure Period (Day)	*Tissues*	*Control*	*Gear Oil-WSF (mg/ml)*	
			1	*2*
4	HP	8.26±2.61	10.24±3.22	11.52±2.48
	G	4.48±1.46	6.20±1.81	7.55±1.69*
	AM	5.21±1.18	6.17±1.86	6.34±1.75
30	HP	7.27±1.52	6.59±1.87	4.23±1.49*
	G	6.25±1.71	5.43±1.03	5.28±1.27
	AM	4.66±1.24	3.26±1.07	3.52±1.08

Values are mean±S D of 5 specimens. Significant response: *P <0.05.

Table 11.12: Protein Acid Content (mg/gm wet weight) in Tissues of *G. divaricatum* Exposed to Xylene

Exposure Period (Day)	*Tissues*	*Control*	*Xylene (mg/ml)*	
			4.25	*8.50*
4	HP	21.69±1.60	23.94±4.56	30.37±2.25**
	G	25.17±3.81	27.28±3.85	28.96±2.95
	AM	18.08±2.37	20.33±1.45	24.96±4.70*
30	HP	18.68±3.78	10.48±1.82**	10.35±1.42**
	G	21.51±5.08	12.21±4.87*	11.34±1.52**
	AM	17.85±3.42	9.60±1.14**	10.59±1.99**

Values are mean±S D of 5 specimens. Significant response: *P<0.05, **P<0.01.

Axiak and George (1987) have also reported decrease in condition index of bivalve, *Venus verrucosa* when exposed to petroleum hydrocarbons. Similarly, Patel and Eapen (1989) observed depletion in condition index of clam, *Anadara granosa* under naphthalene exposure.

Lipophilic compounds such as oil and petroleum hydrocarbons are readily taken up into tissues of bivalves and concentrate to the levels greatly above those in surrounding water. This accumulation of oil and related hydrocarbons in bivalve molluscs has been widely reported (Clement *et al.*, 1980; Awad, 1987; Axiak *et al.*, 1988; Lin-Qin *et al.*, 1990; Lin-Qin and Jia-Xiaoping, 1991; Readman *et al.*,

Table 11.13: Protein Content (mg/gm wet weight) in Tissues of *G. divaricatum* Exposed to Benzene

Exposure Period (Day)	*Tissues*	*Control*	*Benzene (mg/ml)*	
			4.35	*8.7*
4	HP	20.10±2.31	24.12±4.76	27.08±2.60*
	G	24.34±2.12	26.28±1.80	28.24±2.61*
	AM	14.34±2.42	15.33±2.64	18.30±3.09
30	HP	16.25±1.82	15.88±1.80	11.32±1.30**
	G	22.40±3.14	12.72±2.45*	12.63±1.09**
	AM	14.75±2.81	13.11±2.50	12.33±2.33

Values are mean±S D of 5 specimens. Significant response: *P <0.05, **P<0.01.

Table 11.14: Protein Content (mg/gm wet weight) in Tissues of *G. divaricatum* Exposed to Gear Oil-WSF

Exposure Period (Day)	*Tissues*	*Control*	*Gear Oil-WSF (mg/ml)*	
			1	*2*
4	HP	18.25±2.64	20.38±1.50	24.59±2.60**
	G	23.43±2.13	24.11±2.08	24.42±3.49
	AM	14.29±2.46	16.10±2.24	17.20±2.75
30	HP	16.05±6.87	15.30±2.02	12.51±1.47
	G	18.72±5.71	9.33±2.72*	11.95±3.87
	AM	13.39±1.72	9.00±1.62	9.30±3.32*

Values are mean±S D of 5 specimens. Significant response: *P <0.05, **P<0.01.

1992). Accumulation of petroleum hydrocarbons might be disturbing the general osmotic balance in the tissues of *Gafrarium divaricatum* which might be leading to water uptake thereby resulting in increased moisture content of the tissues. Stekoll *et al.* (1980) and Patel and Eapen (1989) have reported increased moisture content in the tissue of marine bivalves as a result of oil and petroleum hydrocarbons exposure.

There is ample evidence to show that the current view of mammalian bioenergetics also applies to marine bivalves. Moreover, the occurence of most of the enzymes which are intermediate of glycolysis and Krebs cycle have been reported in marine bivalves

(Bennet and Nakada, 1968; Engel and Neat, 1970; de Zwaan, 1971a,b; Usuki and Okamura, 1986). It is well known that most of the intertidal bivalves are facultative anaerobics *i.e.*, they can live either aerobically or anaerobically but prefer to use oxygen when available as it allows much more economical use of fuel molecules. However, the condition of environmental or organism level anaerobiasis is very common in intertidal bivalves. The maintenance of high reserves of glycogen is one of the adaptations in intertidal bivalves against anoxic conditions. It is generally accepted that under anaerobic conditions carbohydrates are the sole or main source of energy. The high power output mode during anaerobiasis in marine invertebrates including the intertidal bivalves is linked to classical glycolysis. Furthermore, the increase in anoxia tolerance with larval development appears to go along with shift from functional anaerobiasis to environmental anaerobiasis (Widdows *et al.*, 1989). Therefore, the observed depletion of glycogen in the tissues of hydrocarbon and oil exposed clams might be due to mobilization of glycogen reserves to meet the energy demand. The reduction of glycogen reserves in bivalves under anaerobic conditions has been reported by de Zwaan and Zande (1971) and Badman and Chin (1973). Similarly, utilization of glycogen reserves in bivalves exposed to various pollutants is reported (Dunning and Major, 1974: Theede, 1985, Eapen, 1987; Roy, 1994). Such depletion of glycogen reserves has also been observed in marine crustaceans under pollutant stress (Deshmukh, 1983; Kulkarni, 1983; Reddy and Bhagyalakshmi, 1994). It has been suggested that the histo-pathological damage to gill of *Gafrarium divaricatum* leads to hypoxic condition (Agwuocha *et al.*, 2009).

The reaction step of EMP pathway in which glucose-6-phosphate is converted to pyruvate has been recommended for *Mytilus edulis* both in terms of enzyme activities (Livingstone and de Zwaan, 1983) and intermediates (Ebberink and de Zwaan, 1980). During environmental anoxia, six end products of pyruvate metabolism have been identified in molluscs. Lactate is one of the six end products of anaerobic pyruvate metabolism in marine bivalves. Furthermore, during environmental anoxia, marine bivalves accumulate a variety of organic acids in addition to small amounts of CO_2, opines and lactate. The end product lactate is formed in the cytoplasm. Therefore, the observed changes in lactic acid and pyruvic acid content in tissues of experimental clams suggest disturbed carbohydrate metabolism. The upward trend of lactic acid with respect to pyruvic acid in the tissues of experimental clams also suggests severe respiratory stress leading to anaerobic metabolism. Similar trend of lactic acid and pyruvic acid in tissues of organisms including clams exposed to different pollutants has been reported (Dange and Masurekar, 1982; Eapen, 1989; Ferrando and Andreau-Moliner, 1991; Roy, 1994; Reddy and Bhagyalakshmi, 1994).

Proteins form one of the major fuels in marine clams. The marked depletion in protein content in clam tissues exposed to xylene, benzene and gear oil WSF is symptomatic to severe stress caused by these pollutants which results in increased proteolysis. Such depletion in protein content in the tissues of the clam, *Macoma balthica* exposed to oil has been reported (Stekoll *et al.*, 1980). Furthermore, it is possible that proteins and amino acids from tissues are mobilized under stressful condition of pollutants and released into circulatory system. Eapen and Patel (1989) have reported increased protein levels in serum of clam, *Anadara granosa* exposed to naphthalene. Similarly, Kulkarni (1983) has reported elevation of protein levels in the serum of the crab, *Scylla serrata* exposed to naphthalene. The observed increase in protein of the tissues of experimental clams after 96 hours of exposure in the present investigation also suggests the mobilization of proteins which might have released into the blood of clam during chronic exposure period.

References

Agwuocha, S., Kulkarni, B.G. and Pandey, A.K., 2009. Acute toxicity of xylene, benzene and gear oil–WSF to the intertidal clam, *Garfarium divaricatum* (Gmelin). *J. Appl. Biosci.*, 35: 146–149.

Akbarali, H.B. and Truman, R.R., 1985. Effect of environmental stress on marine bivalve molluscs. In: *Advances in Marine Biology*, (Eds.) J.H.S. Blaxter, Y. Russel and M. Younge. 22: 102–183.

Al–Akel, A.S., Shamsi, M.J.K., Al–Katim, H.F., Chowdhary, M.A. and Ahmed, Z., 1988. Effect of cadmium on cichlid fish, *Oreochromis niloticus*: behaviour and physiological responses. *J. Univ. Kuwait (Sci.)*, 15: 341–346.

AOAC, 2000. *Official Methods of Analysis.* 17th edn. (Ed.) William Hovwitz. Association of Official Analytical Chemists. Washington, D.C.

Augenfield, J.M., Anderson, J.W., Woodruff, D.L. and Webester H.J.L., 1980. Effects of Prudhoe Bay crude oil contaminated sediments on *Protothaca staminea* (Mollusca: Pelecypoda). *Mar. Environ. Res.*, 4: 135–143.

Awad, H., 1987. Aromatic hydrocarbons in some organisms of coastal Red Sea. *Indian J. Mar. Sci.*, 16: 114–116.

Axiak, V. and George, J.J., 1987. Bioenergetic responses of the marine bivalve, *Venus verrucosa* on long–term exposure to petroleum hydrocarbons. *Mar. Environ. Res.*, 23: 33–47.

Axiak, V., George, J.J. and Moore, M.N., 1988. Petroleum hydrocarbons in the marine bivalve, *Venus verrucosa:* accumulation and cellular responses. *Mar. Biol.*, 97: 225–230.

Badman, B.D. and Chin, S.N., 1973. Metabolic responses of the freshwater bivalve, *Pleurobema coccineum* (Conard) to anaerobic conditions. *Comp. Biochem. Physiol.*, 44B: 27–32.

Barker, S.B. and Summerson, W.H., 1941. Colourimetric determination of lactic acid in biological materials. *J. Biol. Chem.*, 138: 535–554.

Bayne, B.L., Widdows, J., Moore, M.N., Salkeld, P., Worrall, C.M. and Donkin, P., 1982. Some ecological consequences of the physiological and biochemical effects of petroleum compounds on marine molluscs. *Phil. Trans. R. Soc. (London)*, 297B: 219–239.

Bennet, R. and Nakada, H.I., 1968. Comparative carbohydrate metabolism of marine molluscs. *Bull. Comp. Biochem. Physiol.*, 24: 787–797.

Biswas, R., 1986. Toxic erffects of an organochlorine insecticide, Thiodon, on marine crab, *Scylla serrata* (Forskal). *Ph.D. Thesis,* University of Bombay, Bombay.

Chaykin, S., 1966. *Biochemistry: Laboratory Techniques*. Wiley-Eastern Pvt. Ltd., New Delhi. 196 p.

Clement, L.E., Stekoll, M.S. and Shaw, D.G., 1980. Accumulation, fractionation and release of oil by the intertidal clam, *Macoma balthica. Mar. Biol.*, 57: 41–50.

Dange, A.D. and Masurekar, V.B., 1982. Naphthalene induced changes in carbohydrate metabolism in *Sarotherodon mossambicus* Peters (Pisces: Cichlidae). *Hydrobiologia*, 94: 163–177.

Deshmukh, N., 1983. Some aspects of toxicity of petroleum hydrocarbons to the speckled prawn, *Metapeneus monoceros* (Fabricus). *Ph.D. Thesis*, University of Bombay, Bombay.

Deshmukh, R.S. 1972. Some aspects of biology of *Meretrix meritrix*. *Ph.D. Thesis*, Marathwada University, Aurangabad.

de Zwaan, A., 1971a. The aerobic carbohydrate metabolism in the sea mussel, *Mytilus edulis* (L.). *Ph.D. Thesis*. University of Utrecht, The Netherlands.

de Zwaan, A., 1971b. Pyruvate kinase in muscle extracts of the sea mussel, *Mytilus edulis* (L.). *Comp. Biochem. Physiol.*, 92B: 7–14.

de Zwaan, A. and van der Zande, T., 1971. The utilization of glycogen and accumulation of some intermediates during anaerobiosis in *Mytilus edulis* (L.). *Comp. Biochem. Physiol.*, 93B: 47–54.

de Zwaan, A. and Mathieu, M., 1992. Cellular biochemistry and endocrinology. In: *The Mussel, Mytillus: Ecology, Physiology, Genetics and* Culture: *Development in Aquaculture and Fisheries Science* (Ed.) E. Gosling. Elsevier Pub., Amsterdam, 25: 223–300.

Dunning, A. and Major, C.N., 1974. The effect of cold seawater extracts of oil fractions upon the blue mussel, *Mytilus edulis*. In: *Pollution and Physiology of Marine Organisms* (Eds.) F.J. Vernberg and W.B. Vernberg. Academic Press, New York. pp. 349–366.

Durairaj, S. and Selvarajan, V.R., 1992. Influence of Quinolphos, an organophosphorus pesticide, on biochemical constituents of the tissues of fish, *Oreochromis mossambicus*. *J. Environ. Biol.*, 13: 181–185.

Eapen, T., 1987. Physiological and biochemical responses of an arcid clam, *Anadara granosa* exposed to naphthalene. *Ph.D. Thesis,* University of Bombay, Bombay.

Ebberink, R. and de Zwaan, A., 1980. Control of glycolysis in the posterior adductor muscle of the sea mussel, *Mytillus edulis*. *J. Comp. Physiol.*, 137: 165–171.

Engel, H.R. and Neat, M.J.C., 1970. Glycolytic and gluconeogenic enzymes in the Quahog, *Mercenaria mercenaria*. *Bull. Comp. Biochem. Physiol.*, 37: 397–403.

Ferrando, M.D. and Andreu–Moliner, E., 1991. Changes in selected biochemical parameters in the brain of fish, *Anguilla anguilla* (L.) exposed to Lindane. *Bull. Environ. Cotam. Toxicol.*, 47: 459–464.

Friedmann, T.K. and Haugen, G.E., 1943. Pyruvic acid. II. The determination of ketoacids in blood and urine. *J. Biol. Chem.*, 47: 415–442.

Khan, R.A. and Nag, K., 1993. Estimation of hemosiderosis in sea birds and fish exposed to petroleum. *Bull. Environ. Contam. Toxicol.*, 50: 125–131.

Kostylev, Eh.F., 1988. Changes in metabolic rates in bivalves under conditions of water pollution with detergents and oil–products. In: *Aquatic Toxicology and Optimization of Bio–production Management Processes in Aquaculture* (Ed.) S.A. Patin. C. Kueh, Moscow. pp. 83–93.

Krishna, P., Ventatarani, L., Ravi, C. and Indira, K., 1987. Metabolic consequences of methylparation exposure in the bivalve, *Lamellidens marginalis. Bull. Environ. Contam. Toxiciol.*, 38: 509–514.

Kulkarni. B.G., 1983. Toxicity studies: effects of some petroleum hydrocarbons on some marine animals. *Ph.D. Thesis*, University of Bombay, Bombay.

Kulkarni. B.G., 1990a. Effect of copper exposure on tissue enzyme activities in the intertidal clam, *Gafrarium divaricatum* (Gmelin). *Nat. Acad. Sci. Letter*, 13: 4–6.

Kulkarni, B.G., 1990b. ATPase activity in the gills and hepatopancreas of the intertidal clam, *Meretrix casta* var. *ovum* (Hanley) expose to naphthalene. *J. Environ. Biol.*, 11: 275–278.

Kulkarni, B.G. and Kulkarni, R.G. 1987. The effect of naphthalene exposure on tissue aminotranferase activities in the clam, *Meretrix casta* var. *ovum* (Hanley). *Poll. Res.*, 7: 9–12.

Lin-Qin and Xin-Xiaoping, 1991. Petroleum hydrocarbons in marine organism from Honghai Bay, north-eastern South China. *Seq. Mar. Sci. Bull.* (Haiyang Tongbao), 10(1): 33–38.

Lin-Qin, Xin-Xiaoping and Lu-Xiaoyu, 1990. Petroleum hydrocarbons in marine animals from the Pearl River estuary. *Mar. Sci.* (Haiyang Kexue), 3: 34–38.

Livingstone, D.L. and de Zwaan, A. 1983. Carbohydratye metabolism in gastropods. In: *The Mollousca. Vol. 1* (Ed.) K.M. Wilbur. Academic Press, New York. pp. 177–242.

Lowry, O.H., Rosenbrough, N.J., Farr, A.L. and Randall, R.J., 1961. Protein measurement with folin–phenol reagent. *J. Biol. Chem.*, 193: 265–275.

Mane, U.H. and Nagabhushanam, R., 1975. Body distribution and seasonal changes in the biochemical composition of the estuarine mussel, *Mytilus viridis* at Ratnagiri. *Riv. J. Hydrobiol.*, 14: 163–175.

Mane, U.H., 1974. Adaptations of the estuarine clam, *Katelysia opima* to the salinity fluctuations. *Rev. Biol.*, 6: 73–107.

Mane, U.H., 1975. Study on the rate of water transport of the clam, *Katelysia opima* in relation to environmental conditions. *Hydrobiologia.* 47 : 439–451.

Mane, U.H., Akarte, S.R. and Kulkami, D.A., 1986. Acute toxicity of fenthion to freshwater lamellibranch mollusc, *Indonaia caeruleus* (Prasad, 1918) from Godavari river at Paithan: A biochemical approach. *Bull. Environ. Contain. Toxicol.*, 37: 622–628.

Moles, A., Babcock, M.M. and Rice, S.D., 1987. Effects of oil exposure on pink salmon, *Oncorhynchus gorbuscha* alevins in a simulated intertidal environment. *Mar. Environ. Res.*, 21: 49–58.

Nagabhushanam, R. and Mane, U.H., 1978. Seasonal variation in the biochemical composition of *Mytilus viridis* at Ratnagiri on the west coast of-India. *Hydrociologia*, 571: 69–72.

Neff, J.M., Hillman, R.E., Carr, R.S., Buhl, R.L. and Laney, J.I., 1987. Histopathological and biochemical responses in arctic marine bivalve molluscs exposed experimentally to spilled oil. *Arctic*, (Suppl. 1): 220–229.

Patel, B. and Eapen, T., 1989. Biochemical evaluation of naphthalene intoxication in tropical arcid clam, *Anadara granosa*. *Mar. Biol.*, 103: 203–209.

Readman, J.W., Fowler, S.W., Villeneuve, J.P., Cattini, C., Oregioni, B. and Mee, L.D., 1992. Oil and combustion product contamination of the Golf marine environment following the war. *Nature*, 358: 662–665.

Reddy, P.S. and Bhagyalakshmi, A., 1994. Changes in oxidative metabolism in selected tissues of crab, *Scylla serrata* in response to cadmium toxicity. *Ecotoxicol. Environ. Saf.*, 29: 255–264.

Rodriguez, G.A., 1981. Osmoregulation and total serum protein in two species of Penaidean shrimps from Pacific coast of Mexico. *J. Crust. Biol.*, 1: 392–400.

Roy, S.S., 1994. Effects of copper on ecophysiological aspects of the clam, *Gafrarium divaricatum. M.Sc. Thesis*, University of Bombay, Bombay.

Sabo, D.J. and Stegeman, J.J., 1977. Some metabolic effects of petroleum hydrocarbons in marine fish. In: *Physiological Responses of Marine Biota to Pollutants*, (Eds.) F.J. Vemberg, A. Calabrese, P.P. Thurberg and V.B. Vemberg. Academic Press, New York, pp. 279–287.

Schoffeniels, E., 1976. Biochemical approaches to osmoregulatory processes in Crustacea. In: *Perspectives in Experimental Biology, Vol. 1.* (Ed.) P.S. Davis. Pergaman Press, Oxford, pp. 107–124.

Seifter, S., Dayton, S., Noovic, B. and Muntwyler, I., 1950. Estimation of glycogen with anthrone reagent. *Arch. Biochem.*, 25: 191–200.

Shankara, C.R., Babu, N.SA., Indira, K. and Rajendra, W., 1989. Metabolic response to methyl parathion toxicity in bivalve, *Lamellidens marginalis. Curr. Sci.*, 58: 767.

Shreedevi, P., Sivaramakrishna, B., Suresh, A. and Ramakrishnaiah, K. 1992. Effect of nickel on some aspects of protein metabolism in the gill and kidney of the freshwater fish, *Cyprinus carpio* L. *Environ. Pollut.*, 77: 59–63.

Stekoll, M.S., Clement, L.S. and Shaw, D.G., 1980. Sublethal effects of chronic oil exposure on the intertidal clam, *Macoma bolthica. Mar. Biol.*, 57: 51–60.

Theede, H., 1985. Some problems of pollution with heavy metals under the environmental conditions of the Baltic Sea. *Symposium Biol. Hung.*, 29: 223–238.

Usuki, I. and Okamura, N., 1986. Glycolytic in oyster gill. *Sci. Rep. Tohuku Univ.* Fourth Ser., 22: 225–232.

Venkatrama, P. and Radhakrishnaiah, K., 1987 Lethal and sublethal effects of copper on the protein metabolism of freshwater teleost, *Labeo rohita. Trends in Life Sci.*, 2(2): 82–86.

Widdows, J., Bayne, B.L., Livingstone, D.R., Newell, R.I.E. and Donkinn, P., 1979. Physiological and biochemical responses of bivalve molluscs to exposure to air. *Comp. Biochem. Physiol.*, 62A: 301–308.

Chapter 12

Prevalence of Invasive Slug *Laevicaulis alte* (Ferrussac, 1821) (Systellommataphora : Veronicellidae) in Bangalore Region

M. Jayashankar[1], A.S. Jeevan Chakravarthy[2], R. Shreyas[2], L.S. Vinesh[3], N. Manjunath[1], P. Ravi[1] and J. Manjunath[1]*

[1]Department of Zoology, Bangalore University, Bangalore – 560 056, Karnataka, India

[2]St. Joseph's College (Autonomous), Bangalore – 560 027, Karnataka, India

[3]Division of Entomology and Nematology, Indian Institute of Horticultural Research, Hesaraghatta, Bangalore – 560 089, Karnataka, India

ABSTRACT

Survey carried out during 2008-10 to record the occurrence of the invasive slug *Laevicaulis alte* revealed its prevalence in 58 (51 per cent) of 114 villages in Bangalore region comprising of Bangalore Rural and Bangalore Urban districts. Highest percentage of *L. alte* occurrence was recorded in Bangalore

* Corresponding author: E-mail: jay_zoology@rediffmail.com

South taluk (78.57 per cent) of Bangalore Urban District and the lowest recorded was in Nelamangala taluk (10 per cent) of Bangalore Rural District. Based on the district wise distribution *L. alte* incidence was noticed in 40 (70 per cent) villages in Bangalore Urban district and 18(32 per cent) villages in Bangalore Rural district. Further spread of the *L. alte* in the region needs to be monitored and reported, as controlling or eradicating invasive species once established is often extremely difficult and costly, while prevention and early intervention have more chances to be successful and cost-effective.

Keywords: *Invasive, Laevicaulis alte, Bangalore.*

Introduction

Climate change and biological invasion are two of the most important ecological issues facing the world today (Kiritani, 2011). Invasive alien species are serious global threats to biological diversity and may disrupt ecological systems, homogenization of biota and extinction (SCOPE, 1996; Tobin, 2007) by colonizing in the introduced area. Such colonization involve four steps *viz.*, introduction, establishment, spread and naturalization (Kiritani, 1998). Invasive malacofauna are reported to devour native plants, modifying the environment, probably also outcompete native gastropods (Leslie *et al.*, 2008). The present article deals with one such invasive slug, *Laevicaulis alte* (Ferrussac) (Gastropoda : Systellommataphora) reported to be pestiferous in different parts of India (Raut and Ghose, 1984; Mahobe and Pachori, 1993; Patnaik and Tripathy, 2001; Kumari and Sheela Thakur, 2005; Kalidas *et al.*, 2006). Evident from the literature perusal the potential costs of such pestiferous slug on agriculture and environment seems potentially immense. Prerequisite to answering the question on pest management is the need for a thorough knowledge of the distribution and level of abundance of the species in the region where the invasive pest has naturalized. Given the fact that reports on such alien species from the region are scarce, also their impact on local environment lacking, the survey on occurrence of invasive slug *L. alte* was undertaken.

Materials and Methods

The Study Organism

Systematic Position of *Laevicaulis alte* (Ferrussac, 1821)

Kingdom	:	Animalia
Phylum	:	Mollusca
Class	:	Gastropoda
Subclass	:	Pulmonata
Order	:	Systellommataphora
Family	:	Veronicellidae
Genus:	:	*Laevicaulis*
Species	:	*alte*

Survey Region

The study was conducted in the Bangalore region (12°58'N, 77°34'E) (Figure 12.6) of Karnataka State. Bangalore region comprises of two districts namely Bangalore Urban and Bangalore Rural. Bangalore Urban district has four taluks *viz.*, Bangalore North, Bangalore South, Bangalore East and Anekal. Bangalore Rural taluk comprises of Hoskote, Devanahalli, Doddaballapura and Nelamangala.

Survey Period and Methodology

Field observations were conducted during the monsoon (July-September) and the period just after the monsoon (October-January). Field surveys were undertaken employing visual search method and questionnaire prepared consulting earlier surveys (Asamoah, 1999, USDA–APHIS, 2005) to gather information about the prevalence of the pestiferous slugs. The interviewees were provided with informational photographs and contact numbers to communicate on the status of the slugs.

The information gathered in trace back investigations was used to identify the potential pathways of slug introduction. In the field, exact location of each sample collection site was recorded with a handheld Garmin e-trex, 12 Channel global positioning system (GPS) device to determine coordinates for each study point and of district and taluk map (Land records of 2001) of Karnataka State Remote

Figure 12.1: *L. alte*-Adult (Dorsal View)

Figure 12.2: Mating Pair (Ventral view exposing the copulatory tube)

Figure 12.3: Mating Pair (Dorsal view)

Figure 12.4: Egg Mass of *L. alte*

Figure 12.5: *L. alte* Adult Attached to Nursery Plant Packet

Figure 12.6: Bangalore Region Showing Two Districts of Urban and Rural each with Four Taluks

Sensing Applications Centre (KSRAC), Government of Karnataka. All points were within the administrative boundary of the locations and the data thus obtained was analyzed and projected using Arc GIS Software (8.1 version).

Assistance regarding confirmation on the species identity was sought from Dr.N.A.Aravind, Malacologist, Ashoka Trust for Research in Ecology and Environment (ATREE), Bangalore.

Results and Discussion

According to the results analyzed during the present survey in Bangalore region, *L. alte* was detected in 58 (51 per cent) of 114 villages in Bangalore region (Figures 12.7 and 12.8). Highest percentage of *L. alte* occurrence was recorded in Bangalore South taluk (78.57 per cent) and the lowest recorded in Nelamangala taluk (10 per cent).The district wise distribution was found to be 40 (70 per cent) villages in Bangalore Urban district and 18(32 per cent) villages in Bangalore Rural district (Figure 12.8). Survey of

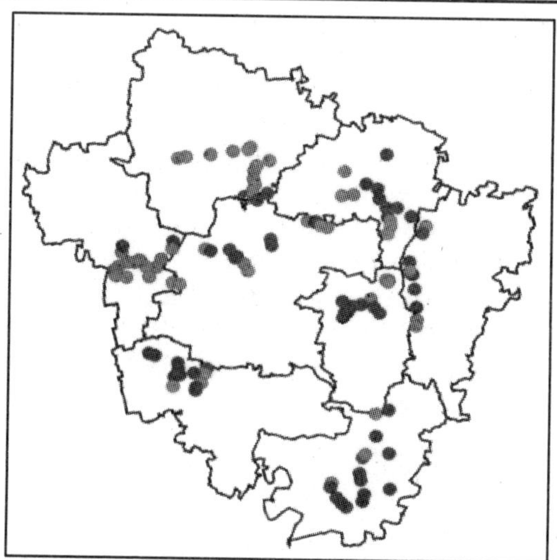

Figure 12.7: The Global Positioning System (GPS) Coordinates of Villages of the Detection Survey in Bangalore Region, Red Spots Indicating Presence of the Slug and Green Indicating Absence

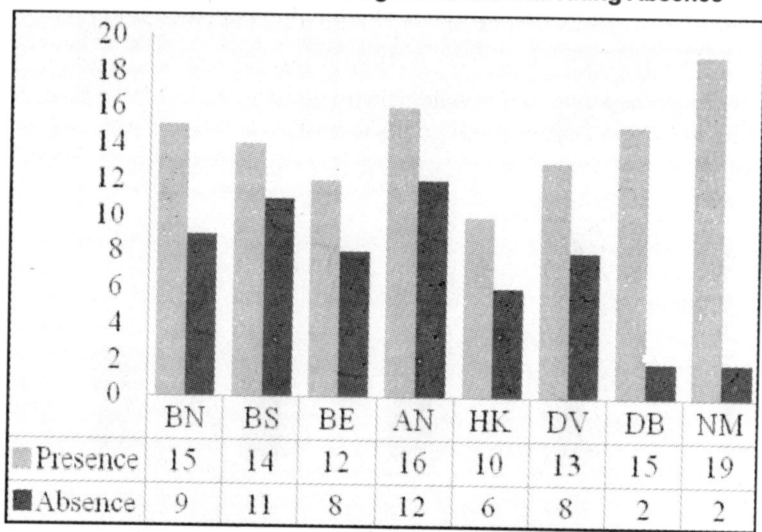

	BN	BS	BE	AN	HK	DV	DB	NM
Presence	15	14	12	16	10	13	15	19
Absence	9	11	8	12	6	8	2	2

Figure 12.8: Prevalence of *L. alte* in Eight Taluks of Bangalore Region

Legends (Study sites) - BN: Bangalore North; BS: Bangalore South; BE: Bangalore East; AN: Anekal; HK: Hoskote; DV: Devanahalli; DB: Doddaballapur and NM: Nelamangala.

gastropods by Kaur and Chabbra (2009) in commercial plant nurseries at Ludhiana revealed two biotypes of *Filicaulis alte*, brown slug with black spots and creamish yellow with black anterior and posterior ends, wrinkled body with irregular polygonal network of brown lines on its dorsal surface. Taxonomic identity of the similar looking invasive slugs still remains a herculean task compelling researchers to be cautious in differentiating closely related species. Careful observations and thorough literature review are necessary in knowing species that are re-described and are new records. Similar to observations made by Panigrahi (2000a) different colour variants of *L.alte* adults ranging from studier black to thin brown slugs were observed during the present survey.

Co-occurrence of four different species of malacofauna *viz., Macrochlamys indica, Cryptozona bistrialis* (endemic snails); *Mariella dussumeri* (endemic slug) and *Achatina fulica* (Invasive snail) along with *L.alte* during the study period is represented in a bar graph (Figure 12.9). During the present survey it was found that, *L.alte-A.fulica* combination co-occured in 46 per cent of villages followed by-*L.alte- C. bistrialis* (31 per cent), *L.alte- M. indica* (20 per cent) and *L.alte - M. dussumeri* in 19 per cent of the total 114 villages surveyed.

Human aided agency was found during the study as main means of dispersal for the slug in the region. Collection and dispersal linked to agri-horticultural goods was often given as a mechanism

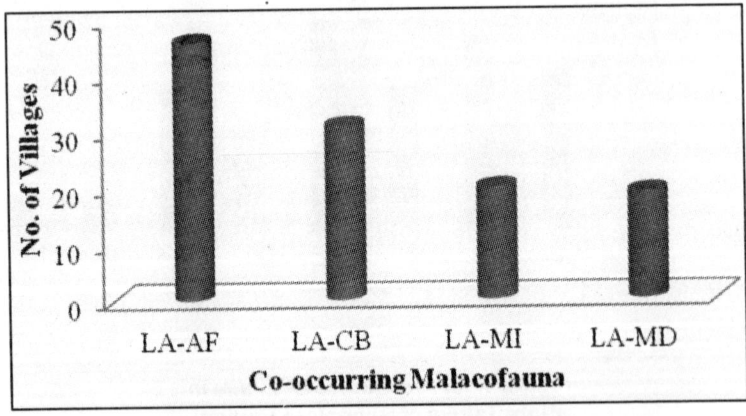

Figure 12.9: Co-occurring Malacofauna in the Study Region

Legends: LA: *Laevicaulis alte*; AF: *Achatina fulica*; CB: *Cryptozona bistrialis*; MI: *Macrochlamys indica* and MD: *Mariella dussumeri.*

for dispersal by respondents during the interviews. But, its impact as pest seems less known and still unaware of in the region. A range of natural enemies have been reported by different workers, frog (Muthumani *et al.*, 1992); mites (Raut, 1996) and glow worm (Panigrahi, 2000b) but no instance of natural enemies parasitizing or predating the slug was noticed during the present field survey. According to Barker (2002), the emergence of gastropods as important crop pests in temperate and tropical regions is attributed to cultivation of new crops, intensification of agricultural production systems, the spread through human trade and adaptation of the species to modified environments. Furthermore, in some crops, pestiferous gastropods have gained significance with decline in the importance of other pest groups, such as insects, for which effective control strategies have been developed.

Study carried out by Kaur and Chhabra (2009) found a positive correlation between population of surface active gastropods including *F.alte* with air and soil temperature and relative humidity (RH) minimum and a negative correlation with RH maximum and soil temperature, indicating the adaptation of the invasive slugs to Indian conditions. Kaur and Mehta (2010) have reported *Filicaulis alte* as an emerging pest in Punjab. Continued movement of agri-horticultural goods and extreme climatic disturbances as witnessed world over may perhaps create more opportunities for invasive pest colonization and establishment. Thus, further spread of the *L.alte* needs to be monitored and reported, as controlling or eradicating invasive species once established is often extremely difficult and costly, while prevention and early intervention have more chances to be successful and cost-effective.

Acknowledgements

The first author expresses his acknowledgements to Dr. M.S. Reddy, Associate Professor and Department of Zoology, Bangalore University for providing necessary facilities. The help provided by Officials at KSRAC is also acknowledged.

References

Asamoah, S.A., 1999. Ecology and Status of the giant African snails in the Bia Biosphere Reserve in Ghana. Final Report (September, 1999): Ecological Studies on the Giant African Snails, Submitted to: The Division of Ecological Sciences, UNESCO (MAB) Young Scientist Research Award Scheme, *Paris Cedex 15 France*, p. 1–31.

Barker, G.M., 2002. Preface. In: *Mollusc as Crop Pests*, (Ed.) G.M. Barker. CABI Publishing, Wallingford, p. 55–114.

Kalidas, P., Rao, C.V., Nasim Ali and Babu, M.K., 2006. New pest incidence on oil palm seedlings in India: A study of black slug (*Laevicaulis alte*). *Planter*, **82** (960): 181–186.

Kaur, H. and Mehta, R., 2010. Survival of Slug, *Filicaulis alte*, exposed to pesticides commonly sprayed on vegetables. *Pesticide Research Journal*, **22(2)**: 151–155.

Kaur, H. and Chabbra, E., 2009. Impact of environmental factors on the population of gastropods in commercial ornamental plant nurseries at Ludhiana, Punjab. *Journal of Research, Punjab Agricultural University*, **46(1/2)**: 60–67.

Kiritani, K., 1998. Exotic insects in Japan. *Entomological Science*, **1**: 291–298.

Kirtani, K., 2011. Impacts of Global warming on *Nezara viridula* and its native congeneric species. *Journal of Asia-Pacific Entomology*, **14**: 221–226.

Kumari, P. and Thakur, Sheela, 2005. Population dynamics of slug pest, *Laevicaulis alte* (Ferussac) in North Bihar. *Journal of Applied Zoological Researches*, **16 (2)**: 177–179.

Leslie, A.R., Mark, F.D. and Robert, O.H.J.R. 2008. Invasive species impact: asymmetric interactions between invasive and endemic freshwater snails. *Journal of North American Benthological Society*. **27(3)**: 509–520.

Mahobe, J. and Pachori, R., 1993. Biology and control of land slug, *Laevicaulis alte* (Ferrustac). *Orissa Journal of Agricultural Research*, **6(1/2)**: 14–18.

Muthumani, K., Jeyakumar, E., Mani, A. and Mathavan, S., 1992. Biological control of veronicellid slug *Laevicaulis alte* (Systellommatophora: Veronicellidae) using frog as a predator. *Environment and Ecology*, **10(3)**: 649–652

Panigrahi, A., 2000a. Colour variants in the garden slug *Laevicaulis alte* (Ferussac). *Environment and Ecology*, **18(1)**: 143–145.

Panigrahi, A., 2000b. *Lampyris* larva (glowworm), the effective predator of the pestiferous slug *Laevicaulis alte* (Ferussac). *Environment and Ecology*, **18(4)**: 1011–1013.

Patnaik, N.C. and Tripathy, M.K., 2001. Incidence and management of the garden slug, *Laevicaulis alte* Ferrusac in Orissa. *Journal of Applied Zoological Researches*, **12(1)**: 57–58.

Raut, S.K., 1996. Factors determining the effectiveness of the mites *Fuscuropoda marginata* in the control of the slug pests *Laevicaulis alte*. Slug & snail pests in agriculture. Proceedings of a Symposium, University of Kent, Canterbury, UK, 24-26 September, p. 247-254.

Raut, S.K. and Ghose, K.C., 1984. Pestiferous Land Snails of India, *Z.S.I. Technical Monograph* **11**: 151. Zoological Survey of India, Calcutta.

SCOPE, 1996. UN/Norway Conference on Alien species. *Newsletter*. **50**: 6–7.

Tobin, P.C., 2007. Space-time patterns during the establishment of a non-indigenous species. *Population Ecology*, **49**: 257–263.

USDA–APHIS, 2005. New Pest Response Guidelines. Giant African Snails: Snail Pests in the Family Achatinidae. USDA – APHIS – PPQ – Emergency and Domestic Programs – Emergency Planning, Riverdale, Maryland. *http://www.aphis.usda.gov/import_export /plants /manuals/index.shtml.*

Chapter 13

Effects of Sublethal Doramectin Administration on Certain Biochemical Parameters in Muscles of *Channa punctatus*

Namita Verma*

Pest and Parasite Research Laboratory, Department of Zoology, Bareilly College, Bareilly – 243 005, U.P., India

ABSTRACT

Channa punctatus (Pisces-Teleostomi) is distributed in the fresh water habitat of different regions of India including Uttar Pradesh. This species is an important member of fresh water ecosystem. Physiological functions including biochemical activity of fishes varies with the changing conditions in the environment and responds immediately to changes in the water quality because of it's intimate contact thorough gill surface. This edible species of economic importance demands thorough toxicological screening under the exposure of doramectin, a pesticide of new generation. Fresh water ecosystem bears the risk of contamination by pesticide residues during monsoon and other seasons. In the present work the activity levels of

* E-mail: namitaverma38@yahoo.com

acid phosphates, alkaline phosphates and amylase were examined in muscles of teleost fish *Channa punctatus* which had been exposed to sublethal concentrations (0.008 ppm, 0.012ppm, 0.016ppm) of doramectin for a period of 30 days. Significant reduction in the activity of acid phosphates and amylase activity and increase in metabolism of alkaline phosphates in *Channa punctatus* was recorded during present study. Alteration in biochemical response is suggestive of environmental stress which is indicator of disturbed water system.

Keywords: *Doramectin, Biochemical parameters, Channa punctatus.*

Introduction

Out of all human needs, food is the most essential need. As the human population increased, agriculture has also increased, but by increase in agriculture, main barrier was the destruction of crops by harmful insects and pests. For protecting the crops, some pesticides are being used to kill and repel the pests. The use of these pesticides is creating very serious health problem for human population, as they make undesirable changes in our land and water resources, making them unfit for use. Only a small fraction of pesticides is used to control the pest while remaining part reaches the air, land surface and water bodies. These pesticides affect the human population directly or indirectly.

Pesticides are the toxic compounds linked to human use that have profound effect on aquatic life and water quality. Unintentional pesticides related fish kills occur through out of the world. Some of these kills have been large, involving thousands of fishes and other aquatic animals. Bioaccumulation is the accumulation of pesticides in animal tissue at levels greater than those in the water or soil to which they were applied. Some fish may concentrate certain pesticides in their body tissue and organs (especially fat) at levels 10 million times greater than in the water. Biomagnification is the accumulation of pesticides at each successive level of the food chain. At each step in the food chain the concentration of pesticides increases. Fish can pass these poisons on to humans.

Doramectin is a member of avermectins which is a group of chemicals discovered in 1975, isolated from culture of the actenomycete *Streptomyces avermitilis* (Campbell, 1989; Fisher and Mrozik, 1992). Doramectin is used for a wide range of ecto and endo parasites of domestic and wild animals. Recently these compounds are introduced as insecticides. A mixture of avermectins B1a and B1b is also used currently as a foliar spray on various crops against a variety of insect pests. Aziz *et al.* (1982) and Taylor and Green (1989) studied a number of possible mechanisms of action of avermectins on neural transmission and chloride channels. Many workers have reported the effect of heavy metals, carbamates, organochlorides, organophosphates and pyrethroids but little work has been done on the effect of avermectins on fresh water fishes.

The insecticides are reported (Banik *et al.*, 1996; Saxena and Seth, 2002; Saxena and Gupta, 2003) to produce toxic effects on the biochemical and haematological parameters of various terrestrial and aquatic animals. The persistence of these insecticides in the aquatic environment is dangerous for the survival of the fish because gill and skin of fish directly absorb these insecticides from surrounding water.

In the present investigations effect of doramectin has been studied on biochemical parameters of fresh water fish *Channa punctatus*.

Material and Methods

Channa punctatus were obtained from local water bodies and acclimated in the laboratory for a maximum period of 30 days and then divided into four treatment and one control groups. Fish were then transferred to different aquaria containing various lethal and sublethal doses of doramectin were maintained up to 30 days. During this period fish were sacrificed after 5, 10, 10, 15, 20, 25 and 30 days for the study of the changes in biochemical constituents due to exposure to doramectin. For the determination of biochemical muscles were collected, weighted and homogenized in glass homogenizer and centrifuged in Zanetaki K-24 refrigerated. The supernatants were separated and used for measuring the activity of enzymes. The activities of acid, alkaline phosphates and amylase were measured by the methods given in Bergmeyer (1971). The experiment was repeated at least four times. The data obtained was analyzed by ANOVA to test its significance.

Results

The pattern of changes in activities of amylase, acid phosphates and alkaline phosphates are given in Tables 13.1–13.3 respectively.

Effect of Doramectin on the Mortality of *Channa punctatus*

When *Channa punctaus* were exposed to doramectin at the concentration of 0.017 ppm and above *i.e.*, 0.018ppm and 0.02ppm mortality was observed. At 0.02 ppm fish died within 20 hours and at 0.018 ppm fish died within 24 hours but at 0.017 ppm fish remained alive up to 10 days. After that mortality was recorded upto 30th day.

When the concentration was less than 0.017 ppm fish survived for 30 days or more. Therefore, three concentrations (0.016, 0.012, 0.008 ppm), which were below 0.017 ppm were selected as sublethal concentrations for studying the effect of doramectin on fish metabolism.

Effect of Doramectin on Biochemical Activities

(1) Amylase Activity

During the present investigations amylase activity in *Channa punctatus* was reduced up to 30 days at all sublethal concentrations of doramectin used. However, there was slight increase in amylase activity after 5 days exposure to 0.012 ppm of doramectin and after that there was continuous reduction up to 30 days. Maximum reduction in (93.43 per cent) amylase activity was recorded at 0.016 ppm after 5 days exposure (Table 13.1). The results obtained were statistically insignificant for the concentrations used but significant for the period of exposure (Table 13.1).

(2) Acid Phosphomonoesterase Activity

The results of ACP activity are summarized in Table 13.2. During the present studies alterations in ACP activity were significant due to concentrations of doramectin and exposure period. Maximum reduction (36.66 per cent) was recorded on 5th day at 0.016 ppm of doramectin and some increase (17.85 per cent) was also recorded on 10th day at 0.008 ppm of doramectin. Besides, these mostly acidphosphomonoesterase activity was reduced but there was no alteration on some days and concentrations. The results obtained were statistically significant for the concentrations and period of exposure (Table 13.2.1)

Table 13.1: Effect of Doramectin on Amylase Activity (Street-close unit/100 ml) in Muscles of *Channa punctatus*

Days	*Concentration*			
	Control	*0.008 ppm*	*0.012 ppm*	*0.016 ppm*
0	30.95			
5	30.47	6.82 (– 77.61)*	32.65 (+ 7.15)	2.00 (– 93.43)
10	30.95	4.55 (– 85.29)	2.70 (– 91.27)	4.08 (– 86.81)
15	29.98	5.13 (– 82.88)	3.85 (– 87.15)	11.11 (– 62.94)
20	30.15	6.98 (– 76.84)	7.10 (– 76.45)	9.08 (– 69.88)
25	29.19	7.50 (– 74.30)	14.29 (– 51.04)	12.53 (– 57.07)
30	30.12	5.26 (– 75.06)	13.11 (– 56.47)	11.11 (– 63.14)

* Per cent change from control value.

Table 13.1.1: ANOVA for Amylase Activity Due to the Effect of Doramectin in *Channa punctatus*

SOV	*DF*	*SS*	*MS*	*F*	$F_{0.05}$
Concentration	3	2154.24	718.08	18.24	3.29
Days	5	137.58	27.51	0.699	2.90
Error	15	590.48	39.36		
Total	23	2882.3			

SOV: Source of variance; DF: Degree of freedom; SS: Sum of square; MS: Variance (SS/DF); F: Ratio b/w variance of source and error;' $F_{0.05}$: Value of F at 0.05 level of significance.

For concentration insignificant at 5 per cent level.

For days significant at 5 per cent level.

(3) Alkaline Phosphomonoesterase Activity

During present studies insignificant increase occurred in ALP activity for the concentrations used but according to exposure period this change was significant (Table 13.3.1). Maximum increase in percentage of ALP activity recorded 100 per cent on 20 day at 0.008 ppm of doramectin (Table 13.3).

Table 13.2: Effect of Doramectin on ACP Activity (µmoles pnp/mg protein/hour) in Muscles of *Channa punctatus*

Days	*Concentration*			
	Control	*0.008 ppm*	*0.012 ppm*	*0.016 ppm*
0	0.30			
5	0.30	0.28 (– 6.66)*	0.28 (– 6.66)	0.19 (– 36.66)
10	0.28	0.33 (+ 17.85)	0.25 (– 10.71)	0.28 (0.0)
15	0.30	0.30 (0.0)	0.22 (– 26.66)	0.33 (+ 10)
20	0.33	0.30 (– 9.09)	0.28 (– 15.15)	0.30 (– 10)
25	0.28	0.28 (0.0)	0.30 (+ 7.14)	0.28 (0.0)
30	0.33	0.30 (– 9.09)	0.33 (0.0)	0.33 (0.0)

* Per cent change from control value.

Table 13.2.1: ANOVA for ACP Activity Due to the Effect of Doramectin in *Channa punctatus*

SOV	*DF*	*SS*	*MS*	*F*	$F_{0.05}$
Concentration	3	0.01	0.003	1.5	3.29
Days	5	0.01	0.002	1.0	2.90
Error	15	0.03	0.002		
Total	23	0.05			

SOV: Source of variance; DF: Degree of freedom; SS: Sum of square; MS: Variance (SS/DF); F: Ratio b/w variance of source and error;' $F_{0.05}$: Value of F at 0.05 level of significance.

For concentration insignificant at 5 per cent level.

For days significant at 5 per cent level.

Structural and Behavioural Changes in *Channa punctatus*

Structural and behavioural changes were observed in *Channa punctatus*. During exposure period fish remained restless. They showed erratic and uncoordinated swimming. Sometimes they tried to jump out of the aquarium. An important structural change was the formation of a hole on the lower side between pectoral girdles. The diameter of hole was increased gradually and in some fishes heart was visible through the hole.The colour of the body was also fade during exposure period.

Table 13.3: Effect of Doramectin on ALP activity (µmoles pnp/mg protein/hour) in Muscles of *Channa punctatus*

Days	Concentration			
	Control	0.008 ppm	0.012 ppm	0.016 ppm
0	0.24			
5	0.24	0.36 (+ 50.0)*	0.24 (0.0)	0.24 (0.0)
10	0.23	0.24 (+ 4.34)	0.36 (+ 56.52)	0.33 (+ 43.48)
15	0.21	0.36 (+ 71.43)	0.24 (+ 14.29)	0.36 (+ 71.43)
20	0.24	0.48 (+ 100)	0.36 (+ 50)	0.33 (+ 37.50)
25	0.26	0.36 (+ 38.46)	0.36 (+ 38.46)	0.48 (+ 84.62)
30	0.26	0.36 (+ 38.46)	0.48 (+ 84.62)	0.36 (+ 38.46)

* Per cent change from control value.

Table 13.3.1: ANOVA for ALP Activity Due to the Effect of Doramectin in *Channa punctatus*

SOV	DF	SS	MS	F	$F_{0.05}$
Concentration	3	0.03	0.01	10	3.29
Days	5	0.01	0.002	2	2.90
Error	15	0.02	0.001		
Total	23	0.06			

SOV: Source of variance; DF: Degree of freedom; SS: Sum of square; MS: Variance (SS/DF); F: Ratio b/w variance of source and error;' $F_{0.05}$: Value of F at 0.05 level of significance.

For concentration insignificant at 5 per cent level.

For days significant at 5 per cent level.

Discussion

Doramectin is a known anthelmintic which is recently introduced as insecticide also. The biochemical changes observed in present studies in *Channa punctatus* are due to exposure to sublethal concentrations of doramectin. These findings are in accordance with the findings of previous workers (Saxena and Seth, 2002; Saxena and Gupta, 2003; Saxena and Gupta, 2005; Sirohi and Saxena, 2006; Saxena and Sirohi; 2007; Maheswaran *et al.*, 2008). Pesticides are known to cause biochemical toxicological changes in fishes. Avermectins, which is a known antiparasitic compound (Singh and Kaushal, 1995) are also being used as pesticide and is becoming a wide spectrum compound against a variety of pests in agriculture and other purposes.

In the present studies doramectin caused mortality in *Channa punctatus* at very low concentrations *i.e.,* 0.02 ppm. Exposure to doramectin caused significant alterations in the activities of acid phosphates, alkaline phosphates and amylase.

Acid and alkaline phosphates are reported (Arora and Kulshreshtha, 1985) to be associated with the transport of metabolites with metabolism of phospholipids, phosphoproteins, nucleotides and carbohydrates and synthesis of protein. Acid phosphates are responsible for the hydrolysis of ester linkage of phosphates which helps in the autolysis of the cell after its death. The alterations in the activity of acid and alkaline phosphates in present investigations indicate of doramectin in phosphate and protein metabolism *of Channa punctatus.*

The activity of amylase is directly related with the carbohydrate metabolism of the animal. Alterations in the activity of amylase clearly indicate the interferance of doramectin in carbohydrate metabolism of fish. Saxena and Sirohi (2007) have also reported similar changes in *Channa punctatus* due to exposure to pyrethroids pesticide.

In the present studies doramectin at sublethal concentrations caused structural and behavioural changes in *Channa punctatus* and also altered the activities of acid phosphates, alkaline phosphates and amylase. It is an indicator of the toxicity caused by this compound in this fish at these levels. This may be harmful for its survival and growth.

Acknowledgements

We are thankful to Dr. K. Singh, Head, Department of Biotechnology and Dr. (Mrs.) Sunita Sharma, Reader, Department of Zoology, Bareilly College, Bareilly for their cooperation.

References

Aziz, M.A., Diallo, S. and Diopp, I.M., 1982. Efficacy and tolerance of ivermectin in human onchocerciasis. *Lancet,* 2: 171–173.

Arora, N. and Kulshreshtha, S.K., 1985. Effect of chronic exposure to sublethal doses of two pesticides on alkaline and acid phosphatase activities in the intestine of a fresh water teleost, *Channa striatus. Acta Hydrochem. Hydrobiol.*, 13: 619–624.

Banik, S., Chakraborty, S. and Choudhury, J. R., 1996. Haemolytic anaemia in *Anabas testudineus* with reference to endosulfan. *U.P.J. Zool.*, 16: 87–88.

Bergmeyer, H.U., 1971. *Methods of Enzymatic Analysis*. Academic Press, New York.

Campbell, W.C., 1989. *Ivermectin and Abamectin. Springer Verlag. N.Y.*

Fisher, M.H. and Mrozik, H., 1992. The chemistry and pharmacology of avermectins. *Annu. Rev. Phamacol. Toxicol.*, 32: 537–553.

Maheshwaran, R., Devapaul. A., Muralidharan, S., Velmurungan, B. and Ignacimuthu, S., 2008. Haematolgoical studies of fresh water fish, *Clarias batrachus* (L.) exposed to mercuric chloride. *International Journal of Integrative Biology*, 2(1): 49–54.

Saxena, K.K. and Seth, N., 2002. Toxic effects of cypermethrin on certain haematological aspects of fresh water fish *Channa punctatus. Bull.Environ.Contam. Toxicol.*, 69: 364–369.

Saxena, K.K. and Gupta, P., 2003. Effect of permethrin on the activity of acid and alkaline phosphomo noesterase in a fresh water fish *Channa punctatus. Proc. Nat. Symp. Bio. Comp. Sci.*, 477–479.

Saxena, K.K. and Gupta, P., 2005. Carbaryl induced haematological changes in *Channa punctatus. Nat Jour Life Sci.*, 2 (1 and 2): 99–104.

Saxena, K.K. and Sirohi, V., 2007. Effect of λ–cyhalothrin on the activities of trypsin and lipase in fresh water fish *Channa punctatus. Journal of Fisheries and Aquatic Science*, 2(2): 168–172.

Sirohi, V. and Saxena, K.K., 2006. Toxic effect of l–cyhalothrin on biochemical contents of fresh water fish *Channa punctatus. J. Fisheries. Aquat. Sci.*, 1(12): 112–116.

Singh, K. and Kaushal, P., 1995. Effect of fenbendazole and ivermectin on carbohydrate metabolism of swine kidneyworm *Stephanurus dentatus. J. Parasitic Diseases*, 19: 191–192.

Taylor, H.R. and Green, B.M., 1989. The status of ivermectin in the treatment of *Onchocerciasis. Am. J. Trop. Med.*, 41: 460–466.

Chapter 14

Heavy Metals (Cu, Zn, Mn, Fe and Pb) in Soft Tissues of Gastropods, *Bursa spinosa* and *Nerita oryzarum* from Uran Coast, Mumbai (India): A Case Study

Sandhya S. Kupekar[1], *B.G. Kulkarni*[2] *and A.K. Pandey*[3]*

[1]*Department of Zoology, Mahatma Phule A.S.C. College, Panvel – 410 206, Mh., India*

[2]*Department of Zoology, Institute of Science, 15 Madam Cama Road, Mumbai – 400 032, Mh., India*

[3]*National Bureau of Fish Genetic Resources, Canal Ring Road, Lucknow – 226 002, U.P., India*

ABSTRACT

Accumulation of five heavy metals like copper (Cu), zinc (Zn), manganese (Mn), iron (Fe) and lead (Pb) in the soft tissues of gastropods, *Bursa spinosa* and *Nerita oryzarum* from three stations along Uran coast (near Mumbai) was studied during 2005-2007. The coast receives discharges from the heavily industrialized and highly populated settlements nearby areas. Bioaccumulation of heavy metals in *Bursa spinosa* followed the

* Corresponding author: E-mail: akpandey_cifa@yahoo.co.in

order Cu>Fe>Zn>Cd>Mn>Pb and *Nerita oryzarum* Fe>Cu>Zn>Mn>Cd>Pb. Seasonal variations in accumulation in these heavy metals in soft tissue of the gastropods were also noticed. Though the observed concentration of heavy metals in both the bivalves were below the recommended limits, environmental surveillance of the creek is required for contaminant-free gastropods.

Keywords: *Bioaccumulation, Heavy metals, Soft tissues, Bursa spinosa, Nerita oryzarum, Uran coast.*

Introduction

Heavy metals accumulation in marine ecosystem is of global concern (Bhattacharya *et al.*, 1994; Frias-Espericueta *et al.*, 1999a, b; Goksu *et al.*, 2005). They enter in the aquatic environment through atmospheric deposition, erosion of geological matrix or due to anthropogenic activities such as industrial effluents, domestic sewage and mining wastes (Kennish, 1992; de Astudillo *et al.*, 2005; Reddy *et al.*, 2007). The detection of pollutants like petroleum hydrocarbons and heavy metals in the marine environment has been investigated by employing bio-indicators/sentinel organisms (Bryan *et al.*, 1979; Kennish, 1992; Phillips and Rainbow, 1993). Gastropods are filter-feeders and thus uptake of heavy metals in these organisms is not only from food and water but also from ingestion of inorganic particulate materials too (Fang *et al.*, 2003; El-Silkaly *et al.*, 2004; Sidoumou *et al.*, 2006). Since they accumulate most of the contaminants at much higher level than those found in the water column as such they are representative of the pollution of area and considered as appropriate indicators as they are spatially distributed, relatively large in size and easy to collect (Anderlini *et al.*, 1982; Turkman and Turkman, 2004). High concentrations of trace metals have been detected in whole soft bodies of several species of marine bivalves in many parts of the world (Ikuta, 1986; Ikuta *et al.*, 1990; Bu-Olayan and Subrahmaniyam, 1997; Goksu *et al.*, 2005).

Due to their economic and ecological importance as well as their sedentary mode of life, molluscs, especially bivalves, play major role in monitoring contaminants throughout the world and are well established as bio-indicators for the concentration of heavy metals in aquatic ecosystems (Powell and White, 1990; Sarkar *et al.*, 1994; Al-Mafda *et al.*, 1998; Gundacker, 1999; Neuberger-Cywiak *et al.*,

2003). Since rapid industrialization and urbanization during the recent years have led to the significant increase in heavy metal pollution in the coastal ecosystems throughout the world (Phillips, 1980; Kennish, 1992; Kavun *et al.*, 2002; Phillips and Rainbow, 1993; Fang *et al.*, 2003), an attempt was made to record the accumulation of heavy metals such as copper (Cu), zinc (Zn), iron (Fe), manganese (Mn), cadmium (Cd) and lead (Pb) in the two species of gastropods, *Bursa spinosa* and *Nerita oryzarum*, inhabiting three different sites of Uran coast (near Mumbai).

Materials and Methods

At the beginning of investigation, the coast of Uran was surveyed for recording intertidal gastropods. This coast is mixture of rock, sand and muddy shores. Gastropods were found attached to the rocks, in the crevices and below large stones. Three sites of Uran coast were selected for the present study (Figures 14.1a, b). Site–I was located along the eastern shore of Bombay Harbour opposite Colaba, the substratum being rocky (Figure 14.2). Site–II was located near Karanja village, the substratum formed of basalt rock. Due to constant movement of ferries, water of nearby areas at this point is muddy showing sporadic oil patches (Figure 14.3). Site–III was located 5 km away from site II and situated at Naval Base, Dronagiri. Here upper part of inter-tidal zone is rocky whereas substratum of remaining part is sandy (Figure 14.4). Since all the above sites were exposed to the environment during neap tide, collection of animals was possible throughout the year.

Bursa spinosa (Lamarck, 1843) and *Nerita oryzarum* (Recluz, 1841), found in intertidal zone, were collected during the low tide from the three selected sites of Uran coast between January to December during 2005-2007. They were of uniform size and their whole soft tissues carefully removed by shelling with plastic knife (Chiu *et al.*, 2000). Samples were dried at 60°C to constant weight. Atomic absorption spectrophotometer (GBC 93ZAA) was used to estimate metals in the samples. Briefly, an accurately weighed dried powdered sample was taken in a beaker. To this, 20 ml 70 per cent HNO_3 was added and subjected to digestion till brown fumes completely disappeared and residue became whitish. 1 ml 30 per cent $HClO_4$ added after the residue was cooled. This was digested for 10-15 minutes to dryness. The dried residue was cooled and the final volume made to 25 ml with 2 M HNO_3. All the reagents were of

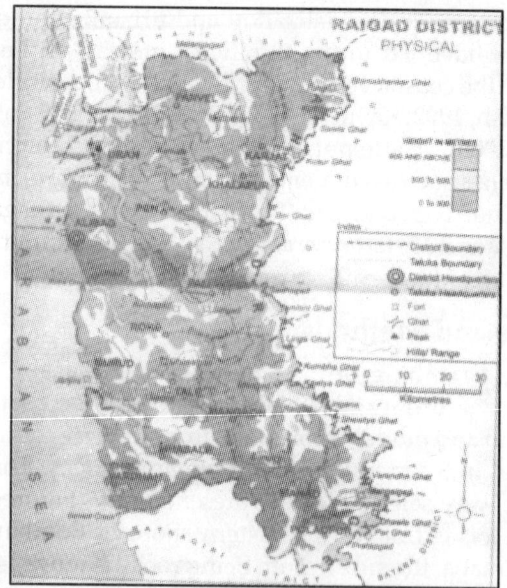

Figure 14.1a: General Map of the Area

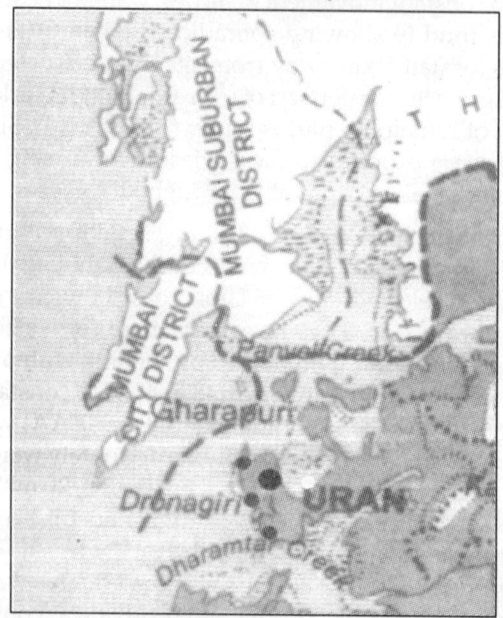

Figure 14.1b: Magnified to show Sampling Sites

Figure 14.2: Site-I Opposite to Colaba, Mumbai

Figure 14.3: Site-II Located Near Karanja Village

Figure 14.4: Site-III Near Naval Base, Dronagiri

analytical grade. Metals were estimated from the samples using acid as a blank. The metal concentration in the tissues of the molluscs was calculated by using standard calibration curve.

Results and Discussion

Bivalve clams are employed as bio-monitor to determine the effect of marine pollution (Al-Mafda *et al.*, 1988; Zorba *et al.*, 1992; Tomazelli *et al.*, 2003; El-Silkaly *et al.*, 2004). The concentration of metals in molluscs not only dependent on levels of the elements in environment but also on other factors *viz.* size, age, growth, sex and reproductive conditions (Phillips, 1980; Phillips and Rainbow, 1993; Boening, 1999; Frias-Espericueta *et al.*, 1999 b; Fang *et al.*, 2003). In the present study, the intertidal clams, *Bursa spinosa* and *Nerita oryzarum* were identified from Uran coast near Mumbai. Both the species have been reported among intertidal molluscan diversity of Mumbai coasts (Apte, 1993; Datta *et al.*, 2010; Pawar, 2012). Bioaccumulation of heavy metals in the soft tissues of *Bursa spinosa* and *Nerita oryzarum* have been summarized in Tables 14.1 and 14.2 while the average values of bioaccumulation are given in Table 14.3. Accumulation of heavy metals in *Bursa spinosa* followed the order Cu>Fe>Zn>Cd>Mn>Pb (Table 14.1) while in *Nerita oryzarum*, the

Table 14.1: Bioaccumulation of Heavy Metals in *Bursa spinosa* (mg/g)

Heavy Metals	*Site*	*Jan*	*Feb*	*Mar*	*Apr*	*May*	*Jun*	*Jul*	*Aug*	*Sept*	*Oct*	*Nov*	*Dec*	*Average*	*Average SD*
Zn	I	0.314	0.311	0.275	0.305	0.348	0.319	0.301	0.316	0.326	0.491	0.312	0.446	0.339	±0.025
	II	0.309	0.309	0.274	0.304	0.339	0.315	0.321	0.212	0.291	0.422	0.362	0.432	0.324	
	III	0.319	0.313	0.279	0.308	0.351	0.321	0.412	0.326	0.336	0.516	0.491	0.502	0.373	
Cu	I	0.557	0.973	0.546	0.748	0.662	0.530	0.682	0.649	0.511	0.414	0.408	0.439	0.593	±0.009
	II	0.542	0.969	0.534	0.742	0.652	0.538	0.689	0.691	0.522	0.410	0.419	0.429	0.595	
	III	0.569	0.979	0.551	0.753	0.670	0.551	0.699	0.703	0.530	0.421	0.441	0.445	0.609	
Mn	I	0.455	0.280	0.213	0.238	0.110	0.453	0.090	0.055	0.032	0.203	0.053	0.155	0.195	±0.012
	II	0.443	0.275	0.195	0.213	0.103	0.445	0.078	0.073	0.029	0.195	0.055	0.130	0.186	
	III	0.473	0.285	0.223	0.248	0.120	0.460	0.103	0.078	0.066	0.218	0.065	0.173	0.209	
Pb	I	0.191	0.152	0.186	0.146	0.169	0.118	0.163	0.163	0.129	0.129	0.208	0.180	0.161	±0.016
	II	0.174	0.146	0.174	0.118	0.152	0.107	0.174	0.118	0.068	0.158	0.186	0.163	0.145	
	III	0.208	0.158	0.197	0.163	0.186	0.124	0.186	0.180	0.146	0.180	0.219	0.174	0.177	
Cd	I	0.241	0.218	0.255	0.250	0.236	0.345	0.132	0.186	0.141	0.145	0.186	0.182	0.210	±0.017
	II	0.239	0.200	0.232	0.236	0.223	0.323	0.095	0.209	0.145	0.141	0.195	0.191	0.202	
	III	0.258	0.223	0.277	0.255	0.250	0.355	0.141	0.214	0.177	0.186	0.255	0.236	0.236	
Fe	I	0.490	0.505	0.524	0.581	0.519	0.552	0.490	0.533	0.533	0.538	0.514	0.548	0.527	±0.010
	II	0.481	0.495	0.519	0.562	0.500	0.538	0.500	0.538	0.529	0.543	0.519	0.538	0.522	
	III	0.500	0.510	0.543	0.595	0.524	0.562	0.505	0.552	0.543	0.557	0.538	0.557	0.541	

Table 14.2: Bioaccumulation of Heavy Metals in *Nerita oryzarum* (mg/g)

Heavy Metals	*Site*	*Jan*	*Feb*	*Mar*	*Apr*	*May*	*Jun*	*Jul*	*Aug*	*Sept*	*Oct*	*Nov*	*Dec*	*Average*	*Average SD*
Zn	I	0.242	0.211	0.364	0.354	0.256	0.311	0.322	0.316	0.412	0.312	0.312	0.311	0.310	±0.023
	II	0.239	0.209	0.358	0.351	0.251	0.309	0.361	0.411	0.316	0.411	0.291	0.322	0.319	
	III	0.248	0.214	0.371	0.356	0.258	0.314	0.411	0.416	0.417	0.512	0.416	0.322	0.355	
Cu	I	0.411	0.755	0.225	0.536	0.534	0.501	0.581	0.672	0.407	0.408	0.456	0.435	0.493	±0.010
	II	0.399	0.744	0.221	0.538	0.511	0.491	0.567	0.662	0.391	0.419	0.443	0.429	0.485	
	III	0.418	0.763	0.229	0.544	0.543	0.505	0.616	0.676	0.417	0.433	0.468	0.447	0.505	
Mn	I	0.255	0.305	0.280	0.385	0.140	0.220	0.078	0.050	0.058	0.065	0.078	0.058	0.164	±0.017
	II	0.250	0.298	0.275	0.373	0.130	0.215	0.055	0.065	0.073	0.223	0.103	0.105	0.180	
	III	0.273	0.310	0.288	0.403	0.143	0.225	0.105	0.078	0.078	0.243	0.105	0.125	0.198	
Pb	I	0.169	0.146	0.146	0.129	0.163	0.158	0.152	0.152	0.174	0.118	0.174	0.163	0.154	±0.021
	II	0.163	0.135	0.135	0.113	0.135	0.146	0.141	0.168	0.146	0.146	0.264	0.169	0.155	
	III	0.225	0.163	0.146	0.158	0.174	0.169	0.163	0.208	0.186	0.203	0.281	0.203	0.190	
Cd	I	0.227	0.209	0.250	0.232	0.223	0.101	0.141	0.113	0.150	0.095	0.178	0.141	0.172	±0.020
	II	0.136	0.195	0.245	0.223	0.205	0.099	0.095	0.101	0.095	0.132	0.155	0.159	0.153	
	III	0.314	0.218	0.250	0.245	0.232	0.111	0.159	0.116	0.159	0.141	0.181	0.186	0.193	
Fe	I	0.486	0.519	0.490	0.505	0.481	0.519	0.495	0.481	0.505	0.519	0.510	0.510	0.502	±0.010
	II	0.481	0.505	0.476	0.495	0.471	0.495	0.495	0.49	0.502	0.51	0.514	0.505	0.495	
	III	0.500	0.519	0.500	0.514	0.495	0.529	0.500	0.495	0.566	0.529	0.519	0.514	0.515	

Table 14.3: Average Values of Heavy Metals (mg/g) Accumulated in the Gastropods at Sites I, II and III

	Zn	*Cu*	*Mn*	*Pb*	*Cd*	*Fe*
Bursa spinosa	0.345±0.025	0.599±0.009	0.197±0.012	0.161±0.016	0.216±0.017	0.530±0.010
Nerita oryzarum	0.328±0.023	0.494±0.010	0.181±0.017	0.166±0.021	0.173±0.020	0.504±0.010

decreasing order was Fe>Cu>Zn>Mn>Cd>Pb (Table 14.2). The concentration of heavy metals showed significant variations between the two species of gastropod (Table 14.3). Seasonal variations in accumulation in these heavy metals in soft tissues of the gastropods were also noticed (Tables 14.1 and 14.2). Accumulation of cadmium (Cd) and lead (Pb) in noticeable amount in *Bursa spinosa* and *Nerita oryzarum* clearly showed that the effluents containing these heavy metals enter into the coastal waters of Uran from the adjacent industries. Bioaccumulation of heavy metals in *Tympanotonus fuscatus* followed the order Cu>Zn>Cd (Daka *et al.*, 2006) whereas in *Donax trunculus* and *Donax faba,* the accumulation was found in order as Zn>Cu>Pb>Cd in two contaminated sites in Gulf of Annaba, Algeria (Beldi *et al.*, 2006) and in the four beaches of Eayong Provice, Thailand (Dungchangwat *et al.*, 2011). In the latter case, the highest concentration of Zn and Cd were recorded in the hot season, Cu in cool season and Pb during the rainy season (Dungchangwat *et al.*, 2011). In the present investigation, higher level of Fe, Cu, Zn and Cd were accumulated by *Bursa spinosa* and *Nerita oryzarum* (Table 14.3). A trend in accumulation of metals in decreasing order Fe>Zn> Mn>Cu>Cd>Hg has also been found in *Cerithidea cingulata, C. obtuse, Telescopiun telescopiun, Thais lacera* and *Nerita articulata* of Sundarban wetland ecosystem of India (Sarkar *et al.*, 2002). Bioaccumulation of heavy metals in *Meretrix meritrix* in estuaries of Sabah, North Borneo showed the decreasing trend like Zn>Cu>Cd>Cr>Pb (Abdullah *et al.*, 2007) whereas in the edible soft tissue of *Perna viridis* inhabiting Mahe estuary, Pondicherry followed the order Zn>Cu>Pb>Ni>Cr (Gopinathan and Sobhana Amma, 2009).

In the present study, essential metals were accumulated in higher concentration than non-essential metals in *Bursa spinosa* and *Nerita oryzarum* as high level of Cu, Zn and Fe were detected than that of Mn, Cd and Pb. Accumulation of high levels of Cu, Zn and Fe in soft-bodied gastropods can be attributed to their metabolic requirement where these metals act as a co-factor in metabolic processes. Similar correlations between trace element accumulations and metabolic activities have been reported in other bivalves too (Gundacker, 1999; Frias-Espericueta *et al.*, 1999a; Ke and Wang, 2001; Wang and Ke, 2002). Further, seasonal variations noticed in accumulation of heavy metals by both the gastropods (Tables 14.1 and 14.2) may probably be associated with food supply, changes in run-off, particularly the material to the sea precipitation and variations related to the reproductive cycle (Fowler and Oregioni, 1976; Lotouche and Mix, 1981; Turkmen and Turkmen, 2004; Beldi *et al.*, 2006; Dungchangwat *et al.*, 2011). It is well known that Cu, Zn and Fe are biologically essential metals and play as cofactor in enzymatic processes (Singh and Steinnes, 1994; Gundacker, 1999), however, their accumulations in higher side affect biological processes in marine vertebrates (Kennesh, 1993; Phillips and Rainbow, 1993). In conclusion, higher level of heavy metals detected in *Bursa spinosa* and *Nerita oryzarum* present in Uran coast may affect the life processes of the gastropopds. Though the observed heavy metals concentrations in both the clams were below the recommended limits, environmental surveillance of the creek for contaminant-free gastropods is required for safe human health (Fang *et al.*, 2003; Modassir and Sivadas, 2003; Babu *et al.*, 2010).

References

Abdullah, M.H., Sidi, J. and Aris, A.Z., 2007. Heavy metals (Cd, Cu, Cr, Pb and Zn) in *Meretrix metretrix* Roding, water and sediments from estuaries in Sabah, North Borneo. *Intl. J. Environ. Sci. Educ.*, 2: 69–74.

Al–Mafda, H., Abdel–Moati, M.A.R. and Al–Glmaly, F.H., 1998. *Pinctada radiata* (pearl oyster): a bio–indicator for metal pollution monitoring in the Qatari waters (Arabian Gulf). *Bull. Environ. Contam. Toxicol.*, 60: 245–251.

Anderlini, V.C., Mohammed, O.S., Zorba, M.A. and Omar, N.A., 1982. Assessment of trace metal and biological pollution in the marine environment of Kuwait. Report 605, Kuwait Institute of Scientific Research (KISR), Kuwait.

Apte, D.A., 1993. Marine Gastropoda of Bombay–a recent survey. *J. Bombay Nat. Hist. Soc.*, 90: 537–539.

Babu, A., Kesavan, K., Annadurai, D. and Rajagopal, S., 2010. *Bursa spinosa*–a mesogastropod fit for human consumption. *Adv. J. Food Sci. Tech.*, 2: 79–83.

Beldi, H., Gimbert, F., Maas, S., Scheifler, R. and Soltani, N., 2006. Seasonal variations of Cd, Cu, Pb and Zn in the edible mollusc, *Donax trunculus* (Mollusca, Bivalvia) from Gulf of Annaba, Algeria. *Afr. J. Agric. Res.*, 1: 85–90.

Bhattacharya, B., Sarkar, S.K. and Maji, P.K., 1994. Bioaccumulation of heavy metals in flora and fauna of Hooghly esturary, east coast of India. *Toxicol. Environ. Chem.*, 42: 123–130.

Boening, D.W., 1999. An evaluation of bivalves as biomonitors of heavty metal pollution in marine water. *Environ. Monit. Assess.*, 55: 459–470.

Bryan, G.W., Waldichuk, M. Pentreath, R.J. and Darracott, A., 1979. Bioaccumulation of marine pollutants. *Phil. Trans. R. Soc. (Lond.)*, 286B: 483–505.

Bu–Olayan, A.H. and Subrahmanyam, M.N.V., 1997. Accumulation of copper, nickel, lead and zinc by snail, *Lunella coronatus* and pearl oyster, *Pinctada radiata* from the Kuwait coast before and after the Gulf war oil spill. *Sci. Total Environ.*, 197: 161–165.

Chiu, S.T., Lam, S.F., Tze, W.L., Chau, C.W. and Ye, D.Y., 2000. Trace metals in mussel from mariculture zones, Hong Kong. *Chemosphere*, 41: 101–108.

Daka, E.R., Oweisana, L.I. and Braide, S.A., 2006. Accumulation of heavy metals from single and mixed metal solution by gastropod mollusc, *Tympanotonus fusatus* Linnaeus a Niger Delta estuary: implications for bio–monitoring. *African J. Biotechnol.*, 5: 1954–1962.

Datta, S.N., Chakraborty, S.K., Jaiswar, A.K. and Ziauddin, G., 2010. A comparative study on intertidal faunal biodiversity of selected beaches of Mumbai coast. *J. Environ. Biol.*, 31: 981–966.

de Astudillo, R., Yen, I.C. and Bekele, I., 2005. Heavy metals in sediments, mussels and oysters from Trinidad and Venezuela. *Intl. J. Trop. Biol.*, 53 (Suppl. 1): 41–53.

Dungchangwat, T., Roachanakanan, R., Naiyanetr, P. and Songchitsomboon, S., 2011. Accumulation of heavy metals in edible Hoi Seab, *Dionax faba* Gmelin, 1789 at beaches of Rayong Province, Thailand. In: *Proceedings of the 37th Congress of Science and Technology of Thailand* (October 10–12, 2011). Faculty of Science, Mahidol University, Bangkok, pp. 1–5.

El–Silkaly, A., Khaled, A. and El–Nemr, A., 2004. Heavy metal monitoring using bivalves from Mediterranean Sea and Red Sea. *Environ. Monit. Assess.*, 98: 41–58.

Fang, Z.–Q., Cheung, R.Y.H. and Wang, M.H., 2003. Heavy metals in oysters, mussels and clams collected from sites along the Pearl river delta, South China. *J. Environ. Sci.*, 15: 9–24.

Fowler, S.W. and Oregioni, B., 1976. Trace metals in mussels from the NW Mediterranean. *Mar. Pollut. Bull.*, 7: 26–29.

Frias–Espericueta, M.G., Ortiz–Arellano, M.A., Osuna–Lopez, J.I. and Ronson–Paulin, J.A., 1999a. Heavy metals in rock oyster, *Crassostrea iridescens* (Filibranchia: Ostreidea) from Mazatlan, Sinaloa, Mexico. *Rev. Biol. Trop.*, 47: 843–849.

Frias–Espericueta, M.G., Osuna–Lopez, J.I., Sandoval–Salazar, G. and Lopez–Lopez, G., 1999b. Distribution of trace metals in different tissues in the rock oyster, *Crassostrea iridescens*: seasonal variation. *Bull. Environ. Contam. Toxicol.*, 63: 73–79.

Goksu, M.Z.U., Akar, M., Cevik, F. and Findik, O., 2005. Bioaccumulation of some heavy metals (Cd, Fe, Zn, Cu) in two bivalvia species (*Pinctatda radiata* Leach 1814 and *Brachidontes pharaonis* Fischer 1870). *Turk. J. Vet. Anim. Sci.*, 29: 89–93.

Gopinathan, K.M. and Sobhana Amma, R., 2009. Bioaccumulation of toxic heavy metals in the edible soft tisuues of green mussel (*Perna viridis* L.) of Mahe Region. Project Report. Submitted to the Department of Science, Technology and Environment (DSTE), Government of Pondicherry, Pondicherry. 30 p.

Gundacker, C., 1999. Tissue–specific heavy metal (Cd, Pb, Cu, Zn) deposition in a natural population of the zebra mussels (*Dreissena polymorpha* Pallas). *Chemosphere*, 38: 3339–3356.

Ikuta, K., 1986. Correlations between ratios of metals concentration in byssuses to those in soft bodies and metal concentration in soft bodies of bivalves. *Bull. Fac. Agric. Miyazaki Univ.*, 33: 265–273.

Ikuta, K., Morikawa, A. and Sakoda, K., 1990. Adequate sample size to means of copper and zinc concentrations in a population of *Crassostrea gigas*. *Bull. Fac. Agric. Miyazaki Univ.*, 37: 7–14.

Ke, C. and Wang, W.X., 2001. Bioaccumulation of Cd, Se and Zn in an estuarine oyster (*Saccrostrea glomerata*). *Aquat. Toxicol.*, 56: 133–151.

Kavun, V.Y., Shulkin, V.M. and Khristoforova, N.K., 2002. Metal accumulation in mussels of the Kuril Islands, north–west Pacific Ocean. *Mar. Environ. Res.*, 53: 219–226.

Kennish, M.J., 1992. *Ecology of Estuaries: Anthropogenic Effects*. CRC Press, Boca Raton, Florida.

Latouche, Y.D. and Mix, M.C., 1981. Seasonal variation in soft tissue weights and trace metal burdens in the Bay mussels, *Mytillus edulis*. *Bull. Environ. Contam. Toxicol.*, 27: 821–828.

Lopez, M.A., Soria, M.L. and Repetto, M., 1989. Heavy metals accumulation in bivalve molluscs in the Huelva estuary. *Bull. Environ. Contam. Toxicol.*, 42: 634–118.

Modassir, Y. and Sivadas, S., 2003. Conservation and sustainable management of molluscan resources along Goa coast, central west coast of India. In: *Recent Advances in Environmental Science* (Fd.) K.G. Hiremath, Discovery Pub. House, New Delhi, pp. 399–421.

Neuberger–Cywiak, L., Achituv, Y. and Garcia, E.M., 2003. Effects of zinc and cadmium on the burrowing behaviour LC_{50} and LT_{50} on *Donax trunculus* Linnaeus (Bivalvia: Donacidae). *Bull. Environ. Contam. Toxicol.*, 70: 713–722.

Pawar, P.R., 2012. Molluscan diversity of mangrove ecosystem of Uran (Raigad), Navi Mumbai, Maharashtra, west coast of India. *Bull. Environ. Pharmacol. Life Sci.*, 1: 55–59.

Phillips, D.J.H., 1980. *Quantitative Aquatic Biological Indicators*. Application Science, Essex, U.K.

Phillips, D.J.H. and Rainbow, P.S., 1993. *Biomonitoring of Trace Aquatic Contaminants*. Elsevier Applied Science, New York.

Powell, M.I. and White, K.N., 1990. Heavy metals accumulation by barnacles and its implication for their use as biological monitors. *Mar. Environ. Res.*, 30: 91–118.

Reddy, M.S., Mehta, B., Dave, S., Joshi, M., Karthikeyan, L., Sarma, V.K.S., Basha, S., Ramachandriah, G. and Bhatt, P., 2007. Bioaccumulation of heavy metals in some commercial fishes and crabs of the Gulf of Cambay, India. *Curr. Sci.*, 92: 1489–1491.

Sarkar, S.K., Bhattacharya, B. and Debnath, S., 1994. The suitability of tropical marine bivalves as biomonitors of heavy metals in deltic Sundarbans, North–east India. *Chemosphere*, 29: 759–770.

Sarkar, S.K., Bhattacharya, B., Debnath, S., Bandopadhaya, G. and Giri, S., 2002. Heavy metals in biota from Sundarban wetland ecosystem, India. Implications to monitoring and environmental assessment. *Aquat. Ecosyst. Hlth. Manage.*, 5: 467–472.

Sidoumou, Z., Gnassia–Barelli, M., Siau, Y., Morton, V. and Romeo, M., 2006. Heavy metal concentration in molluscs from the Senegal coast. *Environ. Intl.*, 32: 384–387.

Singh, B.R. and Steinnes, E., 1994. Soil and water contamination by heavy metals. In: *Soil Processes and Water Quality* (Eds.) R. Lal and B.A. Stewart. Boca Raton, Florida, pp. 233–272

Tomazelli, A.C., Martinelli, L.A., Avelar, W.E.P., de Camargo, P.B., Fostier, A.–H., Ferraz, F.S.B., Krug, F.J. and Santos, D. Jr., 2003. Biomonitoring of Pb and Cd in two impacted watersheds in southeast Brazil using the freshwater mussel, *Anodontites trapesialis* (Lamarck, 1819) (Bivolvia: Mycetopoididae) as a biological monitor. *Brazilian Arch. Biol. Technol.*, 46: 673–684.

Turkmen, A. and Turkmen, M., 2004. The seasonal variation of heavy metals in the suspended particulate material in the Iskenderun Bay (North–eastern Mediterranean Sea, Turkey). *J. Fish. Aquat. Sci.*, 21: 307–311.

Walsh, K., Dunstan, R.H. and Murdoch, R.N., 1995. Differential bioaccumulation of heavy metals and organopollutants in the tissue and shell of the marine gastropods, *Austrocochlea constricta*. *Arch. Environ. Contam. Toxicol.*, 28: 35–39.

Wang, W.X. and Ke, C., 2002. Dominance of dietary intake of cadmium and zinc by two marine predatory gastropods. *Aquat. Toxicol.*, 56: 153–163.

Zorba, A.M., Jacob, P.G., Al–Bloushi, A. and Al–Hafisi, R., 1992. Clams as pollution bioindicators in Kuwait's marine environment: metal accumulation and depuration. *Sci. Total Environ.*, 120: 185–204.

Chapter 15

Advanced Maturation in Indian Major Carps Maintained on Lysine and Methionine Supplemented Semi-balanced Diet

A.K. Pandey[*1], M. Sarkar[2], C.T. Mahapatra[2], G. Kanungo[2] and P.K. Arvindakshan[2]*

[1]*National Bureau of Fish Genetic Resources, Canal Ring Road, Lucknow – 226 002, U.P., India*

[2]*Central Institute of Freshwater Aquaculture, Kausalyaganga, Bhubaneswar – 751 002, Orissa, India*

ABSTRACT

Broodstocks (2+year) of the Indian major carps, *Catla catla*, *Labeo rohita* and *Cirrhinus mrigala* were reared in the ratio of 40:30:30 from January onwards in 0.04 ha ponds with stocking density of 1,500 kg/ha. Fishes of the control group were fed on the 5 mm pelleted semi-balanced diet composed of fish meal 10 per cent, groundnut oil-cake (GOC) 35 per cent, soyabean oil-cake (SOC) 20 per cent, rice bran 24.27 per cent, wheat flour 10 per cent, trace mineral mixture 0.2 per cent, vitamin mixture 0.5 per cent and vitamin C 0.03 per cent (crude protein content 30.65 per cent; crude lipid 12.7 per cent; gross energy 3,800

* Corresponding author: E-mail: akpandey_cifa@yahoo.co.in

kcal/kg) whereas experimental group broodstock were maintained on the diet consisting of groundnut oil-cake (GOC) 35 per cent, soyabean oil-cake (SOC) 30 per cent, rice bran 23.27 per cent, wheat flour 10 per cent, lysine 0.5 per cent, methionine 0.5 per cent, trace mineral mixture 0.2 per cent, vitamin mixture 0.2 per cent and vitamin C 0.03 per cent lacking fish meal (as source of animal protein) but supplemented with lysine and methionine (crude protein 30.58 per cent; crude lipid 12.80 per cent, gross energy 3,820 Kcal/kg). Fishes from both the groups were fed @ 3 per cent of the body weight daily. Gonadal examination of broodstocks revealed that the fishes maintained on control diet matured during late May/early June whereas with the addition of essential amino acids, lysine and methionine in their feed, gonadal maturity among the experimental group fishes (both the sexes) was seen during April. Examination of cleared eggs under microscope revealed the germinal vesicle migration towards periphery in most of the cases. Fishes were bred successfully with ovaprim @ 0.5 ml/kg body weight for females and 0.2 ml/kg body weight for males (single injection). Natural spawning took place in the cemented tank after 6-8 hours of the drug administration. Better fertilization and hatching success were recorded in the carps kept on the semi-balanced diet supplemented with lysine and methionine as compared to those kept on the control diet.

Keywords: *Advanced maruration, Lysine, Methionine, Semi-balance diet, Indian major carps.*

Introduction

The importance of broodstock nutrition in aquaculture has been realized during the recent years (Matty, 1985; Watanabe, 1985; Luquet and Watanabe, 1986; Bromage and Roberts, 1995; Singh *et al.*, 2000, 2002; Pandey *et al.*, 2004). There exist reports on the nutritional requirements during gonadal development of guppy (*Poecilia reticulata*-Dahlgren, 1980), rainbow trout (*Oncorhynchus mykiss*-Takeuchi *et al.*, 1981; Watanabe *et al.*, 1984a; Knox *et al.*, 1988; Waagbo *et al.*, 1989; Washburn *et al.*, 1990; Bromage *et al.*, 1992), Nile tilapia (*Oreochromis niloticus*-Santiago *et al.*, 1983, 1985; De Silva and Radampola, 1991; Gunasekera *et al.*, 1995), coho salmon (*Oncorhynchus kisutch*-Hardy *et al.*, 1984), brook trout (*Salvelinus fontinalis*-McCormick and Naiman 1984), red seabream (*Chrysophrys major*-Watanabe *et al.*, 1984b, c, d, 1985), tilapia (*Oreochromis *mossambicus*-Soliman *et al.*, 1986), dwarf gourami (*Colisa lalia*-Shim *et al.*, 1989), gilthead seabream (*Sparus aurata*-Mourente and Odriozola, 1990a, b), bighead carp (*Aristichthys nobilis*-Santiago *et al.*, 1991), Arctic char (*Salvelinus alpinus*-Svedang, 1991), Indian major carps (*Catla catla* and *Labeo rohita*- Singh *et al.*, 2000) and freshwater catfish (*Heteropneustes fossilis*-Singh *et al.*, 2002). Success of induced breeding depends largely on proper maturity of the broodstocks because the fishes reared without appropriate food supply do not show proper gonadal development (full maturity) and synchronization of breeding (Matty, 1985; Watanabe, 1985; Bromage and Roberts, 1995). Generally, the broodstock diets are formulated with higher level of protein and energy (Halver, 1989; Lovell, 1989; Lall, 1991). Though use of fish meal containing high levels of protein has been the obvious choice in commercial fish feed but its practical acceptance has been limited due to high cost involved and its dubious role in contributing to the nitrogen load in the pond ecosystem (Viola and Lahav, 1991; Chakraborty and Chakraborty, 1998; Stibranyiova and Paraova, 2000). This compels the feed industry to resort to plant ingredients as protein source. Since the oil-cakes and rice bran are poor in lysine and methionine contents (Lim and Dominy, 1989), attempts have been made to balance the diets rich in plant ingredients with the two critical amino acids *viz.* lysine and methionine through commercially available lysomix and methiomix (Pandey *et al.*, 2001, 2004, 2012). The role of micronutrients such as polyunsaturated fatty acids (PUFA), vitamins C and E, carotenoids and various trace elements on gonadal maturation and egg quality of fish has been appreciated during recent past (NRC, 1993; Bromage and Roberts, 1995). Deficiency of any of the ten essential amino acids in fish diets results in reduced growth rate, poor feed conversion and in some cases poor appetite (Cowey and Sargent, 1979; Wilson and Robinson, 1982; Halver 1989). Methionine deficiency has been shown to cause bilateral cataracts in lake- and rainbow trout (Poston *et al*, 1977). We achieved success in advancing maturity by one month in Indian major carps (*Catla catla*, *Labeo rohita*) and catfish, *Heteropneustes fossilis* while maintaining the broodstocks on semi-balanced diet with protein level of 30 per cent and 35.5 per cent, respectively (Singh *et al.*, 2000, 2002). In order to record the effects of essential amino acids on maturity of the Indian major carps under field conditions, the semi-balanced diet was further supplemented with lysine and methionine and the gonadal conditions of both the sexes were as monitored for advanced maturation and commercial seed production of these species.

Broodstock Management

Two 0.04 ha ponds were selected and cleared off the predatory and weed fishes by applying bleaching powder @ 200-300 kg/ha. Lime was added @ 200 kg/ha to adjust pH of the water. Before stocking the broodfish, ponds were fertilized (in split doses) with raw cow dung @ 1000 kg/ha/month, urea @ 25 kg/ha/month and single-super-phosphate @ 30 kg/ha/month to sustain the plankton production. Since the bulk of aquaculture production in India is being contributed by the major carps (Kutty, 2001), the brood fishes (2+years) *of Catla catla* (1.7-1.9 kg), *Labeo rohita* (1.25-2.5 kg) and *Cirrhinus mrigala* (1.0-1.6 kg) were stocked during January @ 1500kg/ ha in the ratio of 40:30:30, respectively. The ingredients like fish meal, groundnut oil-cake, soyabean oil-cake and rice bran, wheat flour, calcium diphosphate, vitamin C, lysomix and methiomix were thoroughly mixed in proper ratio (Table 15.1) and pelleted feeds (5 mm) were prepared.

Table 15.1: Composition of Feeds Used for Rearing of Indian Major Carps

Ingredients	*Control Feed*	*Experimental Feed*
Fish meal (per cent)	10	Nil
Groundnut oil-cake (per cent)	35	35
Soyabean oil-cake (per cent)	20	30
Rice bran (per cent)	24.27	23.27
Calcium di-phosphate (per cent)	0.2	0.2
Vitamin mix (per cent)	0.5	0.5
Vitamin C (per cent)	0.03	0.03
Wheat flour (per cent)	10	10
Lysomix (per cent)	Nil	0.5
Methlomlx (per cent)	NIl	0.5
Composition of feed		
Crude protein (per cent)	30.65	30.58
Crude lipid (per cent)	12.70	12.80
Gross energy(kcal/kg)	3,800	3,820

The crude protein content of feed was analysed by Kjelltech method and the percentage of amino acid were analysed by Amino Acid Analyzer at the Central Institute of Fisheries Technology, Cochin. The gross energy was determined by bomb calorimeter (Parr Instruments, USA). The feed was given early in the morning @ 3 per cent of body weight. Water quality parameters also monitored regularly which ranged optimally- temperature (25-30°C), pH (7.1-8.2), dissolved oxygen (5.5-6.1 mg/l), total alkalinity (45-72 mg/l) and total ammonia (0.12-0.60 mg/l). Sampling was carried out at the regular interval of fifteen days and feed was adjusted accordingly.

Advanced Maturation

Gonadal examination of the broodstocks of the Indian major carps (*Catla catla, Labeo rohita* and *Cirrhinus mrigala*) maintained on the experimental diet showed very good maturity by April in both the sexes (males and females). Examination of cleared eggs under microscope revealed the germinal vesicle migration towards the periphery in most of the cases. All the threes species of Indian major carps were successfully bred with ovaprim (sGnRH+domperidone; Syndell Laboratories, Vancouver) @ 0.5 ml/kg body weight for females and 0.2 ml/kg body weight for males (single injection) during April, May and June. The details of breeding response of the Indian major carps maintained on the control and experimental diets have been summarized in Table 15.2. It was found that the broodstocks maintained on lysine and methionine supplemented diet recorded early maturation as they were bred successfully during late April with better fertilization and hatching success as compared to those maintained on the control diet (Pandey *et al.*, 2006a, b, 2007).

Conclusions

Since protein and lipids are the major components stored in egg yolk, their role in reproduction of fish is expected (Matty, 1985; Watanabe, 1985; Wilson, 1986; Luquet and Watanabe, 1986; Bromage and Roberts, 1995; Singh *et al.*, 2000, 2002; Nandi *et al.*, 2001). Watanabe *et al.* (1984c) showed that cuttlefish meal is a suitable protein source of broodstock diet in red seabream but the eggs produced were not markedly different than the broodstocks receiving different protein levels. Fecundity and oocyte diameter of guppy (*Poecilia reticulata*) did not show significant differences with

Table 15.2: Breeding Responses of the Indian Major Carps with Ovaprim

	Species	*Sets Taken*	*April*		*May*		*June*	
			Control	*Exptl.*	*Control*	*Exptl.*	*Control*	*Exptl.*
Fertilization(%)	*Catla catla*	3	10-15	50-60	35-40	55-65	60-75	85-90
	Labeo rohita	3	10-15	55-60	35-40	60-70	60-75	85-90
	Cirrhinus mrigala	3	20-30	60-65	40-50	68-70	50-65	78-86
Hatching success (%)	*Catla catla*	3	Nil	18-20	22-25	30-32	35-38	78-80
	Labeo rohita	3	Nil	25-30	20-22	40-50	35-45	78-82
	Cirrhinus mrigala	3	Nil	30-32	28-31	46-52	40-46	74-80

changes in dietary protein levels of broodstock (Dahlgren, 1980). Intersetingly, Santiago *et al.* (1991) observed that a higher dietary protein level enhanced reproductive performance in bighead carp (*Aristichthys nobilis*) in terms of total weight of eggs laid by the female at each spawning. Gunsekera *et al.* (1995) recorded early puberty, enhanced oocyte growth and faster maturation in Nile tilapia (*Oreochromis niloticus*) fed higher protein levels (32-40 per cent). A semi-balanced diet with crude protein content about 30 per cent and gross energy about 3,800 kcal/kg gave very good gonadal maturity in *Catla catla* and *Labeo rohita* by the end of May (Singh *et al.*, 2000) while broodstock of *H. fossilis* required a little more protein (35.5 per cent) in the diet to attain maturity by May (Singh *et al.*, 2002). Interestingly, in the present study supplementation of the semi-balanced feed with the essential (critical) amino acids like lysine and methionine advanced maturation of the Indian major carps by one month. Earlier, Somshekarappa *et al.* (1990) could advance maturity in *Labeo rohita* by the second week of April under the agroclimatic conditions of Karnataka using compounded feed consisting of broken rice 20 per cent, rice bran 20 per cent, fish meal 10 per cent, black gram 5 per cent, groundnut oil-cake 15 per cent and deoiled sunflower oil-cake 10 per cent in the diet (crude protein 27-45 per cent, gross energy 3,370 kcal/kg) and human chorionic gonadotropin (HCG). Though Gupta *et al.* (1990) also succeeded in advancing the maturity in *Catla catla* and *Labeo rohita* at Bhubaneswar (Orissa) during May-June through feeding a formulated feed comprising groundnut oil-cake (GOC) 48 per cent, soyabean oil-cake (SOC) 40 per cent, rice bran 5 per cent, fish meal 5 per cent, calcium diphosphate 1.5 per cent, sodium chloride 0.3 per cent and multivitamin mixture 0.1 per cent (having vitamin C 30 mg/kg and vitamin E 200 mg/kg) but the spawning success ranged between 60-85 per cent and rate of fertilization fluctuated widely from 10-95 per cent indicating clear variation in maturity responses in the brood fishes. Interestingly in the present study, breeding responses like fertilization and hatching percentage were better in all the three Indian major carps as compared to the controls. Though the actual mechanism(s) involved in gonadal maturation under the influence of lysine and methionine is not understood but both the essential amino acids help better utilization of feed in Indian major carps and catfish (Pandey *et al.*, 2001, 2004, 2012). The advanced maturation by April will pave the way for rematuration of the Indian major carps in the subsequent 56-64 days and multiple breeding leading to the enhanced seed production for freshwater aquaculture (Bhowmick *et al.*, 1977; Mahanta *et al.*, 1998; Pandey *et al.*, 2007).

Acknowledgements

We are grateful to the Director, Central Institute of Freshwater Aquaculture, Bhubaneswar for extending the necessary facilities to carry out the experiments. The authors are thankful to the farm staff for their help in execution of the work. Financial assistance provided by the Department of Biotechnology, Government of India, New Delhi is acknowledged.

References

Bhowmick, R.M., Kowtal, G.V., Jana, R.K. and Gupta, S.D., 1977. Experiments on second spawning of Indian major carps in the same season by hypophysation. *Aquaculture*, 12: 149–155.

Bromage, N. and Roberts R.J., 1995. *Broodstock Management and Egg and Larval Quality*. Blackwell Science Publications, Oxford, U.K.

Bromage, N., Jones, J., Randall, C., Thrush, M., Davies, B., Springate, J., Duston, J. and Barker, G., 1992. Broodstock management, fecundity, egg quality and the timing of egg production in rainbow trout (*Oncorhynchus mykiss*). *Aquaculture*, 100: 141–166.

Chakraborty, S. C. and Chakraborty, S., 1998. Effect of dietary protein level on excretion of ammonia in Indian major carp, *Labeo rohita*, fingerlings. *Aquacult. Nutr.*, 4: 47–51.

Cowey, C.B. and Sargent. J.R., 1979. Nutrition. In: *Fish Physiology. Vol. VIII. Bioenergetics and Growth* (Eds.) W.S. Hoar, D.J. Randall and J.R. Brett. Academic Press, New York, pp. 1–69.

Dahlgren, B.T., 1980. The effects of three different protein levels on the fecundity in the guppy, *Poecilia reticulata* (Peters). *J. Fish Biol.*, 15: 71–91.

De Silva, S.S. and Radampola, K., 1991. Effect of dietary protein level on the reproductive performance of *Oreochromis niloticus* (L.). In: *Proceedings of the Second Asian Fisheries Forum* (Eds.) R. Hirano and I. Hanyu. Asian Fisheries Society, Manila, Philippines, pp. 559–564.

Gunasekera, R.M., Shim, K.F. and Lam, T.J., 1995. Effect of dietary protein level on puberty, oocyte growth and egg chemical composition in the tilapia, *Oreochromis niloticus* (L.). *Aquaculture*, 134: 169–183.

Gupta, S.D., Reddy, P.V.G.K. and Pani, K.C., 1990. Advance maturity and spawning in Asiatic carps through broodstock management. In: *Carp Seed Production Technology* (Eds.) P. Kesavanath and K.V. Radhakrishnan. Special Publication 2. Asian Fisheries Society: Indian Branch, Mangalore, pp. 34–37.

Halver, J.E., 1989. *Fish Nutrition*. 2nd Edn. Academic Press, New York.

Hardy, R.W., Shearer, K.D. and King, L.B., 1984. Proximate and elemental composition of developing eggs and maternal soma of pen–reared coho salmon (*Oncorhynchus kisutch*) fed production and trace element fortified diets. *Aquaculture*, 43: 147–165.

Knox, D., Bromage, N.E., Cowey, C.B. and Springate, J.C.R., 1988. The effect of broodstock ration size on the composition of rainbow trout eggs (*Salmo gairdneri*). *Aquaculture*, 69: 93–104.

Kutty, M.N., 2001. Diversification of aquaculture. In: *Sustainable Indian Fisheries* (Ed.) T.J. Pandian. National Academy of Agricultural Sciences, New Delhi, pp. 189–212.

Lall, S.P., 1991. Concepts in the formulation and preparation of complete fish diet. In: *Fish Nutrition Research in Asia. Proceedings of the Fourth Asian Fish Nutrition Workshop* (Ed.) S.S. De Silva. Special Publication 1.5. Asian Fisheries Society, Manila, Phillippines, pp.1–12.

Lim, C. and Dominy, W., 1989. Utilization of plant proteins by warmwater fishes. In: *Proceedings of the Republic of China Aquaculture and Feed Workshop* (September 17–30, 1989) (Ed.) D.M. Akiyama. Singapore, pp. 143–161.

Lovell, R.T., 1989. *Nutrition and Feeding of Fish*. AVI, Van Hastrand, Reinnold, New York.

Luquet, P. and Watanabe, T., 1986. Interaction "nutrition–reproduction" in fish. *Fish Physiol. Biochem.*, 2: 121–129.

Mahanta, P.C., Rao, K.G., Pandey, G.C. and Pandey, A.K., 1998. Induced double spawning of an Indian major carp, *Labeo rohita*, in the same breeding season under the agroclimatic conditions of Assam. *J. Adv. Zool.*, 19: 99–104.

Matty, A.J., 1985. Nutrition and aquaculture. *Outlook Aquacult.*, 14: 14–20.

McCormick, S.D. and Naiman, R.J., 1984. Some determinants of maturation in brook trout, *Salvelinus fontinalis*. *Aquaculture*, 43: 269–278.

Mourente, G. and Odriozola, J.M., 1990a. Effect of broodstock diets on lipid classes and their fatty acid composition in eggs of gilthead seabream (*Sparus aurata* L.). *Fish Physiol. Biochem.*, 8: 93–101

Mourente, G. and Odriozola, J.M., 1990b. Effect of broodstock diets on total lipid and fatty acid composition of larvae of gilthead seabream (*Sparus aurata* L.) during yolksac stage. *Fish Physiol. Biochem.*, 8: 103–110.

Nandi, S., Chattopadhyay, D.N., Verma, J.P., Sarkar, S.K. and Mukhopadhyay, P.K., 2001. Effect of dietary supplementation of fatty acids and vitamins on breeding performance of the carp, *Catla catla*. *Reprod. Nutr. Dev.*, 41: 365–375.

National Research Council., 1993. *Nutrient requirement of Warmwater Fish*. National Academy Press, Washington, D.C.

Pandey, A.K., Arvindakshan, P.K., Sarkar, M., Mahapatra, C.T., Kanungo, G., Sahoo, G.C., Muduli, H.K. and Singh B.N., 2001. Role of lysine and methionine in growth of silver carp (*Hypophthalmichthys molitrix*) fingerlings under field conditions. In: *National Seminar on Riverine and Reservoir Fisheries: Challenges and Strategies* (May 23–24, 2001). Central Institute of Fisheries Technology, Cochin, p. CC–8.

Pandey, A.K., Mahapatra, C.T., Sarkar, M., Kanungo, G., Arvindakshan, P.K. and Singh, B.N., 2004. Effect of lysine and methionine on growth and maturation of the Indian catfish, *Heteropneustes fossilis*, under field conditions. *J. Exp. Zool. India*, 7: 91–95.

Pandey, A.K., Singh, B.N., Sarkar, M., Mahapatra, C.T., Kanungo, G. and Arvindakshan, P.K., 2006a. Early maturation of the Indian major carp, *Catla catla*, maintained on semi-balanced feed supplemented with lysine and methioninc. *J. Exp. Zool. India*, 9: 299–303.

Pandey, A.K., Singh, B.N., Sarkar, M., Mahapatra, C.T., Kanungo, G. and Arvindakshan, P.K., 2006b). Advancement of maturity in *Cirrhinus mrigala* maintained on semi–balanced diet supplemented with lysine and methionine. *Proc. Zool. Soc. (Calcutta)*, 59: 209–213.

Pandey, A.K., Singh, B.N., Sarkar, M., Mahapatra, C.T., Kanungo, G. and Arvindakshan, P.K., 2007. Advanced maturation in the broodstock of Indian major carp, *Labeo rohita* (Hamilton–Buchanan), maintained on lysine and methionine supplemented diet. *J. Appl. Biosci.*, 33: 150–152.

Pandey, A.K., Sarkar, M., Mahapatra, C.T., Kanungo, G. and Arvindakshan, P.K., 2012. Effect of dietary lysine and methionine supplementation on growth of *Catla catla* and *Labeo rohita* fingerlings. *J. Exp. Zool. India*, 15: 259–262.

Poston, H.A., Riis, R.C., Rumsey, G.L. and Ketala, G., 1977. The effect of supplemental dietary amino acids, minerals and vitamins on salmonids fed cataractogenic diets. *The Cornell Veterinarian* 67: 472–509.

Santiago, C.B., Aldaba, M.B. and Laron, M.A., 1983. Effect of varying dietary crude protein levels on spawning frequency and growth of *Sarotherodon niloticus* breeders. *Fish. Res. J. Philipp.*, 8 (2): 9–17.

Santiago, C.B., Aldaba, M.B., Abuan, E.F. and Laron, M.A., 1985. The effects of artificial diets on fry production and growth of *Oreochromis niloticus* breeders. *Aquaculture*, 47: 193–203.

Santiago, C.B., Camacho, A.S. and Laron, M.A., 1991. Growth and reproductive performance of bighead carp (*Aristichthys nobilis*) reared with or without feeding in floating cages. *Aquaculture*, 96: 109–117.

Shim, K.F., Landsman, L. and Lam, T.J., 1989. Effect of dietary protein on growth, ovarian development and fecundity in the dwarf gourami, *Colisa lalia* (Hamilton). *J. Aqua. Trop.*, 4: 111–123.

Singh, B.N., Das, R.C., Sahu, A.K., Kanungo, G., Sarkar, M., Sahoo, G.C., Nayak, P.K. and Pandey, A.K., 2000. Balanced diet for the broodstock of *Catla catla* and *Labeo rohita* and induced breeding performance using ovaprim. *J. Adv. Zool.*, 21: 92–97.

Singh, B.N., Das, R.C., Sarkar, M., Kanungo, G., Sahoo, G.C. and Pandey A.K., 2002. Balanced diet for rearing of singhi, *Heteropneustes fossilis* (Bloch), broodfish and induced breeding using ovaprim. *J. Ecophysiol. Occup. Hlth.*, 2: 57–64.

Soliman, A.K., Jauncey, K. and Robberts, R.J., 1986. The effects of dietary ascorbic acid supplementation on hatchability, survival rate and fry performance in *Oreochromis mossambicus* (Peters). *Aquaculture*, 59: 197–208.

Somoshekarappa, B., Chandrashekaraiah, H.N. and Nandeesha, M.C., 1990. Advancement of maturity in an Indian major carp, *Labeo rohita* through improved diet and human chorionic gonadotropin administration. In: *Carp Seed Production Technology* (Eds.) P. Keshavanath and K.V. Radhakrishnan. Special Publication 2. Asian Fisheries Society: Indian Branch, Mangalore, pp. 29–32.

Stibranyiova, I. and Paraova, J., 2000. The effect of lysine, methionine and threonine supplementation in practical diets with graded protein levels on growth, body composition and nitrogen excretion in juvenile African catfish (*Clarias gariepinus*). In: *Proceedings of Fish and Crustacean Nutrition: Methodology and Research for Semi–intensive Pond–based Farming System Workshop* (April 1996). Szarvas, Hungary. 23: 122–130.

Svedang, H., 1991. Effect of food quality on maturation rate in Arctic char, *Salvelinus alpinus* (L.). *J. Fish Biol.*, 39: 495–504.

Takeuchi, T., Watanabe, T., Ogino, C., Saito, M., Nishimura, K. and Nose, T., 1981. Effects of low–protein high calory diets and deletion of trace elements from the fish meal diet on reproduction of rainbow trout. *Bull. Jpn. Soc. Sci. Fish.*, 47: 654–664.

Viola, S. and Lahav, E., 1991. Effects of lysine supplementation in practical carp feeds on total protein sparing and reduction of pollution. *Israeli J. Aquacult.*, 43: 112–118.

Waagbo, R., Thorsen, T. and Sandnes, K., 1989. Role of dietary ascorbic acid on vitellogenesis in rainbow trout (*Salmo gairdneri*). *Aquaculture*, 80: 301–314.

Washburn, B.S., Frye, D.J., Hung, S.S.O., Dorosov, S.I. and Conte, F.S., 1990. Dietary effects on tissue composition, oogenesis and the reprroductive performance of female rainbow trout (*Oncorhynchus mykiss*). *Aquaculture*, 90: 179–195.

Watanabe, T., 1985. Importance of study of broodstock nutrition for further development of aquaculture. In: *Nutrition and Feeding in Fish* (Eds.) C.B. Cowey, A.M. Mackie and J.G. Bell. Academic Press, New York, pp. 395–414.

Watanabe, T., Takeuchi, T., Saito, M. and Nishimura, K., 1984a. Effect of low protein–high calory or essential fatty acid deficiency diet on reproduction of rainbow trout. *Bull. Jpn. Soc. Sci. Fish.*, 50: 1207–1215.

Watanabe, T., Arakawa, T., Kitajima, C. and Fujita, S., 1984b. Effect of nutritional quality of broodstock diets on reproduction of red seabream. *Bull. Jpn. Soc. Sci. Fish.*, 50: 495–501.

Watanabe, T., Ohhashi, S., Itoh, A., Kitajima, C. and Fujita, S., 1984c. Effect of nutritional composition of diets on chemical components of red seabream broodstock and eggs produced. *Bull. Jpn. Soc. Sci. Fish.*, 50: 503–515.

Watanabe, T., Itoh, A., Kitajima, C. and Fujita, S., 1984d. Effect of dietary protein levels on reproduction of red seabream. *Bull. Jpn. Soc. Sci. Fish.*, 50: 1015–1022.

Watanabe,T., Itoh, A., Satoh, S., Kitajima C. and Fujita, S., 1985. Effect of dietary protein levels and feeding period before spawning on chemical components of eggs produced by red seabream broodstock. *Bull. Jpn. Soc. Sci. Fish.*, 51: 1501–1509.

Wilson, R.P., 1986. Protein and amino acid requirements of fishes. *Annu. Rev. Nutr.*, 6: 225–244.

Wilson, R.P. and Robinson, K.H., 1982. *Protein and Amino Acid Nutrition for Channel Catfish.* NAFES, Mississippt State University Information Bulletin No. 25, pp. 1–18.

Chapter 16

Pineal Melatonin and Reproduction in Teleostean Fishes

A.K. Singh* and Ruchi Singh

National Bureau of Fish Genetic Resources, Canal Ring Road, P.O. Dilkusha, Lucknow – 226 002, Uttar Pradesh, India

ABSTRACT

Melatonin, a hormone fundamentally synthesised in the pineal gland contributes to synchronizing behaviours and neuroendocrine regulations with the daily and annual variations of the photoperiod. Studies of the targets and modes of action of melatonin in fish, are of interest to our understanding of time-related neuroendocrine regulation for aquacultural purposes. Melatonin is a much conserved feature in vertebrates that plays a central role in the entrainment of daily and annual physiological rhythms. In this paper, we review the available data on (i) the sites of melatonin production in fish, (ii) the mechanisms that control its daily and annual rhythms of production and (iii) the characterization of its different receptor types, their location and regulation. The *in vivo* and *in vitro* data on melatonin describes reproduction, growth, feeding and behavioural responses. Finally we discuss how manipulation of the photic cues impact on fish circannual clock and annual cycle of reproduction, and how this can be used for aquaculture

* Corresponding author: E-mail: aksingh56@rediffmail.com

purposes. It is possible to manipulate fish gonadal function through exogenous melatonin. However gonadal responses to melatonin vary and depend on time, mode and duration of the administration of the hormone. The effect of melatonin (MLT) implants on the reproductive performance of fish has also been discussed.

Keywords: *Fish, Reproduction, Photoperiod, Melatonin, Circadian, Circannual, Sex differentiation.*

Introduction

The pineal gland is a neuroendocrine structure that synthesizes and releases melatonin with both daily and seasonal rhythms. This peculiar pattern of melatonin release gives the hormone a pivotal function in the synchronization of numerous circadian and annual function with cyclic changes of environmental factors, especially light. Melatonin was isolated in 1958 by Lerner and his associates and its chemical nature was identified as N-acetyl-5-methoxytryptamine (Lerner *et al.*, 1958). Pineal gland is the fundamental seat of melatonin hormone production in animals including fish. However extra source of melatonin production is also known (Falcon, 2010). Melatonin is a naturally occurring compound found in animals, plants, and microbes (Caniato *et al.*, 2003; Tan *et al.*, 2011). In animals, circulating levels of melatonin vary in a daily cycle, thereby regulating the circadian rhythms of several biological functions (Altun and Altun, 2007; Catarina *et al.*, 2010). Melatonin is produced by the photoreceptor cells of the retina and pineal organ (or epiphysis or "Third eye"). The latter is organized as a simplified retina located above the brain hemispheres in a window below the skull and between the two eyes. Photoreceptors detect rapid changes in light environment [which results in the production of a neurotransmitter (glutamate)]. They also provide information on the respective duration of the light (L) and dark (D) phases through the rhythmic production of melatonin. This usually involves circadian clocks. In other words, each photoreceptor cell is a circadian system in which the LD cycle entrains the clock, which in turn drives the expression and activity of enzymes involved in the control of melatonin secretion. Temperature as well as internal chemical transmitters and hormones also modulate the shape of the daily oscillations. These include catecholamines, cortisol, sex-steroids (Falcon, 2010; Singh *et al.*, 2012), and peptides, which might represent elements of a feedback loop. Melatonin levels in the pineal gland and blood are higher during night than during day, whereas a reverse pattern is observed in the retina. Retinal melatonin is produced and metabolized *in-situ*. Pineal melatonin is released in the blood, and the profile of the rhythm varies along the annual cycle. The nocturnal melatonin increase is of high amplitude/short duration during summer, and of low amplitude/long duration during winter (Arendt and Skene, 2005; Falcon, 2010).

Biosynthesis of Melatonin

Melatonin is synthesized from the amino acid Tryptophan. The synthesis takes place primarily in the pineal gland.The sequence of events for the biosynthesis of melatonin is as follows : (1) tryptophan is taken up from the circulation and is converted into serotonin to give N-acetylserotonin; (2) Serotonin is converted into N-acetylserotonin by the enzyme arylalkylamine N-acetyltransferase (AA NAT,EC 2.3.1.87); (3) N-acetylserotonin is converted to melatonin by the enzyme hydroxyl-indole-*O*-methyltransferase (HIOMT,EC2.1.1.4).

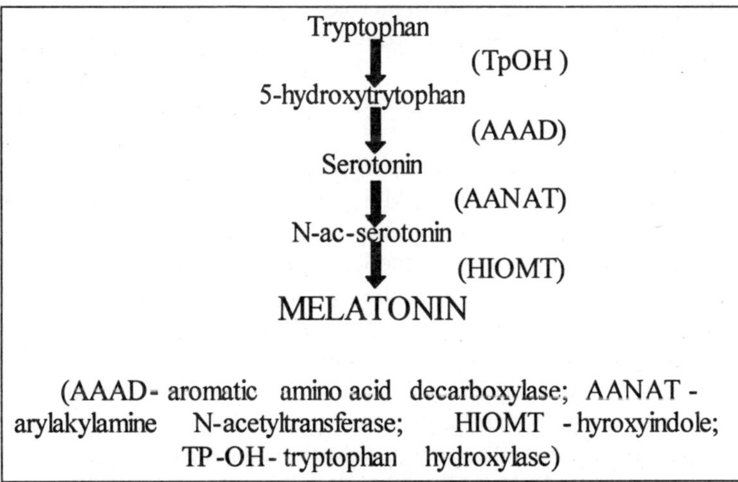

(AAAD- aromatic amino acid decarboxylase; AANAT - arylakylamine N-acetyltransferase; HIOMT - hyroxyindole; TP-OH- tryptophan hydroxylase)

Synthesis and secretion of melatonin is dramatically affected by light exposure to the eyes. The fundamental pattern observed is that serum concentrations of melatonin are low during the daylight hours, and increase to a peak during the dark. The mechanism behind the day night change pattern of melatonin secretion during the dark cycle is the activity of the rate-limiting enzyme serotonin N-acetyltransferase (NAT) in melatonin synthesis which is low during daylight and peaks during the dark phase. In some species, circadian changes in NAT activity are tightly correlated with transcription of the NAT messenger RNA, while in other species, post-transcriptional regulation of NAT activity is responsible. Activity of the other enzyme involved in synthesis of melatonin from serotonin–the methyltransferase–does not show regulation by pattern of light exposure. Two melatonin receptors have been identified from mammals (designated Mel1A and Mel1B) that are differentially expressed in different tissues and probably participate in implementing differing biologic effects. These are G protein-coupled cell surface receptors. The highest density of receptors has been found in the suprachiasmatic nucleus of the hypothalamus, the anterior pituitary (predominantly pars tuberalis) and the retina. Receptors are also found in several other areas of the brain (Margarita *et al.*, 2005).

Biological Action of Melatonin

Many biological effects of melatonin have been reported (Boutin, 2005) while others are due to its role as a pervasive and powerful antioxidant with a particular role in the protection of nuclear and mitochondrial DNA (Falcon *et al.*, 2010). In the mid-70s, it was demonstrated the that the production of melatonin exhibits a circadian rhythm (Oliveira *et al.*, 2010). Many animals use the variation in duration of melatonin production each day as a seasonal clock (Amano *et al.*, 2004). Melatonin has both lipophilic and hydrophilic properties. Once formed it is released into capillaries and can reach all tissues of the body within a very short period (Cardinali and Pevet, 1998). In the circulation,melatonin is partially bound to albumin (Cardinali *et al.*, 1972) and can also bind to haemoglobin (Gilad and Zisapel, 1995). The half- life of melatonin after intravenous infusion is about 30 min (Mallo *et al.*, 1990). However, a recent study indicates that melatonin exhibits a bi-exponential decay with a first distribution half-life of 2 minutes and a second metabolic half-life of 20 min (Claustrat *et al.*, 2005).

Melatonin and Osmoregulation

The action of melatonin in major organs responsible for osmoregulation in fish has been extensively reviewed

(Ewa Kulczykowska, 2001). The influence of melatonin on water/ ion excretion by affecting the circulatory blood homodynamic and by interrelations with other hormones system engaged in water/ion homeostasis have been reported. New data providing the first evidence for the presence of melatonin binding sites in fish gills and kidneys have also ben shown (Ewa Kulczykowska, 2001). It is proposed that new approach that may lead to an improved understanding of the role of melatonin in osmoregulation processes need to be explored.

Melatonin in Reproduction

Melatonin, synthesized in the pinealocytes and secreted into the blood during the scotophase provides an internal environmental signal of alternating photo- and scotophase (diurnal) and of seasonal changes in photoperiod as measured by the duration of the melatonin peak (Bromage *et al.*, 1995). However, the site(s) of action of melatonin on the HPG axis in fishes is not clear. Moreover, experimental evidence based on studies in which melatonin was administered to fish, or of fish subjected to pinealectomy have not provided any clear link between melatonin rhythms and gonadal maturation in fishes. Following melatonin administration to different fish species, some authors report gonadotrophic responses, some authors report antigonadotrophic responses, whereas some were unable to find any measurable response; where responses were noted, for the most part they were relatively small, and often seasonally linked (Sundararaj and Keshavanath, 1976; Nayak and Singh, 1987, Joy and Agha, 1991; Bromage *et al.*, 1995; Singh and Pandey, 1996; Singh and Prasad, 1998).

Recent studies demonstrated that annual changes in the duration of photoperiod play an important role in determining reproductive periodicity in Indian major carp *Catla catla* (Bhattacharya *et al.*, 2005, 2007). Generally it is agreed that the neuroendocrine mechanism by which the photoperiod influences reproduction involves melatonin, which in turn might be used to measure and predict daily and seasonal time (Bromage *et al.*, 2001). However, a possible role of melatonin in the regulation of seasonal reproduction in any tropical or sub-tropical fish remains largely speculative due to lack of comprehensive study in any particular species.

In fish, melatonin is a conservative chemical signal of light-dark cycles of the environment in all the vertebrates. In all the fish species investigated so far, melatonin production is high during the night time and low during the day time. The fish pineal, through the cyclical synthesis and release of melatonin, seems to be involved in the timing and control of a number of rhythmic functions, especially reproduction. Melatonin is important in controlling the reproductive seasonality by stimulating the final stages of sexual maturation and by synchronizing the oocyte maturity with optimal timing of spawning. Melatonin has also been found to affect estradiol levels in mature carp females and to indirectly influence the GtH II secretion *via* hypothalamic stimulatory (GnRH) centers. In Atlantic croaker, *Micropogonias undulates*, melatonin influences LH secretion directly at the pituitary level and indirectly at the brain level. Melatonin has been considered as one of the candidates that mediate the transduction of photoperiodic information to the brain pituitary-gonad axis in gonadal maturation of precocious male masu salmon. On the other hand, a study on the rainbow trout suggested that the pineal organ may not be the site of pacemaker that controls rhythms, and further research is required to study the involvement of other photoperiod-tranducing systems and melatonin (non-pineal origin) in the regulation and expression of circadian rhythm in this species. The response of gonadal activity to surgical removal of pineal organ under both natural and altered photothermal conditions has been reported to indicate the possibility of pineal organ being an important component of the neuro-endocrine-gonadal axis. There are evidences to suggest that the pineal organ and/or melatonin may play a role in the seasonal maturation of gonad in fish and in transduction of photoperiodic information to the brain-pituitary-gonadal axis in teleosts.

Pineal hormone melatonin is now widely recognized to exert profound effects on fish reproduction (Bhattacharya *et al.*, 2005). However, it is reported to help the fish to adapt to changing environmental factors (Falcon *et al.*, 2010). Management of fish reproduction needs adequate basic understanding of factors that control natural reproduction *i.e.*, the reception of external stimuli and the transliteration of message into physiological responses. We have studied the role of melatonin in gonadal development and sex differentiation in *Oreochromis mossambicus* (Singh, 2009; Singh *et al.*, 2012). Effect of melatonin implants on reproduction and growth in fishes has also been reported (Bromage and Randall, 2001). The effect of melatonin implants on the reproductive performance in fish brood stock has been studied (Bromage and Randall, 2001;Alvarino *et al.*, 2002).

Ghosh and Nath (2005) showed that melatonin is effective in inhibiting ovarian vitellogenesis, plasma GtHII levels, and gonadosomatic index (GSI) in a catfish (*Clarias batrachus*). So melatonin may control the reproduction by blocking maturation gonadotropin (LH) release from the pituitary. Opposite effect has been shown where increased LH level in blood, after melatonin injection was observed during the night (dark phase) to mature female carp (Ghosh and Nath, 2005). During the long day the pineal gland and melatonin stimulates the final stages of, sexual maturation by synchronizing full oocyte maturity, with the optimal spawning period and stimulates vitellogenesis, in the post-spawning period in carp (Popek *et al.*, 1991). It has also been reported that melatonin modulates the hypothalamic dopaminergic system rather (Popek *et al.*, 1991) than directly affecting gonads (Popek *et al.*, 1996) or the pituitary gland (Popek *et al.*, 2000), although in other fish species direct melatonin action on the pituitary was observed (Khan and Thomas 1996,). It seems that melatonin can affect reproductive functions most likely by a dopaminergic pathway, which is supported by Falcon *et al.* (2010) who has proved that melatonin is able to inhibit the hypothalamic secretion of dopamine in fish. Dopamine is one of catecholamines produced by hypothalamic aminergic nuclei NRL (nucleus recessus lateralis) and NRP (nucleus recessus posterioris). Dopamine is the most dominant gonadotropin-releasing inhibiting factor (GRIF) (Falcon *et al.*, 2010). The inhibiting effect of dopamine was also proved on LHRH synthesis and secretion (Falcon *et al.*, 2010). Previous studies on the effect of melatonin on the activity of hypothalamic aminergic nuclei in mature female carp, using the Falck-Hillarp histochemical fluorescence technique, has also confirmed the ability of melatonin to inhibit the secretion of catecholamines produced by the hypothalamus (Popek and Epler, 1999). Given that dopamine accounts for 70 per cent of hypothalamic catecholamines (Falcon *et al.*, 2010), these results support the earlier hypotheses concerning the inhibitory role of melatonin on the dopaminergic system in the carp. The results of experiments *in vitro*, where carp females hypothalami were perifused with, or without melatonin addition, also indicated that melatonin inhibited hypothalamic dopamine release (Popek *et al.*, 2005). Discussing the role of melatonin in carp reproduction, it is worth to mention that melatonin stimulates hypothalamic noradrenaline secretion in spawning time, (Ghosh and Nath, 2005). Melatonin affects not only the level of catecholamines in the carp hypothalamus by stimulating or inhibiting their release during the summer, but can also take part in the indirect stimulation of LH release by stimulating and inhibiting the activity of the aminergic system. So the replacement of synthetic drugs by natural melatonin seems to be serious alternative for fisheries. Further research is required to examine the dose and time as well the way of administration dependence of melatonin treatment, on LH secretion, ovulation and spawning efficiency in carp (Popek *et al.*, 2006). There are evidences to suggest that the pineal organ and/or melatonin may play a role in the seasonal maturation of gonad in fish and in transduction of photoperiodic information to the brain-pituitary-gonadal axis in teleosts, however, results of experimental studies in different fish have been identical (Gosh and Nath, 2005; Falcon *et al.*, 2010). Reports are also available that melatonin synchronizes annual variations in fish reproduction and growth, although contradictory results were very often obtained even within the same species.

Intraperitoneal injections of melatonin (20μg/fish for 50 days) into goldfish inhibited the increase in gonad size which accompanied increased daily light exposure in those animals which received placebo injections. The discovery that the melatonin-treated fish had larger pituitary gonadotrophs suggested a possible mode of action of the melatonin (Falcon *et al.*, 2010). The role of melatonin in various biological phenomena can be investigated by studying correlations between melatonin levels, production and receptors and the phenomenon of interest, or by experimental studies involving the removal of the source (s) of melatonin (*i.e.*, usually the pineal organ) and/or by administration of melatonin. Alvaro (2004) found the effect of 18 mg melatonin (MLT) implants on the reproductive performance of turbot (*Scophthalmus maximus* L.) broodstock was determined in 20 females and 12 males. An inhibitory effect of melatonin injections (20 or 100 pg/day) on ovarian weights, pituitary GTH cells and oocyte maturation was found in the catfish *Heteropneustes fossilis*, maintained during the pre-spawning and spawning periods at LD 12 : 12 (Sundararaj and Keshavanath, 1976). On the other hand, there was not any effect of injections of 10 or 100 pg melatonin/day to *H. fossilis* males weighing 30g was found on GSI, seminal vesicle weights, spermatogenesis or number of pituitary GTH cells (Sundararaj and Keshavnath, 1976). The experiment was performed during the pre-spawning period in May-June, and the fish were kept under an LD 12: 12 regime. Further, melatonin administration was found to have no effect on reproduction in *Hetropneustus fossilis* (Garg, 1989).

Administration of melatonin at different doses (25, 50, or 100 µg/100 g body weight) for 1, 15, or 30 days resulted in either pro- or anti-gonadal effects depending on the reproductive seasons. Melatonin secretory cells were immunocytochemically localized only in the end vesicle. Diurnal and seasonal studies indicated that the serum levels of melatonin exhibit a minimum diurnal value in the mid-day of all seasons, but the peak value is attained either at mid-night or just before the onset of light (Maitra *et al.,* 2005). It was suggested that a stimulatory action of melatonin on steroidogenesis of early stage ovarian follicles, shifted to an inhibitory action of the higher concentrations of melatonin in later stages of development, but with a bimodal action of melatonin (related to hormone concentration) being present during the time of maximum follicular growth (Singh *et al.,* 2012). Melatonin-induced changes in ovarian function in the freshwater fish *Channa punctata* the effect of melatonin on reproduction seems to depend on the photoperiod and duration of exposure to melatonin. In fish held either in long days or LL, daily restricted exposure to melatonin tends to inhibit gonads, whereas continuous exposure to melatonin either stimulate or has no effect these results apparently support the view that duration of exposure to melatonin signals darkness (Renuka and Joshi, 2010).

Influences of photoperiod on plasma melatonin profiles and effects of melatonin administration on long-day-induced modification in masu salmon (*Oncorhynchus masou*) have been investigated in order to reveal the roles of melatonin in the regulation of smoltification in salmonids. Under light-dark (LD) cycles, plasma melatonin levels exhibited daily variation, with higher values during the dark phase than during the light phase. The duration of nocturnal elevation under short photoperiod (LD 8:16) was longer than that under long photoperiod (LD 16:8). Melatonin feeding (0.01, 0.1 and 1 mg/kg body weight) elevated plasma levels of melatonin in a dose-dependent manner for at least 7 h but not for 24 h. When masu salmon reared under short photoperiod were exposed to long photoperiod (LD 16:8) and fed melatonin (1 mg/kg body weight) 7 hours before the onset of darkness, a significantly smaller proportion appeared in the melatonin-fed group after 32 days than in the control group. Effect of melatonin on reproductive processes in fish, most often, points to its stimulatory or inhibitory effect depending on the species, season of the year, time of the day, and degree of sexual maturity, but fails to show how and where melatonin affects the reproductive system (Falcon *et al.*, 2010).

Melatonin treatment induced testicular regression also an elevated serum cortisol, effects comparable to those in hamsters exposed to short days. These findings support the hypothesis that timed oral administration of melatonin can mimic the effects of short days and provide a method by which melatonin can be delivered without the potentially confounding effects of handling and injection stress (Falcon *et al.*, 2010). Melatonin injection has been reported to result in an increase in dopamine concentration in the hypothalamus, which indicates that melatonin influenced aminergic nuclei inhibiting secretion of dopamine. Dopamine probably remained unreleased in the hypothalamic tissue, whereas in pinealectomized fish the lack of melatonin influenced continuous dopamine secretion, what was shown as the low dopamine level in the hypothalamus (Popek, 2006).

Conclusion

Several vital biological actions in fish are controlled by the pineal melatonin. In seabass, it has been shown that the feeding rhythm (diurnal vs. nocturnal) strongly influence the daily patterns of digestive function and clock gene expression in the liver (feeding-entrained clock), (Ana Del Pozo *et al.*, 2012).

The gonad development, sex ratio, and sex steroid [(11-ketotestosterone (11-KT); estradiol (E2); and testosterone, (T)] content in muscle extracts of juveniles Senegal sole was found to be influenced by early exposure of different temperature cycles. The results of such study has provided the first insights into the effect of daily thermocycles on sex differentiation in fish, and underline the key role of natural environmental cycles on the control of sex ratios during larval development, which may be applied to the manipulation of sex ratio in aquaculture (Borja *et al.*, 2011).The reproduction rhythms in Senegal sole has been reported to persist even in the absence of the natural lengthening of photoperiod, although it is also reported that photoperiod manipulation alters the seasonal modulation of melatonin, increase sex steroids and vitellogenin at pre-spawning, and slightly advanced the timing of spawning (Catarina *et al.*, 2011). Studies on the effect of melatonin on reproductive processes in fish, most often, point to its stimulatory or inhibitory effect depending on the species, season of the year, time of the day, and degree of sexual maturity, but fail to show how and where melatonin affects the reproductive system.

In vitro and *in vivo* studies have demonstrated stimulatory effects of GnRH-2 on nocturnal melatonin secretion in the sea bass pineal organ and was the first report in a vertebrate species, converging evidence supporting a role of GnRH-2 in the modulation of fish pineal functions (Arianna *et al.*, 2010). Melonin implanted fishes have shown supraphysio-plasma melatonin concentration during three months after implantation. Melonin implanted fish melatonin has shown higher spermatozoa concentrations in fish and gonadal recrudescence started 45 days earlier. However, further researched are advised so as to ascertain the required dose and duration of melatonin implants for amenable reproduction in fishes.The possible implication of pineal melatonin in gonadal differentiation has been explored and proposed as a a new approach to sex control technology in fishes (Sing and Pandey, 1996;Singh, 2005; Singh *et al.*, 2012).

References

Altun, A. and Altun, U.B., 2007. "Melatonin: therapeutic and clinical utilization". *Int. J. Clin. Pract.*, 61 (5): 835–45.

Alvarino, J.M., Raboller, P.G., Olmedo, M.B. Alvarez-Blazquez, E. Ubilla and Peleteiro, J.B., 2002. An effect of melatonin implants on reproduction and growth of turbot broods tock.*Aquaculture international*, 9(6): 477–487.

Álvaro Aguiar Coelho Teixeira; Manuel de Jesus Simões; Valéria Wanderley Teixeira and José Maria Soares Júnior (2004. Evaluation of the implantation in pinealectomized and/or submitted to the constant illumination rats. *International Journal of Morphology*, 22(3): 189–194.

Amano, M., Iigo, M., Ikuta, K., Kitamura, S., Okuzawa, K., Yamada, H. and Yamamori, K., 2004. Disturbance of plasma melatonin profile by high dose melatonin administration inhibits testicular maturation of precocious male masu salmon. *Zoological Science,* 21: 79–85.

Ana Del Pozo, Ander Montoya, Luisa María Vera, Francisco Javier Sánchez–Vázquez, 2012. Daily rhythms of clock gene expression, glycaemia and digestive physiology in diurnal/ nocturnal European seabass. *Physiology and Behaviour* DOI: 10.1016/j.physbeh.

Arendt, J. and fSkene, D.J., 2005. Melatonin as a chronobiotic. *Sleep Med. Rev.,* 9: 25–39

Arianna Servili, Christèle Lethimonier, Jean–Jacques Lareyre, José Fernando López–Olmeda, Francisco Javier Sánchez–Vázquez, Olivier Kah, José Antonio Muñoz–Cueto, 2010. The Highly conserved gonadotropin–releasing hormone–2 form acts as a melatonin–releasing factor in the pineal of a teleost fish, the european sea bass *Dicentrarchus labrax*. *Endocrinology,* 151(5): 2265–75. DOI: 10.1210/en.2009–120.

Bhattacharya, S., Chattoraj, A. and Maitra, S. K., 2005. Ultrastructural and morphometric study of the pineal complex in Indian major carp *Catla catla* in response to continuous light and darkness. *Journal of Endocrinology and Reproduction,* 9: 1–10.

Borja Blanco-Vives, Luisa María Vera, Jesús Ramos, María José Bayarri, Evaristo Mañanós, Francisco Javier Sánchez–Vázquez (2011. Exposure of larvae to daily thermocycles affects gonad development, sex ratio, and sexual steroids in *Solea senegalensis,* kaup. *Journal of Experimental Zoology Part A Ecological Genetics and Physiology,* 315(3): 162–9. DOI: 10.1002/jez.664

Boutin, J.A., Audinot, V., Ferry, G. and Delagrange, P., 2005. "Molecular tools to study melatonin pathways and actions". *Trends in Pharmacological Sciences,* 26 (8): 412–9.

Bromage, N., Porter, M. and Randall, C., 2001. The environmental regulation of maturation in farmed finfish with special reference to the role of photoperiod and melatonin. *Aquaculture,* 197: 63–98.

Bromage, N.R., Randall, C.F., Porter, M.J.R. abd Davies, B., 1995. How do photoperiod, the pineal, melatonin and circannual rhythms interact to coordinate seasonal reproduction in salmond fish? In: Goetz, F., Thomas, O. (Eds.), *Proc. Fifth Intl. Symp. Reprod. Physiol. Fish Symp.*, Austin 95, pp. 164–166.

Caniato, R., Filippini, R., Piovan, A., Puricelli, L., Borsarini, A. and Cappelletti, E.M., 2003. Melatonin in plants. *Advances in Experimental Medicine and Biology*, 527: 593–597.

Cardinali, D.P., Lynch, H.J. and Wurtman, R.J., 1972. Binding of melatonin to human and rat plasma proteins. *Endocrinol.*, 91: 1213–1218.

Cardinali, D.P. and Pevet, P., 1998. Basic aspects of melatonin action. *Sleep Med. Rev.*, 2: 175–190.

Catarina Oliveira, Evaristo Mañanós, Jesus Ramos, Francisco Javier Sánchez–Vázquez, 2011. Impact of photoperiod manipulation on day/night changes in melatonin, sex steroids and vitellogenin plasma levels and spawning rhythms in Senegal sole, *Solea senegalensis. Comparative biochemistry and physiology. Part A, Molecular and integrative physiology*, 159(3): 291–5. DOI: 10.1016/j.cbpa.2011.03.025.

Claustrate, B., Brun, J. and Chazot, G., 2005. The basic physiology and pathophysiology of melatonin, *Sleep Med. Rev.*, 9: 11–24.

Ewa, Kulczykowska, 2001. A review of the multifunctional hormone melatonin and a new hypothesis involving osmoregulation. *Reviews in Fish Biology and Fisheries*, 11 (4): 321–330.

Falcón, J.H., Migaud, J.A., Muñoz-Cueto and M. Carrillo, 2010. Current knowledge on the melatonin system in teleost fish. *General and Comparative Endocrinology*, 165 (3): 469–482.

Garg, S.K., 1989. Effect of pinealectomy, eye enucleation, and melatonin treatment on ovarian activity and vitellogenin level or long photoperiod. *Journal of Pineal Research.*, 7: 91–104.

Ghosh, J. and Nath, P., 2005. Seasonal effects of melatonin on ovary and plasma gonadotropin and vitellogenin levels in intact and pinealectomized catfish, *Clarias batrachus* (Linn). *Indian Journal of Experimental Biology*, 43: 224–232.

Gilad, E. and Zisapel, N., 1995. High-affinity of melatonin to haemoglobin. *Biochem. Mol. Med.*, 56: 115–120.

Joy, K.P. and Agha, A.K., 1991. Seasonal effects of administration of melatonin and 5-methoxytryptophol on the ovarian activity in the catfish, *Heteropneustes fossilis* (Bloch). *Journal of Pineal Research*, 10: 65–70.

Khan, I.A. and Thomas, P., 1996. Melatonin influences gonadotropin II secretion in the Atlantic croaker (*Micropogonias undulatus*). *General and Comparative Endocrinology*, 104: 231–242.

Lerner, A.B., Case, J.D., Takahashi, Y., Lee, Y. and Mori, W., 1958. Isolation of melatonin, the pineal gland factor that lightens melanocytes. *Journal of American Chemical Society*, 80: 25–87.

Maitra, S.K., Chattoraj, A. and Bhattacharyya, S., 2005. Implication of melatonin in oocyte maturation in Indian major carp *Catla catla*. *Fish Physiology and Biochemistry*, 31: 201–207.

Mallo, C., Zaidan, R., Galy, G., Vermeulen, E., Brun, J., Chazot, G. and Claustrat, B., 1990. Pharmacokinetics of melatonin in man after intravenous infusion and bolus injection. *Eur. J. Clin. Pharmacol.*, 38: 297–301.

Nayak, Ñ.Ê. and Singh, Ô.Ñ., 1987. Effect of melatonin and 5–methoxytryptamine on sex steroids and thyroid hormones during the prespawning phase of the annual reproductive cycle in the freshwater teleost, *Clarias batrachus; J. Pineal Res.*, 4: 377–386.

Oliveira, N.J., Duncan, Pousão-Ferreira P., Mañanós, E., F.J. Sánchez-Vázquez, 2010. Influence of the lunar cycle on plasma melatonin, vitellogenin and sex steroids rhythms in Senegal sole, *Solea senegalensis*. *Aquaculture*, 306: (1–4), 343–347.

Popek, W., £uszczek-Trojnar, E., Dr¹g-Kozak, E., Fortuna-Wrońska, D. and Epler, P., 2005. Effect of the pineal gland and melatonin on dopamine release from perifused hypothalamus of mature female carp during spawning and winter regression. *Acta Ichthyologica et Piscatoria*, 35: 65–71.

Popek, W., £uszczek-Trojnar, E., Dr¹g-Kozak, E., Rz¹sa, J. and Epler, P., 2006. Effect of melatonin on dopamine secretion in the hypothalamus of mature female common carp, *Cyprinus carpio* L. *Acta Ichthyol.Piscat.*, 36 (2): 134–141.

Popek, W., £uszczek-Trojnar, E. and Epler, P., 2000. Effects of pineal gland and melatonin on maturation gonadotropin (GtH2) secretion from perifused pituitary glands of mature carp during spawning. *Electronic Journal of Polish Agricultural Universities, Fisheries* 3 (1): http: / / www.ejpau.media.pl / volume3 / issue1 / fisheries / art–01.html

Popek, W., Bieniarz, K. and Epler, P., 1991. Role of the pineal gland in sexual cycle in common carp. Pp. 99.102. *In*: Surowiak J., Lewandowski M.H. (eds.) *Chronobiology and Chronomedicine*. Verlag Peter Lang, Frankfurt.

Popek, W. and Epler, P., 1999. Effects of intraventricular melatonin microinjections on hypothalamic catecholamine activity in carp females during a year. *Electronic Journal of Polish Agricultural Universities, Fisheries* 2 (2): http: / / www.ejpau.media.pl/ volume2 / issue2 / fisheries / art–02.html

Popek, W., Epler, P., Kime, D.E. and Soko³owska-Miko³ajczyk, M., 1996. Melatonin does not affect steroidogenesis or maturation of carp oocytes in vitro during the prespawning period. *Polish Archives of Hydrobiology*, 43: 379–385.

Renuka, B.N. Joshi, 2010. Melatonin-induced changes in ovarian function in the freshwater fish *Channa punctatus* (Bloch) held in long days and continuous light. *General and Comparative Endocrinology*, 165: 42–46.

Singh, A.K., 2005. Endocrine techniques of reproductive management in finfish aquaculture. Lecture Synopsis and laboratory Manual. *National Workshop on Critical Appraisal and Hands on training in Advanced Techniques in Endocrine Research*. Department of Zoology, Aligarh Muslim University, Aligarh, Jan. 16–30, 2005.

Singh, A.K., 2009. Role of pineal melatonin during environmental sex determination in fishes. In: *Recent Advances in Hormonal Physiology of Fish and Shellfish Reproduction*, (Eds.) B.N. Singh and A.K. Pandey. Narendra Publication House, New Delhi, p. 265–271.

Singh, A.K. and Pandey, A.K., 1996. Implication of pineal melatonin in reproductive management and sex differentiation. *Journal of Natcon.*, 10 (1): 61–67.

Singh, A.K. and Prasad, G.C., 1998. Effect of melatonin and other chemical agents on thyroid gland *in vitro. J. Exp. Zool.*, 1(2): 131–141.

Singh, R., Singh, A.K. and Tripathi, Madhu, 2012. Melatonin Induced Changes in Specific Growth Rate, gonadal maturity, Lipid and protein production in Nile Tilapia *Oreochromis niloticus* (Linnaeus 1758). *Asian-Aust. J. Anim. Sci.*, 25(1): 37–43

Sundararaj, B.I. and Keshavanath, P., 1976. Effects of Melatonin and Prolactin treatment on the hypophysial-ovarian system in the catfish, Heteropneustes fossilis (Bloch). *Gen. Comp. Endocrinol.*, 129: 84–96.

Tan Dun-Xian, Rudiger Hardeland, Lucien C. Manchester, Ahmet Korkmaz, Shuran Ma, Sergio Rosales-Corral and Russel J. Reiter, 2011. Functional roles of melatonin in plants, and perspectives in nutritional and agricultural science. *Journal of Experimental Botany*, 63(2): 577–597.

Chapter 17

Utilization of Fish Meal for Sustainable Aquaculture Management and Productivity

Meenakshi Jindal

Department of Zoology and Aquaculture,
CCS HAU, Hisar – 125 004, Haryana, India

ABSTRACT

Fishmeal (FM) belongs to a short list of excellent feedstuffs that provide essential nutrients in a highly digestible and concentrated form. The use of FM in animal diets will remain a core and efficient practice, particularly for young, rapidly growing, and high-producing animals like maturing fish, berried (egg-laden) shrimp, poultry, and lactating dairy cattle. The beneficial effects of eating wholesome foods will increase the worldwide demand for seafood products resulting thereby in increased use of fishmeal. Optimal use of fishmeal in practical aquaculture diets is necessary to minimize feeding costs which can account for 40 per cent or more of operating expenses. The concentration of high-quality nutrients, especially protein, makes FM one of the most sought and expensive feedstuffs. The cost of high-quality FM (65 per cent protein) has ranged from 2.0 to 3.5 times the price of soybean meal. Therefore, fish diets can be made by partial or total replacement of FM with

* Corresponding author: E-mail: meenjind@gmail.com

other plant and animal proteins. All-plant protein-based diets containing soybean meal, cottonseed meal, and middlings from corn and wheat, supplemented with lysine and methionine, have been used successfully to grow juvenile catfish, carp, and tilapia to market size. The aquaculture industry must continue to seek out alternative sources of high-quality plant and animal-based protein ingredients for their feedstuffs. Presently, this is an active area of research in aquaculture nutrition.

Keywords: *Fishmeal, Domestic and farm animals diets, Aquaculture diets.*

Introduction

The usual traditional diets *i.e.*, rice bran and oil cakes do not contain enough nutrients required for good health of the fish. Further, these nutrients are not only low in energy but also yield a lot of wastes in the treated water. Therefore, to reduce pollution and increase fish growth, a nutritionally balanced diet containing fishmeal (FM) is necessary for enhancing not only the growth but also nutritionally rich fish flesh, such fishes also fetch better price in the market.

Fishmeal (FM) is a generic term for a nutrient-rich feed ingredient used primarily in diets for domestic animals, sometimes used as a high-quality organic fertilizer. FM carries large quantities of energy per unit weight and is an excellent source of protein, lipids (oils), minerals, vitamins and very little carbohydrate.

FM is high quality animal protein. It contains Ca, P, K, Na and certain unidentified factors capable of promoting growth in animals and improved hatchability in poultry. 27 per cent of total world fish catch is converted to FM. Pelagic fish sp. Such as Sardines, mackerel and anchovies support the world's FM industry. FM prepared from whole fish appears to be a better protein supplement than other animal protein.

Most commercial FM is made from small, bony, and oily fish that otherwise are not suitable for human consumption and some is manufactured from by-products of seafood processing industries. FM can be made from almost any type of seafood but is generally manufactured from wild-caught, small marine fish that contain a high percentage of bones and oil (Table 17.1). These fishes are considered 'industrial' since most of them are caught for the sole purpose of FM and fish oil production. The supply is presently stable at 6.0 to 6.5 million tons annually.

Table 17.1: Estimated World Fish Feed Production by Species Groups (2000)

Species Group	*Metric Tons (mt) of Feed*	*Per cent of Total*
Cyprinids (Carp)	6,991,000	53 per cent
Salmon/Trout	1,636,000	12.5 per cent
Shrimp	1,570,000	12 per cent
Marine finfish	1,049,000	8 per cent
Tilapia	776,000	6 per cent
Catfish	505,000	4 per cent
Miscellaneous	5,79,000	4.5 per cent
Total	13,106,000	

Approximately 4 to 5 tons of whole fish are required to produce 1 ton of dry fishmeal. The principal FM-producing countries are Chile, China, Thailand, U.S.A., Iceland, Norway, Denmark, and Japan (Table 17.2). Major groups of industrial fish rendered into FM are anchovies, herrings, menhaden, sardines, shads, and smelts (Table 17.3).

Table 17.2: Top Fishmeal Producing Countries

- ☆ Peru (Anchovy): It produces almost one-third of the total World FM supply
- ☆ Chile (Anchovy and Horse mackerel)
- ☆ China (Various species)
- ☆ Thailand (Various species)
- ☆ U.S.A. (Menhaden, Pollock)
- ☆ European Union, others (Various species)
- ☆ Iceland and Norway (Capelin, Herrings, Bluewhiting)
- ☆ Denmark (Pout, Sandeel, Sprat)
- ☆ Japan (Sardine/Pilchard)
- ☆ South Africa (Pilchard)

Table 17.3: Principal Fish Species in Fishmeal

Anchovies (Engraulidae)

- ☆ *e.g.*, Peruvian anchoveta (*Engraulis ringens*);
- ☆ Japanese anchovy (*Engraulis japonicus*).

Herrings, Menhaden, Sardines and Shads (Clupeidae):

- ☆ *e.g.*, Atlantic herring (*Clupea harengus*);
- ☆ Menhaden (*Brevoortia tyrannus* and *B. patronus*);
- ☆ South American and Japanese pilchards (*Sardinops sagax*) and other species; European pilchard (*Sardina pilchardus*); European sprat (*Sprattus sprattus*).

Smelts (Osmeridae)

- ☆ *e.g.*, Capelin (*Mallotus villosus*).

Jacks (Carangidae)

- ☆ *e.g.*, Chilean horse mackerel (*Trachurus murphyi*), Atlantic horse mackerel (*Trachurus trachurus*). Pollock, Cod, and Haddock (Gadidae) *e.g.*, Walleye or Alaska Pollock (*Theragra chalcogramma*);
- ☆ Atlantic and Pacific cods (*Gadus morhua* and *G. cephalus*);
- ☆ Georges Bank haddock (*Melanogramus aeglefinus*); Norway pout (*Trisopterus esmarkii*);
- ☆ Blue whiting (*Micromesistius poutassou*).

Hakes (Merlucciidae) and Sand lances (Ammodytidae)

- ☆ *e.g.*, Hake (*Merluccius sp.*);
- ☆ Hoki (*Macruronus novaezelandie*).
- ☆ Small and lesser sandeels (*Ammodytes marinus* and *Ammodytes tobianus*).

Tunas and Mackerels (Scombridae)

- ☆ *e.g.*, Skipjack tuna (*Katsuwonos pelamis*), Yellowfin tuna (*Thunnus albacares*);
- ☆ Chub mackerel (*Scomber japonicus*), Atlantic mackerel (*S. scombrus*).

Cutlassfishes (Trichiuridae)

- ☆ *e.g.* Largehead hairtail or Atlantic cutlassfish (*Trichiurus lepturus*).

Source: Miles and Chapman, 2006.

Processing of Fishmeal

Fish is a highly perishable raw material, and spoilage will occur if it is not processed in a timely manner. Preservation using ice or refrigerated seawater is common. Cooking, pressing, drying and grinding the fish make FM. There are several processing methods to produce good quality FM, but the basic principle involves separation of the solids from the oil and water. When no oil needs to be removed, such as with lean fish, the pressing stage is often omitted. The general scheme of FM processing is given in Figure 17.1.

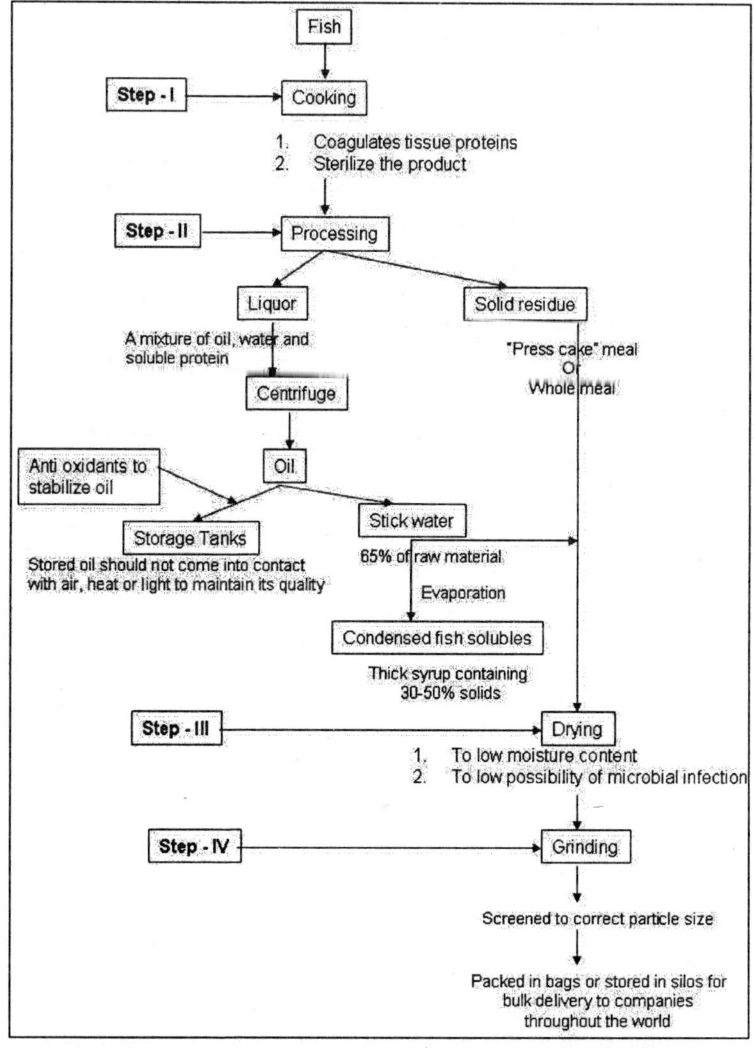

Figure 17.1: General Flow Chart for Fishmeal Preparation

Preservation of Fishmeal

The use of antioxidants in the preservation of FM is essential in order to stabilize its energy value because there are high quantities of PUFAs present in the oil. Without stabilizing FM with antioxidants the available energy content of the meal may be reduced by as much as 20 per cent; the oxygen will modify (damage) the chemical structure of the PUFAs and therefore less energy is available to the animal.

Importance of Fishmeal in Fish Diets

1. Addition of FM to animal diets increases feed efficiency and growth through better food palatability, and enhances nutrient uptake, digestion, and absorption.
2. The balanced amino acid composition of FM complements provides synergistic effects with other animal and vegetable proteins in the diet to promote fast growth and reduce feeding costs.
3. FM of high quality provides a balanced amount of all essential amino acids, phospholipids, and fatty acids, *e.g.*, Docosahexaenoic acid (DHA) and Eicosapentaenoic acid (EPA), for optimum development, growth, and reproduction, especially of larvae and brood stock.
4. The nutrients in FM also aid in disease resistance by boosting and helping to maintain a healthy functional immune system.
5. High-quality FM also allows formulation of nutrient-dense diets, which promote optimal growth.
6. Incorporation of FM into diets of aquatic animals helps to reduce pollution from the wastewater effluent by providing greater nutrient digestibility.
7. The incorporation of high-quality FM into feed imparts a 'natural or wholesome' characteristic to the final product, such as that provided by wild fish.

Protein Quality of Fishmeal

Any complete diet must contain some protein, but the nutritional value of the protein relates directly to its amino acid composition and digestibility (Table 17.4). Proteins are made of amino acids, which are released for absorption into the blood following protein digestion. Animals have requirements for specific amino acids rather than protein. FM and any other feedstuff that contains protein can simply be thought of as a 'vehicle' for providing amino acids to the diet.

Table 17.4: The Percentage Values for the Essential Amino-Acids Composition of Different Feedstuffs

Essential Amino-Acids	*Fishmeal (64.5%)*[1]	*Meat and Bone meal (55.6%)*[1]	*Poultry byproduct meal (59.7%)*[1]	*Blood Meal (89.2%)*[1]	*Soybean Meal (50.0%*[1]
Arginine	3.82	3.60	4.06	3.75	3.67
Histidine	1.45	0.89	1.09	5.14	1.22
Isoleucine	2.66	1.64	2.30	0.97	2.14
Leucine	4.48	2.85	4.11	10.82	3.63
Lysine	4.72	2.93	3.06	7.45	3.08
Methionine + Cystine[2]	2.31	1.25	1.94	2.32	1.43
Phenylalanine + Tyrosine[3]	4.35	2.99	3.97	8.47	4.20
Threonine	2.31	1.64	0.94	3.76	1.89
Tryptophan	0.57	0.34	0.46	1.04	0.69
Valine	2.77	2.52	2.86	7.48	2.55

1: Percentage of total crude protein in feedstuff.

2: Cystine can be synthesized from methionine.

3: Tyrosine can be synthesized from phenylalanine.

The values in parenthesis indicates the percentage of crude protein in the meal

Source: NRC, 1993.

Inclusion Level of Fishmeal

High-quality FM normally contains 60 to 72 per cent crude protein by weight. From a nutritional standpoint, FM is the preferred animal protein supplement in the diets of farm animals and often the major source of protein in diets for fish and shrimp. A typical inclusion rate of FM in terrestrial livestock diets is usually 5 per cent or less on a dry matter basis. Various inclusion levels of FM as protein supplement for different fish species are shown in Table 17.5.

Table 17.5: Various Inclusion Levels of FM for different Fish Species

Fish Species	*Per cent Inclusion Level of Fishmeal*
Typical fish	32 to 45
Shrimps	25 to 42
Carp and Tilapia	5 to 7
Trout, Salmon and Marine fishes	40 to 55

Lipid Contents and Energy in Fishmeal

The lipids in FM not only impart an excellent profile of essential fatty acids but also provide a high content of energy to the diet. Since there is very little carbohydrate in FM, the energy content of FM relates directly to the percentage of protein and oil it contains. The quantity and quality of oil in FM will in turn depend on the species, physiology, sex, reproductive status, age, feeding habits of the captured fish, and the method of processing.

The lipids in FM and fish oil are easily digested by all animals, especially fish, shrimp, poultry, pigs, and ruminants such as cows, sheep, and goats. In these animals the lipid digestibility is 90 per cent or greater. The high digestibility of fish lipids means they can provide lots of usable energy. If a diet does not provide enough energy, the fish or shrimp will have to break down valuable protein for energy, which is expensive and can increase production of toxic ammonia (Figure 17.2).

Fish lipids are highly digestible by all species of animals and are excellent sources of the essential polyunsaturated fatty acids (PUFA) in both the omega-3 and omega-6 families of fatty acids. The predominant omega-3 fatty acids in fishmeal and fish oil are linolenic acid, DHA and EPA. PUFAs, are easily damaged and become rancid when exposed to oxygen, a process known as oxidation and one that releases heat. Both DHA and EPA fatty acids are produced and passed along the food chain by small-size algae and zooplankton, which are consumed by fish.

Mineral Value of Fishmeal

When a sample of feed is taken to the laboratory and analyzed for nutrient content, the procedure involves burning a portion of the sample. Ash is the material remaining after the feed sample is

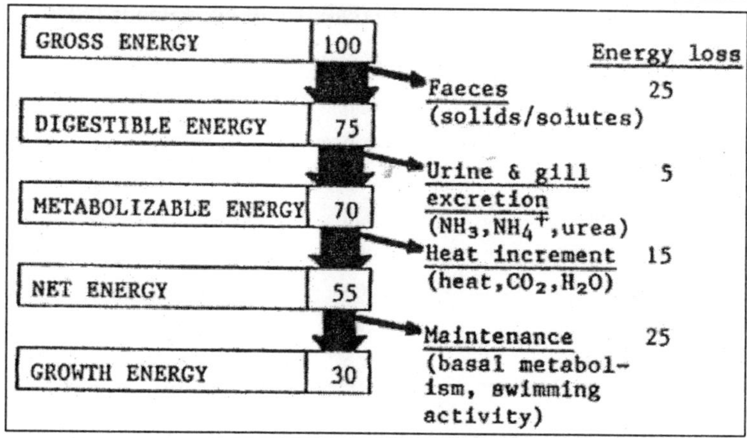

Figure 17.2: Utilization of Dietary Energy by the Fish (after Luquet, 1982)

completely burned. Normally, the ash content of good quality averages between 17 per cent and 25 per cent. More ash indicates a higher mineral content, especially calcium, phosphorus, and magnesium. Calcium and phosphorus constitute the majority of the ash found in FM.

Vitamin Value of Fishmeal

The vitamin content of FM is highly variable and influenced by several factors, such as origin and composition of the FM processing method, and product freshness. The content of fat-soluble vitamins in FM is relatively low because of their removal during extraction of the oil. FM is considered to be a moderately rich source of vitamins of the B-complex especially cobalamine (B_{12}), niacin, choline, pantothenic acid, and riboflavin.

Economic and Environmental Considerations of Fishmeal Utilization

The high quality and concentration of essential nutrients, especially of well-balanced amino acids, essential fatty acids, and energy content makes FM an indispensable ingredient in diets of most aquaculture species and many land-farm animals. Because of its nutrient content, high digestibility and palatability, FM serves as the benchmark ingredient in aquaculture diets (Tables 17.6–17.8).

Table 17.6: Estimated Fishmeal (FM) Use in Aqua Feeds (2000)

Species Group	*Metric Tons (mt) of FM*	*Per cent of Total*
Cyprinids (Carp)	350,000	16.5 per cent
Salmon	454,000	21.5 per cent
Shrimp	372,000	17.6 per cent
Marine fish	415,000	19.6 per cent
Trout	176,000	8.3 per cent
Eels	173,000	8.2 per cent
Flatfish	69,000	3.3 per cent
Other fish	106,000	5.0 per cent

Table 17.7: FM Use by Species Group (per cent in diet and 000mt) Showing an Increase of 38 per cent in FM Use by 2010

Species Group	*2000 (%)*	*2010* (%)*	*2000 (mt)*	*2010* (mt)*
Cyprinids (Carp)	5	2.5	350	675
Salmon	40	30	454	377
Shrimp	20	20	372	485
Marine fish	45	40	415	688
Trout	30	25	176	147
Flatfish	55	45	69	263
Catfish	3	0	15	0

Source: Barlow 2000.

Table 17.8: Predicted Use of FM in Aqua Feeds in 2010

Compared to amount needed if percentage of FM in feeds remains the same as 2000.

	Feed (000mt)	*FM (000mt)*
2000	13,098	2,115
2010 (today's use levels)	37,226	4,081
2010*		2,831

Difference for 2010 at today's use levels and Barlow's prediction–1,250,000mt.

Source: Barlow 2000.

FM belongs to a short list of excellent feedstuffs that provide essential nutrients in a highly digestible concentrated form. The use of FM in domestic and farm animal diets will remain a core and efficient practice, particularly for young, rapidly growing, and high-producing animals like maturing fish, berried (egg-laden) shrimp, poultry, and lactating dairy cattle. The beneficial effects of eating wholesome foods will increase the worldwide demand for seafood products resulting in increased use of fishmeal.

The best approach in feed formulation is to use high-quality feedstuffs to manufacture a diet that meets the nutritional and energy requirements of the aquaculture species in question. Fishmeal contains the profile of amino acids that most closely meets the amino acid requirements of fish. If a portion or all of the fishmeal in a diet can be replaced successfully with other high-quality protein sources, doing so will contribute greatly towards protecting the surrounding environment and promoting a sustainable aquaculture industry.

Replacement of FM with other Plant and Animal Proteins

Optimal use of FM in practical aquaculture diets is necessary to minimize feeding costs which can account for 40 per cent or more of operating expenses. The concentration of high-quality nutrients, especially protein, makes FM one of the most sought and expensive feedstuffs. The cost of high-quality FM (65 per cent protein) has ranged from 2.0 to 3.5 times the price of soybean meal.

Unfortunately, the specific dietary requirements for energy, essential amino acids, fatty acids, and other nutrients are not yet known for many aquaculture species. Also, the digestibility of different feedstuffs has not yet been established for many of the commercially important fish species. Therefore, there has been a reluctance to diminish the total amount of FM used in many of the different aquaculture diets. Simply, over-supplementing the diet with high-quality FM is an easy and very successful way of overcoming lack of knowledge about aquaculture diets in relation to specific nutrient requirements and digestibility of feedstuffs.

New information on nutrient requirements of aquatic organisms coupled with advances in feed technology indicates that species-specific fish diets can be made by partial or total replacement of FM with other plant and animal proteins. All-plant protein-based diets containing soybean meal, cottonseed meal, and middlings from corn and wheat, supplemented with lysine and methionine, have been used successfully to grow juvenile catfish, carp, and tilapia to market size (Jindal and Garg, 2005, Jindal *et al.*, 2007a, b; Jindal 2008, 2011a,b; Jindal *et al.*, 2008a, b, c).

However, larvae and young fish still require fishmeal to grow at an optimal rate. Animal proteins and fats, by-products of the animal-rendering industry, can be used in aquaculture diets because they also provide essential amino acids and fatty acids. These "fishmeal substitutes" will be used more extensively by the aquaculture industry in the future.

Fishmeal is Better Than Plant Protein Sources

1. Better Amino-Acid Profile of Fishmeal

The amino acid profile of FM is what makes this feed ingredient so attractive as a protein supplement (Table 17.4). Proteins in cereal grains and other plant concentrates do not contain complete amino acid profiles and usually are deficient in the essential amino acids lysine and methionine. Soybean and other legume meals, which are widely used in the diets of most farm animals such as pigs and chickens, are a good source of lysine and tryptophan but are limiting in the sulphur-containing amino acids methionine and cystine. An animal's requirement for a limiting amino acid can be met by simply adding more of the protein. However, this would be very costly, and the excess nitrogen in the protein would deleteriously affect water quality. Excess nitrogen arising from the amino acids of proteins is excreted from the fish into the water in the form of ammonia. Ammonia is toxic to fish and must be removed from the water by filtration or water flushing. (Jindal and Garg, 2005; Jindal *et al.*, 2007a, 2010).

2. Better Protein Digestibility of Fishmeal

Plant-based proteins, even when properly processed, are usually not as digestible as FM; and their inclusion rate into the diet is often limited as it results in depressed growth rates and feed intake. Over-all protein digestibility values for FM are consistently above 95 per cent. In comparison protein digestibility for many plant-based proteins varies greatly, for example, from 77 per cent to 96 per cent, depending on the species of plant.

3. Fishmeal as an Excellent Source of Omega-3 Fatty Acids

FM and oil contain more omega-3, than omega-6 fatty acids. In contrast, most plant lipids contain higher concentrations of omega-6 fatty acids. For example, oil extracted from soybeans, corn, or cottonseed is rich in linoleic acid, an omega-6 fatty acid. Some oils, like those from canola and flax seeds contain linolenic acid (of the omega-3 family), however, its conversion into essential DHA and EPA by most animals may be limited.

4. Better Palatability of Fishmeal

FM contains certain compounds that make the feed more acceptable and agreeable to the taste (palatable). This property allows for the feed to be ingested rapidly, and will reduce nutrient leaching. It is thought the non-essential amino acid glutamic acid is one of the compounds that imparts to FM its palatability.

5. Lack of Anti-nutritional Factors (ANF's) in Fishmeal

ANF's are compounds that interfere with nutrient digestion, uptake, or metabolism and can also be toxic. The structural nature of plants is totally different from that of animals. Proteins isolated from plants are associated with indigestible non-structural carbohydrates (oligosaccharides) and structural fiber components (cellulose), which are not associated with animal proteins. It is the presence of these components which are thought to be contributing obstacles to efficient utilization of proteins in many economically plant-based feedstuffs (Jindal *et al.*, 2007a b, 2008a b, 2010). The lack of nutritional inhibitors or ANF's in FM also makes this meal more attractive than plant proteins for use in aquaculture diets. Some examples showing presence of ANF's in plant protein sources are shown below:

1. A naturally occurring ANF in uncooked soybeans is the Kunitz trypsin-inhibitor that prevents the enzyme trypsin from breaking down dietary proteins in the intestine of animals. Other ANF's present in soya products and the process of removing these ANF's are given in Table 17.9.
2. Lathyrogens in chickpeas also disrupt collagen formation. Collagen is the most abundant protein present in animals, making up most connective tissue and providing structural support.

Table 17.9: Anti-Nutritional Factors in Soy Products and Processing Steps to Remove or Inactivate them

Anti-nutrient	*Heat Sensitive*	*Extractable*	*Other Treatment*
Trypsin inhibitors	Yes	No	No
Hemagglutinin	Yes	No	No
Phytic acid	No	No	Phytase
Saponin	No	Yes	No
Phytoestrogen	No	Yes	No
Anti-vitamin	Yes	?	No

3. Thiaminases found in raw fish are known to destroy thiamine (vitamin B_1), and the avidin in egg white binds biotin (another water-soluble vitamin of the B-complex).
4. Gossypol is another ANF found in cottonseed meal/oil that is toxic to animals and lowers fertility in males.

Principle Objectives in the Formulation of Fish Feeds

1. Balancing nutrients in diets by using the minimum amount of FM to meet specific amino acid requirements for fast growth and reproduction and reducing feed costs constitute one of the principal objectives in formulation of fish feeds.
2. Another important aim in feed formulation is to increase dietary nutrient density and digestibility of the feed to increase biological performance and to reduce nutrient leaching and water-quality degradation (Figure 17.3).

Conclusions

Fish fed diets formulated with a high percentage of FM will contain high concentrations of the PUFAs in their tissues, especially of the omega-3 family. PUFAs are essential for human biological functions, particularly the production of prostaglandins. PUFAs and prostaglandins may ameliorate many human health disorders, such as high blood pressure, heart disease, arthritis, migraines, diabetes, and cancer. Incorporation of DHA and EPA found in FM into the diets of fish and other farm animals is an efficient method to ensure a proper concentration of these important omega-3 fatty acids

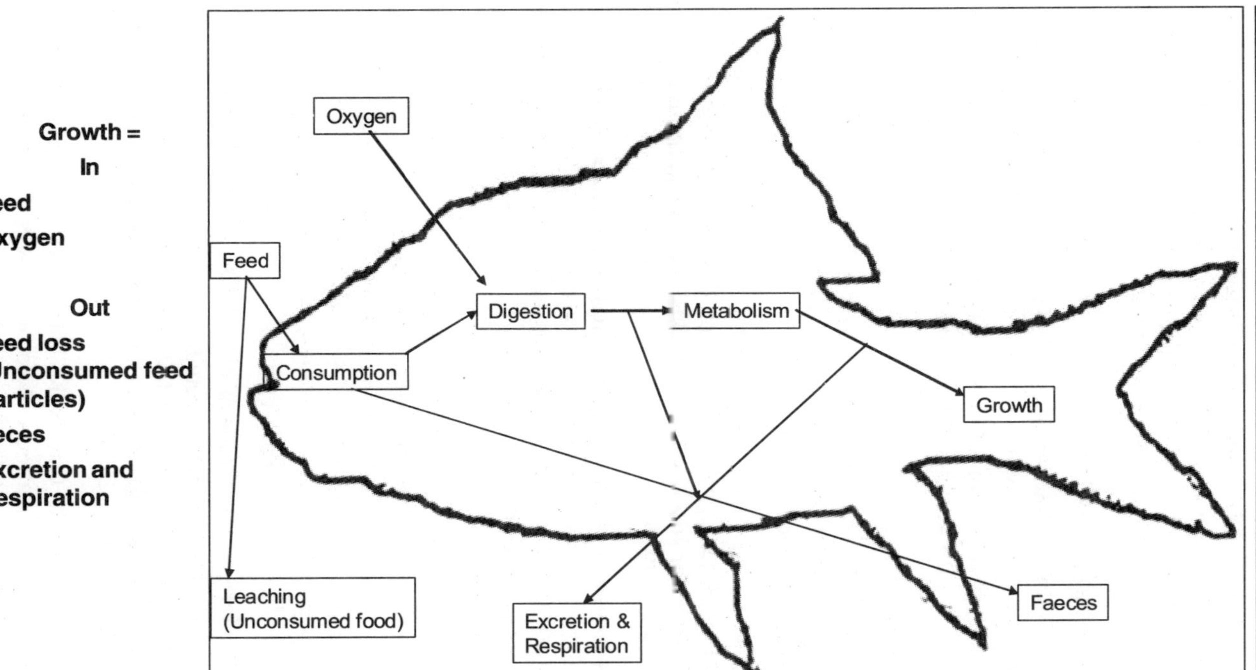

Figure 17.3: Pathway Showing Utilization of Dietary Nutrition (Garg, 2003)

in the human diet. The aquaculture industry must continue to seek out alternative sources of high-quality plant and animal-based protein ingredients for their feedstuffs. Presently, this is an active area of research in aquaculture nutrition.

Acknowledgements

The author acknowledges funding received under the scheme "Women Scientist Scholarship Scheme for Societal Programmes (WOS-B), Department of Science and Technology, Government of India" for the publication of this technical note.

References

Garg, S.K., 2003. *Recent Methods in Fish Nutrition and Biotechnology.* Practical Manual, Department of Zoology and Aquaculture, CCS HAU, Hisar, pp. 1–183.

Jindal, Meenakshi, Garg, S.K. and Yadava, N.K., 2008a. Effect of feeding dietary protein sources on daily excretion in *Channa punctatus* for sustainable aquaculture. In: *Advances in Aquatic Ecology*, 3: 49–60.

Jindal, Meenakshi, Garg, S.K. and Yadava, N.K., 2008b. Effect of replacement of fishmeal with processed soybean on daily excretion of ammonical–nitrogen (NH_4–N) and ortho–phosphate (O–PO_4) in *Channa punctatus* (Bloch). *Pb. Univ. Res. J. (Sci.)*, 58: 25 –33.

Jindal, M., Garg, S.K., Yadava, N.K. and Gupta, R.K., 2007b. Effect of replacement of fishmeal with processed soybean on growth performance and nutrient retention in *Channa punctatus* (Bloch.) fingerlings. *Livestock Research for Rural Development, Volume 19, Article #165.* Retrieved from http: / /www.cipav.org.co/lrrd/ lrrd19/11/jind19165.htm

Jindal, M., Garg, S.K. and Yadava, N.K., 2007a. Effect of replacement of fishmeal with dietary protein sources of plant origin on the growth performance and nutrient retention in the fingerlings of *Channa punctatus* (Bloch.) for sustainable aquaculture. *Pb. Univ. Res. J. (Sci.)*, 57: 131–138.

Jindal, Meenakshi and Garg, S.K., 2005. Effect of replacement of fishmeal with defatted canola on growth performance and nutrient retention in the fingerlings of *Channa punctatus* (Bloch.) *Pb. Univ. Res. J. (Sci.)*., 55: 183–189.

Jindal, Meenakshi, Garg, S.K. and Yadava, N.K., 2009. Effect of feeding defatted canola on daily excretion of ammonical nitrogen (NH_4–N) and ortho–phosphate (O–PO_4) in *Channa punctatus* (Bloch.). *Livestock Research for Rural Development, USA. Vol. 21, Article # 35* Retrieved from http: //www.lrrd.org./ lrrd21/3/jind21035.htm

Jindal, Meenakshi., 2008. Studies on protein requirements of catfish *Clarias batrachus* fingerlings for sustainable aquaculture. In: National seminar on Technical advances in environment management and applied zoology. Jan. 23–25, Department of Zoology, Kurukshetra University, Kurukshetra, pp. 59–62.

Jindal, Meenakshi., 2011a. Influence of different feeding regimes on the growth performance and nutrient retention of walking catfish, *Clarias batrachus.* In: *Prospectives of Animal Ecology and Reproduction* (Editor: Dr. V.K. Gupta) Vol. 7, Daya Publishers, New Delhi. Chapter 12; pp 173–186.

Jindal Meenakshi, Yadava, N.K., Jain, K.L. and Gupta, R.K., 2010. Effect of two dietary protein levels on body weight and composition in *Channa punctatus* (Bloch.) fingerlings. *Turkish journal of Fisheries and Aquatic Sciences,* Vol. 10 (2) pp.203–208.

Jindal, Meenakshi., 2011b. Studies on protein requirements of catfish *Clarias batrachus* for sustainable aquaculture. *Indian J. Fisheries.* 58(2): 95–100.

Luquet, P., 1982. Aspects du metabolisme des poisons particulierement importants pour la qualite de leau. In: *Razionale utilizzo delle risorse idriche in acquacoltura. Atti del Convengo Internazionale Verona, Salone Acquacoltura,* 16 Ottobre.. Pubblicazione curate dal Centro regionale tutela e sperimentazione Pesca e Acquacolture dell' E.S.A.V., Italy. pp. 30–38.

Miles, R.D. and Chapman, F.A., 2006. The benefits of fishmeal in aquaculture diets. In: *Catfish Protein Nutrition,* (Eds.) Robinson, Edwin H. and Menghe, H. Li, pp. 51–59.

NRC (National Research Council), 1993. *Nutrient Requirements of Fish.* National Academy of Science, Washinton, D.C., 102p.

Chapter 18

Feed Supplementation and Water Quality Management for Sustainable Aquaculture

Meenakshi Jindal

Department of Zoology and Aquaculture,
CCS HAU, Hisar – 125 004, Haryana, India

ABSTRACT

The optimization of fish production and also for the sustainability of the system requires research into the feeding techniques, which could promote good growth and also at the same time reduce the quantity of waste output in the environment. Water quality management is vital for sustained high fish production in intensive or semi-intensive aquaculture systems. The artificial feed constitutes the major input component for sustained high production, but also responsible for deterioration of water quality in such systems. The main pollutants in farm effluent water are phosphorous, nitrogen and fecal solids. These can be reduced by using pelleted and water stable feed and also by supplementation of phytase enzyme in feed along with modifying feeding levels and methods.

Keywords: *Artificial feed, Feeding techniques, Fecal solids, Nitrogen, Phosphorous, Water quality.*

* Corresponding author: E-mail: meenjind@gmail.com

Introduction

Fisheries is the most rewarding activity in today's time of socio-economic developments, particularly in developing countries, including India. It being the major potential source of protein rich animal food as well as livelihood for humans constitute one of the major component of global models of world food production. Global fisheries, however, is facing constant decline in fish stocks, both in coastal and inland water resources on account of continuously increasing water pollution. In intensive culture systems, the feed is considered to constitute major input cost and an important source of pollution leading to eutrophication. Fecal solids and excessive presence of phosphorus and nitrogen originating from unconsumed feed and undigested nutrient constituents in feces are major causes of water pollution in culture systems. Undigested protein and unutilizable phosphorus are excreted in feces. The unconsumed feed left at bottom and indigested nutrients in fecal solids release their phosphorus and nitrogen contents after decomposition and increase biological oxygen demand. Reduction in wastage from unconsumed feed is possible by pelletisation of feed and increasing water stability of pellets and devising appropriate feeding strategies and methods.

The strategies for reducing amount of unconsumed feed, excretion level of phosphorus and nitrogen, undigested part of feces, fecal solid discharges *etc.* for controlling pollutants originating from feed are discussed in this communication.

Feed and Feeding Strategy

Ingredient composition and manufacturing procedure of feeds have greater influence on meeting the nutritional requirement of fish, water stability of pellets and digestibility of feedstuffs. Broadcasting of unprocessed feed and open feeding systems with loose feed balls are most inefficient ways of fish feeding and liable to feed loss in large amounts and subsequent cause of unwarranted pollution originating from uneaten feed. For each culturable fish the optimum ingredient composition of feed has to be determined as per requirement of each and every stage of its life cycle. More emphasis in feed formulation has to be given on aspects like gross energy, food conversion ratio, optimum nitrogen and phosphorus levels in preparing low-pollution feed. To control nutrient levels in farm effluents strict restrictions were imposed on feed composition in Denmark and other Scandinavian countries. Danish specification of feed composition prescribed for beyond 1992 is as; feed conversion ratio 1: 1.0; gross energy 6000 kcal/kg; nitrogen 8 per cent maximum (50 per cent protein) and total phosphorus 1.0 per cent maximum.

Apart from ingredient composition of feed, the manufacturing process also affects its nutrient digestibility. Manufacturing by extrusion processing has been reported to increase fish feed digestibility (Figure 18.1). The heat and pressure associated with extrusion processing increases digestibility of the protein and carbohydrate portions and digestible energy of the feed (Jindal *et al.*, 2010; Jindal, 2008). Extruded pellets can absorb high amount of added lipid thus increase nutritional density (Figure 18.2). Other feeding strategies useful in lowering nutrient level in water of fish rearing systems are phase feeding and use of demand feeders.

Fecal Solids

Various undigested organic constituents like protein, lipids, soluble and fibrous carbohydrates *etc.* form solid fecal wastes. These organic solid wastes can be reduced by proper management strategies:

1. Developing diets using ingredients with high nutrient digestibility.
2. Supplementation of diets with microbial enzymes such as amylases and proteases to improve the digestion of carbohydrate and protein resulting in increased nutrient utilization.
3. Supplementation of diets with premix minerals and amino acids. Each kg of which contains Copper–312mg; Cobalt–45mg; Magnesium–2.114g; Iron -979mg; Zinc–2.13g; Iodine 156mg; DL-Methionine–1.92g; L-lysine mono hydrochloride–4.4g; Calcium 30 per cent and Phosphorous–8.25 per cent (Jindal and Garg, 2005; Jindal, *et al.*, 2007a, b, 2010 and Jindal, 2011 a, b).

Phosphorus

Phosphorus is an essential mineral which is available only by diet and its dietary requirement for various fish species ranges between 0.3 to 0.9 per cent of diet. Higher dietary phosphorus

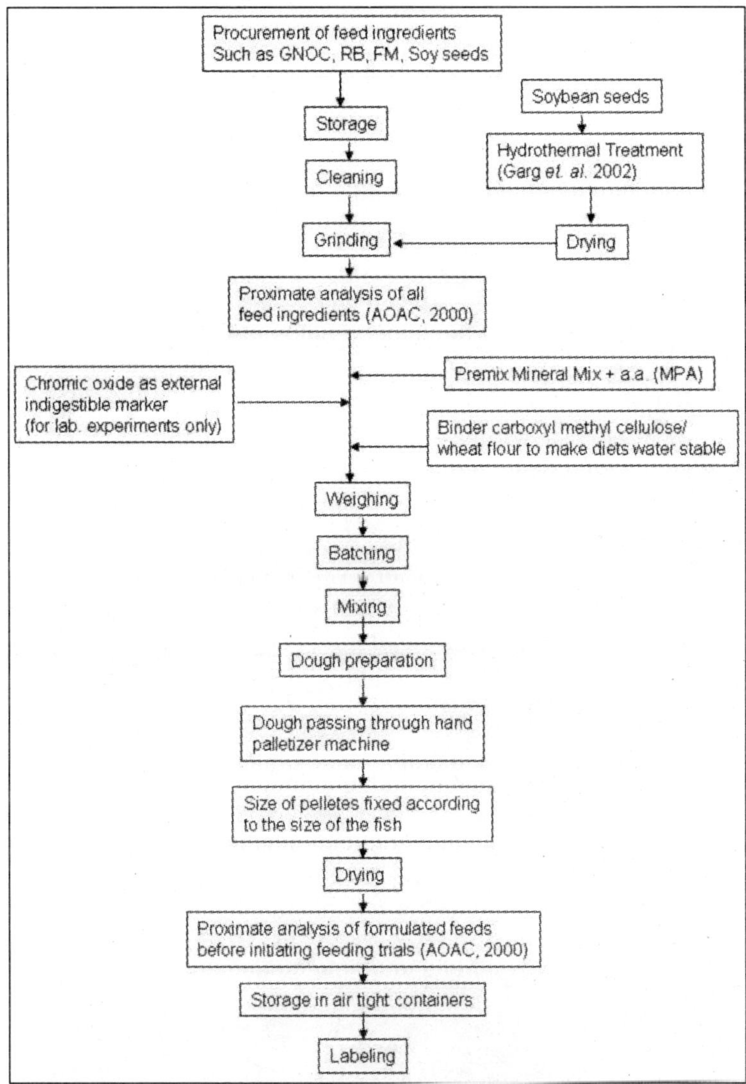

Figure 18.1: Typical Flow Scheme for Manufacturing Catfish Feeds

requirement has been reported for scaly fishes than scale-less one. Fish excrete phosphorus in soluble and particulate forms. The soluble fraction called ortho-phosphate (O-PO_4) is most available for plant growth. However, the main loading of phosphorus to the

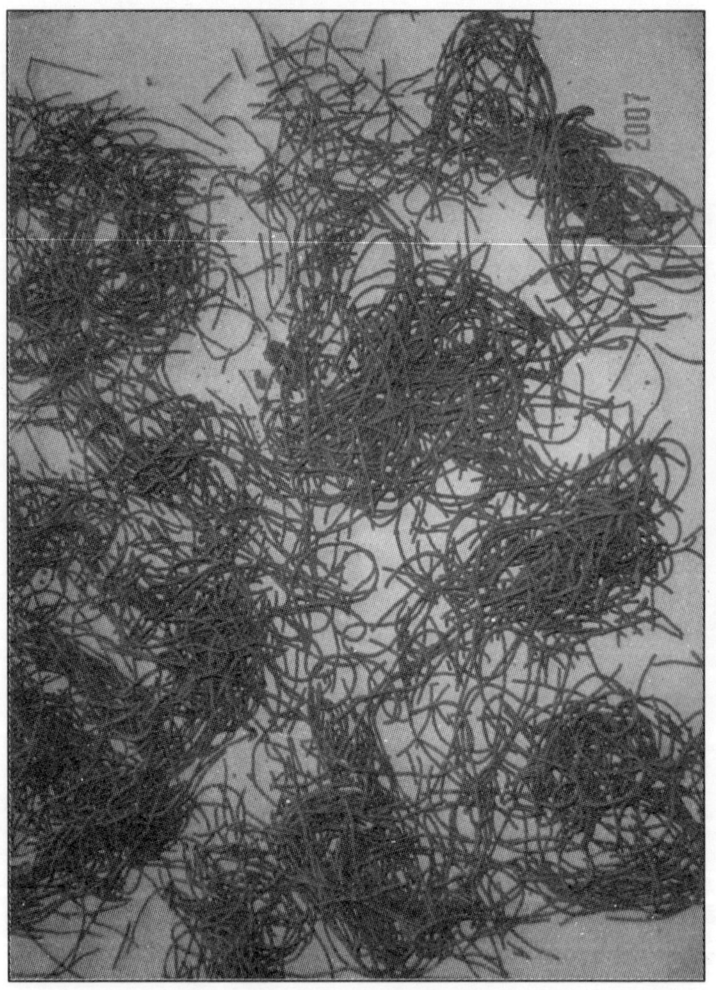

Figure 18.2: Pelleted Feed for Fish

environment was reported to be via faecal pellets (Figures 17.3 and 18.3).

Various management steps in reducing the phosphorus excretion by fish

1. First step in reducing the phosphorus excretion by fish is to assess the accurate level of bioavailability of phosphorus in feed ingredients. It assumes significance because easiest way in reducing phosphorus excretion is to reduce the proportion lost in feces.

   In general feed made from animal by products contain comparatively high phosphorus level coming from bone component. However, phosphorus availability declines as fish bone level in fish meal increases probably due to formation of Ca-P complexes. Adding citric acid to the diet inhibits formation of Ca-P complexes and tends to increase phosphorus bioavailability.

   The feed ingredients produced from plants *i.e.* grains and oilseeds, are relatively low in total phosphorus, but it is stored in seeds as phytate phosphorus which is indigestible in monogastric animals like fishes. However, indigestible phytate phosphorus becomes readily available to fish when phytase enzyme is supplemented in diet (Vielma *et al.*, 2000).

2. Formulation of feed with phosphorus levels at levels required by a fish species at each stage of production. Using most comprehensive approach, 0.66 per cent and 0.55 per cent phosphorus requirements for 200g and 400g rainbow trout respectively were reported.

3. Formulation of feeds to match the available phosphorus level in the feed to the requirement of the fish. When fish are fed with phosphorus deficient diet then phosphorus from body reserves are drawn up and if dietary phosphorus exceeds requirement levels these are absorbed in excess and then excreted in urine (Jindal, 2010).

Nitrogen

In freshwater systems, nitrogen is sometimes a limiting nutrient, so adding it stimulates plant and algal growth. A majority of the excess nitrogen in either tank or pond culture systems originates as ammonia excreted by fish through gills (Kaushik and Cowey, 1991). The ammonia as a waste product, is formed during the breakdown of proteins and excess amino-acids not incorporated into the tissues by the fish (Kibria *et al.*, 1998). Nitrogenous wastes are dietary in origin and depending upon species up to 52–95 per cent of feed nitrogen is excreted as waste (Figures 17.3 and 18.3). Nitrogen excretion is influenced by the quantity and quality of dietary protein. Protein quality is assessed by amino acid composition and digestibility. Dietary protein not digested and absorbed in gastrointestinal tract is excreted in feces. When the concentration of dietary protein is excessive relative to non-protein energy then the portion of dietary protein not used for protein accretion is broken down and used for energy resulting in elevated ammonia production by fish. Fishes are considered metabolically efficient at utilizing protein for energy and higher level of digestible protein relative to digestible energy increased ammonia excretion in rainbow trout (Figure 18.4).

Various Properly Managed Strategies to Lower Nitrogen Excretion

1. One strategy to lower nitrogen excretion involves increasing digestible energy level in the diet relative to digestible protein. Increasing energy density of diets by lipid supplementation to enhance growth and protein retention has been observed in Atlantic salmon (Kaushik and Cowey, 1991).

   Most appropriate approach for increasing nutrient density of feed has been suggested to exclude ingredients with low protein and energy contents.
2. Replacing feed ingredients having low protein digestibility with those having high protein digestibility coefficients is another strategy for lowering nitrogen excretion.
3. Manufacturing procedures like use of extrusion processing of fish diets which has been found to increase digestibility of diets and reduce ammonia excretion is another useful tool for curtailing nitrogen excretion (Jindal, 2007).
4. Reducing nitrogen level will require formulation of feed which is able to support high percentage of nitrogen retention (protein increase) by fishes (Jindal *et al.*, 2007a, b; Jindal *et al.*, 2008, 2010; Jindal, 2008).

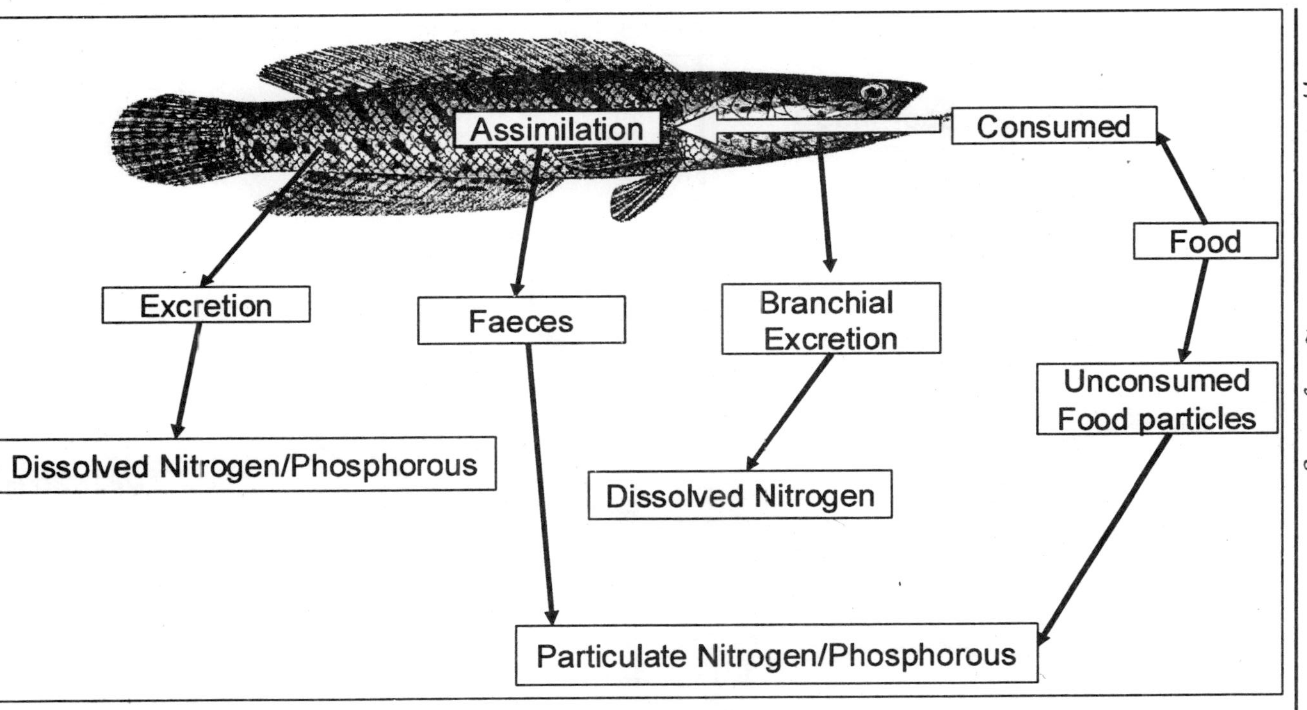

Figure 18.3: Nitrogen and Phosphorous Excretion/Pollution in Aquaculture (Jindal *et al.*, 2008a)

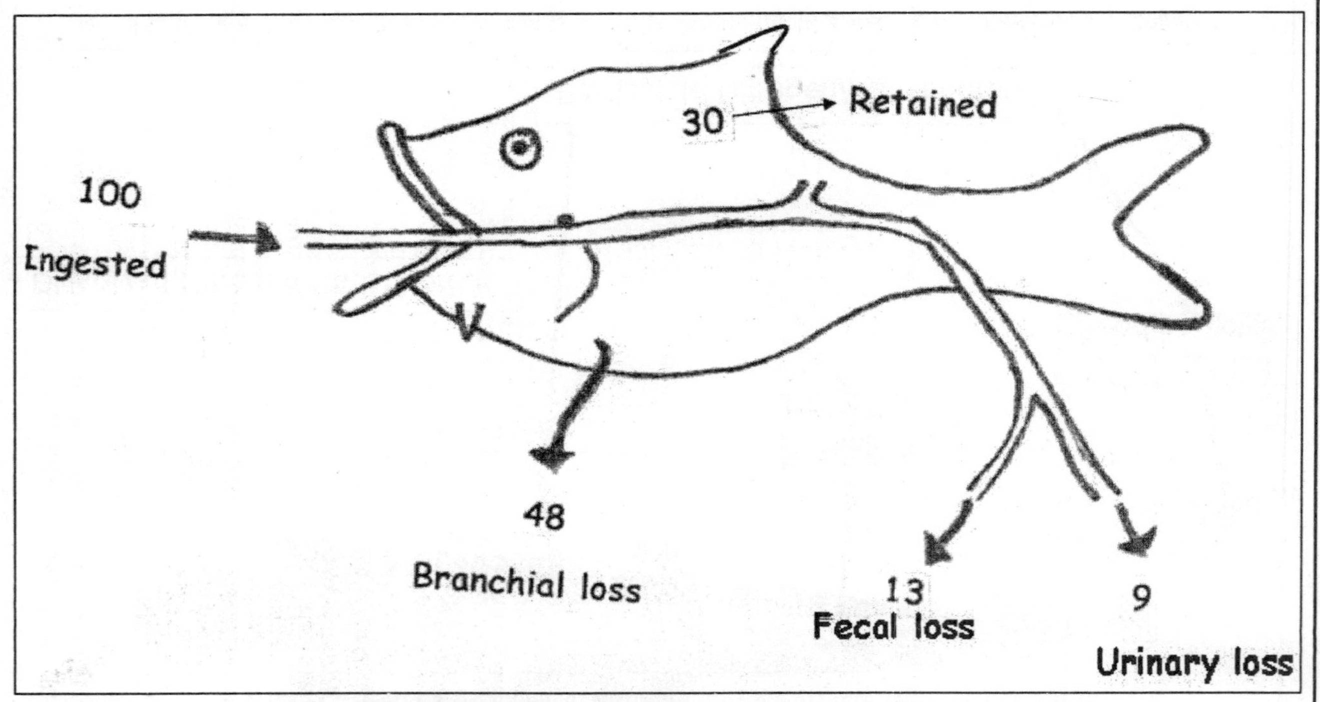

Figure 18.4: Generalized View of Nitrogen Balance in Fish (after Luquet, 1982)

Energy Balance and Dietary Requirements

A schematic representation of the fate of dietary energy in fish is shown in Figure 17.2. It is apparent from the energy balance presented that only a portion of the total chemical energy contained within the food ingested is available to the animal (*i.e.* net energy) for maintenance, activity and growth; 45 per cent of the ingested food energy being lost as undigested food (faeces), metabolic excretion and as heat. Although the energy losses will vary depending on the composition and digestibility of the feed ingredients used, feeding regime, water temperature, fish size and the physiological status of the animal, since all biological systems obey the laws of thermodynamics the energy balance equation can be represented as follows:

$$C = P + R + U + F$$

where,

- ☆ C (for consumption) is the gross energy content of the food ingested,
- ☆ P the energy utilized in growth materials (production),
- ☆ R the net loss of energy as heat (R standing for respiration),
- ☆ U (urinary loss) the energy lost in nitrogenous excretory products,
- ☆ F the energy lost in the faeces (Brafield, 1985).

Using this approach Brett and Groves (1979) have produced the following generalised energy budgets for young carnivorous and herbivorous fish fed on natural foods;

$$\text{Carnivores: } 1\ DOC = 29P + 44R + 7U + 20F$$

$$\text{Herbivores: } 1\ DOC = 20P + 37R + 2U + 41\ F$$

where,

The figures are expressed as a percentage of the ingested food energy.

Conclusions

Thus it is evident that by using appropriate strategies involving diet composition, feed manufacturing processes, feeding strategies *etc.* level of enriching nutrients in fish rearing systems can be significantly reduced without compromising production or profitability.

Acknowledgements

The author acknowledges funding received under the scheme "Women Scientist Scholarship Scheme for Societal Programmes (WOS-B), Department of Science and Technology, Government of India" for the publication of this technical note.

References

AOAC (Association of Official Analytical Chemists), 2000. Official methods of analysis. Association of Official Analytical Chemists, Washington, Sc, USA.

Brafield, A.E., 1985. Laboratory studies of energy budgets. In: *Fish Energetics: New Perspective*, (Eds.) P. Tytler and P. Calow. Croom Helm Ltd., London and Sydney, pp. 257–281.

Brett, J.R. and Groves, T.D.D., 1979. Physiological energetics. In: *Fish Physiology*, (Eds.) W.S. Hoar, D.J. Randall and J.R. Brett. Academic Press, New York, 8: 279–352.

Garg, S. K., 2003. *Recent Methods in Fish Nutrition and Biotechnology: Practical Manual*, Department of Zoology and Aquaculture, CCS HAU, Hisar pp. 1–183.

Garg, S. K., Kalla, A. and Bhatnagar, A., 2002. Evaluation of raw and hydrothermical processed leguminous seeds as supplementary feed for the growth of two Indian Major carp species. *Aquaculture Research*, 33: 151–163.

Jindal, M., Garg, S.K., Yadava, N.K. and Gupta, R.K., 2007b. Effect of replacement of fishmeal with processed soybean on growth performance and nutrient retention in *Channa punctatus* (Bloch.) fingerlings. *Livestock Research for Rural Development, Volume 19, Article #165.* Retrieved from http: //www.cipav.org.co/lrrd/lrrd19/11/jind19165.htm

Jindal, Meenakshi, 2007. Use of supplementary feeds for the development of sustainable aquaculture technology. A report submitted to Science and Society Division, Department of Science and Technology, New Delhi. pp 47

Jindal, Meenakshi, 2008. Influence of supplementary feeds on the growth and excretory metabolite levels in *Heteropneustes fossilis.*

In: *Advances in Aquatic Ecology*, (Ed.) V.B. Sakhare. Daya Publishers, New Delhi, 3: 19–31.

Jindal, Meenakshi and Garg, S.K., 2005. Effect of replacement of fishmeal with defatted canola on growth performance and nutrient retention in the fingerlings of *Channa punctatus* (Bloch.), *Panjab University Research Journal (Science)*, 55: 183–189.

Jindal, Meenakshi, Garg, S.K., and Yadava, N.K., 2008. Effect of feeding dietary protein sources on daily excretion in *Channa punctatus* for sustainable aquaculture. In: *Advances in Aquatic Ecology*, (Ed.) V.B. Sakhare. Daya Publishers, New Delhi, 3: 49–60.

Jindal, M., Garg, S.K., and Yadava, N.K., 2007a. Effect of replacement of fishmeal with dietary protein sources of plant origin on the growth performance and nutrient retention in the fingerlings of *Channa punctatus* (Bloch.) for sustainable aquaculture. *Panjab University Research Journal (Science)*, Vol. 57, pp. 132–138.

Jindal, Meenakshi., 2011a. Influence of different feeding regimes on the growth performance and nutrient retention of walking catfish, *Clarias batrachus*. In: *Perspectives of Animal Ecology and Reproduction*, (Ed.) V.K. Gupta. Daya Publishers, New Delhi, 7: 173–186.

Jindal Meenakshi, Yadava, N.K., Jain, K.L. and Gupta, R.K., 2010. Effect of two dietary protein levels on body weight and composition in *Channa punctatus* (Bloch.) fingerlings. *Turkish journal of Fisheries and Aquatic Sciences*, 10(2): 203–208.

Jindal, Meenakshi., 2011b. Studies on protein requirements of catfish *Clarias batrachus* for sustainable aquaculture. *Indian J. Fisheries*, 58(2): 95–100.

Kaushik, S.J. and Cowey, C.B., 1991. Dietary factors affecting nitrogen excretion by fish. In: *Nutritional Strategies and Aquaculture Waste*, (Eds.) C.B. Cowey and C.Y. Cho, p. 7–19.

Kibria, G., Nugegoda, D., Fairclough, R. and Lam, P., 1998. Can nitrogen pollution from aquaculture be reduced? *NAGA, ICLARM*, 21: 17–25.

Luquet, P., 1982. Aspects du metabolisme des poisons particulierement importants pour la qualite de leau. In: *Razionale utilizzo delle risorse idriche in acquacoltura. Atti del*

Convengo Internazionale Verona, Salone Acquacoltura, 16 Ottobre. Pubblicazione curate dal Centro regionale tutela e sperimentazione Pesca e Acquacolture dell' E.S.A.V., Italy. Pp. 30–38.

Vielma, J., Makinen, T., Ekholm, P. and Koskela, J., 2000. Influence of dietary soybean and phytase levels on performance and body composition of large rainbow trout (*Oncorhynchus mykiss*) and algal availability of phosphorous load. *Aquaculture*, 183: 349–362.

Chapter 19

Alterations in Haematological and Biochemical Parameters of Climbing Perch, *Anabas testudineus* (Bloch), Due to Mercury Intoxication

Kuldeep Kumar* and A.K. Pandey[1]

Central Institute of Freshwater Aquaculture, Kausalyaganga, Bhubaneswar – 751 002, Orissa, India

[1]National Bureau of Fish Genetic Resources, Canal Ring Road, Lucknow – 226 002, U.P., India

ABSTRACT

Alterations in haematological and biochemical parameters of the freshwater air-breathing fish, *Anabas testudineus* exposed to 100, 50, 30 and 10 per cent of LC_{50} concentration of mercuric chloride ($HgCl_2$) were recorded. The treatment elicited variations in the differential leucocytes count causing lymphocytosis in larger cells, lymphoponia in small cells, neutrophilia, monocytosis, eosinophilia and thrombocytopenia in 100 per cent concentration of LC_{50} of $HgCl_2$ on day 4 of the exposure. Similar, but less severe, effects were also found on blood constituents of the fish subjected to other concentrations of the toxicant. Poikilocytosis and vacuolization of cytoplasm as well as nucleus were the other pathological changes induced

* Corresponding author: E-mail: kuldeepkumar.kk@gmail.com

by $HgCl_2$. Blood glucose level of the treated fish increased while the liver glycogen content decreased. Serum protein levels in blood of *Anabas testudineus* varied widely. Though the treatment levels did not show significant differences in the serum protein contents, they were significant between periods ($F_{6, 18}$= 5.340; p< 0.1).

Keywords: *Mercuric chloride, Haematological and biochemical parameters, Anabas testudineus.*

Introduction

Industrial effluents and domestic sewage carry considerable loads of metal residues into the water bodies finally affecting the aquatic biota (Pandey *et al.*, 1999). The heavy metals impair the physiological functions on their entry into the circulation by altering the blood proteins, causing respiratory disorders, hormonal imbalance, reduced resistance to diseases and loss of vigour (Roche and Boge, 1996; Wendelaar Bonga, 1997; Safahieh *et al.*, 2010; Younis *et al.*, 2012). Mercury often undergoes the processes of bioconcentration, bioaccumulation and biomagnification in fish and shellfish tissues (Johnels *et al.*, 1967; Westoo and Norem, 1968). Mercurial pollution in aquatic environment has received attention of several workers due to its adverse effects as revealed by the histopathological changes in the vital organs leading to the reduced fish production (Rucker, 1968; Jayachandran and Paulraj, 1975; Biegert and Valkovic, 1980; McFarlane and Frnzin, 1980; Mahajan and Juneja, 1980; Gill and Pant, 1981; Menezes and Qasim, 1984; Mohan *et al.*, 1986a; Bhattacharya *et al.*, 1987; Paulose, 1989; Dalal and Bhattacharya, 1991; Pandey, 1994; George *et al.*, 1995; Pandey *et al.*, 1996a, b; Roche and Boge, 1996). An attempt has been made to record exhaustively the effects of varying concentrations of mercury on the haematological as well as biochemical parameters of the air-breathing perch, *Anabas testudineus*, a commercially important fish recommended for culture in swampy as well as derelict water bodies (Dehadrai and Kamal, 1993).

Material and Methods

Healthy specimens of *Anabas testudineus* (Bloch) (average length 12.04±2.56 cm; average weight 31.94±2.24 g) were procured from the village ponds located near the Central Institute of Freshwater

Aquaculture, Bhubaneswar (Orissa) and were acclimatized to laboratory conditions for 15 days prior to use. During this period, they were given formulated feed in pelleted form comprising groundnut oil-cake (GOC) 50 per cent, rice bran 23 per cent, fish meal 25 per cent, calcium diphosphate 1.5 per cent, sodium chloride 0.3 per cent, trace mineral mix 0.1 per cent and vitamins mix 0.1 per cent at the rate of 5 per cent of the body weight. The faecal matter was removed daily by siphoning to avoid water contamination. The water quality parameters were analysed as per standard methods (APHA, AWWA, WPCF, 1981). Based on the LC_{50} value of mercuric chloride (0.641) as calculated by Sihna and Kumar (1992), the acclimatized nine fishes of uniform size in each group were exposed to the five concentrations of the toxicant such as–(i) 100 per cent of LC_{50} value, (ii) 50 per cent of LC_{50} value, (iii) 30 per cent of LC_{50} value, (iv) 10 per cent of LC_{50} value and (v) 0 per cent of LC_{50} value (without $HgCl_2$) (control). The concentration of $HgCl_2$ in different treatments was maintained through periodical replenishment of the water.

Fishes from all the groups were sacrificed on day 0, 1, 2, 3, 4, 7, 15 and 30. Blood samples from experimental as well as control fishes were collected according to the methods suggested by Steucke and Schoettger (1967). Thin smears of blood were prepared on alcohol cleaned slides and fixed in methanol after air-drying. The combination of Wright and Giemsa method was used for staining. Blood glucose was analyzed by GOD-PAP method, serum protein by Biurete method while liver glycogen was estimated using anthrone reagent (Oser, 1965). The data were subjected to analysis of variance with F and t-tests to determine the significance of variations in different parameters between the treatments and periods (Fischer and Yates, 1963; Fischer, 1970; Finney, 1971).

Results

Water Quality Regime

The water quality parameters ranged as : temperature 31-33.5°C, pH 7.1-7.3, total alkalinity 120-126 ppm, free CO_2 0.5-0.8 ppm, chloride 14.4-15.2 ppm, hardness ($CaCO_3$) 102.4-120.4 ppm, electrical conductivity 349-407 mmhos/cm, dissolved oxygen (DO) 3.84-4.48 ppm, P_2O_5, 0.02-0.04 ppm, NH_4^+-N 0.04-0.05 ppm and NO_3-N 0.4-0.45 ppm. However, no significant difference could be observed between the control and treatment media during the entire course of investigation.

Differential Leucocyte Counts

The population of different leucocytes of control *Anabas testudineus* ranged between large lymphocytes 9.40±1.52–12.02±2.24 per cent, small lymphocytes 18.20±1.92–21.00±2.55 per cent, neutrophils 19.40±3.01–22.60±1.52 per cent, monocytes 1.40±1.14–3.2±1.48 per cent, eosinophils 1.40±1.34–4.00±1.58 per cent, basophils 0.80±0.84–1.60±1.82 per cent and thrombocytes 41.000±4.47–45.60±3.36 per cent (Table 19.1). The fishes maintained in 100 per cent concentration of LC_{50} $HgCl_2$ exhibited an elevation in the population of large lymphocytes, progressive decline in the number of small lymphocytes, steady increase of neutrophils to a maximum of 34.4 per cent, slight decrease in monocytes (3.8 per cent), increase in eosinophils (6.6 per cent) and marked reduction in thrombocyte population on day 4 of the exposure. In 50 per cent concentration of LC_{50} value, the large lymphocyte population increased to 19.8 per cent, small lymphocytes declined to 10.6 per cent and neutrophils increased to 35.6 per cent till day 7. The monocytes and eosinophils recorded an initially increase of 4.2 per cent and 6.2 per cent, respectively on day 1 while thrombocytes population registered a sharp decrease to 24.0 per cent on day 7. Exposure to 30 per cent concentration of LC_{50} value caused almost similar variations in blood cell populations of *Anabas testudineus*. 10 per cent concentration of LC_{50} of $HgCl_2$, the toxicant elicited an increase in large lymphocytes and neutrophils till day 7. Maximum number of eosinophils were recorded on day 3 of the exposure. The differences in the leucocyte cells on application of mercuric chloride were significant for large lymphocytes both between periods ($F_{6,18}$= 7.371; $p < 0.001$) and treatments ($F_{3,18}$ = 24.742; $p < 0.001$), small lymphocytes-between periods ($F_{6,18}$=7.256; $p < 0.001$) and treatments ($F_{3,18}$=19.709; $p < 0.001$), neutrophils between periods ($F_{6,18}$= 3.033; $p < 0.05$) and treatments (($F_{3,18}$=3.257; $p < 0.05$), eosinophils-between treatments (($F_{3,18}$=3.061; $p < 0.10$) and among thrombocytes between periods ($F_{6,18}$= 2.511; $p < 0.10$) and treatments ($F_{3,18}$=7.193; $p < 0.01$)

Poikilocytosis and vacuolization of cytoplasm and nucleus were the pathological changes in the climbing perch induced by mercuric chloride exposure. The severity was greater on long-term exposure and with higher LC_{50} concentration of $HgCl_2$ like 100 per cent > 50 per cent > 30 per cent >10 per cent. In 100 per cent concentration of LC_{50}, vacuolization in cytoplasm and nuclei was seen on day 4. In

Table 19.1: Effects of Mercuric Chloride Treatment on Differential Leucocytes Counts (per cent) of *Anabas testudineus*

Cells	1 Day	2 Days	3 Days	4 Days	7 Days	15 Days	30 Days
Control (0 per cent concentration of LC_{50} value of $HgCl_2$)							
Lymphocytes (large)	10.0±2.0	9.6±2.19	10.2±2.39	10.6±2.07	9.4±1.52	12.0±2.24	9.6±1.82
Lymphocytes (small)	20.4±2.07	20.4±1.95	18.4±2.70	20.6±1.14	18.2±1.92	21.0±2.55	19.4±2.97
Neutrophils	19.4±3.05	20.2±1.92	20.6±1.67	22.6±1.52	20.6±2.07	19.4±3.21	20.0±2.92
Monocytes	2.4±1.14	2.0±1.58	2.6±1.14	3.2±1.48	2.6±0.89	1.4±1.14	1.4±1.14
Eosinophils	3.8±0.84	3.2±1.30	1.4±1.34	2.2±1.30	3.0±1.58	3.2±1.92	4.0±1.58
Basophils	1.6±1.82	1.4±1.14	1.2±0.84	1.4±1.34	0.8±0.84	1.0±1.22	1.2±1.30
Thrombocytes	42.2±1.92	43.6±4.93	45.6±3.36	41.0±4.47	45.4±3.78	42.0±5.24	44.4±4.22
100 per cent concentration of LC_{50} value of $HgCl_2$							
Lymphocytes (large)	13.8±1.64	16.2±3.27	19.2±2.86	17.8±1.30	–	–	–
Lymphocytes (small)	18.2±2.39	13.4±1.82	10.0±4.36	8.0±2.39	–	–	–
Neutrophils	24.2±1.94	27.4±2.30	30.8±4.02	34.4±3.85	–	–	–
Monocytes	3.2±0.84	3.8±1.48	3.2±1.30	2.6±1.14	–	–	–
Eosinophils	4.8±1.64	5.6±2.07	4.6±1.67	6.6±1.67	–	–	–
Basophils	1.4±1.14	0.8±0.84	1.6±1.14	1.6±0.89	–	–	–
Thrombocytes	34.4±2.51	32.8±5.63	28.6±4.39	283.2±7.01	–	–	–

Contd...

Table 19.1–*Contd...*

Cells	*1 Day*	*2 Days*	*3 Days*	*4 Days*	*7 Days*	*15 Days*	*30 Days*
50 per cent concentration of LC_{50} value of $HgCl_2$							
Lymphocytes (large)	12.6±1.82	15.8±2.59	16.8±1.79	18.2±1.92	19.8±2.39	16.6±1.52	13.8±2.59
Lymphocytes (small)	17.4±3.05	14.2±2.17	14.6±1.52	10.6±1.67	10.6±2.07	16.8±1.92	17.8±2.39
Neutrophils	23.6±2.61	24.8±1.64	29.4±2.07	33.6±1.82	35.6±1.52	29.4±1.95	25.4±2.07
Monocytes	4.2±1.79	3.6±1.14	2.8±0.84	3.2±1.30	2.4±1.14	2.2±0.84	1.2±1.30
Eosinophils	5.0±1.58	5.0±2.55	4.6±1.82	5.6±1.52	6.2±1.48	3.6±2.07	3.4±1.14
Basophils	1.0±1.00	1.2±0.84	1.0±0.71	1.2±0.84	1.4±0.89	1.2±1.30	1.0±1.00
Thrombocytes	36.2±2.59	35.6±5.41	29.6±1.82	29.2±1.64	24.0±3.81	30.2±1.64	37.4±3.36
30 per cent concentration of LC_{50} value of $HgCl_2$							
Lymphocytes (large)	12.6±1.82	15.8±2.59	16.8±1.79	18.2±1.92	19.8±2.39	16.6±1.52	13.8±2.59
Lymphocytes (large)	12.0±2.74	14.2±2.86	15.6±1.95	16.2±1.92	18.2±1.92	13.8±2.28	11.2±2.59
Lymphocytes (small)	18.0±2.55	16.2±2.39	16.4±3.58	14.2±2.86	13.0±2.74	16.8±1.92	18.0±2.92
Neutrophils	23.2±2.39	24.8±3.11	28.6±2.88	31.6±2.07	32.6±1.67	29.6±2.07	20.4±2.07
Monocytes	2.4±1.82	3.4±1.52	1.8±0.84	3.0±1.22	2.2±1.30	1.8±0.84	2.0±1.00
Eosinophils	3.2±1.48	4.8±1.48	3.6±1.82	5.4±1.52	5.0±1.58	3.4±1.14	3.0±1.58
Basophils	0.8±0.84	1.0±0.71	0.8±1.30	1.0±1.22	1.2±0.84	1.0±0.71	1.2±1.10
Thrombocytes	40.4±5.03	35.4±6.73	33.4±8.11	29.0±4.30	29.8±5.40	33.6±6.62	44.2±5.07

Contd...

Table 19.1–*Contd...*

Cells	*1 Day*	*2 Days*	*3 Days*	*4 Days*	*7 Days*	*15 Days*	*30 Days*
	10 per cent concentration of LC_{50} value of $HgCl_2$						
Lymphocytes (large)	10.8±2.39	13.6±2.97	14.4±2.70	15.6±1.95	16.2±2.39	13.2±1.92	9.4±2.07
Lymphocytes (small)	18.8±3.96	17.8±2.95	16.0±2.55	16.4±2.07	15.2±1.92	18.6±2.07	21.2±1.92
Neutrophils	21.6±3.04	23.2±1.92	26.0±2.65	30.2±2.59	30.6±2.07	25.6±1.95	21.2±1.03
Monocytes	2.0±1.58	2.2±1.30	2.8±1.79	2.0±1.58	2.0±1.58	2.4±1.14	2.4±1.67
Eosinophils	3.6±1.82	4.0±1.22	6.2±1.31	4.4±1.67	2.4±1.14	3.2±1.30	1.8±0.84
Basophils	1.0±0.71	0.6±0.89	1.8±1.48	1.0±0.71	1.2±1.30	1.0±1.00	1.8±1.64
Thrombocytes	42.2±9.31	38.6±4.83	32.4±5.86	30.4±2.07	32.4±1.52	35.6±2.88	42.2±5.54

50 per cent concentration of LC_{50} value of $HgCl_2$, vacuolization of cytoplasm and nuclei was recorded on day 7. The same concentration inflicted severe damage in the leucocyte on day 15 as vacuolization exaggerated in cytoplasm and nuclei. The leucocytes enlarged and the cytoplasmic material started disintegrating. Vacuoles were also seen in the erythrocyte cells. Finally, the same concentration caused extrusion of the nuclei on day 30 of exposure. In 30 per cent concentration of LC_{50} value of the toxicant, vacuolization in cytoplasm of erythrocytes was noticed on day 30. Interestingly, no marked pathological changes were observed in the blood cells of *Anabas testudineus* maintained in the 10 per cent concentration of LC_{50} of $HgCl_2$.

Blood Glucose

The blood glucose levels of the control *Anabas testudineus* ranged between 78.29±0.89–85.36±1.41 mg/dl. Exposure of the fish to mercuric chloride caused distinct variations in blood glucose levels (Table 19.2). Exposure to 100 per cent concentration of LC_{50} of $HgCl_2$ elicited 76.93 per cent elevation in blood glucose on day 1. Though the value registered a progressive decline but the level remained high (24.97 per cent over initial value) even on day 4. In 50 per cent LC_{50} concentration too, a similar trend was observed. In 30 per cent and 10 per cent concentrations of LC_{50} of the toxicant, blood glucose level decreased on day 4 and 15. The differences between the blood glucose levels on application of different doses of mercuric chloride were significant ($F_{3,18} = 7.8035, p < 0.01$) as also between periods ($F_{6,18} = 2.8443; p < 0.05$).

Serum Protein

The serum protein content of the control *Anabas testudineus* ranged between 3.96±0.23 and 4.56±0.64 g/dl. In all the concentrations of $HgCl_2$, maximum increase in serum protein level was observed on day 2 followed by a progressive decline on day 4. Serun protein levels again registered an increase on day 7, thereafter, it decreased till the close of the experiment (day 30). Though the response of serum protein levels to $HgCl_2$ treatment was a little late among the fishes exposed to the lower (50, 30 and 10 per cent concentrations of LC_{50} value of $HgCl_2$), the effect was almost similar in all the treatments. The treatment levels of mercuric chloride did not show significant differences in their effects on serum protein

Table 19.2: Effect of Mercuric Chloride Treatment on Certain Blood Parameters of *Anabas testudineus*

Concentrations (per cent)	*0 Day*	*1 Day*	*2 Days*	*3 Days*	*4 Days*	*7 Days*	*15 Days*	*30 Days*
Blood glucose (mg/dl)								
100	79.64±1.83	140.91±1.94	120.89±2.01	109.53±2.08	99.53±2.12	–	–	–
50	79.64±1.83	120.38±1.62	143.14±1.69	100.66±2.43	80.26±1.89	123.31±2.11	97.59±1.86	142.46±1.73
30	79.64±1.83	133.15±1.52	99.50±1.91	90.00±2.11	69.85±1.63	113.89±1.44	76.75±1.73	121.82±1.52
10	79.64±1.83	146.06±1.72	100.42±1.58	91.81±1.78	70.26±2.09	100.56±2.15	90.52±1.35	95.06±1.19
Control	79.64±1.83	78.87±2.39	80.84±1.58	81.84±2.06	85.36±1.41	81.04±1.86	78.29±1.89	80.09±1.46
Serum protein (g/dl)								
100	3.96±0.23	6.04±0.467	7.08±0.37	6.10±0.32	5.93±0.43	–	–	–
50	3.96±0.23	5.78±0.34	6.23±0.42	5.86±0.26	4.80±0.45	5.05±0.47	3.52±0.22	2.94±0.23
30	3.96±0.23	5.67±0.36	6.03±0.30	5.57±0.43	4.02±0.29	4.85±0.31	3.82±0.23	3.22±0.29
10	3.96±0.23	5.08±0.40	5.75±0.60	4.97±0.29	4.01±0.42	5.01±0.34	4.08±0.31	3.91±0.29
Control	3.96±0.23	4.07±0.34	4.14±0.29	4.03±0.29	4.26±0.43	4.05±0.20	4.56±0.64	4.03±0.31

Table 19.3: Effect of Mercuric Chloride Treatment on Liver Glycogen (mg/g) of *Anabas testudineus*

Concentrations (per cent)	*0 Day*	*1 Day*	*2 Days*	*3 Days*	*4 Days*	*7 Days*	*15 Days*	*30 Days*
100	14.54±0.81.52	8.05±0.5811	9.20±0.7416	10.11±0.7919	11.57±0.7207	–	–	–
50	14.54±0.81.52	9.00±0.3937	8.17±0.5783	10.97±0.6629	13.20±0.8700	9.63±0.9539	12.90±0.4873	8.81±0.8373
30	14.54±0.81.52	8.68±0.7321	11.86±1.3509	13.56±0.8455	14.81±0.9492	11.44±0.7316	11.95±0.6857	9.05±0.5120
10	14.54±0.81.52	8.13±0.4527	12.11±0.5182	14.67±0.4206	14.35±0.7535	10.95±0.5547	14.75±0.8032	12.50±0.8396
Control	14.54±0.81.52	15.03±0.4795	15.24±0.3972	15.12±0.5333	14.55±0.5854	15.47±0.5118	15.26±0.7088	15.00±0.8602

content, but they were significant between the periods ($F_{6,18}$= 5.340; p < 0.10).

Liver Glycogen

The liver glycogen content decreased sharply on application of mercuric chloride. The liver glycogen content of the control *Anabas testudineus* ranged narrowly between 14.54±0.82–15.47±0.51 mg/g. Liver glycogen exhibited significant variations in response to $HgCl_2$ exposures (Table 19.3). In 100 per cent concentration of LC_{50} of $HgCl_2$, the liver glycogen content registered decline of 44.63 per cent and 20.42 per cent, respectively, over the initial value on day 1 and 4. In 50 per cent concentration of LC_{50} of value, 43.81 per cent decrease in liver glycogen was recorded on day 2 while 9.21 per cent and 39.40 per cent reduction (over initial value) were observed in the fish on day 4 and 30, respectively. In 30 per cent concentration of mercuric chloride, though liver glycogen initially exhibited a significant decline on day 1 and 2 but normal value was regained by day 7. Thereafter, a progressive declining trend in the liver glycogen content was recorded till close of the experiment (day 30). In 10 per cent concentration of LC_{50} of $HgCl_2$, normal values were recorded in liver glycogen on day 3 and 15 after an initial decline on day 1 and 2. Interestingly, maximum decrease in the liver glycogen value was noticed on day 1 in all the levels of mercuric chloride. The differences between the liver glycogen value on application of different doses of mercuric chloride were significant ($F_{3,18}$ = 13.446; $p < 0.001$) as also between periods ($F_{6,18}$ = 4.169; $p < 0.10$).

Discussion

Though the lethal and sub-lethal effects of metals as well as the deteriorating effects of mercury individually on aquatic organisms have been amply recorded (Webeser, 1973, 1975a, b; Eisler *et al.*, 1979; Aruna and Gopal, 1987; Sultana *et al.*, 1991; Pandey, 1994; George *et al.*, 1995; Pandey *et al.*, 1996a, b; Roche and Boge, 1996; Safahieh *et al.*, 2010), early detection of specific physiological abnormalities provide an indication of exposure prior to manifestation of any gross damage to the tissues. Exposure to mercuric chloride induced variations in differential leucocyte counts and caused lymphocytosis in large cells, lymphopenia in small cells, neutrophilia, monocytosis, eosinophilia and thrombocytopenia in *Anabas testudineus* maintained in 100 per cent concentration of LC_{50}

of the toxicant on day 4. The 50, 30 and 10 per cent concentrations of LC_{50} value of $HgCl_2$ too caused similar effects in lymphocytes, neutrophils, eosinophils and thrombocytes of the climbing perch. Newman and Maclean (1974) reported a dose-dependent threefolds increase in neutrophils and decrease in lymphocytes in *Tautogolabrus adspersus*. Larsson *et al.*(1985) observed markedly enhanced lymphocyte count in perch from a river polluted with cadmium. Exposure of *Puntius conchonius* to organomercurical fungicide induced significant thrombocytopenia, lymphocytosis and neutropenia (Pant and Gill, 1983). Dhanekar *et al.* (1985) reported increase in large lymphocyte, reduction in small lymphocyte and thrombocytes populations as also elevation in monocyte, neutrophil and eosinophil cells in *Heteropneustes fossilis, Channa punctatus* and *Mastacebalus puncalus* on long-term exposure to least effective concentration of mercuric chloride. Our findings with *Anabas testudineus* are in the line of the observations recorded in other telostean species. There are several morphological types of multinuclear leucocytes such as neutrophils, eosinophils and basophils. Only neutrophils and eosinophils possess the ability of phagocytosis. Any inclusions of foreign substances are subject to removal from the systems with the phagocytic ability of these cells. The increase in number of neutrophils and eosinophils in experimental *Anabas testudineus* (Table 19.1) may be due to their demand for removal of foreign substance (mercuric chloride). Thrombocytosis occurred due to toxic action of the chemical on the blood and haemopoisis system. Deformity in the shape of blood cells (poikilocytosis), cytoplasmic deterioration, vacuolization in cytoplasm and nuclei, and extrusion of nuclei have been recorded in *Anabas testudineus* maintained in varying concentrations of $HgCl_2$. Pant and Gill (1983) also reported similar effects in *Puntius conchonius* upon chronic exposure (56 days) by an organomercurial fungicide. Chandra and Pathani (1983) too reported severe corpuscular deformity induced by heavy metals (Hg, Cd and Cr) in *Puntius conchonius*. Murugesan *et al.* (1989) recorded cytopathological changes in erythrocytes of *Heteropneustes fossilis* exposed to textile mill effluents. Vacuolization of cytoplasm and extrusion of nuclei may be explained on the fact that hypertrophied erythrocytes have weak cell membrane which breaks down during the process of smearing releasing the nuclei. Vacuolization in the cytoplasm and nuclei is the earliest sign of damage to cells.

The blood glucose and liver glycogen of *Anabas testudineus* exhibited marked variations due exposures to varying concentrations of mercuric chloride. The blood glucose increased while liver glycogen decreased in response to $HgCl_2$ treatments. It is well established that under exposure to toxicant, many biochemical changes in the fish preceede cellular and systemic dysfunction. Gill and Pant (1981) reported that subtle glucovariations with concurrent hepatic glycogenolysis provide reliable evidence of biochemical stress. They also recorded increase in the blood glucose levels and low liver glycogen in *Puntius conchonius* exposed chronically with $HgCl_2$ upto three weeks. Bhattachacharya *et al*. (1987) observed that mercury caused elevation in blood glucose on day 1 and 2 of exposure. Mercuric chloride treatment to *Anabas testudineus* brought about a significant decline in glycogen content of liver throughout the experimental period (Table 19.2). Hepatic glycogen appears to be the source of hyperglycemia and its depletion corresponded with the increase in blood glucose levels suggesting that some of the hepatic glycogen (via the intermediate, glucose-1-phosphate) is getting converted into glucose. Decrease in glycogen may be attributed to an initial regulatory step which increases intermediary metabolism resulting in protection of the hepatocytes. The interconversion of glycogen and glucose during metal toxicity has been demonstrated by Larsson *et al*. (1976), Sastry and Rao (1981), Sastry and Tyagi (1982) and Choudhury and Nath (1985).

The serum protein levels of *A. testudineus* exhibited wide variatons in response to mercuric chloride treatments. The maximal initial increase on day 2, followed by a decrease on day 4 and again an elevation on day 7 in all the levels of the exposures were registered. However, the minimal level of serum protein was recorded on day 30. Sharma and Davis (1980) reported that methyl mercury administration brought about a distinct decline in the rate of protein synthesis in the liver of *Cyprinus carpio*. Ramalingam and Ramalingam (1982) observed a sharp reduction in liver protein in *Tilapia mossambica* exposed to DDT, malathion and mercury for 1, 7 and 15 days. Ram and Sathyanesan (1984) found reduction in liver and ovarian protein by exposing *Chánna punctatus* to mercury. McKim *et al*. (1970) noted an increase in protein levels after 6 days followed by a decrease after 21 days in brook trout by copper poisoning. Agrawal *et al*. (1982) reported that lithium poisoning brought about a significant increase in total plasma protein after 96 hours of exposure in *Heteropneustes fossilis*. Sharma (1989) reported a decline in serum protein in *Clarias batrachus* in the early phase of exposure due to higher levels of mercuric chloride. Lower levels of serum protein may be associated with its excessive loss due to nephrosis or reduced protein synthesis. When carbohydrate sources of glucose are depleted, proteins are utilized for glucose production (gluconeogenesis) which would evidently lead to a faster and irreversible reduction in serum protein.

Acknowledgement

The authors are grateful to the Director, Central Institute of Freshwater Aquaculture (ICAR), Bhubaneswar for providing necessary facilities to carry out this work.

References

Agrawal, V.P., Sharma, M.L., Wadhwa, S., Gupta, T. and Mishra, B.P., 1982. Lithium induced haematological and biochemical changes in *Heteropneustes fossilis*. *Proc. Symp. Environ. Van. S. P.*, pp. 27–31.

APHA, AWWA, WPCF, 1981. *Standard Methods for the Examination of Water and Wastewater*. 15th edn. American Public Health Association, Washington D.C.

Aruna, D. and Gopal,V.,1987. Toxic effect of sublethal level of mercury on haematological parameters. *Indian J. Environ. Hlth.*, 29: 52–56.

Bhattacharya, T., Ray, A.K. and Bhattacharya, S., 1987. Blood glucose and hepatic glycogen interrelationship in *Channa punctatus* (Bloch). A parameter of nonlethal toxicity bioassay with industrial pollutants. *Indian J. Exp. Biol.*, 82: 25–32.

Biegert, E.K. and Valkovic, V., 1980. Acute toxicity and accumulation of heavy metals in aquatic animals. *Period Biol.*, 82: 25–32.

Boetius, J. 1960. Lethal action of mercuric chloride and phenylmercuric chloride acetate on fishes. *Medd. Dan. Fish. Havunders*, 3: 93–115.

Brouwer, M. and Brouwer–Hoexum, T., 1985. Cupric and mercuric ions decrease red cells deformability. *Environs*, 3:

Chandra, J. and Pathani, S.S., 1983. Corpuscular pathology induced by heavy metals in a freshwater teleost, *Puntius conchonius* (Ham.). In: *Proceedings of the Fourth All India Seminar on Ichthyology*, Abst. No. 2.29, 30 p.

Choudhury, H.S. and Nath, K., 1985. Nickel induced hyperglycemia in the freshwater fish, *Colisa fasciatus*. *Water, Air & Soil Poll.*, 24: 173–176.

Dalal, R. and Bhattacharya, S., 1991. Effects of chronic nonlethal doses of non–metals and metals on hepatic metauothionein in *Channa punctatus* (Bloch). *Indian Exp. Biol.*, 29: 693–694.

Dehadrai, P.V. and Kamal, M.Y., 1993. Role of air-breathing fish culture in rural upliftment. In: *Souvenir: Third Indian Fisheries Forum* (October 14–16, 1993). College of Fisheries, G.B. Pant University of Agriculture and Technology, Pantnagar, pp. 28–33.

Dhanekar, S., Srivastava, S., Rao, K.S. and Pandya, S.S., 1985. Studies on toxic effects of least effective concentration of mercury on fish: A haematological study. *Matsya*, 11: 75–78.

Eisler, R., Rossoll, R.M. and Gaboury G.A., 1979. Fourth Annotated Bibliography on Biological Effects of Metals in Aquatic Environments. U.S. Environmental Protection Agency Res. R. P. EPA–600/3–79–084. Washington, D.C.

Finney, D.J. 1971. *Probit Analyses*. 3rd edn. Cambridge University Press, Cambridge.

Fisher, R.A. and Yates, F.I. 1963. *Statistical Tables*. 6th edn. 146 p.

Fisher, R.A., 1970. *Statistical Methods for Research Workers.* Heafner Press, New York and Coller–McMillan Publisher, London.

George, K.C., Pandey, A.K. and Mohamed, M.P., 1995. Mercury chloride induced renal lesions in the mullet, *Liza parsia* (Hamilton–Buchanan). *J. Mar. Biol. Assoc. India*, 37; 177–184.

Gill, T.S. and Pant, J.C., 1981. Effects of sublethal concentrations of mercury in a teleost, *Punctius conchonius*: biochemical and haematological responses. *Indian J. Exp. Biol.*, 19: 571–573.

Jayachandran, P. and Paulraj, S., 1975. Mercury in fish from Mettur reservoir. *Curr. Sci.*, 44: 828–829.

Johnels, A.G., Westermark, T., Berg, W., Pearson, P.I. and Sjostrand, B., 1967. Pike (*Exos lucius*) and some other aquatic organisms in Sweden as indicators of mercury contamination in the environment. *Oikos*, 18: 323–333.

Larsson, A., Bengtsson, B.E. and Cvanberg, O., 1976. Some haematological and biochemical effects of cadmium on fish. In: *Effects of Pollutants on Aquatic Organisms* (Ed) A.P.M. Lockwood. Cambridge Univ. Press, Cambridge. pp. 35–45.

Larsson, A., Haux, C. and Sjobeck, M., 1985. Fish physiology and metal pollution: results and experiences from laboratory and field studies. *Ecotoxicol. Environ. Saf.*, 9: 250–256.

Mahajan, C.L. and Juneja, C.J., 1980. Combined effects of simultaneous mercury pollution in diet and water on the fish, *Channa punctatus*. *Curr. Sci.*, 49: 715–717

McFarlane, G.A. and Frnzin, W.G., 1980. An examination of Cd, Cu, Hg concentration on liver of northern pike, *Esox lucius* and white sucker, *Catastomus commersoni* from five Lakes near a base metal smalter at Flin Flon, Manitoba, Canada. *Can. J. Fish. Aquat. Sci.*, 37: 1573–1578.

McKim, J.M., G.M. Christensen and E.P. Hunt 1970. Changes in the blood of brook trout (*Salvelinus fontinalis*) after short-term and long-term exposure to copper. *J. Fish. Res. Bd. Can.*, 27: 1883–1889.

Menezes, M.R. and Qasim, S.Z., 1984. Effects of mercury accumulation on the electrophoretic patterns of the serum, haemoglobin and eye lens protein of *Tilapia mossambica* (Peters). *Wat. Res.*, 18: 153–161.

Mohan, C.V., Gupta, T.R.C. and Menon, N.R., 1986. Acute toxicity of mercury on the early developmental stages of *Cirrhinus mrigala* (Ham.). *Indian J. Fish.*, 33: 133–136.

Murugesan, A.G., Muthu, P. and Haniffa, M.A., 1989. Cytopathological changes in erythrocytes of the catfish, *Heteropneustes fossilis* (Bloch) exposed to textile mill effluent. *Curr. Sci.*, 58: 268–269.

Newman, M.W. and Maclean, S.A., 1974. Physiological response of the cunner, *Tautogolabrus adspersus* to cadmium. VI. Histopathology. National oceanic and experiences from laboratory and field studies. *Ecotoxicol. Environ. Saf.*, 4: 250–260.

Oser, B., 1965. *Physiological Chemistry*. 14th edn. McGraw Hill, New York.

Pandey A C, Pandey A K and Das P., 1999. Fish and fisheries in relation to aquatic pollution. In: *Environmental Issues and Resource Management* (Eds.) P. Das, S.R. Verma and A.K. Gupta. Nature Conservators, Muzaffarnagar. pp. 67–112.

Pandey, A.K., 1994. Branchial and renal lesions of the estuarine mullet, *Liza parsia*, exposed to sublethal concentration of mercury. *Proc. Nat. Acad. Sci., India*, 64B: 283–287.

Pandey, A.K., George, K.C. and Mohamed, M.P., 1996a. Histopathological changes in liver of a teleost, *Liza parsia*, induced by mercuric chloride and DDT. *Indian J. Fish.*, 43: 272–284.

Pandey, A.K., George, K.C. and Mohamed, M.P., 1996b Histopathological changes in gill of *Liza parsia* induced by mercuric chloride and DDT. *Indian J. Fish.*, 43: 285–291.

Pant, J.C. and Gill, T.S., 1983. Blood cell dyserasia and biochemical anomalies caused by an organomercurial fungicide in the fish, *Punctius conchonius* (Ham.). In: *Proceedings of the Fourth All India Seminar on Ichthyolog.*, pp. 29–36.

Paulose, P.V., 1989. Histological changes in relation to accumulation and elimination of inorganic and methyl mercury in gills of *Labeo rohita* (Ham.). *Indian J. Exp. Biol.*, 27: 146–150

Ram, R.N. and Sathyanesan, A.G., 1984. Mercuric chloride induced changes in the protein, lipid and cholesterol levels of the liver and ovary of the fish, *Channa punctatus*. *Environ. Ecol.*, 2: 113–117

Ramalingam, K. and Ramalingam, K., 1982. Effects of sublethal levels of DDT, malathion and mercury on tissue proteins of *Sarotherodon mossambicus* (Peters). *Proc. Indian Acad. Sci. (Anim. Sci.)*, 91: 501–505.

Roche, H. and Boge, G., 1996. Fish blood parameters as a potential tool for identification of stress caused by environmental factors and chemical intoxicants. *Marine Environ. Res.*, 41: 27–43.

Rucker, P.R.,1968. Effects of mercury compounds on fish and humans. *Bull. Int. Epizoot.*, 69: 1431–1437.

Safahieh, A., Hedayati, A., Savari, A and Movahedinia, A., 2010. Experimental approaches of haematotoxic and immunotoxic effects of mercury on yellowfin seabream (*Acanthopagrus latus*). *Amer.–Euras. J. Toxicol. Sci.*, 2: 169–176.

Sastry, K.V. and Rao, D.R., 1981. Enzymological and biochemical changes produced by a teleost fish, *Channa punctatus. Toxicology Lett.*, 11: 17–21.

Sastry, K.V. and Tyagi, S., 1982. Toxic effects of chromium in a freshwater teleost fish, *Channa punctatus*. *Toxicology Lett.*, 9: 321–326.

Sharma, S.D., 1989. Haematochemical adversities in *Clarias batrachus* induced by lithium. *Indian. J. Environ. Hlth.*, 31: 354–357.

Sharma, D.C. and Davis, P.S., 1980. Effect of methyl mercury on protein synthesis in liver of European carp, *Cyprinus carpio*. *Indian J. Environ. Hlth.*, 31: 354–357.

Sinha, Y.K.P. and Kuldeep Kumar, 1992. Acute toxicity of mercuric chloride for *Anabas testudineus* (Bloch). *Environ. Ecol.*, 10: 720–722.

Steucke, E.W. and Schoettger, R.A., 1967. *Wintrope M M Clinical Haematology*. Lee Febeger, Philadelphia, U.S.A.

Sultana, R., Uma Devi, V. and Prasad, M.N., 1991. Effect of heavy metals on the respiration of the catfish, *Mystus gulio*. *J. Ecotoxicol. Environ. Monit.*, 1: 234–237.

Webeser, G., 1973. Aquatic mercury pollution: studies of its occurrence and pathologic effects on fish and milk. *Ph. D. Thesis*. Univ. Saskatchewan, Saskatoon, Canada.

Webeser, G., 1975a. Prolonged administration of methyl mercury chloride to rainbow trout (*Salmo gairdneri*) fingerlings. *J. Fish. Res. Bd. Can.*, 32: 2015–2023.

Webeser, G., 1975b. Acute toxicity of methyl mercury chloride and mercuric chloride for rainbow trout (*Salmo gairdneri*) fry and fingerlings. *J. Fish. Res. Bd. Can.*, 32: 2005–2013.

Wendelaar Bonga, S.E., 1997. The stress response in fish. *Physiol. Rev.* 77, 591–625.

Westoo, G. and Norem, K., 1968. Mercury and methyl mercury in fish. The amount of mercury and methyl mercury in fish caught or brought from September 1966 to November 1967. Var Foda (Sweden). *Chem. Abstr.*, 19: 138.

Younis, E.M., Abdel–Warith, A.A. and Al–Asgah, N.A., 2012. Haematological and enzymatic approaches to Nile tilapia, *Oreochromis nilotucus*, during short– and long–term sublethal exposure to zinc. *Afr. J. Biotechnol.*, 11: 4442–4446.

Chapter 20

Eco-morphological, Histochemical and Functional Organization of Integumentary Covering in Vertebrates with Particular Reference to the Teleost Fishes

Iqbal Parwez and Subuhi Abidi*

Department of Zoology, Aligarh Muslim University, Aligarh – 202 002, U.P., India

ABSTRACT

Integument is not merely an external body covering but an extensive interactive interface between the organism and its environment. Its tripod primary functions include lubrication, water proofing and indeed protection. In addition, other functions such as sensory perception, respiration, thermoregulation and also ion-regulation in teleost fishes have also been reported. This entire system is built upon a common basic plan consisting of three primary zones – epidermis, dermis and hypodermis which are finally anchored with the underlying musculature. The epidermis may have three regions *i.e.*, external cuticular covering, mid region of large cells with

* Corresponding author: E-mail: iparwez2002@yahoo.co.in

interspersed mucous cells and posterior most stratum germinativum which largely rests on basement membrane. Presence of lymphatic spaces, pit organs and taste buds have also been reported in the epidermis. There is a prominent layer of chromatophores lined just below basement membrane which is responsible for skin colouration. Dermis consists of two regions, the loosely arranged connective tissue network forming stratum spongiosum and densely packed stratum compactum with the occurrence of blood vessels and nerve endings. The hypodermis is largely made up of adipose cells and like dermis also with the presence of connective tissues, blood vessels and nerve ending and occasional presence of chromatophores. The integument is provided with two types of derivatives- the first type are soft glandular such as mucous cells, club cells, sacciform cells and some complex glands in higher vertebrates such as sweat glands, mammary glands, poison glands etc. The other type are those of hard structures which include scales, plates, spines, beak, teeth, head projections like horn, antlers and foot projections like hoofs, nail, claws etc. Both types of these derivatives are predominantly epidermal in origin though scales may either be epidermal in some and dermal in others. The teeth may have dual origin with enamel portion epidermal and dentine part dermal in origin. The structural and functional details of teleost integument along with an elaborate account of the various secretary glands have been described. Mucus secreted principally by the goblet cells is the most widely occurring secretion in the teleost integument. In term of chemical composition, it is constituted 95 per cent by water and the remaining 5 per cent is in the form of glyco-conjugates where the latter moiety is predominantly protein with only a small lipid fraction. Amongst the glycoproteins, carbohydrates moiety in the form of mucopolysaccharide is 65 per cent and is classified as neutral and acidic mucopolysaccharide which can be easily localized histochemically by the use of Periodic Acid Schiff (PAS) for neutral and Alcian blue (AB) at pH 2.5 for total acidic component. The significance of integumentary system as a sensitive indicator of environmental degradation and also as stress marker has been highlighted in the present write-up.

Keywords: *Integument, Structural, Histochemical, Functional organization, Vertebrates, Teleost fishes.*

(A) Significance and Functions

Skin or Integument provides an all-round covering over the body surface and serves as the major interface between the animal and its environment. It is a multifunctional organ system which shows considerable variations depending on species, the specific region of the body and the type of environment inhabited by the animals. Amongst the broad functions associated with skin, the primary function is that of protection of (a) underlying organs against mechanical abrasion, (b) invasion of pathogen, (c) warding off the predaceous enemies by the presence of scales and (d) release of venom to kill the predator, colour camouflaging to avoid being spotted and release of fright pheromones to alert other or con-species. The other functions are thermoregulation particularly through sweat glands in mammals, cutaneous respiration in amphibians and larval fishes, friction-free locomotion facilitated by the slippery mucous covering and sensory perception due to the presence of taste buds, receptors for touch, pressure and temperature. Most of these integumentary functions can also be ascribed to fish integument in addition to the function of ion-regulation due to the presence of ionocytes in some teleost fishes. Further, the presence of bioluminescent organs or photophores in many mid water or bottom dwelling deep sea fishes or the multiple photophores spread over the body surface may serve the variety of functions of colour camouflage for protection, to illuminate dark waters or to attract the prey. The slippery mucus over the fish integument may aid friction-free locomotion at a great speed with less expenditure of energy. Some studies have also suggested that fish skin is an important site for ammonia excretion particularly in marine teleosts. In short, fish integument performs majority of the functions associated with general vertebrate skin in addition to some of its own specialized functions due to its habitat need.

(B) Typical Vertebrate Integument

Based on the generalized pattern, it consists of three broad regions. An outer *Epidermis* derived from ectoderm, the next underlying layer *Dermis* or *Corium* derived from mesoderm which is followed by the third, the *Subcutaneous Tissue* which may also have a layer of fat called panniculus adiposus or subcutis being made up of several inches of thickness in case of obese individuals. In tetrapods, there may be a periodical shedding or sloughing off the epidermis which may either be partial or complete. The relevant details of each layer are as follows:

1. Epidermis

The epidermis of a typical vertebrate skin may have three distinct cellular regions which may either be single or multilayered. These are the innermost *stratum germinativum* or *malpighian layer* resting on basement membrane, the middle multicellular transitional layer and the outermost *stratum corneum*. In some cases, particularly in the skin of sole and palm of humans, the transitional layer has the additional layer of *stratum granulosum* and *stratum lucidum*. The stratum germinativum derives nourishment from the blood vessels of underlying dermis, divide continually to form new cells which are pushed upward to form multilayered transitional layers. In most of the vertebrates, the cells of stratum germinativum are columnar whereas those in the transitional layer get gradually flattened and are of stratified squamous epithelial types. The outermost layer is made of flattened cells which have lost their nuclei, are dead and cornified hence called stratum corneum which undergoes regular sloughing in many vertebrate groups. However, this may not be the universal feature in all the vertebrates. Majority of the skin derivatives such as feathers, hair, horns, nail, claws and antlers are exclusively epidermal in origin. Some scales are epidermal whereas others are dermal in origin. Teeth are derived both from epidermis (enamel portion) and dermis (dentine). The epidermis may also have colour pigments which may either be contained in certain specialized pigment containing cells called *chromatophores* such as in fishes, amphibians and reptiles or these pigments may be dispersed in the malpighian layer of the epidermis and the degree of pigmentation determines the skin colour. Interestingly, such vertebrate groups which are provided with fur, feather and scales, the pigmentation may be provided to these skin derivatives. The epidermis may also have soft integumentary derivatives in the form of glandular structures such as various kinds of glands *e.g.*, mucous cells/goblet glands, sacciform cells, glandular club cells etc, each with a specific secretion and specialized functions. In addition, it may also have sensory structures such as taste buds and other sensory organs for perception of various kind of stimuli.

2. Dermis

The dermis commonly has an outer vascular *stratum spongiosum* and a deeper thicker *stratum compactum* which are merged with each other. The entire dermal layer is provided with an extensive network of loose connective tissue which contains abundance of blood and lymph vessels, nerve endings, sense organs, and some assorted dermal derivatives.

3. Subcutis

It is the last part of the integument which anchors with dermal collagenous fibres on one side and the underlying muscles below. It is primarily made up of adipose cells with sporadic network of blood vessels, nerve fibres and occasionally chromatophores.

(C) A Brief Comparative Account of Integument in Different Animal Groups

(i) Invertebrates

The invertebrate integument referred to as hypodermis does not correspond to the typical vertebrate integument organization. It has an outer noncellular layer called cuticle which is made up of chitin or calcareous material. This layer may undergo periodical sloughing called ecdysis. Below cuticular layer lies a single layer of columnar cells called hypodermis which is essentially responsible for the secretion of chitin of the cuticular layer. Hence, in broad terms, the cuticular layer may be analogous to stratum corneum though the latter is cellular and hypodermis may be equivalent to a single layer of stratum germinativum.

(ii) The Amphioxus

Its integument partly resembles the invertebrates in as much as it has columnar epithelial cells secreting noncellular cuticle. But in addition, like typical vertebrates, it is made up of some sort of soft connective tissue forming dermis which, however, is without pigment.

(iii) The Cyclostomes

The integument has more advanced features with multilayered epidermis provided at outer surface with stratum corneum which, unlike most other vertebrates, are nucleated and living and secrete cuticular covering. The dermis is like that of typical vertebrates with extensive network of connective tissue, blood vessels, nerves, smooth muscles and chromatophores.

(iv) The Fishes

Here the integument is same as in cyclostomes with no dead corneal layer. The presence of keratin over the epidermis is reported in some fishes but the outer layer cells are not dead. In certain teleosts, during breeding seasons certain cornified areas in the form of pearl organ are visible. The detailed account of fish integument is described in the later part of this chapter.

Fish integument resembles with the typical mammalian integument in as much as it has well-defined epidermis and dermis and the latter shows the distinct regions of stratum spongiosum and stratum compactum. However, the epidermal layer shows the difference for not having the typical stratum germinativum but all epidermis seems metabolically active. Further, fish scales, unlike reptiles, are dermal in origin which may project out through the epidermis. A detailed account of fish integument is given in the subsequent part.

(v) Amphibians

The integumentary structure is typically that of vertebrates and is different from the lower groups to have dead stratum corneum which is more pronounced in the amphibians adapted to terrestrial life. Since amphibian skin also performs respiratory function, its dermis is richly supplied with large number of blood vessels.

(vi) Reptiles

Due to the mostly dry habitat occupied by this group of animals, the reptilian skin is provided with the well-developed stratum corneum which is periodically shed off. The reptilian scales, unlike those in fishes, are derived from stratum corneum and are shed along with the outer layer.

(VII) Birds

The body of the birds are mostly covered with feathers except in the region of beak, legs and feet. The feathers are the derivatives of stratum corneum. The bird skin which is normally thin becomes thick in the exposed region of legs and feet and may also be provided with protective scales. The extraordinary feature of bird skin is the abundance of muscle fibres to control the movement of feathers. The colouration of the bird is generally provided by the pigment present in the feathers and scales which is not confined within a chromatophore.

(VIII) The Mammalian

Integument has already been described in detail under the heading typical integumentary structure.

(D) Integumentary Derivatives

Certain structures which are derived from skin proper *i.e.* epidermis and dermis are called skin or integumentary derivatives which, based on their structural consistency, could be either soft or hard derivatives. The following are some of the generalized features of these integumentary derivatives.

1. In terms of their origin, an overwhelming number of these derivatives originate from the epidermis and very few such as fish scales originate from dermis.
2. Some of these derivatives such as scales etc. may have wider occurrence in different vertebrate groups such as fish, reptiles, birds and mammals whereas others are group-specific such as feathers in birds and hair in mammals.
3. Some of these derivatives may originate from both epidermis and dermis depending upon the group. For example, reptilian scales are epidermal whereas fish scales are dermal in origin.
4. In certain instances, the same integumentary derivative may have dual origin *i.e.*, partly dermal and partly epidermal such as in the case of teeth where enamel portion is derived from epidermis and dentine part from the dermis.
5. Various integumentary glands such as mucous, poison, sweat and sebaceous glands are basically soft and other structures *e.g.* scales, feathers, spines, horns, hoofs, antlers etc. are classified as hard integumentary derivatives.

A brief description of a variety of integumentary derivatives in different groups of animals is given below:

(i) Cyclostomes

There are no hard integumentary derivatives in cyclostomes. However, among soft derivatives are only unicellular *mucous glands*. Hagfishes have peculiar *thread cells* in epidermis projecting spirally coiled fine threads which entangle with mucus secretion forming protective slimy coat. In some cases, other unicellular mesocrine glands such as *granular gland cells* in lamprey and *beaker cells* have been described.

(ii) Fishes

The glandular derivatives include unicellular and multicellular glands which are as follows.

1. Unicellular and multicellular *mucous glands*
2. Multicellular *poison gland*
3. Luminescent *photophores*
4. *Pterygopodial gland*

Both types of mucous glands- unicellular and multicellular are of simple saccular type. The mucus secreted in lungfishes such as South American lungfish, *Lepidosiren* and African lungfish, *Protopterus* forms a dried-up protective cocoon during aestivation. The multicellular poison gland, associated with hollow groove on spines on the fin, tail and gill cover are analogous to hypodermic syringe, protects from enemies. In one species, *Synanceia* from the Indian Ocean, the poison gland is particularly well-developed. In some deep-sea dwelling elasmobranch and teleosts, an extraordinary light producing *luminous gland* called photophores is found which help these fishes to attract their prey. The structure of this gland has been studied in detail in *Porichthys* in which the gland is shown to consist of a lens in the front and a pigmented reflector behind. In certain species of skates and rays, a special pterygopodial gland located on the clasper organ with little known function has been reported.

Among the hard derivatives, the most distinctive derivative in fishes are scales which may not be present in variety of fishes such as catfishes, eel etc. These scales predominantly develop from dermis though in few others, they develop from both layers of skin. Such scales are of four types. The *placoid* found in elasmobranch are most primitive type. The *ganoid* have shiny coating of ganoin found in gar, the *cosmoid* and *cycloid* scales occur in many teleost fishes. In addition, fin rays which are dermal in origin found both in elasmobranchs and teleosts are yet another hard derivatives in fishes. The modification of scales is seen in some fishes such as heavy fin spines in *Squalus acanthias*, the sting of *Dasyatis* and the heavy teeth on the rostrum of sawfish *Pristis*.

It is interesting to note that there has been a general trend in tetrapods to shift from dermal to epidermal origin of scale as has been seen in reptiles, birds and certain mammals.

(iii) Amphibians

Glandular Derivatives (Multicellular Mucous and Granular Glands)

There are abundant flask shaped multicellular simple alveolar mucous glands and granular glands or poison glands both opening at the surface through duct. The granular glands have granular cytoplasm and the secretion is like a milky fluid which in some instances is toxic to the predators. The granular glands are not as numerous as mucous glands and are located at the warts of the toad. The skin of the living amphibians is smooth and lack scales except in a few toads and burrowing limbless caecilians.

(iv) Reptiles

Soft Derivatives (Scent Gland-Only in Few Groups)

Due to the terrestrial mode of life which gives rise to scaly and dried skin, the integument of the reptiles is virtually devoid of any gland. In only exceptional cases where these glands are present, these are basically for reproductive purposes and are called scent gland. The examples are two pairs in crocodile, one on the throat and the other on lower jaw. Some snakes may have these glands in cloacal region and are for defence purpose. In certain turtle, these glands namely musk gland present in the lower jaw and below carapace may also be used for attracting opposite sex.

Hard Derivatives (Scales-Both Dermal and Epidermal)

The only hard integumentary derivatives in the reptilian skin are scales. However, there seems to be inconsistencies in the literature regarding their origin but the overwhelming opinion is in favour of their exclusive epidermal origin. The reptilian scales are broadly put in two categories. One is represented by snakes and lizards in which these scales project backward and overlap the scale behind. Such scales exhibit complete periodic shedding, ecdysis and new set of scales is formed beneath the older ones. The other types of scale are represented by the turtles, crocodiles and alligators where large scale cover the carapace and plastron. Beneath these scales lie the bony plates which may not conform to the pattern of scales. These scales may appear in the form of concentric rings indicating the number of growth period. These scales may get worn off, shed and new scales develop. In crocodiles, these scales on lateral, vertical and tail region may have a pit like depression which may indicate the location of sensory capsule below. Crocodilian scales do not exhibit ecdysis. In literature, dermal scales have been reported in some lizards, snakes and turtles. Modifications of epidermal scales have been observed in lizards and snakes such as the so-called horns of the horned lizard and the rattles of the rattle snake.

(vi) Birds

Glandular Derivatives (Uropygial Gland)

The status of the glands in birds is somewhat similar to those of the reptiles since it does not require keeping the surface moist due to feather covering over the body. However, one gland, so named due to its location at the tail region, the uropygial is bilobed, has separate opening and secretes oil which is taken on beak or bill and is used to preen the feathers. This is to make them impervious and hence this gland is predominantly found in aquatic birds.

Hard Derivatives (Feathers, Epidermal Scales)

The characteristic hard integumentary derivatives in birds are feathers which are exclusive to this group, epidermal in origin and are considered modified reptilian scales. Three main types of these feathers are classified on the basis of structure and distribution. These are hair feathers (filoplume), down feathers (plumulae) and contour feathers (plume). The beautiful colour of the feathers is due to specific pigment granules deposited during their development and the physical phenomenon such as refraction and interference of light rays. Replacement of feathers may occur throughout the life of the bird which may either be gradual throughout the year or seasonal called moulting.

There may be presence of scales on the naked region of birds *i.e.*, legs and toes. In some gallinaceous birds, a bony projection of the tarsometatarsus which is covered with horny scales, used for fighting, may be present. The web on the feet of aquatic birds *i.e.*, geese, ducks and swans may also have the presence of scales.

(vii) Mammals

Glandular Derivatives

(Mammary, Sweat, Sebaceous (Meibomian/Tarsal glands to eyelids, Small glands of Zeis to eyelids, Scent glands, Modified sweat glands, ceruminous or wax producing gland, glands of moll in the eyelids and circumanal glands).

Mammals have very extensive distribution of glands all over the body and some of them such as mammary, sweat and sebaceous glands are characteristic of the class mammals. However, some of the glands may be selectively absent in certain animals and some in certain part of the body. The broad features of these glands are as follows.

Mammary Gland

Characteristic to the class mammals, compound tubular or saccular in structure and apocrine in the nature, secretion in the form of milk for nutrition of young ones and restricted to specific region to the body. In Prototheria (e.g Platypus) these glands are compound tubular, without nipples and open directly on the skin surface. In Metatheria and Eutheria, the glands are compound saccular provided with nipples/teats which are always paired ranging from 1 to 11 pairs and proportional to the number of young ones to be delivered at birth. Interestingly, the nipples and the mammary tissue may be present in both sexes but they are functional only in lactating females although some exceptional cases, the functional mammary glands have also been reported in males.

Sweat Glands

Also called sudoriparous glands present in majority of the mammals except few such as moles, cetaceans (aquatic mammal), scaly anteater, two-toed sloth. These glands are mostly distributed all over the body except the regions like lips, glans penis and nail bed. In some animals, they comprise of specific areas such as in mice, rats and cats present in the underside of the paws, in rabbits around lips, in sheep, goats, cattle, pigs and dogs around the muzzle (snout region). The main function of these glands is temperature regulation through evaporation of sweat and excretion of metabolic waste.

Sebaceous or Oil Glands

Distributed all over except the palm and sole and mostly open into the hair follicle through the duct. Those not connected with hair follicle are present on corner of the mouth, lips, glans penis, labia minora and mammary papillae. The sebaceous gland connected with the eyelids are termed Meibomian gland where the secreted oily substance forms a layer or film between eyelid and eyeball. Some smaller glands with the same function are named as glands of Zeis. The main function of these glands is to keep the hair and skin soft and smooth.

Scent Glands

They are either modified sweat or sebaceous gland and their distribution may vary considerably in different group of animals. Their secretion may either be attracting the same sex, or opposite sex or may serve to repel or frighten the predator.

Modified Sweat Glands

Ceruminous or wax-producing gland present in the external ear passages, structurally coiled hence resembling the sweat gland and secrete fatty wax-like substance called ear-wax. Similarly gland of Moll in the eyelid and ceruminous gland around the anal opening are modified sweat gland.

Hard Derivatives

1. Hairs (Guard hair, under fur, long hair)
2. Head Projections (True horns, Antlers, Rhinoceros horns)
3. Foot Projections (Claws, Nails, Hooves)
4. Scales (Armadillo shell, Scales of Pangolin)
5. Baleen plates (Whales).

Hair

Characteristic of the class mammalia and epidermal in origin. In some mammals, the whole integument may be covered with hair while in others such as whales it may be simply restricted to the snout region. The hair colour is primarily due to the pigment present therein. In some animals, there is periodic shedding of hair.

Head Projection

These are mostly in the form of horns which are found only in class mammalia with the exception of one reptilian example of "horned toad" native to United States. Four different variations of these horns *i.e.* hollow horns in cattle, keratin fibres horns in rhinoceros, pronghorn in antelopes and antlers in deer family have been documented which serve as formidable weapon to fight with the adversary.

Foot Projections

Claws, nails and hoofs are present in the distal end of the digits and grow parallel to the skin and are built upon the same basic plan. The claws in the cat family may be retractile and can be withdrawn back in the sheath when not in use. The group of mammals possessing hoofs are called ungulates.

Scales

They may be present in some mammals and are mostly confined to tail and paw. However, in certain mammals, they extensively cover the body such as in scaly anteater where they are present all over the body except the ventral region. In armadillos scales fuse to form large plates over head, shoulder and ringlike band around mid body region.

(E) Fish Integument

Fish integument, like other vertebrates, also consists of the three classical layers *i.e.* epidermis, dermis and hypodermis or subcutis below which lies the attached muscles. Essentially, epidermis is made up of epithelial cells, some of them may be secretory, dermis consists of collagen fibres and hypodermis is constituted by fat or adipose cells. However, there are some fundamental differences in the fish integument when compared to other vertebrate groups (Figures 20.1A and B)

1. Epidermis

Three distinct regions *i.e.* upper, middle and lower can be identified on the basis of cell size, structure and function. In the mammalian system, the cells of the basal layer of the epidermis are

Figure 20.1: Gross structure of skin of *H. fossilis* showing its various regions (Panel A, C and D stained with haematoxylin and eosin, B with Masson trichrome and E with Alcian Blue, pH 2.5/PAS). Panel (A) and (B) showing distinct regions of epidermis, dermis, hypodermis and underlying muscles, (C) magnified view of epidermis and dermis showing multiple layers of club cells which rest at the basement membrane. Distinct layer of chromatophores (pigment cells) separating epidermis from dermis is visible and two layers of dermis can be clearly demarcated (D) Magnified view of club cells showing shrunk cytoplasm, distinct nucleus and darkly stained eosinophilic area at the base of each club cells. Also note the round shaped unstained mucous cells present between the club cells, germinativum cells can also be noticed (E) Note large population of mucous cells present both at the outermost side which are mostly small and bigger sized ones lying below. Club cells, though visible, are unstained. A clearly stained basement membrane and chromatophore containing layer is also visible (*Unpublished data for author's laboratory*).

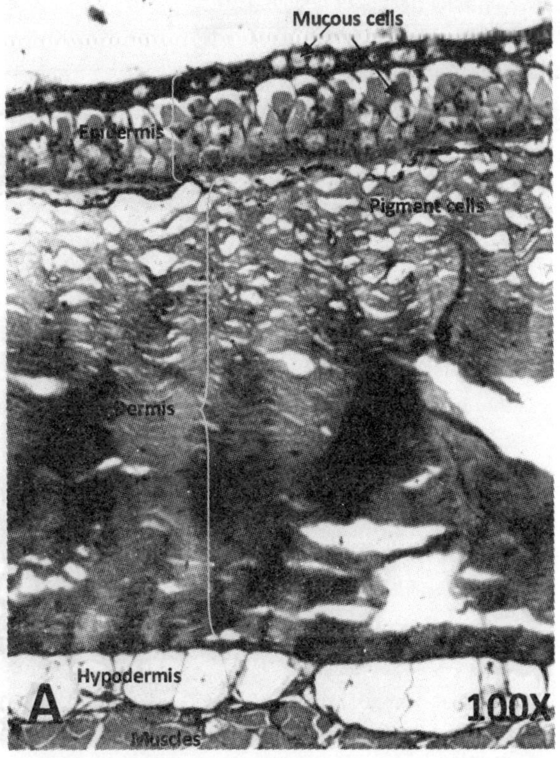

Contd...

Figure 20.1–*Contd...*

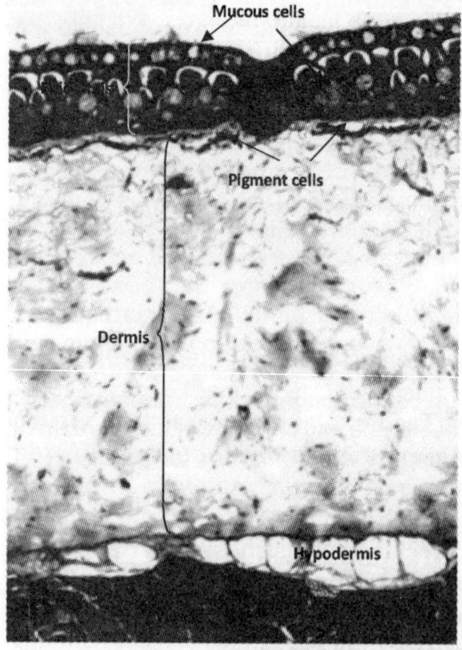

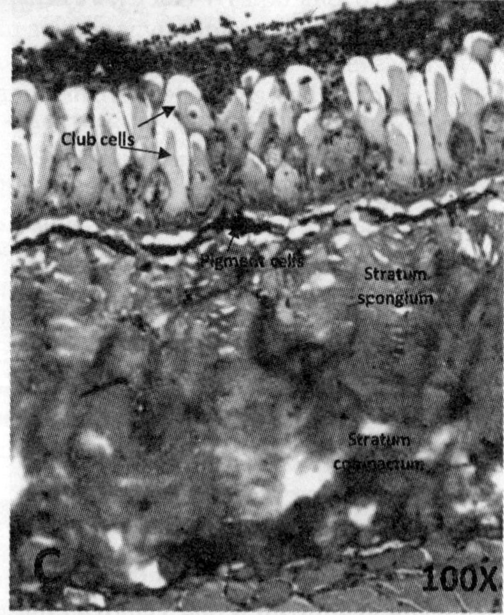

Contd...

Figure 20.1–*Contd...*

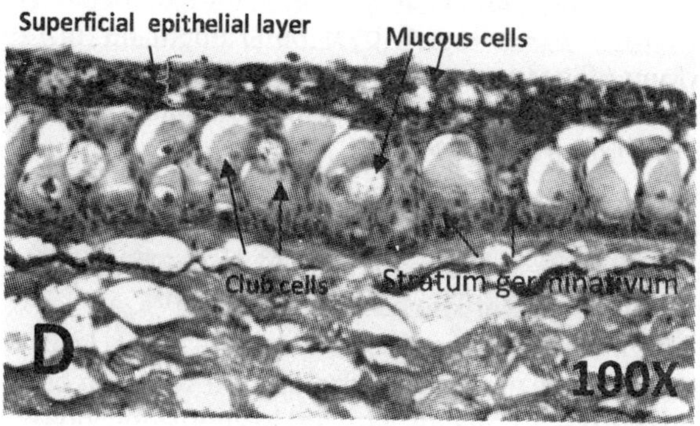

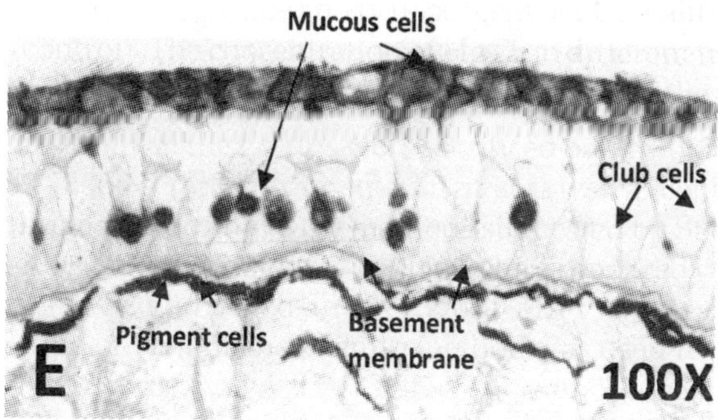

columnar, actively dividing mitotically and pushing the cells upward forming a progressively flattening layer of cells. In teleost, on the contrary, all the cells in the epidermis are metabolically active and dividing and contribute to the overall growth of the epidermis even though the basal layer cells are more active. While in terrestrial vertebrates, the outermost layer is made up of dead, horny cells containing keratin, no such layer of dead cells occur in fishes in general though few exceptions have been reported. Similarly, the term 'cuticle' often referred to the top most layer in some fishes has been reported to be made up of thick mucus layer. Hence, even though the lower most layer of fish epidermis has often been referred to as 'stratum germinativum' but it does not conform to the structural characters of this typical layer in mammals. Similarly, no distinct transitional layer of progressively reducing thickness showing

Figure 20.2: (A) Further magnification of club cells showing shrinkage of cytoplasm and clearly visible stratum germinativum (B)showing blood vessels in dermis region (C) Another magnified view of mucous cells on superficial layer (D), (E) and (F) showing the magnified view taste buds, lymphatic space in epithelium and pit organ (*Unpublished data for author's laboratory*).

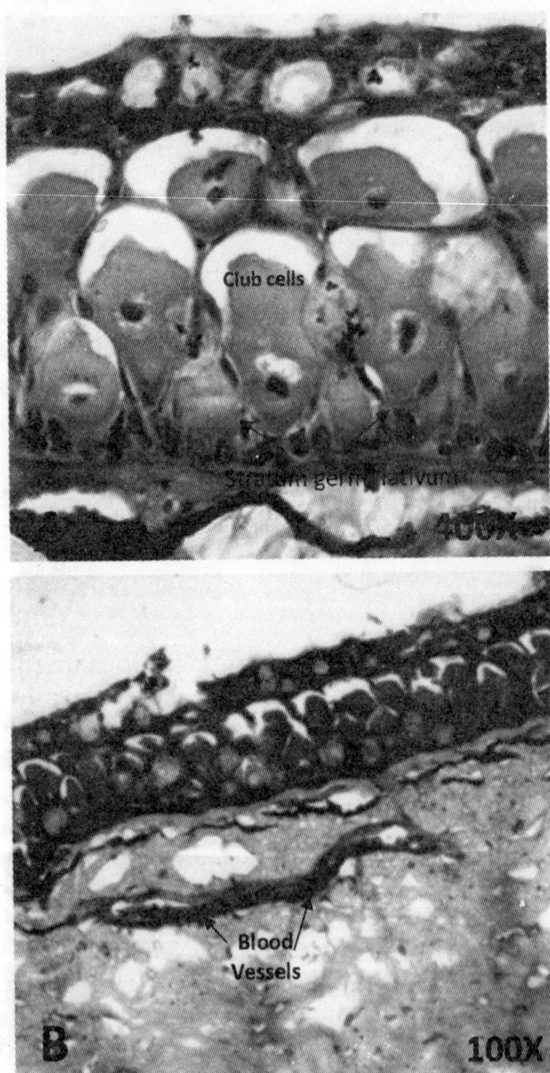

Contd...

Figure 20.2–*Contd...*

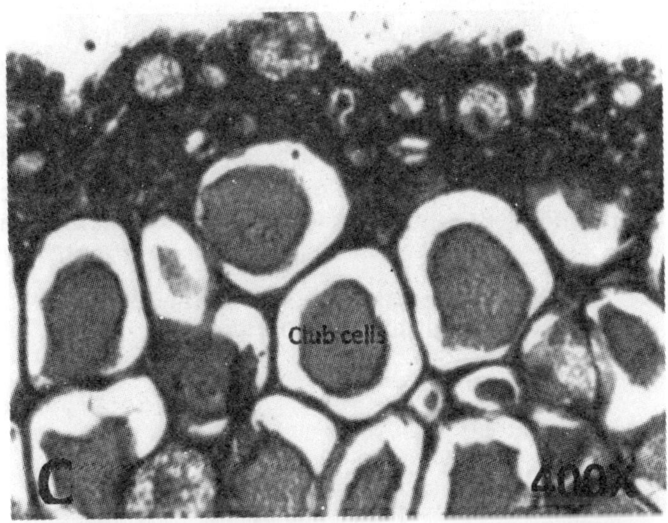

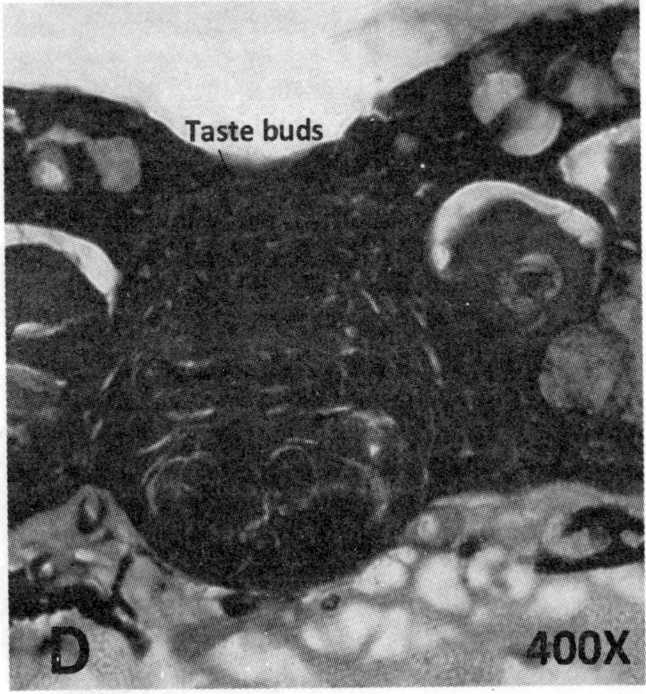

Contd...

Figure 20.2–*Contd...*

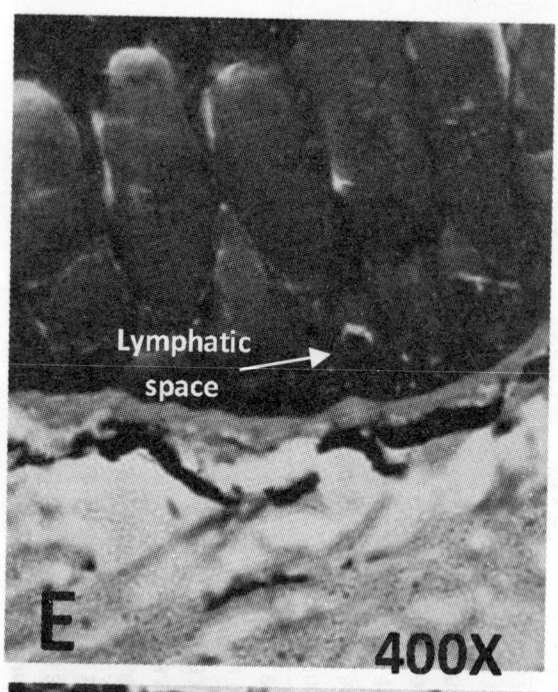

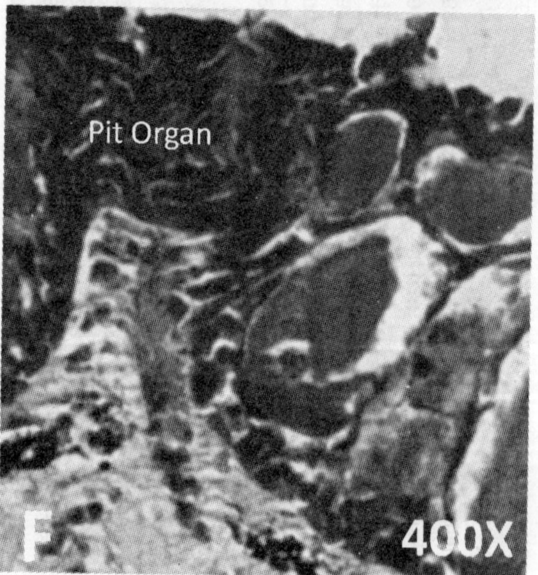

typical stratum granulosum and stratum lucidum and uppermost stratum corneum is distinguishable in fish epidermis (Figures 20.1 C, D and 20.2A).

Under the scanning electron microscope, the surface of the epithelium is characterized by the appearance of concentric ring like structures which often form finger print like pattern. These microridges or micropapillae are attributed to perform mechanical protection or in holding the mucous secretion over the surface layer.

The secretory cells in fish epidermis include goblet cells, sacciform cells, granular cells, club cells, merkel cells, venom cells and some non-secretory but functionally distinct cells such include ionocytes, intrusive cells, rodlet cells, luminescent organs may also be present. However, only some but not all, of these cells may be distributed in different fish species. Some workers have described the presence of lymphatic spaces which are located in the basal layer of epidermis and are characterized by the occurrence 1-2 minute lymphocytes which have deeply stained nuclei and narrow rim by cytoplasm (Figure 20.2E). It has been shown that these lymphatic spaces show enlargement under pathological conditions and are supposedly associated with protection against microorganisms. In certain instances, the specialized sensory organs such as taste buds more clearly seen if stained with Masson trichrome are also visible (Figure 20.2D). Similarly, some other structures in the form of dermal invagination called pit organs are also occasionally observed (Figure 20.2F).

2. Basement Membrane/Basal lamina

The last layer of epidermis finally rests on a basement membrane which separates epidermis from dermis. It is stained positive with PAS and acts as a filtration barrier and attachment site for epidermal epithelial cells (Figure 20.1E).

3. Dermis

In fishes, the dermis, like other typical vertebrates, broadly consists of two layers- the upper stratum spongiosum/laxum and the lower stratum compactum. However, in certain fishes this demarcation may not be clearly visible (Figure 20.1C). Both these layers are made up of collagenous connective tissues with interspersed blood and lymph vessels, nerve ending, sense organs and assortment of epidermal and dermal derivatives (Figure 20.2B).

The upper spongiosum is so named due to loosely arranged connective tissues whereas the compactum consists of densely packed connective tissue fibers which give compact appearance to this layer. There are some vertically arranged strands present at regular intervals in this compactum layer which runs all along up to the end of this layer and are thought to provide reinforcement to this layer. An additional single layer called dermal epithelium consisting of modified fibrocytes located at the depth of stratum compactum has been reported by some workers which are presumed to function as regulatory barrier between stratum compactum and hypodermis. Below the basement membrane and above stratum spongiosum lies the distinct layer of chromatophores which impart integumentary colouration to the fish (Figure 20.1A,B,C,E). The thickness of the dermis varies from fish to fish and region to region including the variations in the width ratio of stratum compactum and spongiosum.

4. Hypodermis/Subcutis

It is regarded as the structural part of the integument proper which lies below dermis and above the body musculature. It is often provided with a deep chromatophore layer above and sporadically occurring blood vessels and nerve ending enclosed in perineurium. It is richly provided with large number of adipose cells which allow considerable flexibility to the integument. Some workers have described the presence of myoseptum which differentiates this layer from underlying musculature.

Secretory Cells in Fish Integumentary Epithelium

The description of secretory cells in the fish integument has given rise to great confusion particularly due to multiplicity of nomenclature of the same cell type. Hence, mucous cells have also been called goblet cells, sacciform cells have been named granular cells, clear mucous cells, white cells etc. Similarly, club cells are also known by giant cells, leyding cells, goblet cells etc. This confusion further deepens when different investigators describe these cells with multiple names as distinct entities. Another interesting aspect is that each species may have only some of the secretory cells and not all thus showing their restricted distribution in different species. A brief overview of the various types of secretory cells in fish integumentary epithelium highlighting their distribution, shape and function has been given in the following text. Broadly, these secretory cells may be classified into two categories one which includes those glands which are more widely distributed in different taxa and groups, such as goblet cells, club cells, sacciform cells, ionocytes, merkel cells, epithelial cells and intrusive cells and whereas the second category of the secretory structures are restricted only to specific taxa such as thread cells (only in hagfish), venom cells (spine-containing fishes) Luminescent organ (deep sea marine fishes) etc.

a. Goblet Cells/Mucous Cells/Mucous Glands

These are most widely occurring glands/cells among fishes present in variety of organs principally, integument, gills and GI tract but have been reported to be absent in *Polydon* and mudskipper *Periophthalmus*. Structurally, these are either unicellular or in some instances multicellular, round to flask shaped, have basally placed nucleus and secrete mucus which is basophilic but remains unstained with routine histological stains such as haematoxylin-eosin. In integument, these are present either at the superficial or middle layer of the epidermis (Figures 20.1E and 20.2C). Since the mucus contains glycoprotein moiety the mucous cells are stained, magenta with PAS and blue with Alcian blue indicating the presence of neutral and acidic glycoprotein. The varying shades of purple suggest the presence of mixture of neutral and acidic glycoproteins.

It has been suggested that two types of goblet cells termed mucous goblet cells and serous goblet cells exist, the latter unlike the former exhibit acidophilic staining and have been reported to occur in variety of fish species like Salmonids, Goby and Coelacanth. Their cytoplasm is made up of large sized granules and number of this type of cells increases with ectoparasite infestation. However, the occurrence of serous goblet cells is less common than mucous goblet cells and they are observed far less in number when both cell types are present together. The population or the number of mucous cells is highly variable which may include seasonal, temperature and reproductive stage or toxicity induced variations.

b. Sacciform Cells/Granular Cells/White Cells

These cells have restricted distribution and have been found to occur in cartilaginous fishes such as rays. Among teleosts, they are confined to gadiformes cods, catfish of the genus *Corydoras* and some salmonids. These are, however, absent in jawless fishes and dipnoans. Under light microscopy, it often becomes difficult to distinguish these cells and is generally characterized by clear to finely granular cytoplasm, basally located nuclei and predominantly proteinaceous secretion. In literature, these cells have often been referred by other names such as granular cells, white cells, clear mucous cells. It is believed that these cells secrete 'noxious substance' which repel or poison the infecting microorganism and predators.

c. Club Cells

Club cells, like many other integumentary cells, are also not universally distributed in all fishes and are predominantly confined to the fishes of superorder ostariophysi. This largely includes families like cypriniformes, siluriformes etc. which put altogether, cover nearly one quarter of known fish species and three-fourth of all freshwater species. Structurally, these cells are large, mostly oval or club shaped and contain 1-2 nuclei, multilayered and mostly located in the middle portion of epidermis. The cytoplasm of these cells is eosinophilic which appears shrunk during the fixation of these cells thus leaving a clear vacuolar space mostly at the apical end (Figures 20.1C,D,E and 20.2A,C). These cells secrete alarm pheromones which serve as anti predator defensive reaction. This secretion may also serve healing function in the event of damaged epithelia.

d. Ionocytes

Ionocytes most widely occur in gill and opercular epithelia but have also been reported in the integument of some euryhaline fishes such as killifish, *Fundulus heteroclitus, Bagarius bagarius,* hagfishes, lamprey (called polyvillous cells) and in elasmobranches. Their characteristic structural features are the same classical characters such as large number of mitochondria and highly branched tubular system. Functionally, they perform ion-absorption in FW and ion-excretion in SW including maintenance of acid-base regulation.

e. Merkel Cells

Merkel cells as secretory cells of integumentary epithelia are not so widely distributed amongst teleost but have been reported in carp, minnow, *Phoxinus phoxinus*. However, they have been reported in lamprey, dipnoans not only in skin but also in oral epithelium. These cells can be recognized by relatively small ratio of cytoplasm to nucleus, presence of dense cytoplasmic granules and peripheral finger-like projections. These cells have been attributed multifunctional role such as tactile reception and responsive to cadmium toxicity as reported in carps.

f. Intrusive Cells

As the name indicates, it is a group of diverse cells which initially do not originate from fish epidermis but are subsequently lodged there. These include (i) chromatophores, (ii) neuromast cells, (iii) leucocytes (iv) macrophages (v) various types of granulocytes (vi) protozoan parasites and (vii) rodlet cells. While the origin and functions of these cells is quite evident and need not be commented upon here, the last of the intrusive cell, the rodlet cell has evolved considerable controversy over the last several decades sharply dividing opinions into two schools. It has been viewed by some that these cells are nothing but a sporozoan parasite *Rhabdospora thelohani* which has got embedded into fish epithelium. This assumption is based on the premise that (a) it is irregularly occurring within fish species (b) its association with certain pathogenic conditions and (c) its prolonged survival beyond the host cells death. However, others believe that these cells are the integral part of the host cells and cite the evidence of (1) wide distribution of these cells in different fish species inhabiting varying temperature and salinities (ii) the presence of these cells in larval and very young fishes (iii) lack of host inflammatory response (iv) presence of desmosomes and functional complexes between these and adjacent cells and more importantly (v) the similarity of the DNA profile between host and these cells. Interestingly, due to overwhelming evidences for and against such assumption, the debate has still not settled with any conclusive view. The distribution of these cells have been reported both in freshwater and marine teleost species exhibiting pear or flask-shaped appearance with displaced nuclei towards wide end and membrane-bound refractile rodlets. Functionally, these cells are suggested to be involved in osmoregulation, sensory reception, lubrication and non-specific stress response.

g. Thread cells

These cells are exclusively restricted in hagfish and are characterized by elongated oval shape basal nuclei, a cluster of central mass of fine granules and peripheral thread like component. Upon being irritated the hagfish discharges copious amount of slime from the thread cells.

h. Venom Cells

These are assemblage of venom cells forming an encapsulated venom gland containing secretary granules and usually located at the base of spine. The discharge of the venom takes place when these glands are pressed and ruptured and the venom passes through the hollow canal of the spine into the wound which is made by the spine itself.

i. Luminescent Organs/Photophores

These organs are well-developed and confined mostly in deep-sea marine fishes. The source of this bioluminescence could be either due to the presence of luminous bacteria in the luminescent organ or the light generated by self-luminous cells of the fish. These luminescent organs may be epidermal in origin and are made of modified epidermal cells or these cells coupled with the reflector organ remain embedded in the dermis.

(F) Histochemical Studies of Integumentary Epithelium

As is evident from the literature, the mucus containing cells are most predominant glandular derivatives in terms of their most wide spread distribution in various vertebrate groups. In fact, these glandular structures also have their wide distribution in majority of vital organs of the body such as GI tract, respiratory tract, reproductive tract, oculo-rhino-otolaryngeal tracts in addition to integumentary epithelium. Hence, study of mucus has gained considerable momentum amongst clinicians, biochemists, histologists, physical chemists and biologists. This prompted Sir Frances Avery Jones to remark "we cannot do without mucus, irritating though it may be, if it is too much or too little and predictably fatal if absent altogether in many body systems where it plays a vital role". In terms of evolutionary history the three-pronged function of mucus includes lubrication, water proofing and indeed protection. Be it an innocuous earth worm where it may provide its permeable skin the passage of O_2 and CO_2 or the GI tract where it acts as the skin of the gut acting more effectively than the skin of the body to provide protection to gastric mucosa against powerful proteolytic enzymes.

Mucous Composition

The initial efforts to isolate and characterize qualitative and quantitative component of mucus proved quite difficult since its gel-like structure did not lend itself to electrophoretic separation and the inability to obtain contaminant-free mucus proved quite a difficult task. Hence, the available histochemical methods to assess the chemical moiety of the mucus assumed considerable significance.

Broadly, mucus consists of 95 per cent water and remaining 5 per cent is made up of organic and inorganic constituents. The organic constituents are typically made up of carbohydrates, proteins

Figure 20.3: Photomicrographs of catfish *H. fossilis* skin showing mucous cells containing different types of glycoproteins differentiated on the basis of various histochemical stains (A) Skin stained with Alcian blue (pH 2.5) showing acid mucopolysaccharides (B) Skin stained with Alcian blue (pH 1.0) showing sulphated mucopolysaccharides (C) Stained with Periodic Acid Schiff showing mucous cells with neutral mucopolysaccharides (D) Combination of both Alcian blue (pH 2.5) and PAS showing acid, neutral and mixture mucopolysaccharides (E) Magnified view of acidic mucous cells (F) acidic and mixture mucous cells (G) Predominantly mucous cells with the mixture of acidic and neutral moiety (*Unpublished data for author's laboratory*).

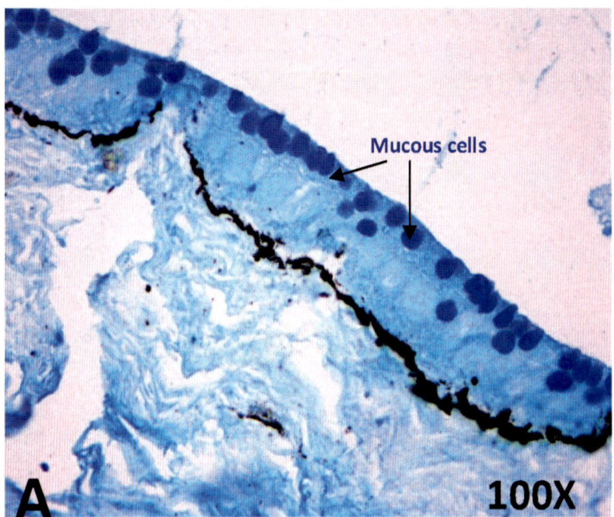

Contd...

Figure 20.3–*Contd...*

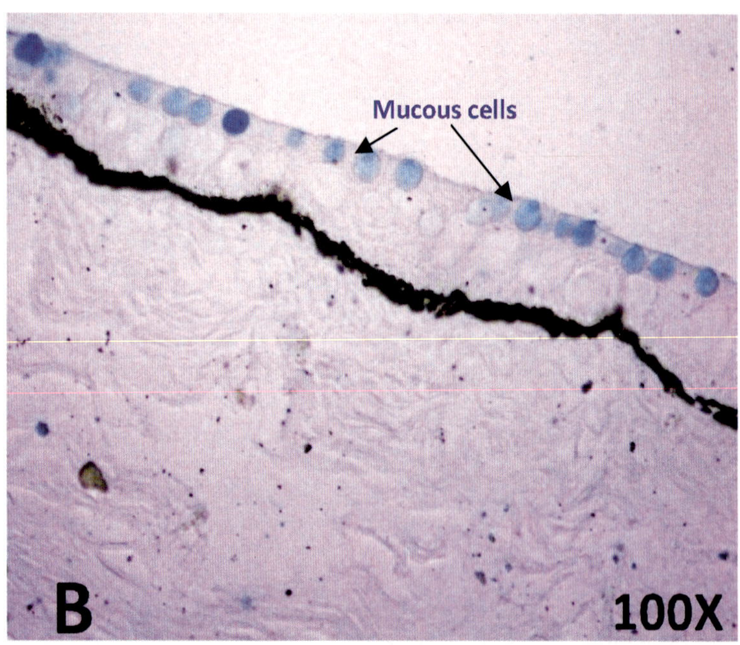

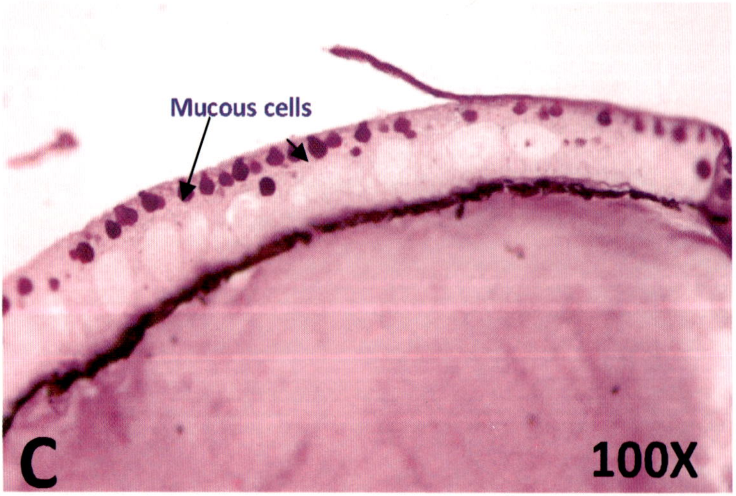

Contd...

Figure 20.3–*Contd...*

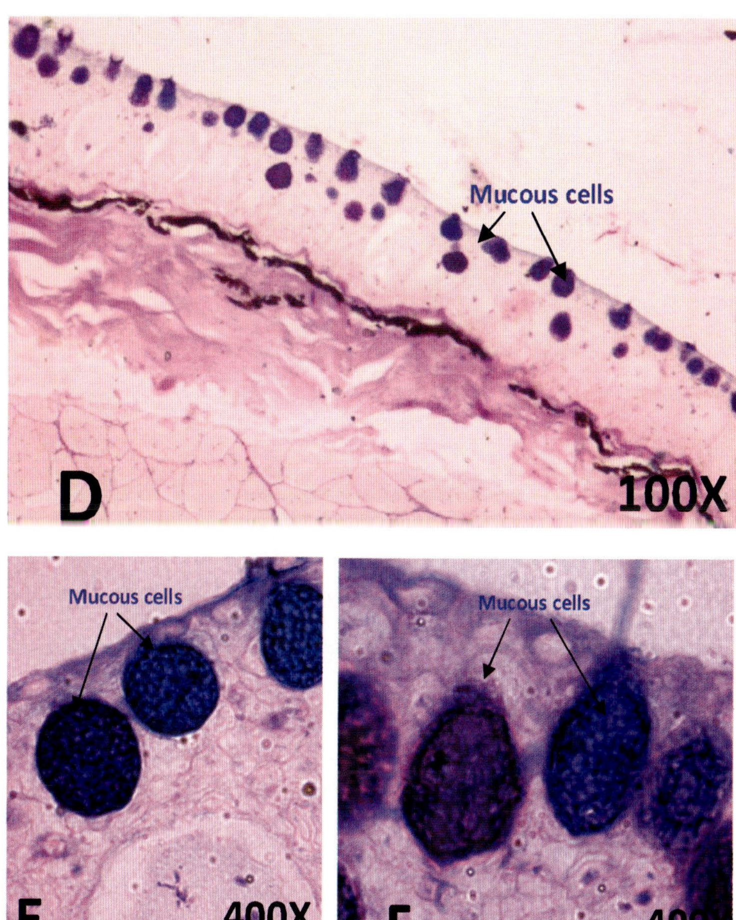

Contd...

and lipids either in free or conjugated forms called glycoconjugates. Among the conjugated moieties, glycoproteins are most predominant which are covalently bound with each other through disulphide bonds and are responsible for jellying properties. These mucus glycoproteins are characterized by high carbohydrate to protein ratio which distinguishes mucus glycoproteins from other glycoproteins.

Figure 20.3–*Contd...*

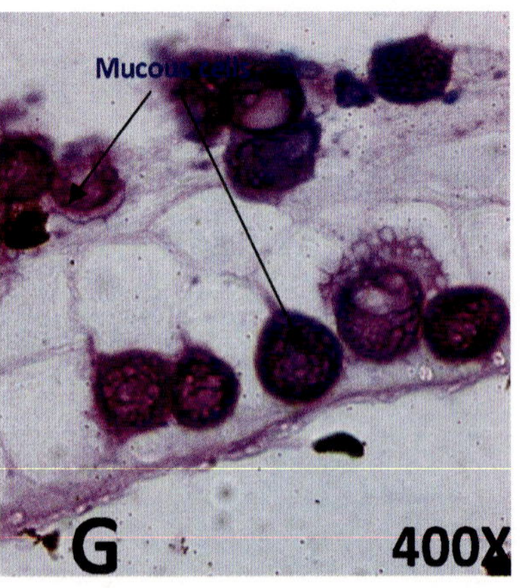

Histochemical Localization of Various Mucopolysaccharides

Based on the chemical entity of various carbohydrate constituents of mucus contained in integumentary goblet cells, the following three types of mucous cells are clearly distinguishable in various group of animals including teleosts.

1. Neutral mucopolysaccharide – can be specifically localized by Periodic Acid Schiff (PAS) staining where these cells stain magenta (Figure 20.3C).
2. Acidic mucopolysaccharides – can be specifically stained with Alcian Blue (AB) (pH 2.5) which stains both carboxylated as well as sulphated acidic mucopolysaccharides. However, in order to distinguish these two types, AB at pH 1.0 or pH 0.5 is used which only stains sulphated acidic mucopolysaccharides (Figures 20.3A, B). Both carboxylated as well as sulphated acidic mucopolysaccharides can be distinguished in the same slide by the use of Acid fuchsin and AB (pH 2.5) combination where carboxylated and sulphated mucous cells stain blue and purple respectively.
3. Acidic and neutral mucopolysaccharides moieties within the same mucous cells- There are large number of mucous cells which have a mixture of both neutral and acidic

mucopolysaccharides within the same cell which can be stained by the combination of PAS and AB (pH 2.5) and yield varying shades of purple depending upon the relative contents of both these moieties (Figure 20.3D).

Concluding Remarks and Future Perspectives

The integumentary system is an extensive interface between the organism and its surrounding environment. Hence, it is not merely body covering but the first line of defence, protection and much more in terms of its multifunctionality. Considering its multifaceted significance, it is somewhat surprising that the studies on the integumentary system are mostly focussed on its histological characteristics and its potential applications in many other aspects of organism's adaptability and successful survival in its environment has been scantly explored. Hence, studies dealing with histological and physiological changes in the integumentary components as indices of environmental degradation hold tremendous promise. The exact nature of chemical moieties of secretion of various glandular structures such as club cells, sacciform cells, merkel cells, ionocytes etc are virtually non-existent. It is hoped that this small write-up on this very important aspect will act as a trigger for more fruitful studies in this area.

Acknowledgements

The authors are grateful to the financial assistance provided by University Grants Commission, New Delhi and Council of Scientific and Industrial Research, New Delhi through their financial support in the forms of major research projects Nos.34-488/2008(SR) and 37(1421)/10/EMR-II respectively. Some of the data included in this write-up on the catfish, *Heteropneustes fossilis* is from the author's laboratory and still not published. One of the authors (SA) is grateful to UGC, New Delhi for project fellowship to her.

References

Bansil, R. and Turner, B.S., 2006. Mucin structure, aggregation, physiological functions and biomedical applications. *Current Opinion in Colloid and Interface Science*, 11: 164–170.

Datta Munshi, J.S. and Choudhary, S., 1996. *Ecology of H. fossilis*. Narendra Publishing House, p. 17–25.

Datta Munshi, J.S., 2009. *The Life of Air Breathing Fishes Vol. II.* Narendra Publishing House, p. 77–91.

Fletcher, T.C., Jones, R. and Reid, L., 1976. Identification of glycoproteins in goblet cells of epidermis and gill of plaice (*Pleuronectes platessa* L.), flounder (*Platichthys flesus* L.) and rainbow trout (*Salmo gairdneri* Richardson). *Histochem. J.*, 8: 597–608.

Fletcher, T.C. and Grant, P.T., 1968. Glycoproteins in the external mucus secretions of the plaice, *Pleuronectes platessa* and other fishes. *Biochem. J.*, 106: 12.

Gona, O., 1979. Mucus glycoproteins of teleostean fish: A comparative histochemical study. *Histochemical Journal*, 11: 709–718.

Gonas, O., 1979. A comparative study of fish mucus cells. *Anat. Rec.*, 193: 552.

Jakowska, S., 1963. Mucus secretion in fish. *Ann. NY Acad. Sci.*, 106: 458–62.

Jones, R. and Reid, L., 1978. Secretory cells and their glycoprotiens in health and disease. *Br. Med. Bull.*, 34: 9–16.

Jones, R. and Reid, L., 1993. The effect of pH on alcian Blue staining et epithelial acid glycoproteins. II. Human branchial submucosal gland. *Histochem. J.*, 5: 19–27.

Jones, R. and Reid, L., 1993. The effect of pH on alcian Blue staining et epithelial acid glycoproteins,I. Sialomucins and sulphomucins (singly or in simple combinations). *Histochem. J.*, 5: 9–18.

Ojha, J. and Munshi, J.S.D., 1974b. Morphometric Studies on the gill and skin dimensions in relation to body weight in a fresh water mussel, *Macrognathus aculeatus*. *Zoo. Anz.*, 193, 364–381.

Sadovy, R., Randall, J.E. and Rasotto, M.B., 2005. Skin structure in six dragonet species (Gobiesociformes; Callionymidae) interspecific differences in glandular cell types and mucus secretion. *J. Fish. Bio.* 66: 1411–1418.

Shephard, K.L., 1993. Mucus on the epidermis of fish and its influence on drug delivery. *Advanced Drug Delivery Reviews*, 403–417.

Shephard, K.L., 1994. Function for fish mucus. *Rev. Fish Biol.*, 4: 401–429.

Thronton, D.J. and Sheehan, J.K., 2004. From mucins to mucus toward a more coherent understanding of this essential barrier. *Proc. Am. Thorac. Soc.*, 1, 54–61.

Whitear, M., 1986. The skin of fish including cyclostomes-epidermis. In: *Biology of the Integument*, (Eds.) J. Bereiter-Hahn, A.G. Matolsty and K.S. Richards. Spinger Verlag, Berlin, Heidelberg, New York, Tokyo, p. 8–38.

Zaccone, G., 1972. Comparative histochemical investigations on the mucous cells of the branchial epithelium of *Mugilcephalus* L. and *Anoptichthys jordani* Hubbs and Innes. *Acta histochem.*, 44, 106–15.

Chapter 21

Alterations in Haematological Parameters of an Indian Major Carp *Labeo rohita* Exposed to Dye Effluent

V. Sasikalamathi, M. Ramesh* and R.K. Poopal
Unit of Toxicology, Department of Zoology, School of Life Sciences,,
Bharathiar University, Coimbatore – 641 046, T.N., India

ABSTRACT

Fingerlings of *Labeo rohita* were exposed to three different concentrations (1 per cent–treatment I, 2 per cent–treatment II, and 5 per cent–treatment III) of dye effluent and the changes in the hematological parameters studied at the end of 24 and 96h exposure period. During above treatments, haemoglobin (Hb), haematocrit (Hct), erythrocyte count (RBC) and mean cellular volume (MCV) were decreased in all the Treatments at the end of 24 and 96 h exposure period. In contrast to above a significant increase in WBC count, MCV and MCH (in Treatment III, 24h) were noticed during above exposure period. However, MCHC value was found to be biphasic showing a significant decrease in Treatment I and II at the end of 24h and a significant increase in Treatment III at the end of 24h and in all the Treatments at the end of 96h exposure period. The alterations in

* Corresponding author: E-mail: mathanramesh@yahoo.com

haematological parameters can be used as biomarkers in monitoring the toxicity of dye effluent in aquatic environment. However, more detailed studies on use of specific biomarkers to monitor the toxicity of dye effluents on aquatic organisms are needed.

Keywords: *Dye effluent, Toxicity, Haematology, Labeo rohita.*

Introduction

Natural water bodies are polluted by improperly treated or untreated industrial effluents. The industrial wastes usually contains high quantities of dissolved and suspended solids, organic and inorganic chemicals, high BOD and COD, oils and grease, besides toxic metals (Muley *et al.*, 2007). The presence of some heavy metals in these effluents (either in free form in the effluents or adsorbed in the suspended solids) have been found to be poisonous depending on the dose and exposure duration (Kupchella and Hyland, 1989; Tamburlini *et al.*, 2002). These chemicals are not only poisonous to humans but also found to cause serious injuries to the aquatic life and they may result in food contamination (Novick, 1999; WHO, 2002; Hoshina *et al.*, 2008). Wastewater from dyeing and finishing processes contains a large amount of organic substances such as thickening agents and dyes (Sostar and Schneider, 1998). The wastewater generated from dye and dye intermediate industries mainly have intense colour having various shades of red, blue green, brown and black and usually have high level of COD, BOD, acidity, chlorides, sulphates, phenolic compounds and also various heavy metals *viz.* copper, cadmium, chromium, lead, manganese, mercury, nickel, zinc etc. (Lanciotti *et al.*, 2004). Of the 700,000 tons of dyes produced annually worldwide, approximately 10 to 15 per cent of the dye is disposed off in effluent from dyeing operations. In textile dyeing units, it is estimated that 10 to 15 per cent of the dyes are passed into the effluent during the dyeing process (Sumathi *et al.*, 2001; Cavas and Gozukara, 2003). The discharge of highly coloured effluents containing dyes can be damaging to the receiving bodies (Nigam *et al.*, 1996). Contamination of surface waters with highly potent substances at low concentrations has also been related to alter reproductive performance (Robinson *et al.*, 2003), impairment of immune mechanisms (Hoeger *et al.*, 2004) in aquatic organisms.

There have been increasing trends towards use of bio monitoring to assess the water quality of resources/receiving water bodies (Mwinyihija *et al.*, 2006). Fish are one of the most widely distributed organisms in an aquatic environment and being susceptible to environmental contamination may reflect the extent of the biological effects of environmental pollution in waters (Ramesh *et al.*, 2009). Fish live in very intimate contact with their environment, and are, therefore, very susceptible to physical and chemical changes which may be reflected in their blood components (Wilson and Taylor, 1993). Blood is known to exhibit pathological changes before the onset of any external symptoms of toxicity (Zutshi *et al.*, 2010). Changes in blood parameters are often sensitive to environmental or physiological changes. Furthermore, these parameters are easily measurable and provide an integrated measure of the physiological changes in organisms (Remyla *et al.*, 2008).

The haematological variables such as haemoglobin (Hb), hematocrit (Hct), red blood cell (RBC) count, white blood cell (WBC) count, and haematological indices such as mean cellular volume (MCV), mean cellular haemoglobin (MCH) and mean cellular haemoglobin concentration (MCHC), parameters widely used to assess the toxic stress induced by environmental contaminants (Saravanan *et al.*, 2011).

Rapid industrialization in India has resulted in the significant increase in the liquid waste (spent wash or effluent), which is usually discharged into open land or into nearby natural water which results in a number of environmental problems such as threat to plants and animal lives, surface water logging, ground water contamination and salinizing good quality land (Ramona *et al.*, 2001). India is among one of the major producers of dyes and dyes intermediate from Asian region and meets the requirement of the world at large (Mwinyihija *et al.*, 2006). There are about 900 dyes and dyes intermediate producing industries in India. Most of them are in small-scale unorganized industrial sector, and about five thousand dyes and dyes intermediate are in commercial production. Since the conventional wastewater treatment processes are not completely effective in removing dyes from effluent water, the unused dyes are discharged into aquatic environments with other chemical components in the effluent (Cavas and Gozukara, 2003). Hence in the present study an attempt was made to study the effect of dye effluent toxicity on certain haemtological parameters of a fresh water fish *Labeo rohita* in order to use these biomarkers in the field of environmental toxicology.

Materials and Methods

Experimental Fish and Water

Specimens of *Labeo rohita*, a fresh water fish belonging to the family Cyprinidae (Hamilton) of either sex were selected for the present study. They were obtained from Tamilnadu Fisheries Development Corporation Limited, Aliyar Fish Farm, Aliyar, Tamil Nadu, India. They were safely brought to the laboratory in well packed polythene bags containing aerated water. Fish were stocked in large cement tank (60'X37'X47') disinfected with potassium permanganate and washed thoroughly prior to introduction of fish (to prevent fungal infection). Fish were acclimatized to laboratory conditions for about 15 days before the commencement of the experiment. Media were changed frequently to avoid fungal growth and contamination by metabolites. These stocks were well aerated with aerators.

During acclimatization, fish were fed, *ad libitum* with rice bran and ground nut oil cake once in a day. Feeding was given at least one hour prior to replacement of water. The feeding was withheld for 24 h before the commencement of the experiment to keep the experimental animals more or less in the same metabolic state. During acclimatization, the fish stock was maintained at natural photoperiod and ambient temperature. Water was replaced every 24 h after feeding in order to maintain a healthy environment for the fish during both acclimatization and experimental period. This ensures sufficient oxygen for the fish and the environment is devoid of any accumulated metabolic state. From the stock fish with an average length of 6–7 cm and weighing 4–5 g were segregated and transferred to clean rectangular glass aquarium tanks (75 × 35 × 37 cm) of 150 L water capacity. Tap water free from chlorine was used for the present study. The hydrobiological features such as temperature, pH, dissolved oxygen, total alkalinity, salinity and total hardness, were estimated for each set of experiment as these factors have a significant influence on the biodegradability and toxicity of pollutants (Table 21.1).

Table 21.1. Physico-chemical parameters of water used for the present investigation

Physico-chemical Parameters	*Values*
Temperature	28.0±1.0°C
pH	7.2±0.07
Dissolved oxygen	6.8±0.04 mg/L
Total alkalinity	19.3±7.0 mg/L±
Total hardness	18.0±0.5 mg/L
Salinity	0.2±0.02 ppt
Calcium	4.1±0.3 mg/L
Magnesium	2.3±0.6 mg/L

Values are mean±S.E. of five individual observations

Dye Effluent

Dye effluent used in the present study was collected from Tirupur District, Tamilnadu, India. The effluent was collected at the point of discharge in plastic cans and brought to the laboratory and kept immediately in deep freezer to avoid any microbial decomposition. The physico-chemical parameters of the dye effluent were estimated following the method of APHA (1998) (Table 21.2).

Table 21.2: Physico-chemical Properties of Dye Effluent Used for the Present Investigation

Parameters	*Values*
pH	9.32
TDS (Total dissolved solvent)	4917 mg/l
Colour	2168 pt/co (Platinum/Cobalt)
Turbidity	66.8 NTU (Nephaloturbidity)
Total hardness	350 mg/l
Chloride	3248 mg/l
COD	724 mg/l
BOD	240 mg/l
FRC (Free residual chlorine)	Nil
DO	0.6 mg/l
Sulphate	544 mg/l

Assessment of Effluent Toxicity

For the assessment of dye effluent toxicity, 8 circular plastic tubs with 10 litres of water were taken. Then to each tub, different concentrations of effluent (*i.e.* 100 per cent, 75 per cent, 50 per cent, 25 per cent, 10 per cent, 5 per cent, 1 per cent 0.5 per cent) were added. A control tub with 10 litres of water was also taken. Ten healthy fish, with an average length of 7.0 cm and average weight of 5.0 gms were selected and introduced into each tub. The manifestation and survival time of fish was observed in each concentration according to the method of Wurhman (1950) and graphically expressed as survival curve. The manifestation time, the time at which fish lose their sense of balance and float on their sides were determined and presented as manifestation curve (Warren, 1900). Based on the method of Powers (1971), the relationship between the reciprocal of survival time and per cent effluent concentration was determined and graphically expressed as fatality curve.

Acute Toxicity Studies

Based on survival time three effluent concentrations 1 per cent, 2 per cent and 5 per cent were taken for acute studies. Four glass tanks with 80 litres of water were taken and marked as Treatment I, Treatment II, Treatment III and control. Then, 40 fishes were introduced into each tank. The above said effluent concentrations (1 per cent, 2 per cent and 5 per cent) were added to their respective glass tanks and a common control was maintained. At the end of 24 and 96 h, fish from control and experimental fish were taken and sacrificed for dye effluent treated haematological studies.

Preparation of Samples

Fish from control and dye effluent treated glass tank were sacrificed and blood was drawn by cardiac puncture using plastic disposable syringe fitted with 26 gauge needle. The syringe and needle were prechilled and moistured with heparin (Beparine R heparin sodium, IP 1000 IU ml1 derived from beef intestinal mucosa containing 0.15 per cent w/v chlorocresol IP preservative), an anticoagulant manufactured by Biological E Limited, Hyderabad, India. It was transferred into small vials, previously rinsed with heparin. MCV, MCH and MCHC were calculated by standard formulas.

Haematological Studies

Haemoglobin content of the blood was estimated by Cyanmethemoglobin method (Drabkin, 1946). Haematocrit was estimated by microhalematocrit (capillary) method. Erythrocytes and Leucocytes were counted by the method of Rusia and Sood (1992) using haemocytometer. MCV, MCH and MCHC were calculated by standard formulas.

$$\text{MCV (cubic micra)} = \frac{\text{HCT (\%)}}{\text{RBC (million} \times \text{cu} \times 10^6)} \times 100$$

$$\text{MCH (picograms)} = \frac{\text{Hb (g/dl)}}{\text{RBC (million} \times \text{cu} \times 10^6)} \times 100$$

$$\text{MCHC (g/dl)} = \frac{\text{Hb (g/dl)}}{\text{HCT (\%)}} \times 100$$

Statistical Analysis

The data were analyzed statistically at $P < 0.05$. To test their significance the t values were calculated between control and experiment value by Student's t-test.

Results

Behavioural toxicology is a tool for hazard assessment of water pollution. In the present study, the behavioural pattern of fish was dependent upon the effluent concentration. In 100 per cent, 75 per cent, 50 per cent and 25 per cent effluent concentration, the fish exhibited hyper excitability, enhanced opercular movements, erratic and jerky movements and vigorous activity before they become lethargic. In 10 per cent, 5 per cent, 1 per cent and 0.5 per cent fish were found at the bottom of the tank at the initial stage, and subsequently they came up to the surface and remained in the surface thereafter.

Changes in the haemoglobin content of fish *Labeo rohita* exposed to Treatment I (1 per cent), Treatment II (2 per cent) and Treatment III (5 per cent) of dye effluent are presented in Table 21.3. A significant decrease in haemoglobin content was observed in all the Treatments at the end of 24 h exposure period. However, at the end of 96 h exposure period haemoglobin content was increased in 1 per cent treatment when compared to control group showing a per cent increase of 0.38. All the values were significant at 5 per cent level except 1 per cent treated fish at the end of 96h treatment.

Table 21.3: Changes in the Haemoglobin Content of *Labeo rohita* Treated with Different Concentrations of Dye Effluent for 24 and 96h

Exposure Period (in hrs)	*Haemoglobin Content (g/dl)*			
	Control	*Treatment I 1 per cent*	*Treatment II 2 per cent*	*Treatment III 5 per cent*
24	5.051±0.021 (–7.78)	4.658±0.023* (–40.55)	3.002±0.021* (–15.17)	4.284±0.526*
96	5.127±0.290 (+0.38)	5.146±0.014 (–19.65)	4.120±0.013* (–4.17)	4.913±0.056*

Values are Means±SE of three individual observations.

(+) indicates per cent increase over control.

(–) indicates per cent decrease over control.

(*) Values are significant at 5 per cent level.

Table 21.4: Changes in the Hematocrit Values of *Labeo rohita* Treated with Different Concentrations of Dye Effluent for 24 and 96h

Exposure Period (in hrs)	*Haematocrit (per cent)*			
	Control	*Treatment I 1 per cent*	*Treatment II 2 per cent*	*Treatment III 5 per cent*
24	14.90±0.031	13.92±0.020* (–6.71)	8.84±0.040* (–40.94)	12.54±1.508* (–15.84)
96	14.80±0.031	15.00±0.031 (+1.35)	11.84±0.040* (–20.00)	14.52±0.159 (–1.89)

Values are Means±SE of three individual observations.

(+) indicates per cent increase over control.

(–) indicates per cent decrease over control.

(*) Values are significant at 5 per cent level.

Haematocrit value of fish *Labeo rohita* exposed to Treatment I (1 per cent), Treatment II (2 per cent) and Treatment III (5 per cent) of dye effluent are shown in Table 21.4. In all the Treatments, haematocrit levels were found to be decreased from that of control group except in Treatment I at the end of 96 h (+1.35). The values of Treatment I (1 per cent), Treatment II (2 per cent) and Treatment II (5 per cent) of 24 h exposure period were significant at 5 per cent level, whereas Treatment I and III at the end of 96 h were not significant at 5 per cent level.

Table 21.5. Changes in the Erythrocyte Count of *Labeo rohita* Treated with Different Concentrations of Dye Effluent for 24 and 96h

Exposure Period (in hrs)	*Erythrocytes (million/cu.mm)*			
	Control	*Treatment I 1 per cent*	*Treatment II 2 per cent*	*Treatment III 5 per cent*
24	0.432±0.018	0.396±0.046* (–8.33)	0.334±0.036* (–22.68)	0.280±0.010* (–35.18)
96	0.422±0.012	0.314±0.0496* (–25.59)	0.338±0.0392* (–19.90)	0.370±0.067* (–12.32)

Values are Means±SE of three individual observations.

(–) indicates per cent decrease over control.

(*) Values are significant at 5 per cent level.

Table 21.6. Changes in the Leucocyte Count of *Labeo rohita* Treated with Different Concentrations of Dye Effluent for 24 and 96h

Exposure Period (in hrs)	*Leucocytes (1000/cu.mm)*			
	Control	*Treatment I 1 per cent*	*Treatment II 2 per cent*	*Treatment III 5 per cent*
24	27.00±0.612	35.74±3.584* (+32.37)	37.74±2.230* (+39.77)	42.20±2.004* (+56.30)
96	23.50±0.314	32.00±0.422* (+36.17)	32.44±0.517* (+38.04)	24.55±0.247* (+4.47)

Values are Means±SE of three individual observations.

(+) indicates per cent increase over control.

(*) Values are significant at 5 per cent level.

The different concentration of dye effluent has caused a decrease in the erythrocyte count of fish *Labeo rohita* when compared to control groups (Table 21.5). With Treatment I, II and III a per cent decrease of 8.333 per cent, 22.685 per cent and 35.185 per cent were noted at the end of 24 h, whereas in 96h treatment a per cent decrease of 25.592 per cent, 19.905 per cent and 12.322 per cent were recorded during above said Treatment periods. All the values were significant at 5 per cent level.

Table 21.6 describes the WBC count in *Labeo rohita* exposed to dye effluent. Overall increase in WBC counts were observed in all the Treatments; at the end of 24 h exposure period the increase in WBC count were found to be 32.22 per centm 39.77 per cent and 56.30 per cent in Treatment I, Treatment II and Treatment III, respectively. Likewise, the WBC counts were also increased after 96 h exposure period showing a per cent increase of 36.17 per cent, 38.04 per cent and 4.47 per cent in Treatment I, Treatment II and Treatment III respectively. The values were significant at 5 per cent level.

Table 21.7: Changes in the MCV Value of *Labeo rohita* Treated with Different Concentrations of Dye Effluent for 24 and 96h

Exposure Period (in hrs)	*Mean Cellular Volume (cubic micra)*			
	Control	*Treatment I 1 per cent*	*Treatment II 2 per cent*	*Treatment III 5 per cent*
24	347.40±14.74	283.82±14.30* (−18.34)	205.09±8.599* (−40.96)	444.28±46.213* (+27.88)
96	352.01±11.06	292.51±7.19* (−16.90)	273.42±10.91* (−22.32)	311.76±14.32* (−11.43)

Values are Means±SE of three individual observations.

(+) indicates per cent increase over control.

(–) indicates per cent decrease over control.

(*) Values are significant at 5 per cent level.

Changes in the MCV values in *Labeo rohita* exposed to dye effluent are shown in Table 21.7. The MCV values decreased (except Treatment III of 24 h) throughout the study period when compared to their control groups. The MCV levels of fish decreased to 18.34 per cent and 40.96 per cent at the end of 24 h exposure period respectively in Treatment I and II. In case of Treatment III there was a per cent increase (27.88 per cent) noted at 24 h. After 96h exposure also MCV value was found to be decreased in all the treatments. All the values were significant at 5 per cent level.

The data shown in Table 21.8 demonstrate that in all the Treatments (1 per cent, 2 per cent and 5 per cent) MCH values decreased at the end of 24 and 96h exposure period (except in Treatment III at the end of 24h) showing a per cent decrease of 19.35 per cent and 40.83 per cent at the end of 24h exposure period and 16.58 per cent, 21.28 per cent, 15.26 per cent in Treatment I, Treatment II and Treatment III) at the end of 96 h, respectively. All the values in Treatment I. II and III were significant at 5 per cent level.

Table 21.8: Changes in the MCH Values of *Labeo rohita* Treated with Different Concentrations of Dye Effluent for 24 and 96h

Exposure Period (in hrs)	*Mean Cellular Haemoglobin (picograms)*			
	Control	*Treatment I 1 per cent*	*Treatment II 2 per cent*	*Treatment III 5 per cent*
24	117.79±5.148	94.99±5.109* (–19.35)	69.69±3.104* (–40.83)	152.91±17.195* (+29.82)
96	120.32±2.757	100.36±2.512* (–16.58)	94.70±4.052* (–21.28)	101.90±4.649* (–15.26)

Values are Means±SE of three individual observations.

(+) indicates per cent increase over control.

(–) indicates per cent decrease over control.

(*) Values are significant at 5 per cent level.

Changes in the MCHC values of fish *Labeo rohita* exposed to different concentrations of dye effluent for 24 and 96 h are presented in Table 21.9. The MCHC values were found to be decreased at the end of 24h exposure period showing a per cent decrease of 2.42 per cent and 0.69 per cent in Treatment I and II respectively. In treatment the MCHC value was found to be increased (+0.69). In 96h exposure period the MCHC value was found to be increased in all the treatments. Statistical analysis indicates that all the values of 24 h and 96 h were not significant at 5 per cent level (except Treatment II of 24 h).

Table 21.9: Changes in the MCHC Values of *Labeo rohita* Treated with Different Concentrations of Dye Effluent for 24 and 96h

Exposure Period (in hrs)	*Mean Cellular Haemoglobin Concentration (g/dl)*			
	Control	*Treatment I 1 per cent*	*Treatment II 2 per cent*	*Treatment III 5 per cent*
24	34.299±0.373	33.467±0.038 (–2.42)	33.967±0.115 (–6.96)	34.536±0.302 (+0.69)
96	33.812±0.152	34.312±0.097* (+1.47)	34.622±0.112* (+2.39)	33.841±0.111 (+0.08)

Values are Means±SE of three individual observations.

(+) indicates per cent increase over control.

(–) indicates per cent decrease over control.

(*) Values are significant at 5 per cent level.

Discussion

Effluents are the main source of direct and continuous input of pollutants in the aquatic ecosystems and hence the study of the effects of effluents exposure on organisms has a high ecological relevance. In recent years, disposal of textile dyeing effluents into the aquatic environment has become an important problem mainly due to increasing varieties and quantities of dyes used (Cavas and Gozukara, 2003). However due to the unknown and complex nature of effluents it is virtually impossible to relate observed effects to specific pollutants or even a class of pollutants. The lethal effects of various industrial effluents on fish were recorded by many authors (Baskaran *et al.*, 1989; Varadaraj and Subramanium, 1991).

In the present study, the toxicity of dye effluent on the mortality of fish was dose dependent. Similarly observations have also been made by Jayachandran and Chokkalingam (1987), Subramaniam *et al.* (1988), Varadaraj and Subramanium (1991) in fish reared in tannery effluent. In the present study mortality of the fingerlings of *Labeo rohita* may be due to toxic chemicals found in the dye effluent. With the increase in concentration of effluent the survival time of the fish was reduced (Metcalf and Eddy, 1916). The manifestation and maximum survival time of *Labeo rohita* when exposed to 100 per cent effluent was 18 min. In 75 and 25 per cent effluent concentration they showed a survival time of 24 and 38 min. In concentrations of

20, 15, 10, 5, 3 and 1 per cent effluent concentration the fish survived up to 3 to 5 days. The fish survived for a short period in higher concentration of dye effluent. This may be due to a stage of exhaustion during which adaptive mechanisms may by inadequate and homeostasis might have not regained leading to death of fish as suggested by Roborts (1978).

Further the abnormal behavioural responses like increased ventilation and excursion of fish in the dye effluent may be in response to stress produced by the effluent and the above behavioural aspects of the fish depended on the effluent concentrations. The above observation was supported by the work of Srivastava and Srivastava (1979) on *Channa punctatus*. Eisler (1970) reported that toxicity of a toxin is found to be variable due to large number of environmental factors like temperature, salinity or hardness of the medium. The toxic effect of pollutants may be decreased or even increased by various water quality factors such as pH, temperature, hardness and dissolved oxygen content of the water. In the present study, high mortality in the higher concentration of effluent may be either due to in the direct effect of metal salts present in the effluent, or soaking of gills by suspended solids of the effluent, leading to death as supported by Metcalf and Eddy (1916) and Herbert and Merkens (1961).

According to Mason (1981) temperature is important because it not only influences the metabolic activity and behaviour of organisms, but it may also alter the physical and chemical status of the pollutant. Holden (1973) suggested that hardness and pH which is usually interrelated in natural waters and it is not possible to state which of the two factors are responsible for affecting toxicity. In the present study the pH of the dye effluent was found to be 9.32 indicating the toxicity to aquatic organisms. Solids concentration is another important characteristic of wastewater (Lee and Lin, 1999). In the present study the TDS of dye effluent was found to be 4917 mg/l. The high levels of BOD are indications of the pollution strength of the wastewaters. They also indicate that there could be low oxygen available for living organisms in the wastewater when utilizing the organic matter present.

High COD levels imply toxic condition and the presence of biologically resistant organic substances (Sawyer and McCarty, 1978). The concentrations of COD observed in surface waters range from 20 mg/l O_2 or less in unpolluted waters to greater than 200 mg/l O_2 in waters receiving effluents (Sahan *et al.*, 2007). The most important measure of water quality is the dissolved oxygen (Peirce *et al.*, 1997). The low level of DO recorded could result in the non-maintenance of conditions favourable to the aerobic organisms. Low dissolved oxygen content in the medium considerably increases the toxicity of many poisons to fish. When dissolved oxygen content of water goes below a critical level, mortality of fish may occur (Ellis, 1973). In the present study, the dissolved oxygen content was minimum which may be the reason for the sudden death in 100 per cent effluent. In the present study also high level of BOD, COD and low level of DO in dye effluent indicates that the dye effluent is highly toxic to aquatic organisms.

Untreated waste water has severe toxic effects on fish (Moneim *et al.*, 2008). Fish respond to toxic agents similarly to higher vertebrates, which can allow the assessment of substances that are potentially hazardous to humans (Lemos, 2007). The uses of fish as biomarkers of the effects of pollution are of increasing importance and can permit the early detection of aquatic environment pollution (Van der Oost and Beyer, 2003). Fish haematology is gaining increasing importance in fish culture because of its value in monitoring the health status of fish (Hrubec *et al.*, 2000). Knowledge of the haematological characteristics is an important tool that can be used as an effective and sensitive index to monitor physiological and pathological changes in fish (Kori-Siakpere *et al.*, 2005).

Hb seems to be the best blood indicator of environmental stress. Buckley *et al.* (1976) reported that prolonged reduction in haemoglobin content is deleterious to oxygen transport and degeneration of the erythrocytes could be ascribed as pathological conditions in fishes exposed to toxicants. The decrease in haemoglobin concentration signifies that the ability of fish to provide sufficient oxygen to the tissues is considerably restricted and will result in a decrease of physical activity (Joshi *et al.*, 2002). Reduction in TEC count and Hb content of fish from polluted lakes may due to the inhibition of aerobic glycolysis curtailing synthesis of iron and haemoglobin via the lowered energy status in fish (Zutshi *et al.*, 2010). Joshi *et al.* (2002) reported that inhibition of erythropoiesis and increase in the rate of erythrocyte destruction in haematopoietic organs is the cause of decrease in RBC count. The decrease in the haemoglobin and haematocrit values in the fish could also be attributed to the lysing of erythrocytes (Kori-Siakpere *et al.*, 2008). Depression in haematocrit and haemoglobin values coupled with decreased and deformed erythrocytes are obvious signs of anaemia (Flos *et al.*, 1987). Male fathead minnows (*Pimephales promelas*) showed decreased haematocrit values after exposure to high sewage effluent concentrations when compared to control fish (Hemming *et al.*, 2001).

The decrease in haemoglobin content, haematocrit value and erythrocyte number observed in the present investigation may be due to the disruptive action of the dye effluent on the erythropoiectic tissue as a result of which the viability of the cells might be affected. Decreases in RBC count, Ht level and Hb content are symptoms of anaemia (Koprucu *et al.*, 2006). The decrease in Hct, Hb and RBC count levels may be indicators of anaemia and reduction in RBC count caused either by the inhibition of erythropoiesis or by the destruction of red cells by the pharmaceutical drugs in the fish *Cyprinus carpio* (Saravanan *et al.*, 2011). There are various factors to cause anaemia in animals, such as haemorrhage, haemolysis and renal diseases (Roberts, 1978). Further, in the present study impaired osmoregulation and gill damage during dye effluent exposure resulted in haemodilution, which lead to a decrease in the number of red blood cells through haemolysis.

Blood cells including WBCs are frequently used as indicators of health status in fish because WBCs are key components of innate immune defence and leukocytes are involved in regulation of immunological function in the organisms (Ballarin *et al.*, 2004). Increase in TLC was a result of direct stimulation for its defence from diseases due to the presence of polluted substances (Zutshi *et al.*, 2010). As a protective response of the body during stress, TLC increases through stimulation of leukopoietic process and enhanced release of leukocytes to the blood circulation (Das *et al.*, 2006). The increase in the number of leucocytes (leucocytosis) is a normal reaction of the fish body, against infections of foreign substances, which can alter the normal physiological processes in fish (Nussey *et al.*, 1995). In the present study the significant increase in the number of WBC count indicate the stress condition of the fish caused by the dye effluent which might have produced hypoxia and gill damage in fishes.

The erythrocyte index such as MCV and MCH have a wide range of physiological variation (Motlagh *et al.*, 2010). Lethtinum *et al.* (1990) reported alterations in PCV, MCV, and MCH values in fish exposed to effluents from mills with six different bleaching processes. A significant decrease in MCV indicates that the erythrocytes have shrunk, either due to hypoxia, stress or impaired water balance, (Larsson *et al.*, 1985). In the present study the decrease in MCV value was the result of large numbers of immature red blood cells being released into the circulation system. The increase in MCV is attributed to the direct effect of catecholamines, cortisol, and glucose on adenylate cyclase activities in red blood cells, as a response to acute hypoxic stress (Nikinmaa, 1992; Perry *et al.*, 1996). In the present study the significant increase of MCV in Treatment III at the end of 24h exposure period indicates the swelling of red blood cells or increase in the erythrocyte volume due to toxic stress of the dye effluent.

The significant change in the MCH may be due to the reduction in cellular blood iron, resulting in reduced oxygen carrying capacity of blood and eventually stimulating erythropoiesis (Hodson *et al.*, 1978). In the present study the significant decrease of MCH during effluent treatment indicates red blood cells impairment due to toxic stress of dye effluent. Reddy and Bashamohideen (1989) reported that the increase in MCH in carp may be due to symptoms of hypochromic microcytic anaemia, accompanying iron deficiency in the body. In the present study the significant increase of MCH during effluent treatment indicates the swelling of red blood cells or increase in the erythrocyte volume due to toxic stress of dye effluent. Further stress related increase in the erythrocyte volume may be another reason as suggested by Jastrzebska and Protasowicki (2005). MCHC is a good indicator of RBC swelling (Wepener *et al.*, 1992b). The variation in MCHC clearly indicates that the concentration of haemoglobin in the RBCs is reduced in fish thereby indicating anaemic condition (Zutshi *et al.*, 2010). Similarly, MCHC showed fluctuating results (Zutshi *et al.*, 2010). Significant increase of MCHC value may be due to congenital sphaerocytosis as suggested by Sobecka (2001). The increase in MCHC suggested that the anaemia was the macrocytic nesmochromic type. The decrease in MCHC observed in the present study indicates a decrease in Hb synthesis or red blood cell swelling. In contrast the significant increase in mean cell haemoglobin concentration (MCHC) indicates that the concentration of haemoglobin in the red blood cells were much higher in the exposed fish than in the control fish. Alterations in MCV, MCH and MCHC also clearly indicate that the fish are under chemical stress, which leads to a pathological condition in the tissues.

From the present study, it is concluded that, dye effluent has a profound influence on the haematological profiles of fish. These parameters could be effectively used as potential biomarkers of effluent toxicity to the freshwater fish in the field of environmental bio monitoring.

References

APHA (American Public Health Association), 1998. *Standard Methods for the Examination of Water and Wastewater*, 20th edition. American Public Health Association, Washington, DC.

Ballarin, L., Dalloro, M., Bertotto, D., Libertini, A., Francescon, A. and Barbaro, A., 2004. Haematological parameters in *Umbriana cirrosa* (Teleostei, Scianidae): a comparison between diploid and triploid specimen. *Comp. Biochem. Physiol. A.*, 183: 45–51.

Baskaran, P., Palanichamy, S. and Arunachalam, S., 1989. Effects of textile dye effluent on feeding energetic body composition of the freshwater fish, *Oreochromis mossambicus*. *J. Ecobiol.*, 1(3): 203–214.

Buckley, J. A., Whitmore, C. M. and Matsuda, R. I., 1976. Changes in blood chemistry and blood cell morphology in coho salmon, *Oncorhynchus kisutch* following exposure to sublethal levels of total residual chlorine in municipal wastewater. *J. Fish. Res., Board Canada*, 33: 776–782.

Cava, T. and Ergene–Göükara, S., 2003. Micronuclei, nuclear lesions and interphase silver–stained nucleolar organizer regions (AgNORs) as cyto–genotoxicity indicators in *Oreochromis niloticus* exposed to textile mill efuent. *Mutation Res.*, 538: 81–91.

Das, P. C., Ayyappan, S. and Jena, J. K., 2006. Haematological changes in the three Indian major carps, *Catla catla* (Hamilton), *Labeo rohita* (Hamilton) and *Cirrhinus mrigala* (Hamilton) exposed to acidic and alkaline water pH. *Aquaculture*, 256: 80–87.

Drabkin, D. L., 1946. Spectrometric studies, XIV–The crystallographic and optimal properties of the haemoglobin of man in comparison with those of other species. *J. Biol. Chem.*, 164: 703–723.

Eisler, R., 1970. Acute toxicities of organochlorine organophosphorus pecticides to estuarine fish. *Tech. Paper Bur. Sport Fish Wildl.*, 46: 1–12.

Ellis, A. E., 1977. The leucocytes of fish. A review. *J. Fish. Biol.*, 11: 453–91.

Flos, R., Tort, L. and Balacch, J., 1987. Effects of zinc sulphate on haematological parameters in the dogfish (*Scyliohinus canicula*) and influence off MS 222. *Mar. Environ. Res.*, 21: 289–298.

Hemming, J. M., Waller, W. T., Chow, M. C., Denslow, N. D. and Venables, B., 2001. Assessment of the estrogenicity and toxicity of a domestic wastewater efuent owing through a constructed wetland system using biomarkers in male fathead minnows (*Pimephales promelas Rafinesque*), 1820. *Environ. Toxicol. Chem.*, 20: 2268–2275.

Herbert, W. M. and Merkens, J. C., 1961. The effect of suspended mineral solids on the survival of trout. *Int. Air Wat. Poll.*, 5: 46–55.

Hodson, P. V., Blunt, B. R. and Spray, D. J., 1978. Chronic toxicity of water borne lead and dietary lead to rainbow trout (*Salmo gairdneri*) in lake Ontario water. *Water Res.*, 12: 869–878.

Hoeger, B., Van den Heuvel M. R., Hitzfeld, B. C. and Dietrich, D. R., 2004. Effects of treated sewage efûuent on immune function in rainbow trout (*Oncorhynchus mykiss*). *Aquat. Toxicol.*, 70: 345–355.

Holden, A. V., 1973. The effect of pesticides on fish in freshwater. *Proc. Soc. Lond. B.*, 80: 383–394.

Hoshina, M. M., De Franceschi de Angelis D. and Marin–Morales M. A., 2008. Induction of micronucleus and nuclear alterations in fish (*Oreochromis niloticus*) by a petroleum renery efuent. *Mutation Res.*, 656: 44–48.

Hrubec, T. C., Cardinale, J. L. and Smith, S. A., 2000. Haematology and plasma chemistry reference intervals for cultured tilapia (*Oreochromis hybrid*). *Vet. Clin. Pathol.*, 29(1): 7–12.

Jastrzebska, E. B. and Protasowickiz, M., 2005. Effects of cadmium and nickel exposure on hematological parameters of common carp, *Cyprinus carpio* L. *Acta Ichthyol. Piscat.*, 35(1): 29–38.

Jayachandran, K. P. S. and Chockalingam, S., 1987. Effect of tannery effluent on the respiratory parameters and biochemical constituents in an air breathing fish. *Channa punctatus* (Bloch). *Comp. Physiol. Ecol.*, 12: 197–200.

Joshi, P. K., Bose, M. and Harish, D., 2002. Changes in certain hematological parameters in a siluroid cat fish, *Clarias baterchus* (L.) exposed to cadmium chloride. *Poll. Res.*, 21(2): 129–131.

Koprucu, S. S., Koprucu, K., Ural, M., Ulspir, S. and Pala, M., 2006. Acute toxicity of organophosphorous pesticide diazinon and its effects on behaviour and some hematological parameters of fingerling European catfish, *Silurus glanis* (L.), *Pest.Biochem. Physiol.*, 86: 99–105.

Kori–Siakpere, O., Ake, J. E. G. and Idoge, E., 2005. Haematological characteristics of the African snakehead, *Parachacnna obscura*. *Afri. J. Biotechnol.*, 4(6): 527–530.

Kori–Siakpere, O. and Oghoghene, U. E., 2008. Sublethal haematological effects of zinc on the freshwater fish, *Heteroclarias* sp. (Osteichthyes: Clariidae). *Afri. J. Biotechnol.* 7(12): 2068–2073.

Kupechella, C. E. and Hyland M. C., 1989. Environmental Science, Allyn and Baron, London.

Lanciotti, E., Galli, S., Limberti, A. and Givannelli, L., 2004. Ecotoxicological evaluation of wastewater treatment plant effluent discharge: A case study in Parto (Italy) *Ann Ig.*, 16: 549–58.

Larsson, A., Haux, C. and Sjobeck, M. L., 1985. Fish physiology and metal pollution: results and experiences from laboratory and field studies. *Ecotoxicol. Environ. Saf.*, 9: 250–281.

Lee, C. C. and Lin, S. D., 1999. Handbook of Environmental Engineering Calculations, McGraw Hill, New York.

Lemos, C. T., Rödel, P. M., Terra, N. R., Oliveira, N.C.D. and Erdtmann B., 2007. Riverwater genotoxicity evaluation using the micronucleus assay in fish erythrocytes. *Ecotoxicol. Environ. Saf.*, 66: 391–401.

Lethtinum, K. J., Kierkegard, A., Jakobson, E. and Wandell, A., 1990. Physiological effects in sh exposed to efuents from mills with six different bleaching processes. *Ecotoxicol. Environ. Saf.*, 19: 33–46.

Mason, C. F., 1981. In "Biology of water pollution". Longman Group Limited. London.

Metcalf and Eddy, 1961. Report to the Spring Creek Committee upon Disposal of Sewage and Industrial Wastes on the Spring Creek Watershed, Centre County, Pennsylvania, Stream, 4.20.3.

Moneim, A. A. M., Shabana, N. M. A., khadre, S. E. M. and Abdel–Kader, H. H., 2008. Physiological and histopathological effects in catfish (*Clarias lazera*) exposed to dyestuff and chemical wastewater. *Int. J. Zool. Res.*, 4 (4): 189–202.

Motlagh, S. P., Zarejabad, A. M., Nasrabadi, R. G., Ahmadifar, E. and Molaee, M., 2010. Haematology, morphology and blood cells characteristics of male and female Siamese fighting fish (*Betta splendens*). *Comp. Clin. Pathol.* DOI: 10.1007/s00580–010–1058–6.

Muley, D. V., Karanjkar, D. M. and Maske, S. V., 2007. Impact of industrial effluents on the biochemical composition of fresh water fish *Labeo rohita*. *J. Environ. Biol.*, 28(2): 245–249.

Mwinyihija, M., Meharg, A., Dawson, J., Strachan, N. J. C. and Killham, K., 2006. An ecotoxicological approach to assessing the impact of tanning industry effluent on river health. *Arch. Environ. Hlth.*, 50: 316–24.

Nigam, P., Banat, I. M. and Singh., Marchant., 1996. Microbial process for the decolourization of texile effluent containing azo, diazo and reactive dyes. *Process Biochem*. 31: 435–442.

Nikinmaa, M., 1992. Membrane transport and control of haemoglobin oxygen affinity in nucleated erythrocytes. *Physiol. Rev.*, 72: 301–321.

Novick, R., 1999. Overview and the Health in Europe in the 1990s, World Health Organization, Europe Regional Office, Copenhagen, EUR/ICP/EH/CO 02 02 05/6, pp. 20.

Nussey, G., Van Vuren, J. H. J. and Preez, H. H. D., 1995. Effect of copper on the haematology and osmoregulation of the

Mozambique tilapia, *Oreochromis mossambicus* (Cichlidae). *Comp. Biochem. Physiol.*, 11(3): 369–380.

Peirce, J. J., Weiner, R. F. and Vesilind, P. A., 1997. Environmental Pollution and Control, Butterworth- Heinemann, Woburn, MA, 4th Eds. USA. pp. 57–74.

Perry, S. F., Reid, S. G. and Salama, A., 1996. The effects of repeated physical stress on the b–adrenergic response of the rainbow trout red blood cell. *J. Exp. Biol.*, 199: 549–562.

Powers, E. B., 1971. The gold fish, *Carassius carassius* as a test animal in the study of toxicity. *Illions Biol. Monoger.*, 4: 127–193.

Ramesh, M. and Saravanan, M., 2008. Some haematological and biochemical effects of chlorpyrifos in a freshwater fish, *Cyprinus carpio. IJIB.*, 3(1): 80–83.

Ramona, A., Biswas, A. K., Kundu, S., Saha, J. K. and Yadav, R. B. R., 2001. Efficacy of distillery effluent on seed germination and seedling growth in mustard, cauliflower and radish. *Proc. Natl. Acad. Sci. India.*, 71b (II): 129–135.

Reddy, P. M. and Bashamohideen, M. D., 1989. Fenvalerate and cypermethrin induced changes in the hematological parameters of *Cyprinus carpio. Acta. Hydrochimi. Hydrobiol.*, 17: 101–107.

Remyla, S. R., Ramesh, M., Sajwan, K. S. and Senthil Kumar, K., 2008. Influence of zinc on cadmium induced haematological and biochemical responses in a freshwater teleost fish *Catla catla. Fish Physiol. Biochem.*, 34: 169–174.

Roberts, R. J., 1978. Fish pathology. 1st Ed, Bailliere Tindall, London, ISBN: 0702006742.

Robinson, C. D., Brown, E., Craft, J. A., Davies, I. M., Moffat, C. F., Pirie, D., Robertson, F., Stagg, R. M. and Struthers, S., 2003. Effects of sewage efuent and ethynyl oestradiol upon molecular markers of oestrogenic exposure, maturation and reproductive success in the sand goby (*Pomatoschistus minutus*, Pallas). *Aquat. Toxicol.*, 62: 119–134.

Rusia, V. and Sood S. K., 1992. Routine hematological tests. In: 'Medical laboratory technology' Kanai, L. Mukerjee. (Ed.), Vol. I. Fifth reprint. *Tata McGraw Hill Publishing Company Limited*, New Delhi, pp. 252–258.

Sahan, A., Altun, T., Çevik, Fatma., Cengizler, I., Nevþat, E. and Genç, E., 2007. E.U.Comparative study of some haematological parameters in European eel (*Anguilla anguilla* L., 1758) caught from different regions of Ceyhan River (Adana, Turkey). *J. Fish. Aquat. Sci.*, 24(1–2): 167–171.

Saravanan, M., Karthika, S., Malarvizhi, A. and Ramesh, M., 2011. Ecotoxicological impacts of clofibric acid and diclofenac in common carp (*Cyprinus carpio*) fingerlings: Haematological, biochemical, ionoregulatory and enzymological responses. *J. Hazard. Mater.*, 195: 188 –194.

Sawyer, C. C. and McCarty, P. L., 1978. Chemistry for Environmental Engineers, McGraw Hill, New York., pp. 331–514.

Sobecka, E., 2001. Changes in the iron leveling the organs and tissues of wells catfish, *Silurus glanis* L. caused by nickel. *Acta. Ichthyol. Piscat.*, 31(2): 127–143.

Sostar, S. and Schneider, R., 1998. Guar gum as an environment–friendly alternative thickener in printing with reactive dyes, Dyes Pigments. 39: 211–221.

Srivastava, G. J. and Srivastava, O. P., 1979. Urea–induced histopathology in skin and gastric lining of teleost *Channa punctatus* (Bloch). *Ind. J. Exptl. Biol.*, 17(8): 840.

Subramanian, A., Tanabe, S. and Tatsukawa, R., 1988. Estimating some biological parameter of Baird's beakesd whales using PCBs and DDE as tracers. *Mar. Pollut. Bull.*, 19: 284–287.

Sumathi, M., Kalaiselvi, K., Palanivel, M. and Rajaguru, P., 2001. Genotoxicity of textile dye effluent on fish (*Cyprinus carpio*) measured using the comet assay. *Bull. Environ. Contam. Toxicol.*, 66: 407–414.

Tamburlini, G., Ehrenstein, O. V. and Bertollini R., 2002. Children's health and environment: A review of evidence, In: Environmental issue report No. 129, WHO/European Environment Agency, WHO Geneva., p. 223.

Van Der Oost, R., Beyer, J. and Vermeulen, N. P. E., 2003. Fish bioaccumulation and biomarkers in environmental risk assessment: a review. *Environ. Toxicol. Pharmacol.* 13: 57–149.

Varadaraj, G. and Subramanian, M. A., 1991. Toxic effects of dissolved effect of paper and pulp mill effluent on different parameters of bioenergetics in the fingerlings of *Oreochromis anossambicus*. *Env. Eco.*, 9(4): 857–859.

Warren, E., 1900. On the reaction of *Daphnia magna* (Stratus) to certain changes in its environment. *Quast. J. Micer. Sci.*, 43: 199–204.

Wepener, V., Van Vuren, J. H. J. and Du Preez, H. H., 1992b). Effect of manganese and iron at neutral and acidic pH on the haematology of the banded tilapia (*Tilapia sparrmanii*). *Bull. Environ. Contam. Toxicol.*, 49: 613–619.

WHO., 2002. Water Pollutants: Biological Agents Dissolved Chemicals, Non–dissolved Chemicals, Sediments, Heat, WHO CEHA, Amman, Jordan.

Wilson, R. W. and Taylor, E. W., 1993. The physiological responses of freshwater rainbow trout, *Onchorynchus mykiss*, during acute exposure. *J. Comp. Physiol*. 163b: 38–47.

Wurhman, E., 1950. Surquelgues principles dela toxicologiedes poisen. *Bull. Dv. E. D. E.*, 15.

Zutshi, B., Prasad, S. G. R. and Nagaraja, R., 2010. Alteration in hematology of *Labeo rohita* under stress of pollution from Lakes of Bangalore, Karnataka, India *Environ. Monit. Assess.*, 168: 11–19.

Chapter 22

Anti-Hepatotoxic Efficacy of Vitamins ('B' Complex and C) on Ibuprofen Induced Liver Damage in Rats

K. Prema[1], B. Kalaivani[2] and G. Vanithakumari[1]*

[1]*Division of Endocrinology, Department of Zoology, Bharathiar University, Coimbatore – 641 046, Tamil Nadu, India*

[2]*Department of Zoology, Sri Sarada College for Women (Autonomous), Salem – 636 016, Tamil Nadu, India*

ABSTRACT

Non steroidal anti-inflammatory drugs (NSAIDs) are in common use world wide. These drugs may sometimes be used in high or toxic doses by mistake. In this study, we investigated the effects of two different doses of Ibuprofen on liver. Fifty albino adult male wistar rats, weighing 200gm to 250 gm were divided equally in to five groups. The rats in the control group (n–10) were received 0.9 per cent saline orally. The other group received Ibuprofen at a low dose of 1.5 mg/100g body weight/day. The third group received ibuprofen at a dose of 3mg/100g body weight/day. The fourth group received vitamin supplementations (vitamin B complex + vitamin 'C') at a dosage of 0.05ml 100g body weight along with high dose. Final group

* Corresponding author: E-mail: kalai03win@yahoo.co.in

rats received ibuprofen at a dosage of 3mg/100g body weight/ day respectively for 7 days and the animals were kept for further seven days without giving any drug a withdrawal group. No significant changes were observed in the histopathology of the liver tissue of the control rats. In low dose treatment group, enlargement of the peritoneal tract (nearly 10 folds) especially, enlargement of the central vein was observed. Whereas, in high dose treatment group non specific hepatitis, severe periportal inflammation and portal triaditis was also observed. Vitamin supplementations to high dose drug treated animals could not restore the liver histology to normal control levels. Withdrawal of high dose of Ibuprofen for further 14 days caused restorative changes in the liver histology. Hence ibuprofen induced toxic effects to the hepatocytes and vitamin supplementation served as an partially restorative agents.

Keywords: *Liver, Ibuprofen, Hepatotoxicity, Vitamin B complex, Vitamin 'C'.*

Introduction

Non-steroidal anti-inflammatory drugs (NSAID)s are chemically diverse group of substances that exert several distinct pharmacological properties such as the ability to increase inflammation, ability to relieve moderate pain (analgesic), ability to decrease elevated body temperature associated with fever (antipyretic) and the ability to decrease blood clotting by inhibiting platelet aggregation (anticoagulant) (Narma and Mauex, 1989). Ibuprofen is a new generation non- steroidal anti-inflammatory agent, which is classified under non-norcotic analgesics (Klineknecht, 1985).

Several studies have indicated that NSAIDs can generally prevent prostaglandin synthesis from arachidonic acid by inhibiting the activity of the prostaglandin synthesis enzyme cyclooxygenase. The inhibition of cyclooxygenase activity removes the influence of cyclooxygenase products on cellular events (Brune, 1984). Prostaglandins are formed from dietary essential fatty acids (principally arachidonic acid) esterified to phospholipids and in some instances to triglycerides. These products have some potent biological activities affecting cell function in every organ. The high levels of NSAIDs also inhibit the activities of various enzymes, the proteoglycon synthesis from chondrocytes, the ionic exchange rate and the processes depend on prostaglandins (Rubatelli *et al.*, 1979;

Skoutakis et al.,1988 and Kayaalp, 1998). Ibuprofen is extremely metabolized in the liver. Therefore, the present study aimed to investigate the anti-hepatotoxic efficacy of vitamins (B complex and C) on ibuprofen induced liver damage in rats.

Materials and Methods

Drugs

The drug Ibuprofen from Micro Labs Limited, SIPCOT Industrial Complex, Hosur, was used for the experiment. The drug Becousules syrup and vitamin 'C' were from Omni product drugs Ltd, Bhosari, Pune, India.

Animals

The experiments were performed in Wistar strain rats (200g) provided by Tamil Nadu Agricultural University, Coimbatore. They were housed in a clean and well ventilated animal house with a constant 12 hrs light and darkness schedule, They were provided with free access to standard rat pellets and water throughout the 2 weeks acclimatization periods. They were then carefully monitored until the end of the treatment. The experiments were approved by the Animals Care Committee of the Institute (722/02/a/CPCSEA).

Animal Treatment

The animals were weighed and divided into following five groups each consisting of 5 animals subjected to specific regimen of treatment. Group I:*Control group*: the rats received 0.9 per cent saline orally, Group II :*low dose*– the animals received ibuprofen at a dosage of 1.5mg/100g body weight orally for 7 consecutive days, Group III: *High dose*– the animals received ibuprofen at a dosage of 3mg/100g body weight/day for seven consecutive days, Group IV:*High dose* + *vitamin supplementation* (B complex and Vitamin C – rats received vitamin B complex + vitamin 'C' at a dosage of 0.05mg/100g body weight after high dose of treatment, Group V: *High dose withdrawal*– the rats received ibuprofen at a dosage of 3mg/100gm body weight orally for seven consecutive days followed by withdrawal period of further seven days.

Experimental Procedure

Twenty four hours after the last treatment schedule, the animals were sacrificed by decapitation. The liver was removed, washed in

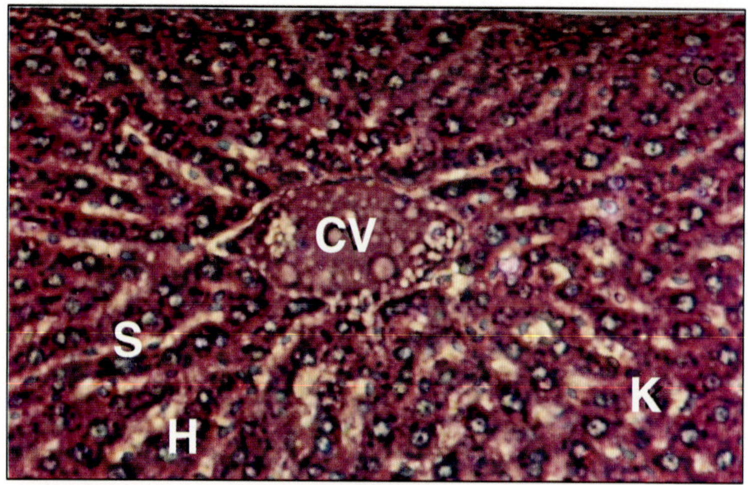

A: H and E 200 X

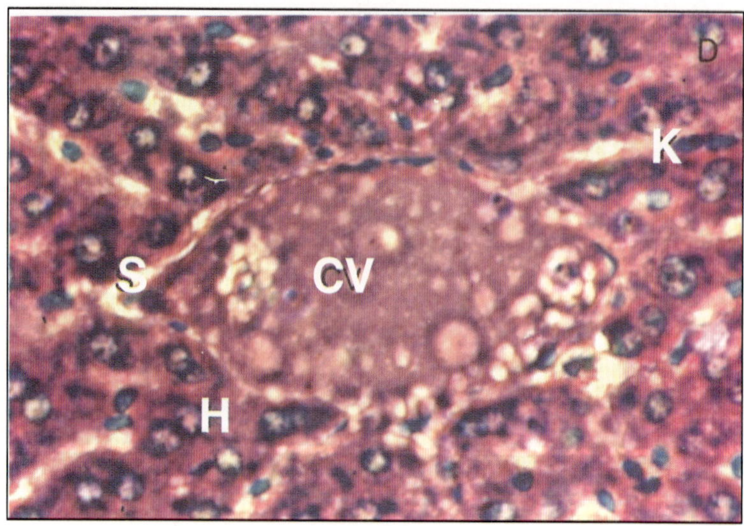

B: H and E 400 X

Figure 22.1: Group I–Control Group
CV: Central Vein; S: Sinusoid; H: Hepatocyte; K: Kupffer cell

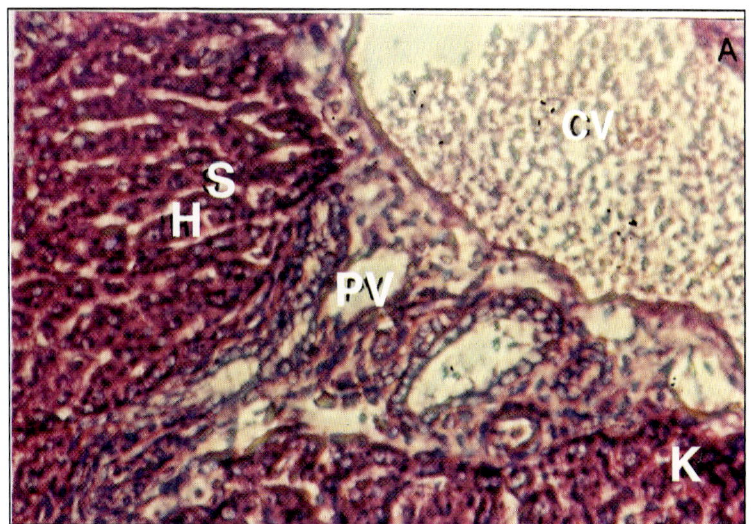

A: H and E 200 X

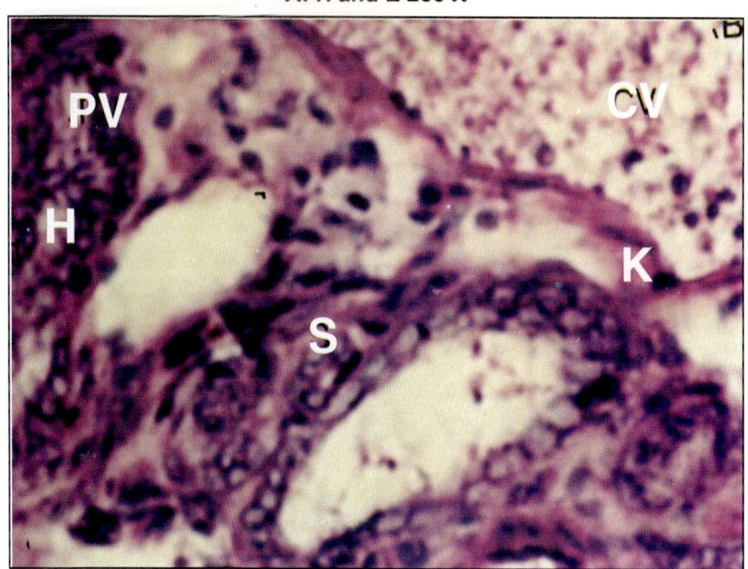

B: H and E 400 X

Figure 22.2: Group II–Low Dose Group

P.V.: Portal vein; S: Sinusoid; H: Hepatocyte; K: Kupffer cell

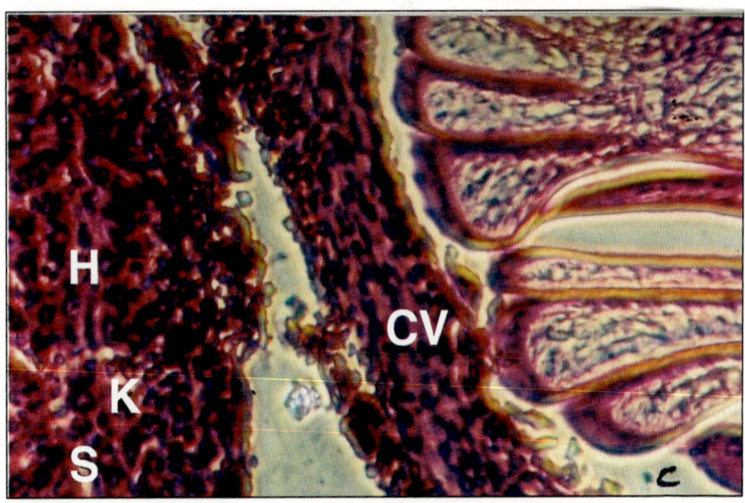

A: H and E 200 X

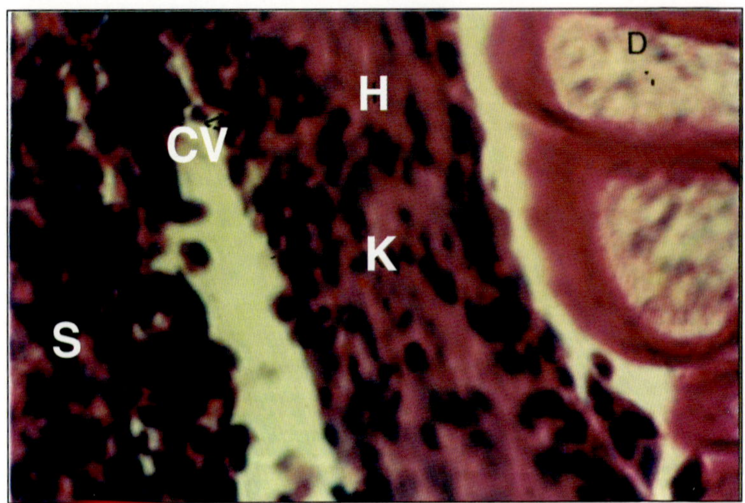

B: H and E 400 X

Figure 22.3: Group III–High Dose Group

H: Hepatocyte; S: Sinusoid; CV: Central vein; K: Kupffer cell; S: Sinusoid

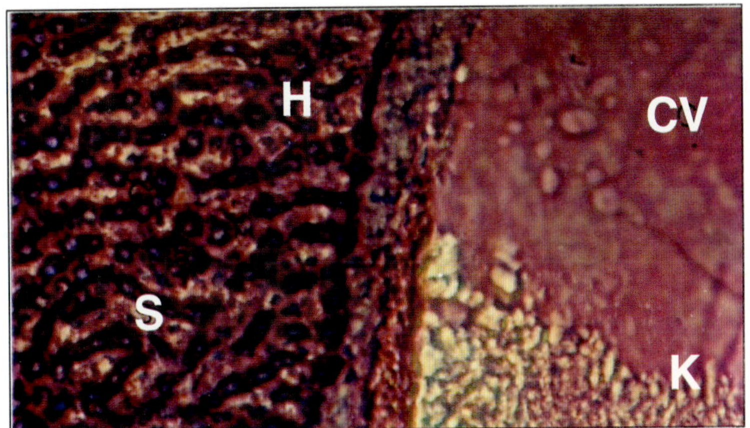

A: H and E 200 X

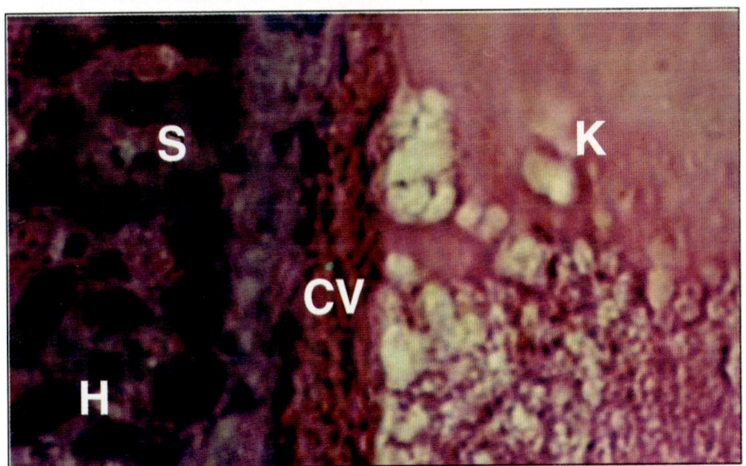

B: H and E 400 X

Figure 22.4: Group IIV–High Dose + Vitamin Supplementation Group
K: Kupffer cell; S: Sinusoid; H: Hepatocyte; PT: Portal triaditis

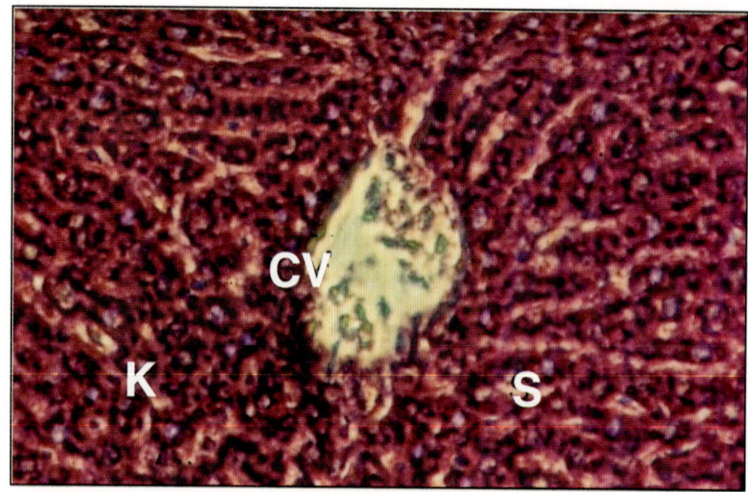

A: H and E 200 X

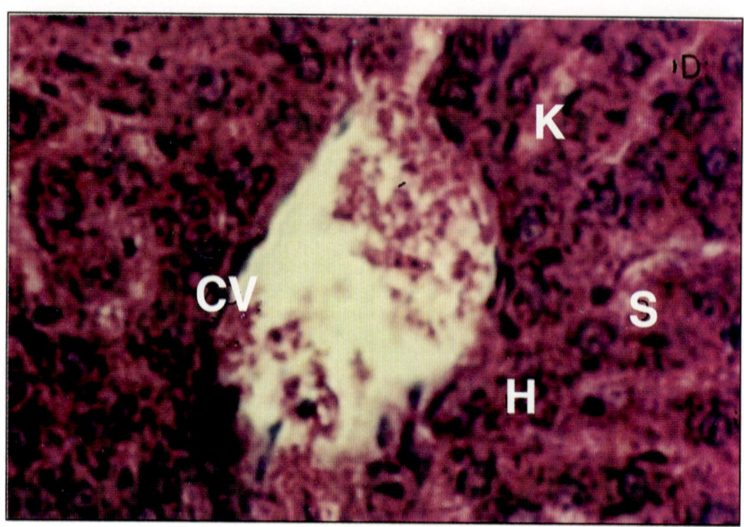

B: H and E 400 X

Figure 22.5: Group V–High Dose Withdrawl Group
K: Kupffer cell; S: Sinusoid; C.V.: Central Vein; H: Hepatocyte

0.9 per cent saline and cleared off adhering fat and connective tissue masses. Tissue samples were taken and that was fixed in a buffered formalin. The tissue was immediately processed for histological studies.

Results and Discussion

The liver plays a central role in metabolizing most drugs into forms that can be readily eliminated from the body. Non- steroidal anti – inflammatory drugs (NSAIDs) which are often used for the relief of non specific fever (Radwan, 2000), continue to be important for the palliation of pain (Simon, 1994). They are the most frequently used medications for the treatment of a variety of common chronic and acute inflammatory conditions (Manoukian et al., 1996). Ibuprofen is a new generation of non-steroidal anti-inflammatory drug, which is classified under non- norcotic analgesic (Klineknecht, 1985). It is sought as a safer, more effective alternative to either corticoids or aspirin for the use with rheumatoid arthritis and osteoarthritis. It is a propionic acid derivative.

Histological examination of the liver of control group (group I) revealed no significant deviation from the normal histological structures. The changes in histologic structures of the liver was significantly affected by the low and high dose ibuprofen treatments. The association of NSAIDs with liver disease is poorly documented. Reports on hepatic injury have ranged enzyme elevation to severe and fulminanat hepatitis (Manoukian and Carson, 1996). Histologically the liver of low dose (group II) treated animals showed enlargement of the portal tract, especially the central vein nearly ten folds. There was moderate periportal inflammation of both lymphocytes and polymorphic leucocytes, which infiltrated into the central vein. Mild kupffer cells hyperplasia was also seen. These findings corroborates with Aydin et al (2003). Prominence of kupffer cells is a common feature of hepatic injury and may occur also in generalized infections associated with histololytic activity unrelated to primary liver disorders (Patrik and Mc Gee, 1988). Similarly Gokcimen et al (2000) and 2001 suggest that a high dose of diclofenac sodium causes irreversible cell death, as well as cell damage, fibrous tissue proliferation and acute hepatitis.

The liver metabolism of drugs occurs in two stages – phase 1 reactions convert the parent drug to a metabolite by oxidation, reduction or hydrolysis. Phase II reactions create a polar excretory product by coupling the drug or metabolite with an endogenous substrate (e.g., glucuronide sulfate). Phase I oxidation occurs primarily in the hepatic monooxygenase system entering on the heme protein cytochrome P450. This system is under genetic control and is also highly sensitive to induction by many factors (e.g.,Insecticides, herbicides, smoking, caffeine) (Merk Manual, 2004).

Microscopic examination of high dose (group III) drug treated animals revealed non specific hepatitis, severe periportal inflammation and portal triaditis. Pollard (1985) has shown the loss of cytoplasmic matrix, filling up of sinusoids with dense cloudy material associated with altered microsomal function and subsequent toxicity to liver in caffeine treated rats. Further there was variable degree of severe portal fibrosis. Same trend was observed by Kappus (1986), Brune and Lindner, (1992). They found that the non- steroidal anti-inflammatory drug diclofenac causes a rare but potentially fatal hepatotoxicity that may be associated with the formation of reactive metabolites. These products presumably were formed via the hepatic enzyme p – 450 (CYP) catalysed oxidation of diclofenac to benzoquinone that are trapped by GSH (Glutathione) conjugation. The effects of NSAIDs in therapeutic doses on oxidative stress have been established (Kappus, 1986) but there is little knowledge on the effects of high doses of these drugs.

Vitamin supplementation (group IV) to high dose drug treated animals could not restore the liver histology to normal control levels. The central vein enlargement as well as portal vein triaditis persisted. Amyeloid also formed round or oval discrete inclusions (globular amyeloid) within the sinusoids and portal tracts. Withdrawal (V) of the drug brought about restoration of the liver histology. The central vein was somewhat reduced to its normal size. So also the hepatocytes and kupffer cell number showing the transient effect of the drug.

Finally it is concluded that, in the present study, dose dependent changes were seen in the liver tissue samples. The present investigation in the liver was in agreement with ibuprofen induced hepatotoxicity that may appear in acute or chronic liver disease. We observed meaningful histopathologic liver changes accompanying acute hepatitis and irreversible cellular changes with the use of high dose ibuprofen compared to low dose treatment. Partial restoration by vitamin supplementation was observed when compared to withdrawal group.

References

Aydin, G., Gokcelman, A., Onco, M., Cleck, E., Karahan, N. and Gokalp, O., 2003. Histopathologic changes in liver and renal tissues induced by different doses of diclofenac sodium in rats. 27: 1131 – 1140.

Brune, K., 1984. The concept of inflammatory mediators. In: Discoveries in Pharmacology. Vol. 2., Haemodynamics. Hormones and inflammation. *Parnhan.*, 94: 467 – 521.

Brune, K. and Linder, J., 1992. Side effects of anti–inflammatory drugs. In: Inflammation and drug therapy series.

Gokcimen, A., Akdogan, M. and Karaoz, E., 2000. Structural and biochemical changes in liver and renal tissues induced by an acute high dose of diclofenac sodium in rats. *Bio Med Res.*, 11: 293 – 302.

Gokcimen, A., Aydin, G., Karaoz, E., Malas, M.A. and Oncu, M., 2001. Effect of Diclofenac sodium administration during pregnancy in the postnatal blood. *Fetal Diagn Ther.*, 16. 417 – 422.

Hep C Net (2003. Drugs and liver damage at WWW. Hep C net.

Kappus, H., 1986. Overview of enzyme systems involved in bioreduction of drugs and redox cycling. *Biochem Pharmacol.*, 35: 1 – 6.

Kayaalp, S.O., 1998. Rasonel tedavi yonunden tibbi farmakoloji. *Medicine Pharmacology*, 8; Baski Ankara.

Klineknecht, D., 1985. Interstitial nephritis, the nephritic syndrome and chronic renal failure to non–steroidal anti–inflammatory drugs. *Semin Nephrol.*, 15: 228 – 235.

Manoukian, A.V. and Carson, L., 1996. Non steroidal anti–inflammatory drug induced hepatic disorders. Incidence and Prevention. *Drug Safe.*, 15: 64–71.

Merk manual of Diagnosis and Therapy (2003. Sec 4, Chap 43. Drugs and Liver. Drug metabolism at http: / / Merk.com.

Patrick, R.S. and Mc Gee, J.O.D., 1998. Liver injury from drugs and toxins. In: Patrick, R.S. and Mc Gee, J.O.D. (ed)., Biopsy pathology of liver, 2 nd ed. Champ and Hall, London, pp. 163 – 190.

Radwan, M.A. and Zidovudine (2000. Diclofenac and ketoprofen pharmacokinetic interations in rats. *J Pharm Pharmacol.*, 52: 665–669.

Rubatelli, F.F., Chiozza, M.L., Zanardo, V., and Cantrutti, F., 1979. Effect of maternal treatment with indomethicin. *J Pediatr.*, 94: 161 – 165.

Simon, L.S., 1994. Actions and toxic effects of the non–steroidal anti–inflammatory drugs. *Curr Opin Rheumatol.*, 6: 238 – 251.

Skoutakis, V.A., Carter, C. A., Mickle, T.R., Smith, V. H., Arkin, C. R., Alissandratos, J. and Petty, D.E., 1988. Review of diclofenac and evaluation of its place in therapy as a non – steroidal anti–inflammatory agent. *Drug Intell Clinn Pharm.*, 22: 850–859.

Chapter 23

Rescue of Antioxidant Vitamins (A and C) on Ciprofloxacin Induced Nephrotoxicity in Albino Rats

D. Srinivasan[1], B. Kalaivani[2] and G. Vanitha Kumari[1]*

[1]Division of Endocrinology, Department of Zoology, Bharathiar University, Coimbatore – 641 046, Tamil Nadu, India

[2]Department of Zoology, Sri Sarada College for Women, Autonomous, Salem – 636 016, Tamil Nadu, India

ABSTRACT

There is an increasing concern that environmental chemicals are having adverse effects on liver, kidney, and reproduction as well as ecological effects. The present study is an attempt to investigate the probable adverse effects of ciprofloxacin, an anti-inflammatory drug used extensively in the present days, on the kidney of albino rats on their enzymatic, non enzymatic antioxidant systems and also histoarchitecture. The animals were divided into three groups of 5 rats. Group I *Control* received saline and oil orally; another batch of controls for vitamin supplementation received gingelly oil orally. Group II *Low dose*: 4.2mg of ciprofloxacin/kg body weight/day. Group III *High dose*: 6.7mg of ciprofloxacin/kg body weight/day.

* Corresponding author: E-mail: kalai03win@yahoo.co.in

Group IV *High dose + Vitamin A*: 6.7mg. of ciprofloxacin followed by 7.5mg.of vitamin A/Kg. body weight/day. Group V *High dose + Vitamin C*: 6.7mg. of ciprofloxacin followed by 500mg. of vitamin C/Kg. body weight/day. VI *High dose withdrawal*: 6.7mg. of ciprofloxacin as an oral dose and were allowed a withdrawal period of the drug for further seven days. In conclusion, the results of present study suggest that neuroprotective effect of antioxidants probably by savenging and sequestering free radicals directly or by improving endogenous antioxidants.

Keywords : *Ciprofloxacin, Nephrotoxicity, Kidney histology, Vitamin A, Vitamin C.*

Introduction

Nephrotoxicity is caused by several xenobiotic substances damaging renal proximal tubule, the portion of the nephron with greater sensitivity to nephritic effects. Some of the chemicals which cause damage to the proximal tubule are antimicrobial agents such as cisplatin and industrial chemicals such as cadmium, hexavalent chromium, mercury and palladium (Cristofori *et al.*, 2007).

Ciprofloxacin is a widely prescribed anti-inflammatory drug belonging to a class of quinolone antibiotics. The first 1 – cyclo propyl fluroquinolone ciprofloxacin was synthesized in 1983 (Wentland, 1990). It is used to treat infections of the lower respiratory tract, the abdomen, the skin, the bones and joints, the urinary tract, including cystis. It is also prescribed for severe sinus or bronchial infections, infectious diarrhoea, typhoid fever, infections of the prostate gland and some sexually transmitted diseases like gonorrhoea.

Oxidative stress has been shown to play a very crucial role in some disease state like liver cirrhosis, atherosclerosis, cancer, diabetes etc. Free radical that generate inside, the body is responsible for oxidative stress and compounds that can scavenge free radicals have great potential in ameliorating these disease processes (Wilson, 1988). Antioxidants have ability to protect the human body against damage by reactive oxygen species (Lollinger, 1981).

So, the present study is an attempt to investigate the probable adverse effects of ciprofloxacin, an anti-inflammatory drug used extensively in the present days, on the histoarchitecture of the kidney.

Materials and Methods

Animals

Wistar strain *albino rats* weighing about 180-200g, readily amenable to ease of handling and restraint during various dosing parameters, have been selected for this study. The rats were procured from Tamil Nadu Agricultural University, Coimbatore, and acclimatized to our animal house conditions for 2 weeks. The animals were housed in a well ventilated, temperature and humidity controlled animal house with constant 12±1 hours light and dark schedule. They were provided with standard diet and clean water *ad libitum*. They were then carefully monitored until the end of the experiment. The experiments were approved by the Animals Care Committee of the Institute (722/02/a/CPCSEA).

Experimental Protocol

The animals were divided into seven groups as follows. The animals were treated with Ciprofloxacin and vitamins supplementation at 12 hours interval for seven consecutive days (Short duration).

Control Group

Received 0.9 per cent saline orally

1. *Low dose*:Received 4.2 mg of ciprofloxacin/kg body weight/ day
2. *High dose*: Received 6.7 mg of ciprofloxacin/kg body weight/day
3. *High dose + Vitamin A supplementation*: Received 400 mg of ciprofloxacin followed by 7.5 mg of vitamin A/kg body weight/day.
4. *High dose + Vitamin C supplementation*:Received 6.7 mg of ciprofloxacin followed by 8.3 mg of vitamin C/kg body weight/day.
5. *High dose + Vitamin E supplementation*:Received 6.7 mg of ciprofloxacin followed by 10 mg of vitamin E/kg body weight/day.
6. *High dose withdrawal*:Received 6.7 mg of ciprofloxacin as an oral dose and were allowed a withdrawal period of the drug for further seven days

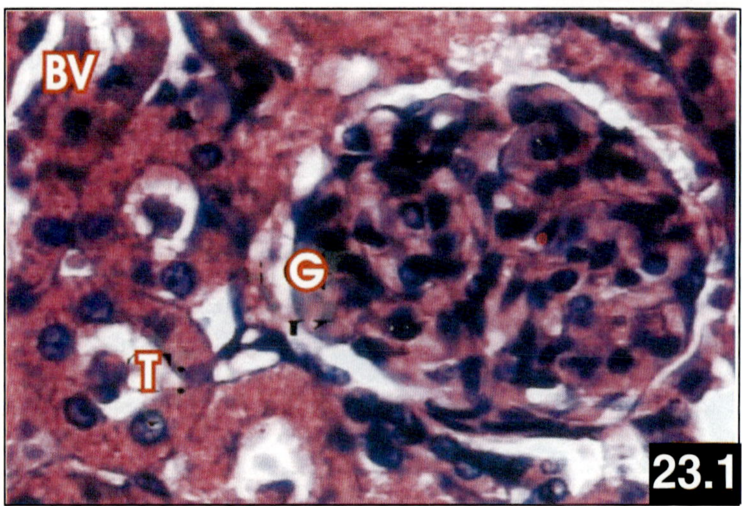

Figure 23.1: Control Group (400X)
BV: Blood Vessel; G: Glomerulus; T: Tubule

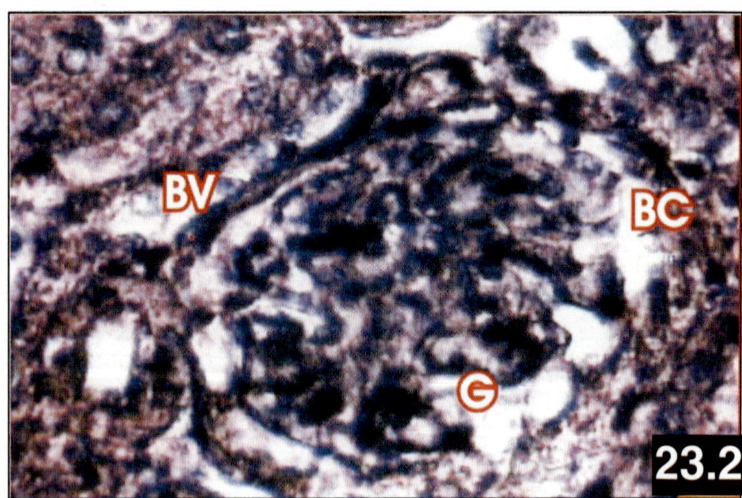

Figure 23.2: Ciprofloxacin Low Dose Treatment Group (400X)
BV: Blood Vessel; G: Glomerulus; BC: Bowman's Capsule

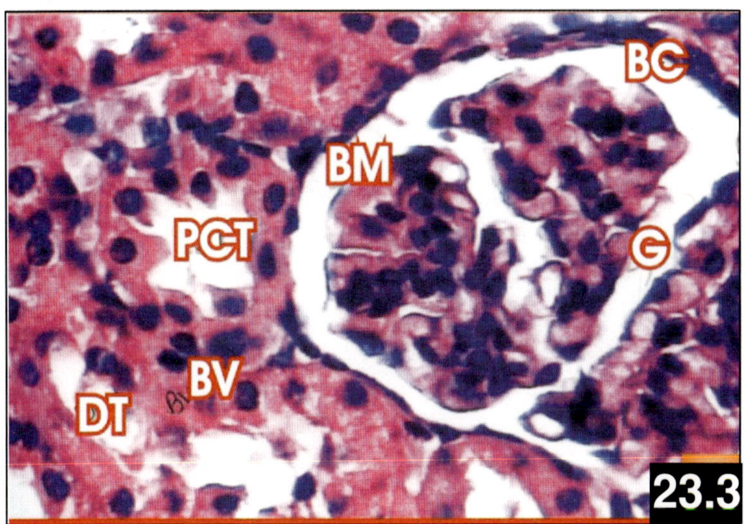

Figure 23.3: Ciprofloxacin High Dose Treatment Group (1000X)

BV: Blood Vessel; G: Glomerulus; BC: Bowman's Capsule; BM: Basement Membrane; PCT: Proximal Convoluted Tubule; DT: Diatal Tubule

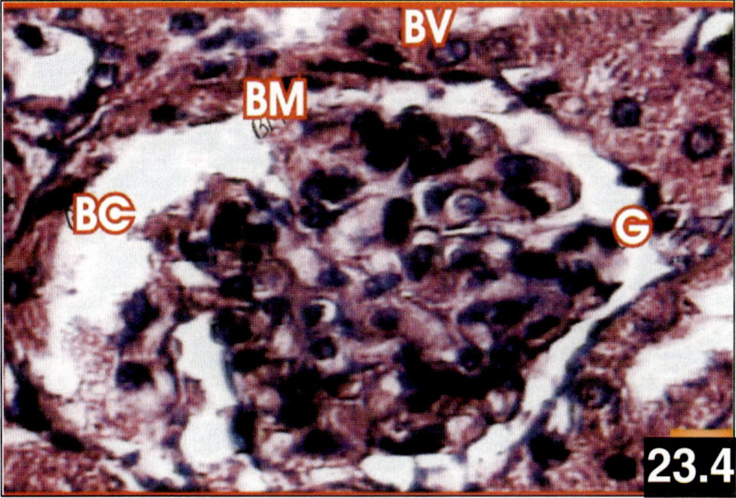

Figure 23.4: Ciprofloxacin High Dose + Vitamin A Treatment Group (400X)

BV: Blood Vessel; G: Glomerulus; BC: Bowman's Capsule; BM: Basement Membrane

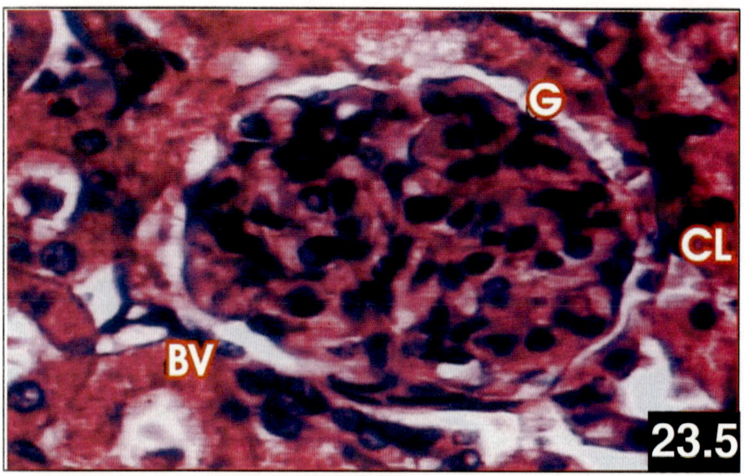

Figure 23.5: High Dose + Vitamin C Treatment Group (400X)
BV: Blood Vessel; G: Glomerulus; CL: Capillary Lumen

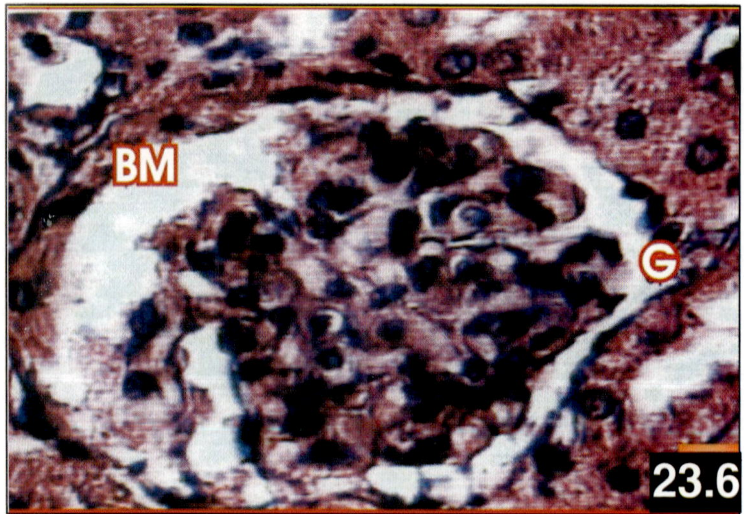

Figure 23.6: High Dose + Withdrawal Group (400X)
G: Glomerulus; BM: Basement Membrane

Sample Collection and Histological Study

The animals were weighed before and after treatment. Twenty-four hour after the last treatment schedule the animals were sacrificed by decapitation method. Serum was separated immediately after sacrifice and centrifugation of blood at 3000 X g for 30 minutes and stored at–20 C until for hormone assay. The insulin hormone test is based on a solid phase enzyme liked immunosorbent assay. The kidney was dissected out and weighed. The kidney was then fixed for histological studies in freshly prepared Bouin's fluid, dehydrated in a graded ethanol series, cleared in benzene and embedded in paraffin wax. Tissues were sectioned at 6 um and the sections were stained with periodic acid – schiff (PAS) and counter stained with haematoxylin.

Results and Discussion

Nephrotoxicity occurs as a disturbance in renal function due to various adverse drug reactions, inadequate elimination of radioactive contrast materials and chemicals. It is of great concern in patients with renal failure. Nephrotoxicity may limit the clinical usefulness of many diagnostic and therapeutic agents, recognition of factors associated with higher risk for renal injury is of great importance. However, the end point of nephrotoxicity is always cell death; therefore it is important to identify the mechanism in addition to the site of action, in order to formulate a strategy for damage prevention. The strategies aimed at ameliorating the nephrotoxicity are of clinical interest.

Histological examination of the kidney tissues in the control groups (Figure 23.1) shows normal histological structures. Histologically, the kidney of low dose (Figure 23.2) Ciprofloxacin treated animals did differ significantly from the control. The total number of glomerulus were 72 of which 24 became slightly changed (66.67 per cent). The lobular assumed a club shaped and 67 per cent of glomerulus was affected. The interstitial tissues became edematous causing separation of the tubules.

The effects of NSAIDs in therapeutic doses on oxidative stress have been established (Kappus, 1986), but there is little knowledge on the effects of high doses of these drugs (Akdogan and Akkus, 2001). The kidney is also affected by the toxic effects of NSAIDs (Palmer, 1995). Microscopic examination of kidney of high dose

Ciprofloxacin treated rats (Figure 23.3), revealed hydropic changes in the proximal convoluted tubules and the distal tubules. A few tubules were showing coagulation necrosis. The glomerulus showed degenerative changes and may loosing their basement membrane.

Ciprofloxacin promotes oxidative stress (OS) by interacting with the cell membrane which triggers the generation of reactive oxygen species (ROS). Enhanced production of ROS also causes oxidative modification of cellular macromolecules such as protein, DNA and lipid. The excess of ROS can be detoxified by the endogenous antioxidant enzymes such as superoxide dismutase (SOD) and catalase (CAT). Endogenous antioxidant system of the body sometimes becomes insufficient because of the continuous exposure to ciprofloxacin which is unavailable. Therefore the use of exogenous antioxidants could be one of the effective mean to ameliorate the ciprofloxacin induced nephrotoxicity. Although, in animal and in vitro studies antioxidants such as vitamin C (Tiwari *et al.*, 1982), vitamin E (Podatawaka *et al.*, 1991) and quercetin have shown their protective efficacy on lindane induced toxicity.

Coagulation necrosis is the most common pattern of necrosis and is characterized by conversion of the cell to an acidophilic opaque tembstone. This usually occurs with loss of nucleus but with preservation of basic cellular shape (Bonventre, 1993), which may be a result of severe ischemia. The NSAIDs adversely change the kidney functions (Goodman and Gillman's, 1995) and may play a role in the induction of membranous nephropathy (Radford *et al.*, 1996).

Supplementation with vitamin A and C (Figures 23.4 and 23.5) restore effectively the histoarchitecture of the kidney. They had brought about significant effect of glomerulus effected by 18 per cent when compared to the 70 per cent of the glomerulus of high dose group. There was increased compactness of tubules and nephrons. Hygroscopic changes were minimal, mesangial cell has increased in glomerulus. Withdrawal (Figure 23.6) of the drug led to partial recovery of the structural changes in high dose Ciprofloxacin treated rats.

Thus in the present study, it was observed that ciprofloxacin exerted an adverse effect on the kidney by way of inducing oxidative stress within the tissues. Vitamins are well established to act as anti-oxidants in many systems. In this study also their supplementation proved to be effective in restoring the balance of the tissue and lowering the ciprofloxacin toxicity. Drug withdrawal was not very effective like vitamin supplementations in the restorative activity suggesting a need for an extended period of drug withdrawal needed by the kidney.

References

Akdogan, M. and Akkus, S., 2001. Investigation of erythrocyte antioxidant enzymes activities in high dose diclofenac sodium applied rats. *Biomed. Res.*, 12: 53 – 57.

Bonventre, G.K., 1993. Effects of acute oral administration of cadmium chloride on uptake of element and control of lipoprotein in hepatic and renal nuclear fractions of rats. *Indian. J. Exp. Biol.*, 26: 449 – 452.

Clive, D.M. and Stoff, J.S., 1984. Renal syndromes associated with non–steriodial anti–inflammatory drugs. *N. Engl. J. Med.*, 310: 563 – 572.

Dixon, R.L. and Hall, J.L., 1982. Reproductive toxicity. In: Methods in toxicology, Hayes, A.W. (eds), New York, Raven press, 21: 213.

Emancipator, M., 1992. In vitro activity of Biochemical constituents against *Brucella melitersis*. *Eur. J. Clin. Microbiol.*, 3: 371.

Goodman and Gilman's (1995. Diclofenac sodium. The Pharmacological Basis of Therapeutics. Ninth Edition, New York, p 637.

Kappus, H., 1986. Overview of enzyme systems involved in bioreduction of drugs and redox cycling. *Biochem. Pharmacol.*, 35: 1 – 6.

Manoukian, A.V. and Carson, J.L., 1996. Non–steroidal anti–inflammatory drug – induced hepatic disorders. Incidence and prevention. *Drug saf.*, 15: 64 – 71.

Murray, M.D. and Brater, D.C., 1993. Renal toxicity of non–steriodal anti–inflammatory drug. *Annu. Rev. Pharmacol. Toxicol.*, 32: 435 – 465.

Palmer, B.F., 1995. Renal complications associated with use of non–steroidal anti–inflammatory agents. *J. Invest. Med.*, 43: 516 – 553.

Radford, M.G. Jr. Holley, K.E., Grande, J.P., Larson, T.S., Wagoner, R.D., Donallio, J.V. and McCarthy, J.T. 1996. Reversible membranous nephropathy associated with the use of non–steroidal anti–inflammatory drugs. *J. Am. Med. Assoc.*, *276*: 466 – 469.

Radwan, M.A., 2000. Zidovudine, diclofenac and ketoprofen pharmacokinetic interaction in Rats. *J. Pharm. Pharmcol.* *52*: 665 – 669.

Simon, L.S., 1994. Actions and toxic effects of the non–steroidal anti–inflammatory drugs. *Curr. Opin. Rheumatol.*, *6*: 238 – 251.

Tischer, B. K. and Brenner, O.G., 1994. Photodermatol photoimmunol photomed oct., 10 (5): 202 – 5.

Wentland, M.P., 1990. Structure, activity, relationship of fluroquinolones. In: Siporin, C., Heifetz, C.L., Domagala, J.M. (eds). The new generation of quinolones. New York, Marcel Dakkar, pp 1 – 43.

Chapter 24

Weight Reducing Effect of Triphala and Trikatu on High Fat Diet Induced Obesity in Albino Rats

B. Kalaivani[*1] and *G. Vanithakumari*[2]

[1]*Department of Zoology, Sri Sarada College for Women (Autonomous), Salem – 636 016, Tamil Nadu, India*

[2]*Division of Endocrinology, Department of Zoology, Bharathiar University, Coimbatore – 641 046, Tamil Nadu, India*

ABSTRACT

The aim of the present study was to investigate the weight reducing effect of triphala and trikatu. This was compared with standard herbal (ayurslim) and synthetic (sibutrex10™) anti-obesity drugs. Albino rats were divided in to three groups. Group I : Control group–the rats received normal standard diet orally, Group II : Short duration group–The animals were fed individually with triphala, trikau and their combination and ayurslim and sibutrex10™ at 12 hours intervals for thirty consecutive days. Group III : Long duration group–The animals groups were fed individually with triphala, ayurslim and sibutrex10 ™ at12 hours intervals for sixty consecutive days. Group II and Group III were further sub-divided into six sub groups, each group consisting of five animals. 1. Obesity

* Corresponding author: E-mail: kalai03win@yahoo.co.in

control–The animals received vanaspathi (hydrogenated oil), 200mg/100g body weight/day as an oral dose. 2. Obesity + Triphala–The animals received 200mg of vanaspathi + 8.3 mg of triphala/100g body weight/day, as an oral dose. 3. Obesity + Trikatu–The animals received 200mg of vanaspathi + 4.16 mg of trikatu/100g body weight/day, as an oral dose. 4 Obesity + Triphala and Trikatu–The animals received 200mg. of vanaspathi + 8.3 mg of triphala and 4.16 mg of trikatu/100g body weight/day, as an oral dose. 5. Obesity + Ayurslim–The animals received 200mg of vanaspathi + 13.3 mg of ayurslim/ 100g. body weight/day, as an oral dose, 6. Obesity + Sibutrex10™–The animals received 200mg of vanaspathi + 0.17 mg. of sibutrex10™/100g body weight/day, as an oral dose. The present study was an attempt to investigate the weight reducing effect of triphala, trikatu and their combination on high fat diet induced obesity. Obesity had caused an increase in body weight and organ weights. On Triphala and trikatu combination was more effective in reducing the bodyweight like that of ayurslim and sibutrex10™.

Keywords: *Obesity, Body weight, Liver weight, Liver index, Kidney weight, Kidney index, Heart weight, Heart index, Triphala, Trikatu, Ayurslim, Sibutrex10™.*

Introduction

Obesity is a global nutritional concern. The increasing prevalence of overweight, obesity and its consequences promoted the World Health Organization (WHO) to designate obesity as a global epidemic (WHO, 1998). According to WHO (2003), there are over million obese adults globally. Low levels of physical activity and sedentary lifestyles have generally been implicated in the world wide trend of weight gain (Popkin *et al.*, 1995).

Many types of treatment approaches are present to treat obesity. This usually involves a combination of diet, exercise and other lifestyle modifications, but the specific method of treatment depends on the severity of obesity. Medications may be prescribed in addition to an exercise programme and low calorie diets. Weight loss drugs that suppress appetite, reduce food intake, increase energy expenditure and/or affect nutrition partitioning metabolism have potential efficacy but are unfortunately frequently accompanied by adverse side effects (Haller and Bencowitz, 2000).

The ideal drug that effectively produces fat loss in an obese individual prevents regain and has no side effects, is not yet available. Therefore, supplementation with safe and natural products, in addition, to a healthy diet and exercise may be helpful. Plants have been the major source of drugs in Indian systems of medicine and other ancient systems in the world. Medicinal plants have the advantages of having little or no side effects. Some of them are being used in traditional systems of medicine from hundreds of years in many countries of the world.

So, the weight reducing effect of some herbal drugs–Triphala, trikatu were studied compared with standard herbal (ayurslim) and synthetic (sibutrex10™) anti-obesity drugs.

Materials and Methods

Animals

Adult female albino rats of Wistar Strain weighing about 180 – 200g were procured from Kerala Agricultural University, Kerala and acclimatized to our animal house conditions for 2 weeks. The animals were housed in well-ventilated animal house with constant 12±1 hours light and dark schedule. They were provided with standard diet and clean water *ad libitum*.

Experimental Design

The following experimental design was as per the guidelines of SPCA and was approved by the Institutional Animal Ethical Committee (IAEC) (No. 722/02/a/CPCSEA).

The animals were divided into three groups as follows.

Group I : Control Group

The rats received normal standard diet orally.

Group II : Short Duration Group

The animals groups were fed individually with triphala, ayurslim and sibutrex10™ at12 hours intervals for thirty consecutive days.

Group III : Long Duration Group

The animals groups were fed individually with triphala, ayurslim and sibutrex10™ at12 hours intervals for sixty consecutive days.

Group II and Group III were further sub-divided into six sub groups, each group consisting of five animals. The animals received the following regimen of treatment, all the treatments were constructed on the basis of adult dosage prescribed by the physicians to humans and extrapolated to animal body weight.

1. Obesity Control

The animals received vanaspathi (hydrogenated oil), 200 mg/ kg body weight/day as an oral dose.

2. Obesity + Triphala

The animals received 200 mg of vanaspathi + 8.3 mg of triphala/ 100 g body weight/day, as an oral dose.

2. Obesity + Trikatu

The animals received 200 mg of vanaspathi + 4.16 mg of trikatu/ 100 g body weight, as an oral dose.

3. Obesity + Triphala and Trikatu

The animals received 200 mg of vanaspathi + 8.3 mg of triphala and 4.16 mg of trikatu/100 g body weight, as an oral dose.

4. Obesity + Ayurslim

The animals received 200 mg of vanaspathi + 13.3 mg of ayurslim/100 g body weight, as an oral dose.

5. Obesity + Sibutrex10™

The animals received 200 mg of vanaspathi + 0.17 mg of sibutrex10™/100 g body weight, as an oral dose.

Suitable controls were maintained for each duration of treatment. However, as there was no obvious difference in any parameter among control groups, a common control was employed in the present study.

Animal Sacrifice

The rats selected for experiments were weighed before and after treatment. Twenty four hours after the last treatment schedule the rats were sacrificed by decapitation method. Blood was collected and was kept at room temperature. Then serum was separated immediately by centrifugation of blood at 3000xg for 30 minutes and stored at 20°C until utilized for hormonal assays and biochemical parameters.

Statistical Analysis

The data were statistically analysed and expressed as mean±standard error of mean (SEM). All the means observed by every treatments were compared by ANOVA (Alder and Rossler, 1977) and ranked by using Duncan's Multiple Range Test (DMRT) (Duncan, 1995) for analyzing the significance at 1 per cent and 5 per cent level. The following package was used to calculate ANOVA and DMRT. Irristat version 3/393, Biometrics unit, International Rice Research Institute, P.O. Box, 933, Manilla, Philippines.

Results and Discussion

Genotype and diet play important roles in human obesity. High dietary fat consumption may lead to obesity and high serum triglyceride levels, which are both linked to an increased risk of cardiovascular disease (Grundy and Denke, 1990). Some studies indicate that replacement of saturated fat with polyunsaturated fat lowers serum cholesterol level and fat deposition (Beynen and Katan, 1984).

Obesity in the mouse is well studied and provides a useful mammalian model system for studying the interaction of dietary fat and genetic factors where, contrary to humans, environmental influences can be closely controlled and specific inbred or selected strains are available (Eisen *et al.*, 1996). Therefore, models of obesity induced by high fat intake may mimic the human condition more closely (Messerli *et al.*, 1981). Indeed, there is a direct relationship between the fat content in the diet and body fat (Faust, 1980; Oscai, 1982).

Many workers have shown that dietary fat has pronounced effect on increase in the body weight and consequent adiposity in humans and animals compared to dietary carbohydrates (Portillo *et al.*, 1999, Fitzgerald *et al.*, 2001). In this investigation, a high fat diet of vanaspathi for short and long duration was shown to increase the bodyweight (Table 24.1 and Figure 24.1) by several folds. This may be due to the high energy dietary fat that promotes the body fat storage and it has been established that the diets consistently containing high level of fat lead to increased calorie intake and therefore body weight and adiposity (Anai *et al.*, 1999). Obesity is an abnormal condition accumulating lipid in adipose tissue. In the present study, obesity induction had raised the perirenal and

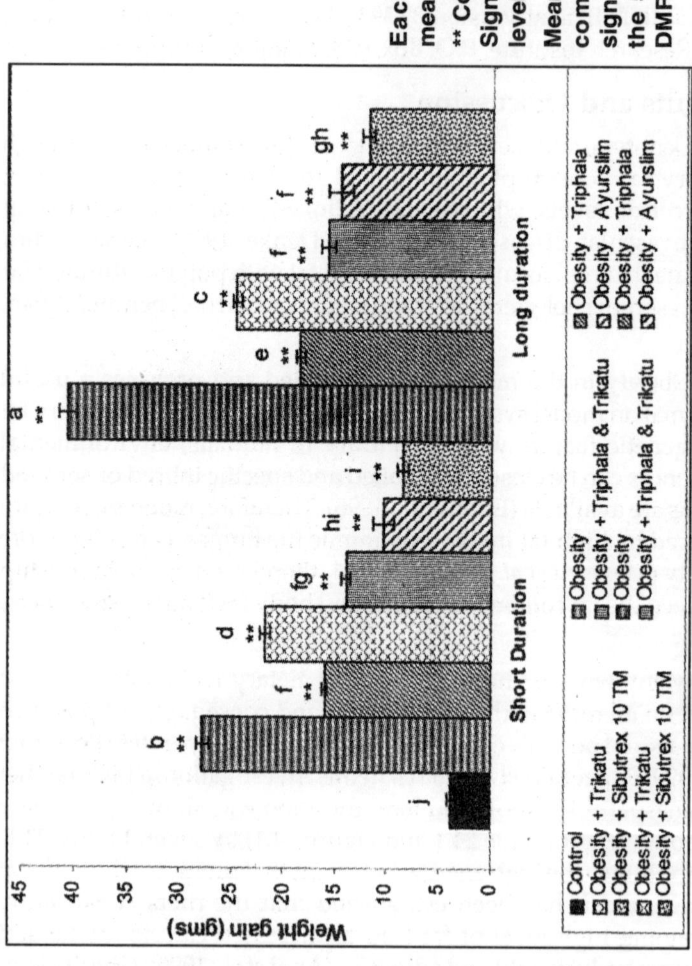

Each value is the mean±SE of five animals

**** Control Vs Treatment, Significant at 1 per cent level by ANOVA**

Means±SE followed by a common letter are not significantly different at the 5 per cent level by DMRT

Figure 24.1: Effect of Triphala, Trikatu and their Combination and Standard Anti-Obesity Drugs on Weight Gain in Albino Rats

abdominal fat pad weights (Table 24.2) when compared to controls in a duration dependent manner. Further, this visceral fat accumulation was seen to be associated with hyperinsulinemia as well as increase in glucose levels suggestive of insulin resistance and impaired glucose tolerance as well as accumulation of lipids. Similar observations have been made by Bjorntorp (1999) in overweight and obese subjects.

Table 24.1: Effect of Triphala, Trikatu and their Combination and Standard Anti-Obesity Drugs on Body Weight of Female Albino Rats

Treatments	*Body Weight (in grams)*	
	Initial	*Final*
Group I–CONTROL	189.2±0.33	193.0±0.28[i]
Group II–SHORT DURATION		
Obesity	198.8±0.28	226.6±0.61**[b]
Obesity + Triphala	199.0±0.28	212.8±0.33**[f]
Obesity + Trikatu	199.6±0.46	218.2±0.33**[d]
Obesity + Triphala and Trikatu	201.4±0.83	208.2±0.33**[fg]
Obesity + Ayurslim	200.8±0.71	205.8±0.33**[hi]
Obesity + Sibutrex10™	200.8±0.66	206.2±0.52**[i]
Group III–LONG DURATION		
Obesity	198.4±0.36	239.8±0.33**[a]
Obesity + Triphala	201.4±0.22	221.0±0.49**[e]
Obesity + Trikatu	198.8±0.33	228.0±0.28**[c]
Obesity + Triphala and Trikatu	202.0±0.39	215.4±0.22**[f]
Obesity + Ayurslim	199.4±0.61	209.2±0.33**[f]
Obesity + Sibutrex10™	198.6±0.22	205.4±0.22**[gh]

Each value is the Mean±SE of five animals

** Control Vs Treatment, significant at 1 per cent level by ANOVA

Means±SE followed by a common letter are not significantly different at the 5 per cent level by DMRT (a, b, c etc).

Effect of Obesity on Liver Weight and Liver Volume

Burnetto *et al.* (2000) found a strong correlation between liver volume and visceral fat accumulation. Moreover, reduction of hepatomegaly accompanied by a specific reduction of ALT concentrations has been demonstrated in overweight patients who lose weight (Palmer and Schaffner, 1999). Cellular swelling is the first manifestation for cell injury, resulting due to shift of extra cellular water into the cells. When all cells in an organ are affected it causes some fall or increased turgor, resulting in increase in organ weight. This pattern of hydropic change or vacuolar degeneration and swelling is reversible (Tishler and Brenner, 1994). Fatty liver, a condition resulting from accumulation of increased amount of triglycerides contribute to increase in size (Flores, 2002).

Table 24.2: Effect of Triphala, Trikatu and their Combination and Standard Anti-Obesity Drugs on Fat Pad Weights of Albino Rats

Treatments	*Fat Pad Weights (in grams)*	
	Abdominal Fat Pad Weights	*Perirenal Fat Pad Weights*
Group I–CONTROL	26.340 ± 0.19^{e}	41.600 ± 0.29^{ef}
Group II–SHORT DURATION		
Obesity	$38.580 \pm 0.17^{**b}$	$57.960 \pm 0.25^{**a}$
Obesity + Triphala	$26.180 \pm 0.17^{**e}$	$43.340 \pm 0.20^{**d}$
Obesity + Trikatu	$29.820 \pm 0.08^{**c}$	$48.760 \pm 0.61^{**b}$
Obesity + Triphala and Trikatu	$25.060 \pm 0.23^{**f}$	$41.040 \pm 0.27^{**fg}$
Obesity + Ayurslim	$20.940 \pm 0.26^{**g}$	$42.060 \pm 0.22^{**ef}$
Obesity + Sibutrex10™	$19.680 \pm 0.07^{**h}$	$39.020 \pm 0.23^{**hi}$
Group III–LONG DURATION		
Obesity	$39.660 \pm 0.16^{**a}$	$57.080 \pm 0.29^{**a}$
Obesity + Triphala	$27.320 \pm 0.28^{**d}$	$42.660 \pm 0.37^{**de}$
Obesity + Trikatu	$30.060 \pm 0.13^{**c}$	$47.640 \pm 0.33^{**c}$
Obesity + Triphala and Trikatu	$25.180 \pm 0.31^{**f}$	$37.980 \pm 0.23^{**i}$
Obesity + Ayurslim	$20.960 \pm 0.36^{**g}$	$40.920 \pm 0.24^{**fg}$
Obesity + Sibutrex10™	$18.000 \pm 0.24^{**i}$	$40.080 \pm 0.28^{**gh}$

Each value is the mean±SE of five animals

** Control Vs Treatment, Significant at 1 per cent level by ANOVA

Means±SE followed by a common letter are not significantly different at the 5 per cent level by DMRT

In the present study, obesity induction brought about a marked increase in liver weight (Table 24.3) as well as liver index with inflammation and congestion as observed histologically. This increase in weight may be due to accumulation of inflammatory cell infiltration as a result of hepatocellular degeneration. The observed küpffer cell hyperplasia, enlargement of central vein and dilation of sinusoids due to lipid secretions might have resulted in an increase in cell mass which was reflected as a weight gain of the liver. Whereas, Sophia and Manoharan, (2007) reported lowered body weight and liver weight in diabetic animals. Increased breakdown of glycogen and pronounced gluconeogenesis in diabetes might be responsible for reduction in liver weight in diabetic animals.

Table 24.3: Effect of Triphala, Trikatu and their Combination and Standard Anti-Obesity Drugs on Liver Weight and Liver Index in Albino Rats

Treatments	*Liver Weight (gms)*	*Liver Index (per cent)*
Group I–CONTROL	5.320±0.08 f	2.756±0.04 cd
Group II–SHORT DURATION		
Obesity	6.420±0.11 **c	2.792±0.03 **c
Obesity + Triphala	5.720±0.05 **e	2.688±0.03 **de
Obesity + Trikatu	6.120±0.03 **d	2.746±0.02 **cd
Obesity + Triphala and Trikatu	5.320±0.08 **f	2.552±0.03 **g
Obesity + Ayurslim	5.340±0.02 **f	2.596±0.02 **fg
Obesity + Sibutrex10™	5.460±0.05 **f	2.652±0.03 **ef
Group III–LONG DURATION		
Obesity	7.600±0.08 **a	3.166±0.03 **a
Obesity + Triphala	5.360±0.05 **f	2.422±0.03 **h
Obesity + Trikatu	6.820±0.03 **b	2.992±0.02 **b
Obesity + Triphala and Trikatu	6.080±0.07 **d	2.820±0.03 **c
Obesity + Ayurslim	6.320±0.03 **c	3.020±0.02 **b
Obesity + Sibutrex10™	5.680±0.03 **e	2.760±0.02 **cd

Each value is the Mean±SE of five animals

** Control Vs Treatment, significant at 1 per cent level by ANOVA

Means±SE followed by a common letter are not significantly different at the 5 per cent level by DMRT (a, b, c etc).

Effect of Obesity on Kidney Weight and Kidney Volume

Kidneys from obese subjects exhibit substantial focal glomerulosclerosis and other morphologic changes similar to those observed in diabetic nephropathy (Kasiske and Crosson, 1986). Study conducted by Heneger *et al.* (2001) reported that a high fat diet, fed to dogs caused significant histologic, biochemical and functional changes in the kidney. Renal damage or alterations in glomerular function affect the kidney's ability to remove metabolic substances from the blood into the urine. They control a number of functions including acid base balance, sodium excretion, urine concentration or dilution, water balance and excretion and small molecule metabolism (Nankivell, 2000).

Renal damage can be due to several different segments of the nephron, renal microvasculature or interstitium. The presence of conditions such as older age, diabetes mellitus, hypertension and congestive heart failure, have a significant influence on the patient's ability to recover from the toxic effects (Wali and Henrich, 2002). In the present study, kidney weight and kidney index (Table 24.4) were increased after obesity induction in a duration dependent manner. The reason might be due to expansion of Bowman's capsule, cell proliferation in the glomeruli, thickening of glomerular and tubular basement membranes and increased mesangial matrix in many glomeruli as observed in the present study. Similar results were observed by Heneger *et al.* (2001) in obese dogs.

Effect of Obesity on Heart Weight and Heart Volume

Increased body weight is associated with a number of cardiovascular abnormalities. Cardiac weight increases with increasing body weight, suggesting cariac work. Heart weight as a percentage of body weight, however, is lower than that in normal weight control group. The increased cardiac work associated with overweight may produce cardiomyopathy and heart failure in the absence of diabetes, hypertension or atherosclerosis (Bray, 2004). In the present study also obesity induction had caused a significant increase in the heart weight and heart index (Table 24.5) in a duration dependent manner. This indicates increased cardiac work. Similarly, Adaramoye *et al.* (2005) reported that cholesterol administered five times a week for 8 consecutive weeks resulted in a significant increase in the relative weight of the heart. The exact mechanism by which obesity increases cardiovascular risk is unknown, but they are multifactorial (Krauss *et al.*, 1998).

Table 24.4: Effects of Triphla, Trikatu and their Combination, Ayurslim, Sibutrex10™ on Kidney Weight and Kidney Index in Albinio Rats

Treatments	*Kidney Weight (gms)*	*Kidney Index (per cent)*
Group I–CONTROL	1.120±0.04[cd]	0.592±0.02[bc]
Group II–SHORT DURATION		
Obesity	1.560±0.07**[a]	0.756±0.03**[a]
Obesity + Triphala	1.200±0.09**[bcd]	0.534±0.05**[c]
Obesity + Trikatu	1.240±0.07**[bcd]	0.570±0.03**[bc]
Obesity + Triphala and Trikatu	1.120±0.04**[cd]	0.492±0.02**[c]
Obesity + Ayurslim	1.400±0.13**[abc]	0.692±0.06**[ab]
Obesity + Sibutrex10™	1.200±0.09**[bcd]	0.502±0.05**[c]
Group III–LONG DURATION		
Obesity	1.640±0.07**[a]	0.748±0.03**[a]
Obesity + Triphala	1.240±0.07**[bcd]	0.586±0.03**[bc]
Obesity + Trikatu	1.560±0.07**[a]	0.726±0.03**[a]
Obesity + Triphala and Trikatu	1.080±0.04**[d]	0.526±0.02**[c]
Obesity + Ayurslim	1.440±0.07**[ab]	0.694±0.03**[ab]
Obesity + Sibutrex10™	0.960±0.07**[d]	0.462±0.07**[c]

Each value is the Mean±SE of five animals

** Control Vs Treatment, significant at 1 per cent level by ANOVA

Means±SE followed by a common letter are not significantly different at the 5 per cent level by DMRT (a, b, c etc).

Effect of Triphala and Trikatu on Body Weight, Fat Pad and Organ Weights

Epidemiological studies support the view that consuming diets rich in fruits and vegetables is associated with a reduced incidence of chronic pathologies such as diabetes, obesity and cardiovascular diseases (Joffe and Robertson, 2001). Triphala is a traditional ayurvedic herbal formulation, consisting equal parts of three medicinal plants namely *Terminalia chebula, Terminalia bellerica* and

Emblica officinalis. Triphala is considered as a tridoshic rasayan, having balancing and rejuvenating effects on the three constitutional elements that govern human life *i.e.*, vata, pitta and kapha by Charka (1500 B.C) in the Charka Samhita (Sharma and Dash, 1998).

Table 21.5: Effects of Triphala, Trikatu and their Combination and Standard Anti-Obesity Drugs on Heart Weight and Heart Index of Albino Rats

Treatments	*Heart Weight (gms)*	*Heart Index (per cent)*
Group I–CONTROL	0.740±0.02[abc]	0.350±0.009[g]
Group II–SHORT DURATION		
Obesity	0.800±0.03**[ab]	0.464±0.01**[ab]
Obesity + Triphala	0.680±0.05**[bcd]	0.394±0.004**[def]
Obesity + Trikatu	0.600±0.03**[cde]	0.408±0.009**[cde]
Obesity + Triphala and Trikatu	0.680±0.06**[bcd]	0.368±0.02**[efg]
Obesity + Ayurslim	0.560±0.08**[de]	0.344±0.03**[g]
Obesity + Sibutrex10™	0.620±0.03**[cde]	0.356±0.003**[fg]
Group III–LONG DURATION		
Obesity	0.860±0.02**[a]	0.486±0.02**[a]
Obesity + Triphala	0.520±0.03**[e]	0.420±0.003**[cd]
Obesity + Trikatu	0.700±0.06**[bcd]	0.436±0.006**[bc]
Obesity + Triphala and Trikatu	0.500±0.03**[e]	0.382±0.005**[dg]
Obesity + Ayurslim	0.740±0.04**[abc]	0.368±0.02**[efg]
Obesity + Sibutrex10™	0.560±0.04**[de]	0.356±0.01**[fg]

Each value is the Mean±SE of five animals

** Control Vs Treatment, significant at 1 per cent level by ANOVA

Means±SE followed by a common letter are not significantly different at the 5 per cent level by DMRT (a, b, c etc).

Triphala contains about 20 per cent tannins of both condensed and hydrolysable type. Other constituents identified in the fruit include lipids, sitosterol, saponins, gallic and ellagic acids, cardiac glycoside and various carbohydrates (Reddy *et al.*, 1990). Fruits of triphala are claimed to have various biological activities such as, excerting a marked heart- protective and cardio –tonic effects, improves digestion, improves liver function and is hepatoprotective

(Gulati *et al.*, 1995). Triphala has been reported to possess anti-oxidant rich herbal formulation (Srikumar *et al.*, 2006) that improves mental faculties (Antarkar *et al.*, 1980) and also assists in the weightloss (Hashimoto and Nakajima, 1997).

Trikatu is another traditional ayurvedic herbal formulation used in the present study for the probable anti-obesity effect. This herbal formulation consists of 3 herbs in equal amounts *Piper nigrum* (black pepper), *Piper longum* (long pepper) and *Zingiber officinale* (ginger). Trikatu administration to obese rats in the present study had brought about a reduction in body weight to a lesser extent than triphala. The possible loss in body weight may be due to piperine, one of the active ingredients of *Piper nigrum*. Piperine has been used in one of the weight loss supplement formula (lean system 7). It is known for the i) availability of many nutrients ii) increased gastrointestinal blood flow and secretion of digestive enzymes (Johri and Jutschi, 1992), iii) increased thyroid peroxidase, which is needed for thyroid hormone synthesis and iv) it promotes thermogenesis by uncoupling oxidative phosphorylation and increasing the use of adenosine tri phosphate for energy and heat (Reanmonokol, 1988). *Piper longum* (pippali) has heat generating property and has been referred by Charak as the best digestive stimulant (Seth, 1996).

The dried fruits of *Terminalia chebula* and *Terminalia belerica* are known to reduce lipid deposits in many organs (Sivarajan, 1994). Many reports are available on effects of saponins on adipose tissue weight and diameter in adipose cell size. The saponins increase the delaying of the intestinal absorption of dietary fat inhibition of pancreatic lipase, examples are tea saponins (Hans, 2001), caffeine and cholinergic acid of green coffee beans (Shimoda *et al.*, 2006), saponins of *Discorea nipponica* (Kwon *et al.*, 2003) and saponins in ginseng (Yun *et al.*, 2004).

Properties of weight loss drugs might be either they showed suppression of the appetite, reduction of food intake, increase in energy expenditure and/or affect nutrition partitioning (I Ialler and Bencowitz, 2000). Triphala combination appears to speed up gastric emptying and increase the intestinal motility, relieves dyspepsia, lower blood glucose and lipids and appears to be cytoprotective to the gastric mucosa. All these properties appear to help in the lowering of body weight to a limited extent (www. Himalaya healthcare.com).

Trikatu with its constituents on the other hand, appears to be a thermogenic leading to high energy expenditure. So, when two herbal formulations were combined, the mixed properties of these herbs appear to be more effective in reducing the body weight by stimulating easy digestion, utilization for glycogen storage and lowering of fat synthesis. Thus, it appears to be equivalent to the anti-obesity drug ayurslim in the body weight lowering effect.

In this study, triphala and trikatu were combined and administered to the obese rats, there was potentiation of both the herbal formulations as a better body weight reduction was seen compared to individual treatments. Similarly liver weight and liver index, kidney weight and kidney index, heart weight and heart index were also reduced. This might be due to the weight reducing effect of triphala and trikatu. Triphala, trikatu and their combination in obese animals had effectively decreased the perirenal and abdominal fat pad weights back to control levels both at short and long duration of their treatments like standard anti-obesity drugs. This effect may be due to the sitostrol and saponin contents in triphala, which are known anti-hypercholesterolemic and triglyceridemic agents.

Effect of Ayurslim and Sibutrex10™ on Body Weight, Fat Pad and Organ Weights

Ayurslim, the ayurvedic anti-obesity formulation is a combination of *Garcinia cambogia* contains hydroxyl citric acid reported to inhibit lipogenesis (Heymsfield, 1998), *Commiphora wightii* known hypolipidemic and hypocholesterolemic compound (Satyavati *et al.*, 1969; Singh *et al.*, 1994), *Gymnema sylvestre* contains gymnemic acid- reduces the cravings for sweets and controls blood sugar levels (Prakash *et al.*, 1986; Ninomia and Imoto, 1995) and lowers the cholesterol levels), and *Trigonella foenum graecum* effectively reduces weight (*Handa et al.*, 2005).

The weight reducing effect of ayurslim might be due to the action of hydroxyl citric acid of *Garcinia cambogia*. Hydroxy citric acid is a competitive inhibitor of ATP citrate lyase, an extra mitochondrial enzyme involved in the initial steps of *de novo* synthesis (and Jena *et al.*, 2002). In addition, there is increased production of hepatic glycogen in the presence of HCA, which may activate glucoreceptors leading to a sensation of fullness and reduced appetite (Lowenstein, 1971; Triscari and Sullivan, 1984). HCA exerts all the above effects without stimulating central nervous system like other dietary stimulants such as ephedra, caffeine or phenylpropanalamine (CNN. Com. Health, 2003). Ramos *et al.* (1995) and others (Sullivan *et al.*, 1983) have shown HCA to reduce body weight in overweight adults when it is taken with meals in humans. Ayurslim acts by inhibiting appetite, increased digestive stimulation, increasing glycogen deposition and lowering lipid synthesis.

Sibutramine has a dual mode of action. It reduces food intake and attenuates the fall in metabolic rate which occurs during weight loss (Finger, 2002). Weight loss with sibutramine has been reported to be attributable to enhanced satiety, as well as to decreased hunger and food consumption (Rolls *et al.*, 1998). Synthetic anti-obesity drug sibutramine is more thermogenic and similar to trikatu in its action. Sibutramine action is by two ways for inducing weight loss, the first is satiety and the second is thermogenesis (Padwal *et al.*, 2003).

Ayurslim and sibutrex10™ had caused a marked reduction in organ weight like that of triphala and trikatu group. The constituent *Garcinia cambogia*, *Commiphora mukul* and *Trigonella foenum* might be the causative factors for reducing the visceral fat and abdominal fats, ultimately reducing fat content in liver. Likewise, sibutramine also decreases abdominal fat, total subcutaneous fat as was reported by Van Gaal *et al.* (1998) corroborating the present observation.

Conclusion

Administration of triphala and trikatu combination was more effective in its restoration compared to individual administration of triphala, trikatu and ayurslim. When compared to ayurslim, triphala and trikatu were more effective. Obesity induced changes takes place through lipid peroxidation. The herbal formulations were reported as well known antioxidants. Reduction in lipid peroxidation and increase in antioxidants as observed in the present study might be the reason for weight reduction.

References

Adaramoye, O.A., Nwaneri, V.O., Anyanwu, K.C., Farombi, E.O. and Emerole, G.O., 2005. Possible anti–atherogenic effect of kolaviron (a garcinia cola seed extract) in hypercholesterolemic rats. *Clinical and Experimental Pharmacology and Physiology.*, 32 (1–2): 40.

Anai, M., Funaki, M., Ogihara, T., Kanda, A., Onishi, Y., Sakoda, H., Inukai, H., Nawano, M., Fukunshima, Y., Yazaki, Y., Kikuchi, M. and Oka, Y. Asano, T., 1999. Enhanced insulin – stimulated activation of phosphatidylinositol 3 – kinase in the liver of high – fat fed rats. *Diabets.*, 48: 158 – 169.

Antarkar, D.S., Vaidya, K.S., Doshi, J.C., Athavale, A.V., Vinchoo, K.S., Natekar, M., Tathed, P.S., Ramesh, V. and Kale, N., 1980. *Ind. J Med Res.*, 72: 588 – 593.

Beynen, A.C. and Katan, M.B., 1984. Relations between the response of serum cholesterol to dietary cholesterol and to the type of the dietary fat, protein and fiber in hypo – and hyperresponsive rats. *Nutr Rep Intl.*, 29: 1369 – 1381.

Bjorntorp, P., Holm, G. and Rosmond, R., 1999. Hypothalamic arousal, insulin resistance and type 2 diabetes mellitus. *Diabetic med.*, 16, 373–381. Black and white pre adolescent girls. *Int. J Eat Disord.*, 18: 59–69.

Bray, G.A., 2004. Medical consequences of obesity. *J. Clin Endocrinol metab.*, 89(6): 2583 – 2589.

Eisen, E.J., Jones, E.E., Rajendran, K.G. and Hall, I.H., 1996. Effect of trimethylamine – carbomethoxyborane on growth traits and lipid metabolism in lines of mice selected for high and low fat content. *Growth Dev Agcing.*, 60: 7 – 14.

Finger, N., 2002. Sibutramine: its mode of action and efficacy. *Int J Obes Relat Metab Discord.*, 26 (Suppl. 4): 529 – 33.

Fitzgerald, S.M., Henegar, J.R., Brands, M.W., Heneger, L.K. and Hall, J.E., 2001. Cardiovascular and renal responses to a high fat diet in Osborne – Mendel rats. *Am J Physiol Regulatory Integrative Comp Physiol.*, 281: R 547 – R552.

Flores, O.J., 2002. Lactate – Dehydrogenase isoenzymes test. In: Gale Encyclopedia of Medicine. The Gale Group Available at www.healthatoz.com

Grundy, S.M. and Denke, M.A., 1990. Dietary influences on serum lipids and lipoproteins. *J Lipid Res.*, 31: 1149 – 1172.

Gulati, R. K., Agarwal, S. and Agarwal, S.S., 1995. Hepatoprotective studies on *Phyllanthus emblica* Linn. and quercetin. *Indian journal of Experimental biology*, 33: 261 – 268.

Haller, C.A. and Bencowitz, N.L., 2000. Adverse cardiovascular and central nervous system events associated with dietary supplement containing ephedra alkaloids. *N. Engl. Med.*, 1833 – 8.

Han, L.K., Kimura, Y., Kawashima, M., Takaku, T., Taniyama, T., Hayashi, T., Zheng, Y.N. and Okuda, H., 2001. Anti–obesity effects in rodents of dietary tea saponin, a lipase inhibitor. *Int J obes Relat Metab Disord.*, 25: 1459 – 1464.

Hashimoto, M. and Nakajima, Y., 1997. *Jpn. Kokai, Tokkyo. Koho. J.P.*, 90: 227 – 398.

Henegar, J.H., Bigler, S.A., Henegar, L.K., Tyagi, S.C. and Hall, J.E., 2001. Functional and structural changes in the kidney in the early stages of obesity. *J Am Soc Nephrol.*, 12: 1211 – 1217.

Heymsfield, S.B., Allison, D.B., Vasseli, J.R., Pietrobelli, A., Green field, D. and Nunez, C., 1998. Garcinia cambogia (hydroxyl citric acid) as a potential anti–obesity agent. *JAMA.*, 280: 1596 – 1600.

Jena, B.S., 2002. Chemistry and biochemistry of (–)– hydroxy citric acid from Garcinia. *J. Agri Food Chem.*, 50(10): 10 – 22.

Johri, R.K. and Zutshi, U., 1992. An ayurvedic formulation 'trikatu and its constituents. *J. Ethopharm.*, 37: 85 – 91.

Kasiske, B.L. and Crosson, J.T., 1986. Renal disease in patients with massive obesity. *Arch. Intern. Med.*, 146: 1105 – 1109.

Krauss, P., Panasenko, O.M. Sdviger, A.G. and Sergienko, V.I., 1998. The rabbit liver in experimental atherosclerosis secretes oxidized lipo proteins. *Bull Exp Mol Biol Med.*, 113: 81.

Messerli, F. H., Christie, B., De Carvalho, J.G., Aristimuno, G.G., Suarexz, D.H., Dreslinski, G.R., and Frohlich, E.D., 1981. Obesity and essential hypertension Haemodynamics, intravascular volume, sodium excretion and plasma rennin activity. Arch Intern Med., 141: 81 – 85.

Nankivell,J., Brain (2001. Creatinine clearance and the assessment of renal function. *Asst Prescer.*, 24: 15–17.

Oscai, L.B., 1982. Dietary induced severe obesity: a rat model Am J physiol Regulatory Interactive comp physiol., 242: R 212 – R 215.

Palmer, M. and Schaffner, F., 1990. Effect of weight reduction on hepatic abnormalities in overweight patients. Gastroenterology., 99: 1408 – 13.

Popkin, B.M., Paeratakul, S., Zhai, F. and Keyou, G.A., 1995. A review of dietary and environmental correlates of obesity with emphasis on developing countries. *Obes Res.*, 3: 1438–535.

Portillo, M.P., Simon, E., Garcia – Calonge, M.A. and Del Barrio, A.S., 1999. Effects of high fat diet on lipo lysis in adipocytes from visceral and subcutaneous WAT. *Eur J Nutr.*, 38: 177 – 182.

Reanmonokol, W., 1998. Effects of piperine bioenergetic function of isolated rat liver mitochondria. *Biochem Pharmacol.*, 37: 753 – 757.

Reddy, V.R.C., Kumari, R., Reddy, S.V., Anzeem, M.A., Prabhakar, M.G., Appa Rao, A.V.N., 1990. *Fitotherapia.*, 61: 517 – 525.

Seth, S., 1996) editor – Guggul (*Commiphora mukul*). Herbs for health and beauty. Bombay, India. Book House Publishers. 113.

Sharma, R.K. and Dash, B., 1998. "Charka Samhita" Vol. II, Chowkamba Sanskrit series office, India.

Sivarajan, V.V., 1994. Ayurvedic Drugs and their plant sources. Lebanon, New Hampshire: International Science publisher.

Tishler, C. and Brenner, B.M., 1994. In: Renal pathology with clinical and pathological correlations (eds) 2nd edn. J. b. Lippin Cott, Philadelphia, pp 75 –106.

WHO (1998. World Health Report, Life in the 21st century: A vision for all. *Geneva*, pp 132.

Chapter 25

Antioxidant Activity of Genistein on Polychlorinated Biphenyl (Aroclor 1260) Induced Toxicity in TM3 Leydig Cell Line

Yasemin Tunali Ayun, Banu Orta and Melike Erkan*

Department of Biology, Faculty of Sciences, Istanbul University, 34134, Vezneciler, Istanbul Turkey

ABSTRACT

Polychlorinated biphenyls are organic chlorinated compounds produced for industrial purposes. Their production has been prohibited in many countries of the world after they were determined to be permanent in nature. These polychlorinated biphenyls threaten the health of humans and the environment. They accumulate in different places in nature and the living organism due to their chemical stability and lipophillic properties. To date, polychlorinated biphenyls have been detected to be carcinogenic, immunosupressive, teratogenic and endocrine impairing.

Genistein is an isoflavone among natural phytoestrogens that is found in soy and soy products. Isoflavones may prevent oxidative DNA damage by affecting free radicals and antioxidant enzymes. Genistein is known to be a compound

* Corresponding author: E-mail: merkan@istanbul.edu.tr

that shows the maximum antioxidant activity among isoflavones. Thus putting forward their protective effect against polychlorinated biphenyls-induced oxidative damage is important.

Leydig cells were exposed to different doses of (10^{-16}–10^{-6} M) PCB (Aroclor 1260) under basal and LH-stimulated conditions for 6, 12 and 24 hours, and an antioxidant, Genistein (10μM) additionally. Steroidegenic enzymes (3β-HSD and 17β-HSD), lipid peroxidation, hydroxyl radical and enzymatic antioxidants (SOD, CAT, GPx and GST) were measured in Leydig cells at the end of applications. The results indicated that steroidogenic enzymes and enzymatic antioxidants decreased and lipid peroxidation and hydroxyl radical increased in a dose and time dependent manner. These findings indicate that PCBs directly affected testosterone biosynthesis by inhibiting steroidogenic enzymes and antioxidant enzymes and stimulated oxidative damage by increasing the amount of hydroxyl radical and lipid peroxidation in the cell. Additionally Genistein was shown to have a healing effect on Aroclor 1260 (A1260) –induced toxicity as a strong antioxidant.

Keywords: *PCBs, Aroclor 1260, Genistein, Steroidogenic enzymes, Antioxidant system, Lipid peroxidation, Hydroxyl radical, Leydig cell line.*

Introduction

In the past years, there has been an increased interest in assessing the relationship between impaired male fertility and environmental factors. Human male fertility is a complex process and, therefore, a great variety of sites may be affected by exogenous noxae (Pflieger-Bruss *et al.*, 2004). Environmental pollutants with endocrine activity, so called "endocrine disruptors", are especially discussed as a possible cause of this detrimental development (Sharpe, 1995). Endocrine disrupting chemicals may interfere with the production, secretion, transportation, metabolism, receptor binding, mediation of effects, and excretion of natural hormones which regulate developmental processes and support endocrine homeostasis in the organism (Kavlock *et al.*, 1996).

Polychlorianted biphenyls (PCB) belong to the synthetic chlorinated organic compounds class. These compounds are environmental pollutants and potentially fertility-related toxic substances (Lloyd *et al.*, 1976). PCBs that were first synthesized in

1881 have begun to be produced for industrial purposes in the 1930s (Hutzinger *et al.*, 1974; De Voogt and Brinkman, 1989; Jones *et al.*, 1992; ATSDR, 1997; Hansen, 1999). Aroclor 1260 is a commercial compound produced for industrial purposes and contains 60 per cent chlorine (Quensen III *et al.*, 1990; Hansen, 1999; ATSDR, 2000). These lipid-soluble compounds are quite stable chemically and physically and may accumulate in the food chain (Kimbrough, 1995).

Polychlorinated biphenyls are carried to every living organism on earth, including humans, by entering the food chain (Carpenter *et al.*, 1998). Polychlorinated biphenyls have been detected to have immunosupressive, endocrine disruptive (Apostoli *et al.*, 2003), neurotoxic (Rogan and Gladen, 1992), carcinogenic (Oakley *et al.*, 1996), teratogenic effects, (Ahlborg *et al.*, 1994) and to lead to behavioural disorders (Shain *et al.*, 1986). Endocrine disruptive effects of these substances affect the male reproductive system and cause a reduction in reproductive capacity, testicular spermatozoa concentration, weight of testes and accessory reproductive organs, testosterone synthesis in Leydig cells and steroidogenic enzyme activities (Sridhar *et al.*, 2004; Murugesan *et al.*, 2005a, b; 2007a, b; 2008).

PCBs are known to cause DNA damage directly and/or by forming reactive oxygen species like other environmental pollutants (Sandal *et al.*, 2008). It is known that a sufficient amount of antioxidants should be consumed to prevent oxidative stress caused by these free radicals and to minimize their effect. Genistein is an isoflavone among natural phytoestrogens found in soy and soy products. Isoflavones may prevent oxidative damage by directly affecting free radicals and antioxidant enzymes. Dietary isoflavones increase the resistance against low density lipoprotein (LDL) oxidation (Djuric *et al.*, 2001). Differences in antioxidant activities of isoflavones depend on the relationship between their chemical structures and reactive oxygen species (Toda and Shiratalki, 1999). Genistein is known as the compound that shows the maximum antioxidant activity among isoflavones (Knight and Eden, 1996).

In vivo studies done in recent years show the protective effect of vitamin C and vitamin E against oxidative stress caused by PCBs in testicular cells (Senthil Kumar *et al.*, 2004; Murugesan *et al.*, 2005a). A number of *in vitro* studies about PCB-induced oxidative toxicity in testicular Leydig cells are quite few (Murugesan *et al.*, 2008).

Additionally, recent *in vitro* studies indicate the protective role of Daidzein, an isoflavone, against PCB-induced oxidative toxicty in mice testicular cells (Zhang *et al.*, 2008).

This study was conducted to put forward the protective effect of an antioxidant, Genistein, against oxidative injury stimulated by Aroclor 1260 in TM3 Leydig cells by measuring steroidogenic enzymes, enzymatic antioxidants, lipid peroxidation and reactive oxygen species.

Materials and Methods

The cells were treated with SFM containing 6M (10^{-6} M A1260), 8M (10^{-8} M A1260), 10M(10^{-10} M A1260), 12M (10^{-12} M A1260), 16M (10^{-16} M A1260) Aroclor 1260 (mixed in 0.08 per cent DMSO) for 6, 12 and 24 hours without or with LH stimulation (100 ng/ml). The control Leydig cells were treated with DMSO, the vehicle used. The most active antioxidant dose of Genistein (10 μM) was used in experiments. Prepared A1260 and A1260+ Genistein solutions were sterilized with a 0.2μm milipore filter. The treated Leydig cells were used for the measurement of steroidogenic enzymes, antioxidant systems, lipid peroxidation and reactive oxygen species. All experiments were repeated three times in duplicates.

Cell Viability Assay

For MTT cell viability assay, five different doses of A1260 and A1260 + Genistein prepared in medium not containing serum in 96 well plates so as to be 5000 cells for each well were incubated in CO_2 incubator at 37°C for 6, 12 and 24 hours under basal and LH-stimulated conditions. After test times had been completed, 10 μl MTT I solution was added into each well and plates were incubated in CO_2 incubator at 37°C for 4 hours. Each well was added 100 μl MTT II solution (SDS) and stored in CO_2 incubator for one night. Optic density of the cells was read in an ELISA device at 540 wave length at the end of this duration. Viability ratios of test cells were expressed as a per cent by accepting control cells viability not treated with test substance as 100 per cent (Mossman, 1983).

Assay of steroidogenic enzymes

3β-HSD Enzyme Activity

The activity of 3β-HSD in cultured Leydig cells was determined by the method described by Bergmeyer (1974). In brief, the cultured

Leydig cells were sonicated in ice-cold Tris–HCl buffer (pH 7.2) and centrifuged at 16 000 g for 5 min at 4 °C. The supernatant was used as enzyme extract for the assay of 3β-HSD. The reaction mixture contained 0.6 ml pyrophosphate buffer (100 mM), 0.2 ml NAD (0.5 mM), 2 ml distilled water, and 0.1 ml dehydroisoandrosterone (0.1 mM). The absorbance at 340 nm was measured immediately after the addition of enzyme extract at 20-s intervals for 5 min in a spectrophotometer against blank.

17β-HSD Enzyme Activity

The activity of 17β-HSD in Leydig cells was determined by the method described by Bergmeyer (1974). In brief, the cultured Leydig cells were sonicated in ice-cold Tris–HCl buffer (pH 7.2) and centrifuged at 10 000 g for 5 min at 4 °C. The supernatant was used as enzyme extract for the assay of 17β-HSD. The reaction mixture contained 0.6 ml pyrophosphate buffer (100 mM), 0.2 ml NADPH (0.5 mM), 2 ml distilled water, and 0.1 ml of 1,4-androstenedine-3,17-dione (0.8 mM). The absorbance at 340 nm was measured immediately after the addition of enzyme extract at 30-s intervals for 5 min in a spectrophotometer against blank.

Leydig Cellular Antioxidant System

After 6, 12 and 24-h A1260 and A1260 + Genistein treatment with or without LH stimulation, the cultured Leydig cells were sonicated in ice-cold Tris–HCl buffer (pH 7.4), centrifuged and the supernatant was collected and used for measuring the following biochemical parameters.

Determination of Lipid Peroxidation

Protein content was determined by the method of Lowry *et al.* (1951). The level of lipid peroxidation was measured by the method of Devasagayam and Tarachand (1987). In brief, the reaction mixture consisted of 1.0 ml of 0.15 M Tris–HCl buffer (pH 7.4), 0.3 ml of 10 mM KH_2PO_4, and 0.2 ml cell extract in a total volume of 2 ml. The tubes were incubated at 37 °C for 20 min with constant shaking. The reaction was stopped by the addition of 1 ml 10 per cent trichloroacetic acid. The tubes were shaken well, 1.5 ml thiobarbituric acid (TBA) added and were heated in a boiling water bath for 20 min. The tubes were centrifuged and the colour developed was measured at 532 nm. The malondialdehyde (MDA) content of the sample is expressed as nanomoles of MDA formed per milligram protein.

Determination of Reactive Oxygen Species

Hydroxyl radical production was quantified by the method of Puntarulo and Cederbaum (1988). In brief, to a 1 ml cell extract, 0.2 ml of 1 M phosphate buffer, 0.1 ml of each of magnesium chloride, sodium azide, DMSO, and NADPH were added and incubated for 10 min at 37°C. The reaction was arrested by adding 1.5 ml trichloro acetic acid, boiled for 30 min, and read at 570 nm, against the reagent blank. The hydroxyl radical content of the samples is expressed as mmol/min per mg protein.

Antioxidant Enzymes Assay

Determination of Superoxide Dismutase (EC: 1.15.1.1, SOD)

The activity of SOD was assayed according to the method of Marklund and Marklund (1974). Briefly, to the assay mixture containing 0.05 ml cell extract, 1,4 ml Tris–HCl buffer (pH 8.2), 0.05 ml of 2 mM pyrogallol. The samples were immediately read at 470 nm against blank at intervals of 30 sec, for 3 min on a spectrophotometer. The enzyme activity is expressed as units per milligram protein.

Determination of Catalase (EC 1.11.1.6, CAT)

The activity of CAT was assayed by the method of Sinha (1972). Briefly, the assay mixture contained 0.4 ml of 0.2 M H_2O_2 and 0,5 ml sodium phosphate buffer. Subsequently, 0.1 ml cell extract was added to initiate the reaction. Then, 2 ml dichromate–acetic acid reagent was added to arrest the reaction. To the control tube, the enzyme was added after the addition of the dichromate–acetic acid reagent. The tubes were then heated for 10 min, allowed to cool, and the green colour that developed was read at 590 nm against blank containing all components except the enzyme in a spectrophotometer. The activity of CAT is expressed as units per milligram protein (1 U is the amount of enzyme that utilizes 1 mmol hydrogen peroxide/min).

Determination of Glutathione Peroxidase (EC 1.11.1.9, GPx)

The activity of GPx was determined by the method of Hafeman *et al.* (1974). Briefly, an assay mixture containing 0.5 ml of sodium phosphate buffer (0.4 M), 0.25 ml sodium azide (10 mM), 0.5 ml reduced glutathione (4 mM), and 0.6 ml distilled water and 0.15 ml cell extract was prepared. The tubes were incubated at 37°C for 5 min. After incubation 0.5 ml H_2O_2 were added (1.25 mM) and incubated at 37°C for 3 min. Then, the reaction was terminated by the addition of 1 ml 10 per cent TCA. To determine the residual glutathione content, the supernatant was removed after centrifugation and 4.0 ml disodium hydrogen phosphate (0.3M) solution and 1 ml of the dithio-bis-nitrobenzoic acid (DTNB) reagent were added. The resulting colour was read at 412 nm against a reagent blank containing only phosphate solution and DTNB reagent in a spectrophotometer. The enzyme activity was expressed as units/mg protein (one unit is the amount of enzyme that converts one mole GSH to GSSG in the presence of hydrogen peroxide/min).

Determination of Glutathione-S-transferase (EC 2.5.1.1.8, GST)

This enzyme was assayed by the method of Habig *et al.* (1974). Briefly, to the assay mixture containing 0.4 ml potassium phosphate buffer, 0.1 ml cell extract, 1.2 ml distilled water, and 0.1 ml 1-chloro-2,4-dinitrobenzene (CDNB) were added and incubated in a water bath at 37 °C for 10 min. After incubation, 0.1 ml of 30 mM reduced glutathione was added. The optical density was measured immediately against a reagent blank at 340 nm at intervals of 30 s, for 3 min in a spectrophotometer. The activity of GST is expressed as units/mg protein (1 U is the amount of enzyme that conjugates 1 nmol CDNB with GSH/min).

Statistical Analysis

Data are presented as the mean±S.E.M. cell viability, amount of lipid peroxidation, amount of hydroxyl radical, SOD, catalase, GPx and GST antioxidant enzyme activity, 3β-HSD, and 17β-HSD enzyme activity were observed in the test group versus time and the administered group. Then, Kruskal Wallis, Mann-Whithey U and Wilcoxon test was applied to calculate statistical analyses.

Results

Effect of Aroclor 1260 and Aroclor 1260 + Genistein on Cell Viability

Viability ratio (per cent) of TM3 Leydig cells treated with Aroclor 1260 and Aroclor 1260 + Genistein under LH stimulated and basal

Figure 25.1: Dose- and Time-Dependent Effects of (A) PCB (Aroclor 1260) and (B) PCB + Genistein (Aroclor 1260 + Genistein) on Basal and LH Stimulated Leydig Cell Viability *in vitro*. Each bar denotes mean±S.E.M. Significance at $p < 0.001$ (*).

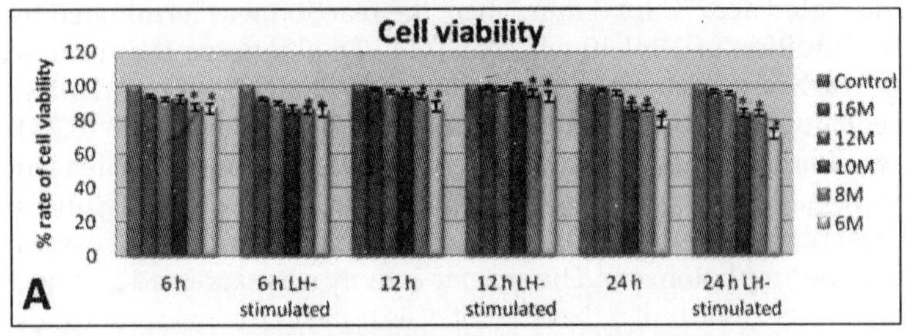

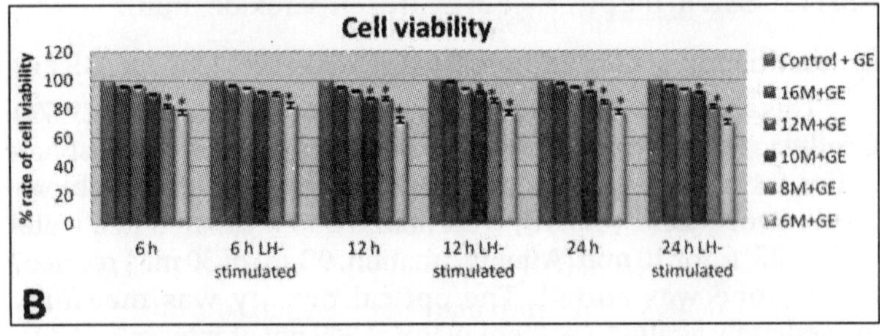

condition at the end of 6, 12 and 24 hours in control and experimental groups are given in Figure 25.1. The viable cells were significantly reduced only due to 8M and 6M A1260 and A1260 + Genistein treatment in both basal and LH-stimulated conditions when compared with respective control. However, other treatments did not alter the cell viability.

Leydig Cellular Steroidogenic Enzymes

Figures 25.2 and 25.3 show the effect of Aroclor 1260 and Aroclor 1260 + Genistein on Leydig cellular steroidogenic enzyme activities, such as 3β-HSD and 17β-HSD, under basal and LH-stimulated conditions. When compared with the control, 3β-HSD and 17β-HSD activities decreased significantly in a dose and time dependent manner under basal and LH-stimulated conditions in Aroclor 1260 exposed Leydig cells. When A1260 alone doses were compared with Aroclor 1260 + Genistein doses, a significant increase occurred at 6, 12 and 24 hours in Genistein added doses in terms of amount of 3β-HSD and 17β-HSD enzymes.

Figure 25.2: Dose- and Time-Dependent Effects of (A) PCB (Aroclor 1260) and (B) PCB + Genistein (Aroclor 1260 + Genistein) on Basal and LH Stimulated 3β-HSD Activities of Leydig Cells *in vitro*. Each bar denotes mean±S.E.M. Significance at $p < 0.05$ (*).

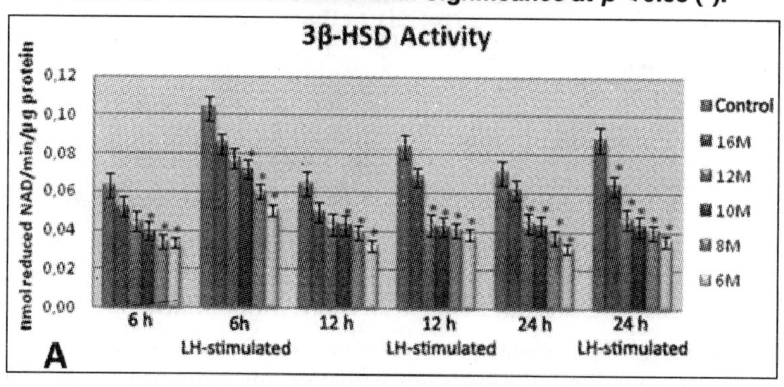

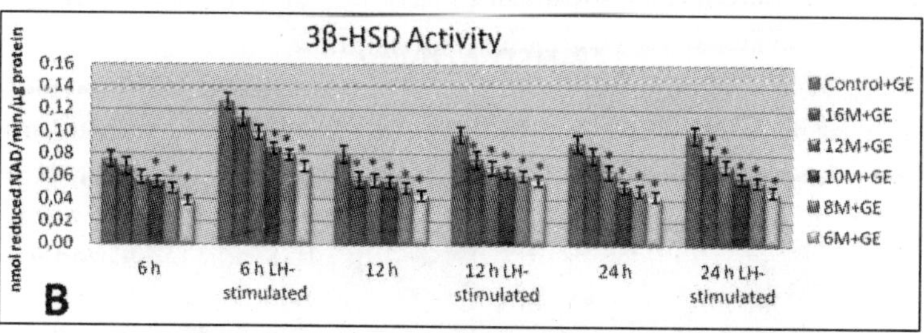

Lipid Peroxidation and Hydroxyl Radical

Figure 25.4 shows levels of lipid peroxidation and hydroxyl radical in A1260 and A1260+ Genistein administered groups. While levels of lipid peroxidation and hydroxyl radical significantly increased in A1260 exposed groups compared to the control, they decreased significantly in all doses in A1260+Genistein exposed groups compared to groups where A1260 was administered alone.

Leydig Cellular Antioxidant Enzymes

Figures 25.5 and 25.6 show antioxidant enzyme levels in Leydig cells that A1260 and A1260+Genistein administered groups for 6,12 and 24 hours under basal and LH stimulated conditions. Antioxidant enzymes, SOD, CAT, GPx and GST, decreased significantly in a dose and time dependent manner compared to the control. When Genistein administered groups and only A1260

Figure 25.3: Dose- and Time-Dependent Effects of (A) PCB (Aroclor 1260) and (B) PCB + Genistein (Aroclor 1260 + Genistein) on Basal and LH Stimulated 17β-HSD Activities of Leydig Cells *in vitro*. Each bar denotes mean±S.E.M. Significance at $p < 0.05$ (*).

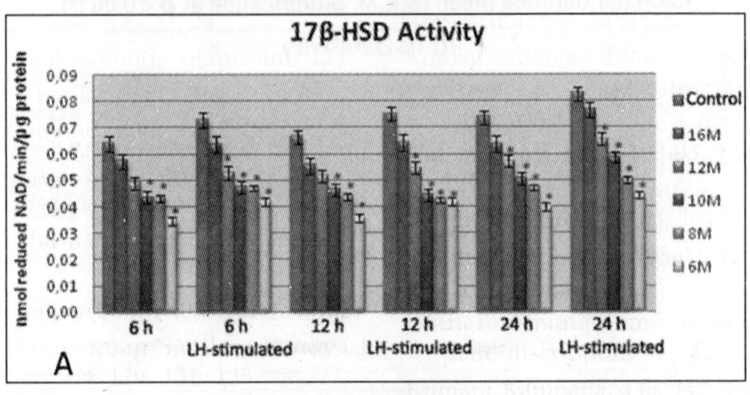

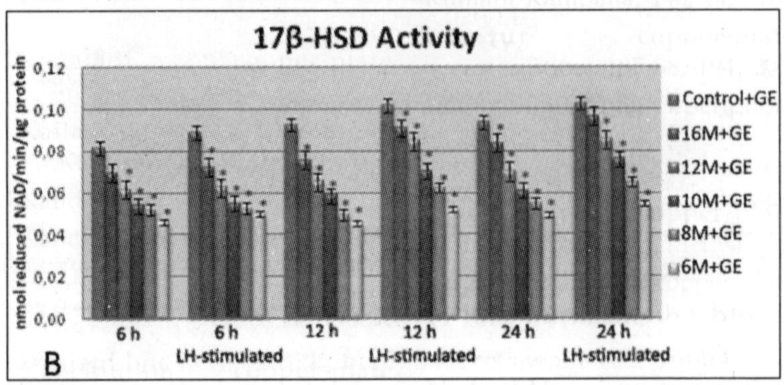

administered groups were compared, while SOD and GPx enzyme amounts significantly increased in all doses and CAT enzyme amount significantly increased in 6 and 12 hour administered groups, a significant difference could not be found in GST enzyme amounts.

Discussion

The present study demonstrates that PCB inhibits basal and LH-stimulated testosterone production in Leydig cells. Disruption of androgen biosynthesis in the Leydig cells has been associated with diminished LH receptors, activities of steroidogenic enzymes,

Figure 25.4: Dose- and Time-Dependent Effects of (A, C) PCB (Aroclor 1260) and (B, D) PCB + Genistein (Aroclor 1260 + Genistein) on Basal and LH Stimulated Lipid Peroxidation and Hydroxyl Radical of Leydig Cells *in vitro*. Each bar denotes mean±S.E.M. Significance at $p < 0.01$ (*).

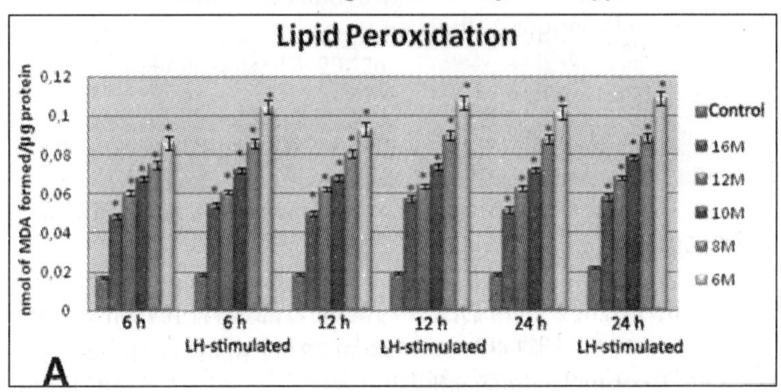

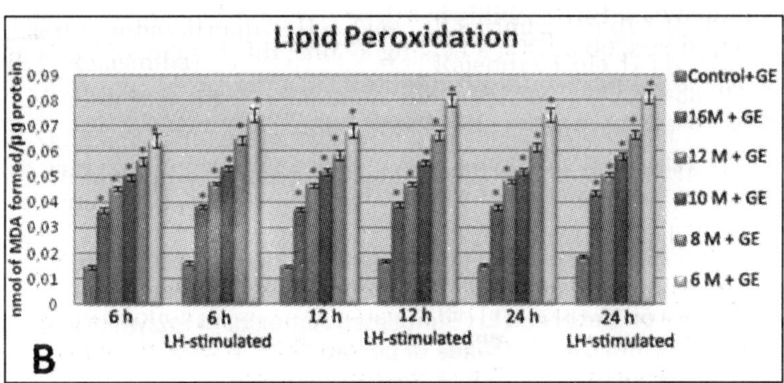

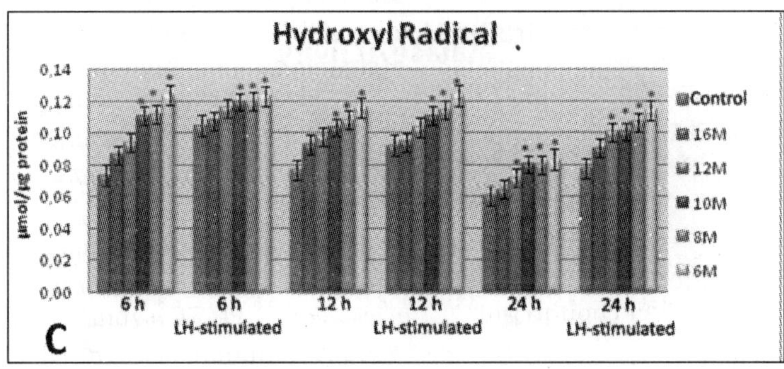

Contd...

Figure 25.4–*Contd...*

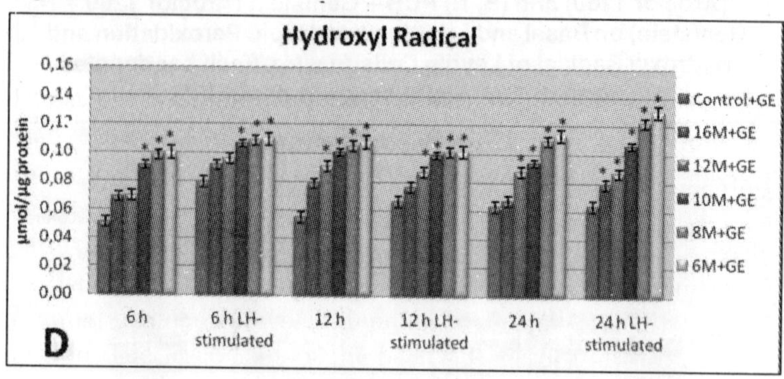

such as 3β-HSD and 17β-HSD, enzymatic antioxidants such as SOD, CAT, GPx and GST by acting directly on Leydig cells (Johansson, 1989; Kovacevic *et al.*, 1995; Murugesan *et al.*, 2007b, 2008). While PCBs decreased cell viability in a dose and time dependent manner in high doses, observing a healing against this toxicity of PCBs in Genistein administered groups indicates the protective role of Genistein.

3β-HSD and 17β-HSD enzymes are important enzymes in steroidogenesis that is testosterone biosynthesis. 3β-HSD converts

Figure 25.5: Dose- and Time-Dependent Effects of (A, C) PCB (Aroclor 1260) and (B, D) PCB + Genistein (Aroclor 1260 + Genistein) on Basal and LH Stimulated Antioxidant Enzymes such as SOD, and CAT Activities of Leydig Cells *in vitro*. Each bar denotes mean±S.E.M. Significance at $p < 0.05$ (*).

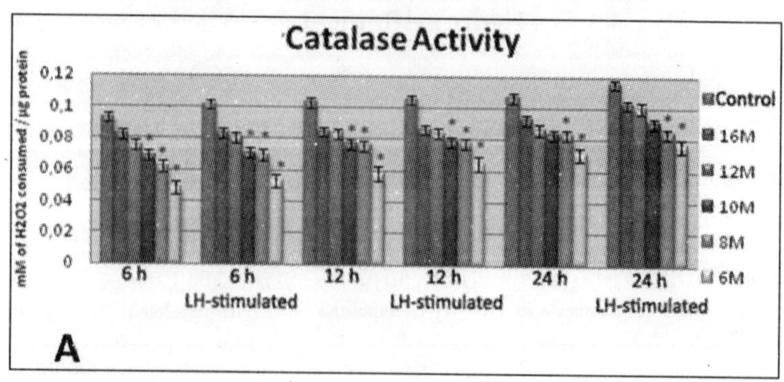

Contd...

Figure 25.5–*Contd...*

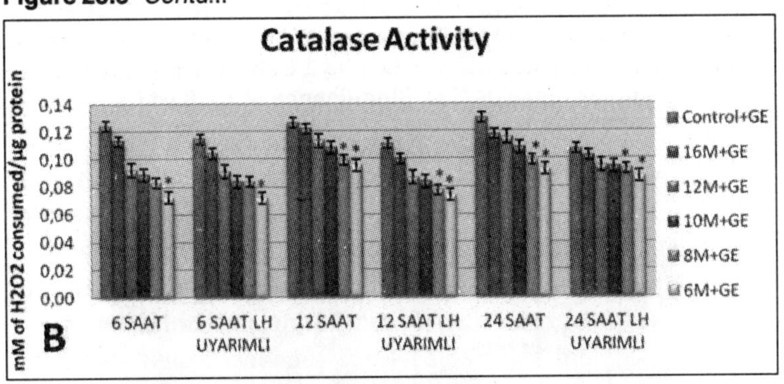

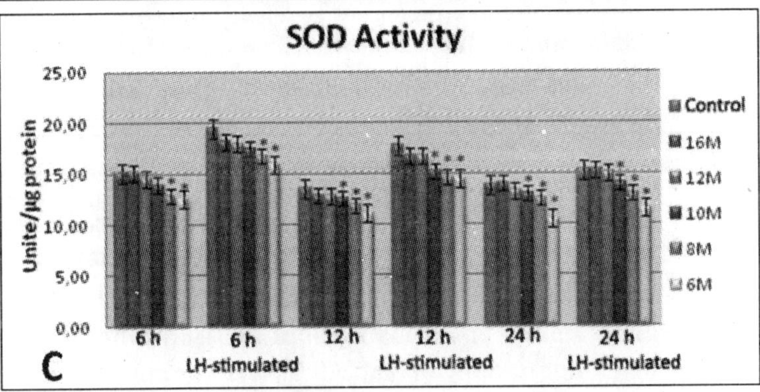

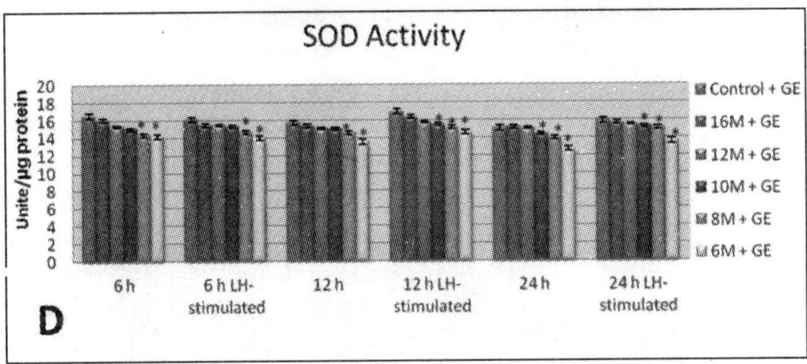

pregnenolone to progeterone, 17α hydroxypregnenolone to 17α progesterone, dehydroepiandrosterone to androsterone. 17β-HSD is responsible for convertion of androstenodione to testosterone (Payne, 2007). A1260 reduces 3β-HSD and 17β-HSD enzyme activities in a dose and time dependent manner as other PCBs. On the other

Figure 25.6: Dose- and Time-Dependent Effects of (A, C) PCB (Aroclor 1260) and (B, D) PCB + Genistein (Aroclor 1260 + Genistein) on Basal and LH Stimulated Antioxidant Enzymes such as GPx, and GST Activities of Leydig Z Cells *in vitro*. Each bar denotes mean±S.E.M. Significance at $p < 0.05$ (*).

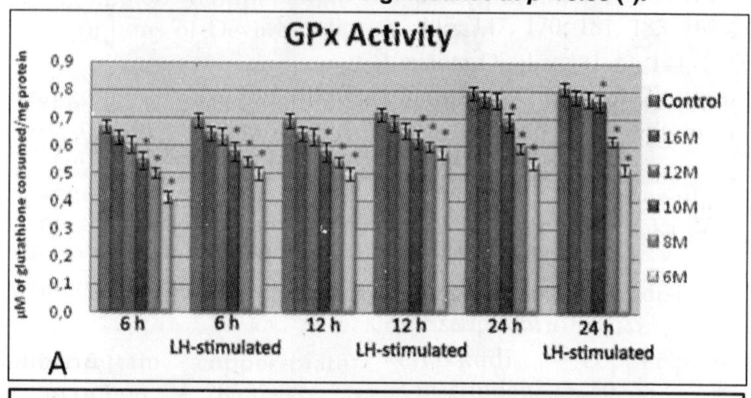

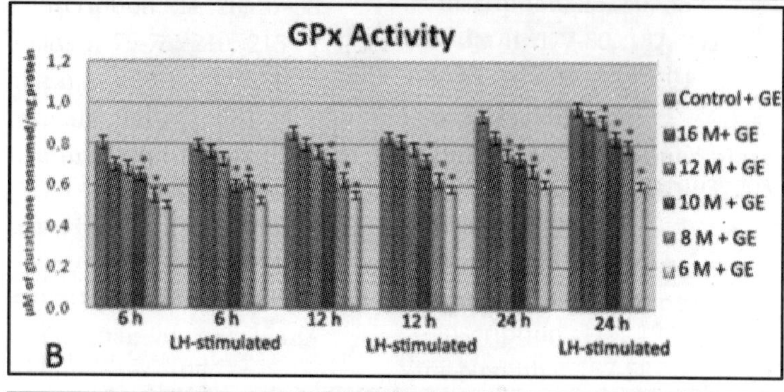

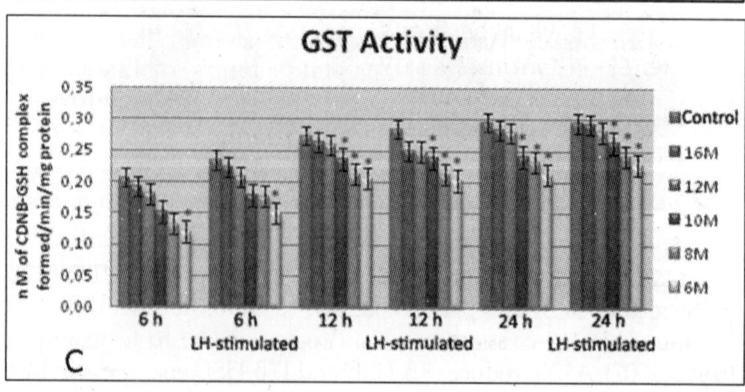

Contd...

Figure 25.6–*Contd...*

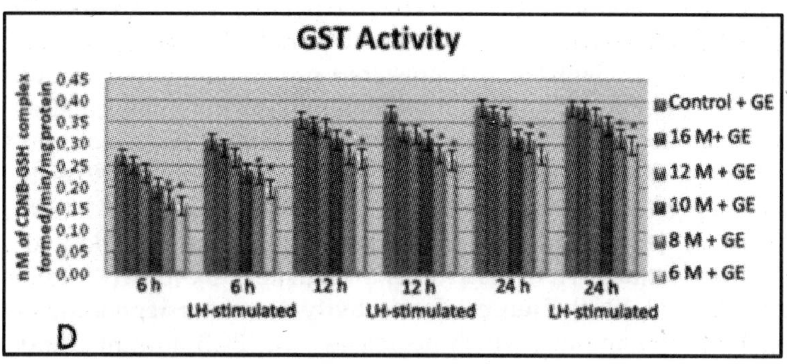

hand, Genistein shows a protective effect on 3β-HSD and 17β-HSD enzyme activities in A1260 administered groups in all doses.

Oxidative damage harms important cellular molecules like proteins and DNA (Halliwell and Gutteridge, 1990; Dalle-Donne *et al.*, 2003). This is caused by reactive oxygen species (ROS) (Cai *et al.*, 1997). Lipid peroxidation products are formed through oxidation of membrane lipids by ROS and membrane integrity is impaired (Halliwell and Gutteridge, 1990). While lipid peroxidation increases in A1260 administered Leydig cells as in A1254 (Murugesan *et al.*, 2008), this damage significantly decreases in groups where additional Genistein is applied. Administration of Genistein in addition to PCBs suppresses lipid peroxidation by inhibiting lipid peroxidation product formation (Oda ve dig., 1987).

ROS lead to reduction of antioxidant enzyme activities, inhibition of protein synthesis, DNA damage and finally cell death (Halliwell and Gutteridge, 1990). PCBs may interact with hydrogen peroxide to form hydroxyl radical (Kamohara *et al.*, 1984). While hydroxyl radical significantly increases in all doses in A1260 administered groups as in A1245 administered studies (Murugesan *et al.*, 2007b, 2008), reduction of hydroxyl radical in Genistein administered groups puts forward the protective role of Genistein.

Various intracellular enzymatic and non-enzymatic antioxidants are produced in order to prevent detrimental effects caused by free radicals. This system is defined as an antioxidant defense system. Superoxide dismutase (SOD) catalyzes conversion of superoxide radical to hydrogen peroxide and molecular oxygen and thus inhibits lipid peroxidation (Murugesan ve dig., 2008). While the amount of the SOD enzyme decreases in high doses in A1260 administered groups, the amount of the SOD enzyme increases in A1260 and Genistein administered groups compared to only A1260 administered groups. Another antioxidant enzyme, catalase, is a haameoprotein containing four heme groups and is found in peroxisomes. Catalase enzyme converts hydrogen peroxide to water and oxygen (Goldman and Blobel, 1978; Murugesan *et al.*, 2008). This enzyme is inhibited by superoxide radical (Kono and Fridovich, 1982) and requires NADPH in order to regulate active form (Kirkman and Gaetani, 1984). Thus catalase activity decreases depending on NADPH the amount which decreases with PCB administration (Twaroski *et al.*, 2001a, b). A1245 reduces amount of catalase enzyme in Leydig and Sertoli cells (Senthilkumar *et al.*, 2004; Murugesan *et al.*, 2007b, 2008,). A1260 also reduces catalase activity significantly like other PCBs. Glutathione peroxidase (GPx) enzyme is a cytosolic enzyme containing four selenium atoms. It plays a role in reduction of hydrogen peroxide and lipid peroxides in presence of reduced glutathione (GSH) (Delibas and Özcankaya, 1995; Simsek, 1999; Burçak and Andican, 2004; Murugesan *et al.*, 2008). Reduction of glutathione levels and increase in peroxide levels were observed to reduce GPx enzyme activity (Twaroski ve dig., 2001b). GST enzyme, another intracellular antioxidant enzyme, constitutes a defense mechanism by showing a selenium-independent GPx activity against lipid peroxides, mainly arachidonic acid and lineolathidroperoxydes. Glutathione neutralizes electrophilic regions of foreign bodies by binding SH groups of cysteine and enables its water-solubility (Simsek, 1999; Banudevi *et al.*, 2006). Glutathione is known to directly interact with free radicals and suppress them (Chausseud, 1979). A1260 reduces GST enzyme activity in a dose and time dependent manner. That Genistein does not have a protective effect against this reduction indicates that Genistein does not affect glutathione levels.

Genistein has an antioxidant and protective effect by scavenging free radicals against reactive oxygen species and inhibiting expression of stress response-related genes (Ruiz-Larrea *et al.*, 1997; Zhou and Lee 1998; Sarkar and Li, 2002). Oxidative stress activates binding NF-κB to DNA (Toledano ve Leonard, 1991). Genistein shows an antioxidant effect through inhibition of NF-κB which is induced by oxidative stress (Davis *et al.*, 2001; Sarkar and

Li, 2002). This effect is mostly effective against oxidative damage in membrane lipids and lipoprotein particles (Wiseman *et al.*, 2004).

In conclusion, Aroclor 1260, a commercial compound of PCBs, was shown to induce cytotoxicity in TM3 Leydig cells by forming lipid peroxidation and oxidative stress. Additionally, it was shown to reduce testosterone production by reducing 3β-HSD and 17β-HSD enzyme activities that have important roles in testosterone biosynthesis. Genistein may protect Leydig cells against A1260-induced injury by reducing lipid peroxidation products and hydroxyl radical, increasing antioxidants and steroidogenic enzyme activities. Thus, a sufficient amount of Genistein may aid reformation of antioxidant systems and correct negative effects of endocrine disruptors which are environmental pollutants on the male reproductive system.

Acknowledgements

This work was supported by Istanbul University Scientific Research Projects (Project Number:2858).

References

Ahlborg, U.G., Becking, G.C., Birnbaum, L.S., Brouwer, A., Derks, H.J.G.M., Feeley, M., Golor, G., Hanberg, A., Larsen, J.C., Liem, A.K.D., Safe, S.H., Schlatter, C., Woem, F., Younes, M. and Yrjanheikki, E., 1994. Toxic equivalency factors for dioxin–like PCB's. *Chemosphere*, 28: 1049–1067.

Apostoli, P., Mangili, A., Carasi, S. amd Manno, M., 2003. Relationship between PCBs in blood and D–glucaric acid in urine. *Toxicology Letters*, 144: 17–26.

Atsdr, 1997. Toxicological Profile for Polychlorinated Biphenyls (update). Atlanta: US Department of Health and Human Services.

Atsdr, 2000. Toxicological Profile for Polychlorinated Biphenyls (PCBs). Atlanta: US Department of Health and Human Services.

Banudevi, S., Krishnamoorthy, G., Venkataraman, P., Vignesh, C., Aruldhas, M.M. and Arunakaran, J., 2006. Role of α–tocopherol on antioxidant status in liver, lung and kidney of PCB exposed male albino rats. *Food and Chemical Toxicology*, 44: 2040–2046.

Bergmeyer, M.U., 1974. Steroid dehidrogenase. In: Bergmeyer H.U. editor, *Methods of Enzymatic Analysis*. New York: Academic Press, 476–477.

Burçak, G. and Andican, G., 2004. Oksidatif DNA hasari ve yaþlanma, *Cerrahpasa. Journal of Medicine*, 35: 159–169.

Cai, Q., Rahn, R.O. and Zhang, R., 1997. Dietary flavonoids, quercetin, luteolin and genistein, reduce oxidative DNA damage and lipid peroxidation and quench free radicals. *Cancer Letters.* 119(1): 99–107.

Carpenter, D.O., Arcaro, K.F. and Bush, B., 1998. Human health and chemical mixtures: an overview. *Environmental Health Perspective*. 106: 1263–1270.

Chausseud, L.F., 1979. The role of glutathione and glutathione–S–transferase in the metabolism of chemical carcinogens and other electrophilic agents. *Advances of Cancer Research*, 29: 175–274.

Dalle-Donne, I., Giustarini, D., Colombo, R., Rossi, R. and Milzani, A., 2003. Protein carbonylation in human diseases. *Trends Molecular Medicine*, 9: 169–176

Davis, J.N., Kucuk, O., Djuric, Z. and Sarkar, F.H., 2001. Soy isoflavone supplementation in healthy en prevents NF–kappaB activation by TNF–alpha in blood lymphocytes. *Free Radic Biol Med.*, 30: 1293–1302

De Voogt, P. and Brinkman, U.A.Th., 1989. *Production, properties and usage of polychlorinated biphenyls*, In: Kimbourgh, R, Jensen, A.A. (Eds),: Halogenated Biphenyls, Terphenyls, Naphthalenes, Dibenzodioxins and Related Products. Elsevier, Amsterdam, 3–45.

Delibas, N. and Özcankaya, R., 1995. Serbest radikaller, SDÜ Tip Fakültesi Dergisi 2(3): 11–17.

Devasagayam, T.P. and Tarachand, U., 1987. Decreased lipid peroxidation in the rat kidney during gestation. *Biochemical and Biophysical Research Communications*, 198: 273–80.

Djuric, Z., Chen, G. and Doerge, D. R., 2001. Effects of soy isoflavone supplemantation on markers of oxidative stress in men and women., *Cancer Letters*, 172: 1–6.

Goldman, B.M. and Blobel, G., 1978. Biogenesis of peroxisomes: Intracellular site of synthesis of catalase and uricase. *Proceedings in National Academic Science*, 75(10): 5066–5070.

Habig, W.H., Pabst, M.J. and Jakoby, W.B., 1974. Glutathione–S–transferase. the first enzymatic step in mercapturic acid formation. *The Journal of Biological Chemistry*, 249: 7130–7139.

Hafeman, D.G., Sunde, R.A. and Hoekstra, W.G., 1974. Effect of dietary selenium on erythrocyte and liver glutathione peroxidase in the rat. *Journal of Nutrition*, 104: 580–587.

Halliwell, B. and Gutteridge, J.M.C., 1990. Role of free radicals and catalytic metal ions in human disease: an overview. *Methods in Enzymology*, 186: 1–85.

Hansen, L.G., 1999. *The Ortho Side of PCB's Occurrence and Disposition.* Kluwer Academic Publishers, Boston, London.

Hutzinger, O., Safe, S. and Zitko, V., 1974. *The Chemistry of PCB's.* CRC Press, Cleveland.

Johansson, B., 1989. Effects of polychlorinated biphenyls (PCBs) on *in vitro* biosynthesis of testosterone and cell viability in mouse leydig cells. *Bulletin of Environmental Contamination and Toxicology*, 42(1): 9–14.

Jones, K.C., Sanders, G., Wild, S.R., Burnett, V. and Johnston, A.E., 1992. Evidence for a decline of PCBs and PAHs in rural vegetation and air in the United Kingdom. *Nature*, 356: 137–140.

Kamohara, K., Yagi, N. and Itokawa, Y., 1984. Mechanism of lipid peroxide formation in polychlorinated biphenyls (PCB) and dichlorodiphenyltrichloroethane (DDT)–poisoned rats. *Environmantal Research*, 34: 18–23.

Kavlock, R. J., Daston, G. P., Derosa, C., Fenner–Crisp, P., Gray, L. E., Kaattari, S., Lucier, G., Luster, M., Mac, M. J., Maczka, C., Miller, R., Moore, J., Rolland, R., Scott, G., Sheehan, D. M., Sinks, T. and Tilson, H. A., 1996. Research needs for risk assessment of health and environmental effects of endocrine disruptors: a report of the U.S. EPA–sponsored workshop. *Environmental Health Perspectives*, 104: 715–740.

Kimbrough, R.D., 1995. Polychlorinated biphenyls (PCBs) and human health: an update. *Critical Review in Toxicology*, 25: 133–163.

Kirkman, M.N. and Gaetani, G.F., 1984. Catalase: A tetrameric enzyme with four tightly bound molecules of NADPH. *Proceedings of the National Academy of Sciences of United States of America*, 81: 4343–4347.

Knight D. C. and Eden, J. A., 1996. A review of the clinical effects of phytoestrogens. *Obstet. Gynecol.*, 87: 897–904.

Kovacevic, R., Vojinovic–Miloradov, M., Teodorovic, I. and Andric, S., 1995. Effect of PCBs on androgen production by suspension of adult rat leydig cells *in vitro*. *Journal Steroid of Biochemistry Molecular Biology*, 52: 595–597.

Lloyd, J.W., Moore, Jr. R.M., Woolf, B.S. and Stein, H.P., 1976. Polychlorinated biphenyls. *Journal of Occupational Medicine*, 18: 109–113.

Lowry, O.H., Rosebrough, N.J., Farr, A.L. and Randall, R.J., 1951. Protein measurement with folin phenol reagent. *Journal Of Biological Chemistry*, 193: 265–275.

Marklund, S. and Marklund, G., 1974. Involvement of the superoxide anion radical in the autooxidation of pyrogallol and a convenient assay for superoxide dismutase. *European Journal of. Biochemistry*, 47: 469–474.

Mossman, T., 1983. Rapid colourimetric assay for cellular growth and survival: application to proliferation and cytotoxicity assays. *Journal of Immunological. Methods*, 65: 55–63

Murugesan, P., Kanagaraj, P., Yuvaraj, S., Balasubramanian, K., Aruldhas, M.M. and Arunakaran, J., 2005a. The inhibitory effects of polychlorinated biphenyl (Aroclor 1254) on Leydig cell LH receptors, steroidogenic enzymes, and antioxidant enzymes in adult rats. *Reproductive Toxicology*, 20: 117–126.

Murugesan, P., Muthusamy T., Balasubramanian, K. and Arunakaran, J., 2005b. Studies on the protective role of vitamin C and E against polychlorinated biphenyl (aroclor 1254)–induced oxidative damage in leydis cells. *Free Radical Research*, 39(11): 1259–1272.

Murugesan, P., Muthusamy T., Balasubramanian, K. and Arunakaran, J., 2007a. Effects of vitamins C and E on steroidogenic enzymes mRNA expression in polychlorinated bipenyls (Aroclor 1254) exposed adult rat Leydig cells. *Toxicology*, 232: 170–182.

Murugesan, P., Balaganesh, M., Balasubramanian, K. and Arunakaran, J., 2007b. Effects of polychlorinated biphenyl (Aroclor 1254) on steroidogenesis and antioxidant system in cultured adult rat Leydig cells. *Journal of Endocrinology*, 192, 325–338.

Murugesan, P., Muthusamy T., Balasubramanian, K. and Arunakaran, J., 2008. Polychlorinated biphenyl (Aroclor 1254) inhibits testosterone biosynthesis and antioxidant enzymes in cultured rat Leydig cells. *Reproductive Toxicology*, 25: 447–454.

Oakley, G.G., Devanaboyina, U., Robertson, L.W. and Gupta, R.C., 1996. Oxidative DNA damage induced by activation of polychlorinated biphenyls (PCBs): implications for PCB induced oxidative stress in breast cancer. *Chemical Research in Toxicology*, 9: 1285–1292.

Oda, H., Yamashita, K., Sasaki, S., Horio, F. and Yoshida, A., 1987. Long–term effects of dietary polychlorinated biphenyl and high level of vitamin E on ascorbic acid and lipid metabolism in rats. *Journal of Nutrition*, 117: 1217–1223.

Payne, A.H., 2007. *Steroidogenic Enzymes in Leydig Cells*, The Leydig Cell in Health and Disease. Humana Press Inc., Totowa, New Jersey, 3–30.

Pflieger–Bruss, S., Schuppe, H. C. and Schill, W. B., 2004. The male reproductive system and its susceptibility to endocrine disrupting chemicals. *Andrologia*, 36 (6): 337–345.

Puntarulo, S. and Cederbaum, A.I., 1988. Effects of oxygen concentration on microsomal oxidation of ethanol and generation of oxygen radicals. *Biochemical Journal*, 251: 787–794.

Quensen III, J.F., Stephen, A.B. and James, M.T., 1990. Dechlorination of Four Commercial Polychlorinated Biphenyl Mixtures (Aroclors) by Anaerobic Microorganisms from Sediments. *Applied And Environmental Microbiology*, 56(8): 2360–2369.

Rogan, W.J. and Gladen, B.C., 1992. Neurotoxicology of PCBs and related compounds. *Neurological Toxicology*, 13: 27–36.

Ruiz–Larrea, M.B., Mohan, A.R., Paganga, G., Miller, N.J., Bolwell, G.P. and Rice–Evans, C.A., 1997. Antioxidant activity of phytoestrogenic isoflavones. *Free Radic Res* 26: 63–70.

Sandal, S., Yilmaz, B. and Carpenter, D.O., 2008. Genotoxic effects of PCB 52 and PCB 77 on cultured human peripheral lymphocytes. *Mutation Research/Genetic Toxicology and Environmental Mutagenesis*, 654: 88–92.

Sarkar, F.H. and Li Y., 2002. Mechanisms of cancer chemoprevention by soy isoflavone genistein. *Cancer and Metastasis Reviews*, 21: 265–280.

Senthilkumar, J., Banudevi, S., Sharmila, M., Murugesan, P., Srinivasan, N., Balasubramanian, K., Aruldhas, M.M. and Arunakaran, J., 2004. Effects of Vitamin C and E on PCB (Aroclor 1254) induced oxidative stress, androgen binding protein and lactate in rat Sertoli cells. *Reproductive Toxicology*, 19: 201–208.

Shain, W., Overmann, S.R., Wilson, L.R., Kostas J. and Bush B., 1986. A congener analysis of polychlorinated biphenyls accumulating in rat pups after perinatal exposure. *Archives of Environmental Contamination and Toxicology*, 15: 687–707.

Sharpe, R.M., 1995. Another DDT connection. *Nature*, 75: 538–539.

Sinha, A.K., 1972. Colourimetric assay of catalase. *Analytical Biochemistry*, 47: 389–394.

Sridhar, M., Venkataraman, P., Siva, D., Arun Kumar, A., Aruldhas, M.M. and Srinivasan, N., 2004. Impact of polychlorinated biphenyl (Aroclor 1254) and vitamin C on antioxidant system of rat ventral prostate. *Asian Journal of Andrology*, 6: 19–22.

Simsek, F., 1999. Serbest Oksijen Radikalleri, Antioksidanlar Velipit Peroksidasyonu. *Pediatri Dergisi*, 8(1): 42–47.

Toledano, M.B. and Leonard, W.J., 1991. Modulation of transcription factor NF–cB binding activity by oxidation–reduction *in vitro*. *Proceedings of the National Academy of Sciences*, 88: 4328–4332.

Toda, S. and Shirataki, Y., 1999. Inhibitory effects of isoflavones on lipid peroxidation by reactive oxygen species. *Phytotherapy Research*, 13: 163–165.

Twaroski, L.W., O'brien, M.L., Larmonier, N., Glauret, H.P. and Robertson, L.W., 2001a. Polychlorinated biphenyls–induced effects of metabolic enzymes. Ap–1 binding, vitamin E and oxidative stres in the rat liver. *Toxicology and Applied Pharmacology*, 171: 85–93.

Twaroski, L.W., O'brien, M.L., Larmonier, N., Glauret, H.P. and Robertson, L.W., 2001b. Polychlorinated biphenyl (PCB) congeners on hepatic glutathione, glutathione–related enzymes and selenium status: implications for oxidative stress. *Biochemical Pharmacology*, 62: 273–278.

Wiseman, H., Casey, K., Bowey, E.A., Duffy, R., Davies, M., Rowland, I.R., Lloyd, A.S., Murrary, A., Thompson, R. and Clarke, D.B., 2004. Influence of 10 wk of soy consumption on plasma concentrations and excretion of isoflavonoids and on gut microflora metabolism in healthy adults. *American Journal of Clinical Nutrition*, 80(3): 692–699.

Zhang, D., Mi, Y., Wang, K., Zeng, W. and Zhang, C., 2008. Attenuating effects of daidzein on polychlorinated biphenyls–induced oxidative toxicity in Mouse testicular cells. *J. Zhejiang Univ. Sci. B.*, 9(7): 567–571.

Zhou, Y. and Lee, A.S., 1998. Mechanism for the suppression of the mammalian stress response by genistein, an anticancer phytoestrogen from soy. *J Natl Cancer Inst.*, 90: 381–388.

Chapter 26

Protective Role of Vitamin E on Polychlorinated Biphenyl (Aroclor 1260) Induced Oxidative Damage and Changes in Steroidogenic Enzymes in TM3 Leydig Cell Line

Yasemin Tunali Ayun* and Melike Erkan

Department of Biology, Faculty of Sciences, Istanbul University, 34134, Vezneciler, Istanbul Turkey

ABSTRACT

Polychlorinated biphenyls are aromatic chlorinated compounds, which were first manufactured in the 1930s for industrial purposes, and are formed as a result of chlorine ions in different numbers (ranging between 1 and 10) attached to a phenyl molecule at different configurations (ortho, meta, para). Their production has been prohibited completely in many countries and also its use has been restricted in some countries across the world after it was found that they cause environmental contamination when they are deposited in to the food chain due to their lipophilic features and chemical stability and they began to threaten human health. However, industrial equipment and industrial products containing polychlorinated biphenyl are still in use in many countries.

* Corresponding author: E-mail: yastun@istanbul.edu.tr

It was determined that free radicals caused by polychlorinated biphenyls are essential in every type of damage in an organism. Such damage in an organism, which has been exposed to these matters, may be minimized if the organism eats an antioxidant matter in sufficient amount to protect itself against this damage. Therefore, vitamin E, which is a chain breaking antioxidant to prevent radical reactions from occurring, protects the cell against oxidative damages.

Leydig cells were exposed to PCB (Aroclor 1260) for 6, 12 and 24 hours under basal and LH-stimulated conditions at different doses (ranging between 10^{-8} M and 10^{-6} M) and, in addition to these doses, to vitamin E (50μM) as an antioxidant. After these applications, steroidogenic enzymes (3β-HSD and 17β-HSD), lipid peroxidation, hydroxyl radical and enzymatic antioxidants (SOD, CAT, GPx and GST) in Leydig cells were measured. The results indicated that steroidogenic enzymes and enzymatic antioxidants decreased depending on dose and time while lipid peroxidation and hydroxyl radical increased depending on dose and time. These findings indicate that PCBs directly affect testosterone biosynthesis in Leydig cells by inhibiting steroidogenic enzymes and antioxidant enzymes and encourage oxidative damage by increasing hydroxyl radical and lipid peroxidation amount. Furthermore, it was evidenced that vitamin E has a healing effect on toxicity caused by Aroclor 1260 (A1260).

Keywords: *PCBs, Aroclor 1260, Vitamin E, Steroidogenic enzymes, Antioxidant system, Lipid peroxidation, Hydroxyl radical, TM3 Leydig Cell Line.*

Introduction

In the last 50 years, it has been discussed that the essential problem in the male reproductive system is the continuous decrease in number of sperms and the reason for this may be environmental (Sharpe and Shakkebaek, 1993). Endocrine disruptors are chemical substances existing all over and they have a bad effect on the endocrine system (Pflieger-Bruss *et al.*, 2004). The endocrine and reproductive effects of these chemicals are believed to be due to their ability to: (a) mimic the eect of endogenous hormones; (b) antagonize the eect of endogenous hormones; (c) disrupt the synthesis and metabolism of endogenous hormones; (d) disrupt the synthesis of hormone receptors (Amaral-Mendes, 2002). PCBs are one of the known endocrine disruptors.

Polychlorinated biphenyls (PCB) are industrial compounds, which were used prevalently from the 1930s until the mid-1970s (Hutzinger *et al.*, 1974; De Voogt and Brinkman, 1989; ATSDR, 1997, 2000; Hansen, 1999). Their chemical and physical stability and their low inflammableness and electrical conductivity are their positive features (WHO, 1993; Hansen, 1999). Due to these features, they are mostly used in capacitors and transformers, which have been designed as closed systems, as fluid in hydraulic and vacuum pumps, in furniture, internal decoration and building construction works, as plasticizer in coatings and printing inks, as an additive for pesticides, and used in the manufacturing of glues, linings, cutting oils, hydraulic oils, immersion oils, carbon-free copy papers and electricity insulation fluids to prevent fire and energy loss (WHO, 1993; Safe, 1994; Eisler and Belisle, 1996; De Vos *et al.*, 2003; Seyran and Erisir, 2008). Polychlorinated biphenyls are the most insistent compounds environmentally among all known chemicals. It was found that they cause environmental contamination when they are deposited in to the food chain, due to their lipophilic features and chemical stability and this threatens human health (Lilienthal *et al.*, 2000).

PCBs, which are very insistent environmentally, have high toxicity. It was detected that PCBs have effects suppressing immune system, disrupting the endocrine (Apostoli *et al.*, 2003), and they have neurotoxic (Rogan and Gladen, 1992), carcinogenic (Oakley *et al.*, 1996) and teratogenic (Ahlborg *et al.*, 1994) effects and in addition, they cause behavioural disorders (Shain *et al.*, 1986). PCBs have endocrine disruptive effects. These effects can be seen decreased in reproductive capacity, testicular spermatozoa concentration, weight of testis and secondary reproductive organs, and testosterone synthesis and steroidogenic enzyme activities in Leydig cells (Sridhar *et al.*, 2004; Murugesan *et al.*, 2005a, b; Murugesan *et al.*, 2007a; Murugesan *et al.*, 2008). A number of *in vitro* studies conducted until now on PCBs, and its effects on Leydig cell, which has a direct relationship with sperm formation in testes and the success of reproduction, and is responsible for a reduced testosterone production (Johansson, 1989; Kovacevic *et al.*, 1995; Andric *et al.*, 2000a, b; Murugesan *et al.*, 2007b; Murugesan *et al.*, 2008). Furthermore, any study conducted with Aroclor 1260, which is a polychlorinated biphenyl mixture in testes, is not available.

Antioxidants should be eaten in sufficient amount to prevent oxidative stress caused by free radicals and to minimize their effect (Coskun, 2005). Vitamin E is a dietary factor, which is necessary for reproduction in humans and animals, and is an antioxidant for all mammals (Murugesan *et al.*, 2007b). *In vivo* and *in vitro* studies have evidenced that Vitamin E, which is a chain breaking antioxidant to prevent radical reactions from occurring, protects the cell against oxidative damages by inhibiting oxidant inducer lipid peroxidation in Leydig cells. Moreover, it has a significant protective effect on testosterone production ability (Senthil Kumar *et al.*, 2004; Murugesan *et al.*, 2007b; Murugesan *et al.*, 2008).

This study was conducted with Aroclor 1260 alone and with vitamin E to discover its effects on testis TM3 Leydig cells, to detect cell vitality, lipid peroxidation, hydroxyl radical as a reactive oxygen species, enzymatic antioxidants to find the oxidative damage, and to display vitamin E's protective role against such damages.

Materials and Methods

Polychlorinated biphenyl doses to be administered to the TM3 Leydig cell line were specified by considering the doses to the extent that the organisms have been exposed to polychlorinated biphenyls. Two different doses for Aroclor 1260 in a medium containing 0,08 per cent DMSO, 10^{-6}M (6M) and 10^{-8}M (8M), were produced and the medium containing 0,08 per cent DMSO was administered to the control group. α-tocopherol, which is most active and abundant form of vitamin E, was used in the experiments (Kosowski and Clouatre, 2009; Simone and Palozza, 2009). The dose level to be administered was specified as 50μM that is *in vitro* antioxidant dose (Takacs *et al.*, 2001). The prepared A1260 and A1260 + Vitamin E solutions were sterilized with the help of 0.2 μm milipore filter. Then, they were administered to TM3 Leydig cells for 6, 12 and 24 hours under basal and Luteinizing Hormone (LH) stimulated conditions. To provide LH-stimulated conditions and non- stimulated conditions, the cells were incubated with a serum-free medium. The treated Leydig cells were used to measure steroidogenic enzymes, lipid peroxidation, hydroxyl radical and antioxidant enzymes.

Cell Viability Assay

For MTT cell viability assay, two different doses of A1260 and A1260 + Vitamin E prepared in medium not containing serum in 96 well plates so as to be 5000 cells for each well were incubated in CO_2 incubator at 37°C for 6,12,24 hours under basal and LH-stimulated conditions. After test times had been completed, 10 µl MTT I solution was added into each well and plates were incubated in CO_2 incubator at 37°C for 4 hours. Each well was added 100 µl MTT II solution (SDS) and stored in CO_2 incubator for one night. Optic density of cells was read in ELISA device at 540 wave length at the end of this duration. Viability ratios of test cells were expressed as per cent by accepting control cell viability not treated with test substance as 100 per cent (Mossman, 1983).

Assay of Steroidogenic Enzymes

3β-HSD Enzyme Activity

The activity of 3β-HSD in cultured Leydig cells was determined by the method described by Bergmeyer (1974). In brief, the cultured Leydig cells were sonicated in ice-cold Tris–HCl buffer (pH 7.2) and centrifuged at 16 000 g for 5 min at 4 °C. The supernatant was used as enzyme extract for the assay of 3β-HSD. The reaction mixture contained 0.6 ml pyrophosphate buffer (100 mM), 0.2 ml NAD (0.5 mM), 2 ml distilled water, and 0.1 ml dehydroisoandrosterone (0.1 mM). The absorbance at 340 nm was measured immediately after the addition of enzyme extract at 20-s intervals for 5 min in a spectrophotometer against blank.

17β-HSD Enzyme Activity

The activity of 17β-HSD in Leydig cells was determined by the method described by Bergmeyer (1974). In brief, the cultured Leydig cells were sonicated in ice-cold Tris–HCl buffer (pH 7.2) and centrifuged at 10 000 g for 5 min at 4 °C. The supernatant was used as enzyme extract for the assay of 17β-HSD. The reaction mixture contained 0.6 ml pyrophosphate buffer (100 mM), 0.2 ml NADPH (0.5 mM), 2 ml distilled water, and 0.1 ml of 1,4-androstenedine-3,17-dione (0.8 mM). The absorbance at 340 nm was measured immediately after the addition of enzyme extract at 30-s intervals for 5 min in a spectrophotometer against blank.

Leydig Cellular Antioxidant System

After 6, 12 and 24-h A1260 and A1260 + vitamin E treatment with or without LH stimulation, the cultured Leydig cells were sonicated in ice-cold Tris–HCl buffer (pH 7.4), centrifuged and the supernatant was collected and used for measuring the following biochemical parameters.

Determination of Lipid Peroxidation

Protein content was determined by the method of Lowry *et al.* (1951). The level of lipid peroxidation was measured by the method of Devasagayam and Tarachand (1987). In brief, the reaction mixture consisted of 1.0 ml of 0.15 M Tris–HCl buffer (pH 7.4), 0.3 ml of 10 mM KH2PO4, and 0.2 ml cell extract in a total volume of 2 ml. The tubes were incubated at 37 °C for 20 min with constant shaking. The reaction was stopped by the addition of 1 ml 10 per cent trichloroacetic acid. The tubes were shaken well, 1.5 ml thiobarbituric acid (TBA) added and were heated in a boiling water bath for 20 min. The tubes were centrifuged and the colour developed was measured at 532 nm. The malondialdehyde (MDA) content of the sample is expressed as nanomoles of MDA formed per milligram protein.

Determination of Reactive Oxygen Species

Hydroxyl radical production was quantified by the method of Puntarulo and Cederbaum (1988). In brief, to a 1 ml cell extract, 0.2 ml of 1 M phosphate buffer, 0.1 ml of each of magnesium chloride, sodium azide, DMSO, and NADPH were added and incubated for 10 min at 37°C. The reaction was arrested by adding 1.5 ml trichloro acetic acid, boiled for 30 min, and read at 570 nm, against the reagent blank. The hydroxyl radical content of the samples is expressed as mmol/min per mg protein.

Antioxidant Enzymes Assay

Determination of Superoxide Dismutase (EC: 1.15.1.1, SOD)

SOD activity was measured as the inhibition of autoxidation of pyrogallol, according to the method of Marklund and Marklund (1974). Activity was monitored at 440 nm for 180 s (s). Data is expressed as U of SOD/mg protein.

Determination of Catalase (EC 1.11.1.6, CAT)

CAT activity was measured according to the method of Sinha (1972) as the rate constant of hydrogen peroxide (H_2O_2) decomposition. Activity was monitored at 590 nm for 60s. Data is expressed as U of CAT/mg protein.

Determination of Glutathione Peroxidase (EC 1.11.1.9, GPx)

GPx activity was measured according to the method of Hafeman *et al.* (1974) by using H_2O_2 as a substrate and as the rate constant of glutathione decomposition. Activity was monitored at 412 nm for 60s. Data is expressed as µM of glutathione consumed/mg protein.

Determination of Glutathione-S-transferase (EC 2.5.1.1.8, GST)

The activity of glutathione S-transferase (GST) was measured according to *Habig et al.* (1974) using 1-chloro-2,4-dinitrobenzene (CDNB) as a substrate. GST activity of was expressed as µmol GS-DNB min^{-1} per mg protein using an extinction coefficient of 9.6 mM cm^{-1}.

Statistical Analysis

Data are presented as the mean±S.E.M. Cell viability, formed MDA amount, the amount of hydroxyl radical, SOD enzyme activity, catalase enzyme activity, GPx enzyme activity, GST enzyme activity, 3β-HSD enzyme activity and 17β-HSD enzyme activity were observed in the test groups versus time and the administered dose. Then, Kruskal Wallis test was applied in multi-comparisons. Mann-Whithey U and Wilcoxon test were applied in dual comparisons to calculate statistical analyses. $p<0.001$, $p<0.01$ and $p<0.05$ significance levels were referenced to interpret the results.

Results

Effect of Aroclor 1260 and Aroclor 1260 + Vitamin E on Cell Viability

Figure 26.1 shows viability rates for TM3 Leydig cells treated with Aroclor 1260 and Aroclor 1260 + Vitamin E in per cent for the control and test groups under basal and LH-stimulated conditions at the end of 6, 12 and 24 hours. The viable cells were significantly reduced due to 8M and 6M A1260 and A1260 + vitamin E treatment in both basal and LH-stimulated conditions when compared with respective control. However, other treatments did not alter the cell viability.

Leydig Cellular Steroidogenic Enzymes

Figures 26.2 and 26.3 shows the effect of Aroclor 1260 and Aroclor 1260 + vitamin E on Leydig cellular steroidogenic enzyme

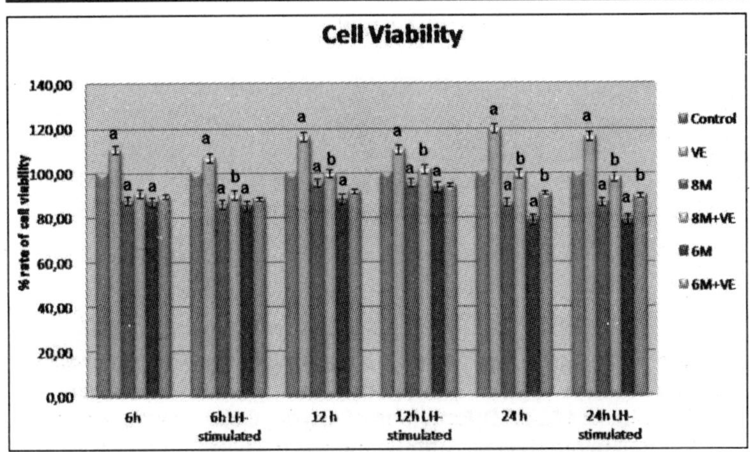

Figure 26.1: Dose- and Time-Dependent Effects of PCB (Aroclor 1260) and PCB + Vitamin E (Aroclor 1260 + Vitamin E) on Basal and LH Stimulated Leydig Cell Viability *in vitro*. Each bar denotes mean±S.E.M. Significance at $p < 0.001$, a: compare to control, b: compare to PCB-exposed dose.

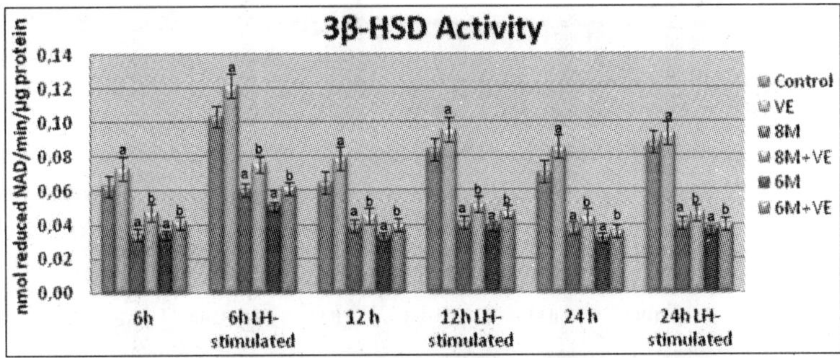

Figure 26.2: Dose- and Time-Dependent Effects of PCB (Aroclor 1260) and PCB + Vitamin E (Aroclor 1260 + Vitamin E) on Basal and LH Stimulated 3â-HSD Activities of Leydig Cells *in vitro*. Each bar denotes mean±S.E.M. Significance at $p < 0.05$, a: compare to control, b: compare to PCB-exposed dose.

activities, such as 3β-HSD and 17β-HSD, under basal and LH-stimulated conditions. When compared with control, 3β-HSD and 17β-HSD activities decreased significantly in a dose and time dependent manner under basal and LH-stimulated conditions in

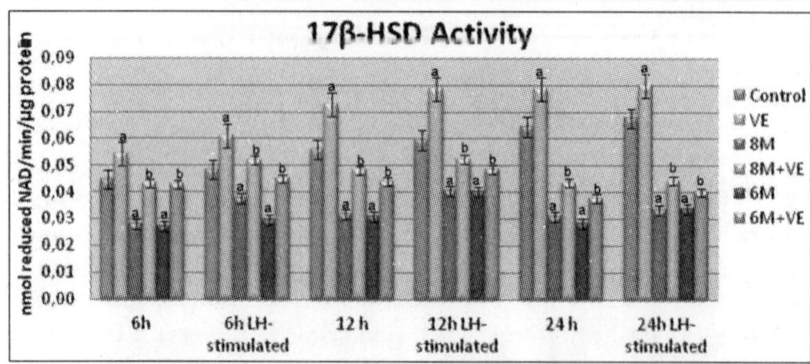

Figure 26.3– Dose- and Time-Dependent Effects of PCB (Aroclor 1260) and PCB + Vitamin E (Aroclor 1260 + Vitamin E) on Basal and LH Stimulated 17â-HSD Activities of Leydig Cells *in vitro*. Each bar denotes mean±S.E.M. Significance at $p < 0.05$, a: compare to control, b: compare to PCB-exposed dose.

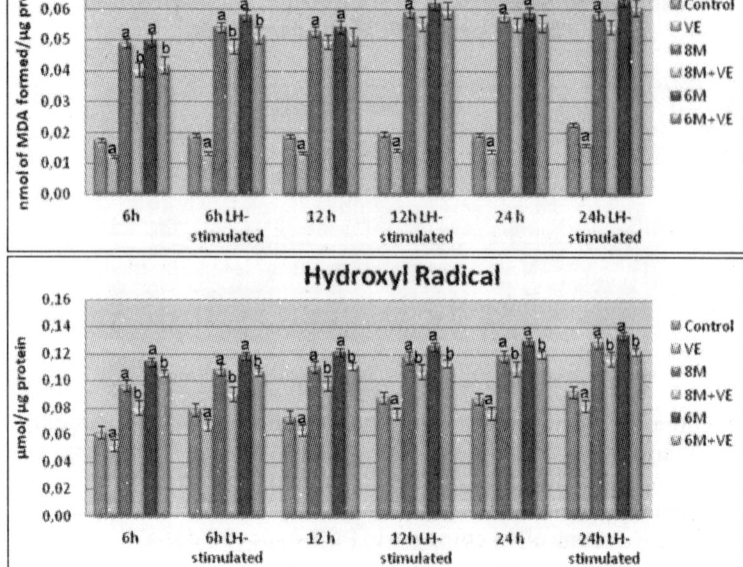

Figure 26.4– Dose- and Time-Dependent Effects of PCB (Aroclor 1260) and PCB + Vitamin E (Aroclor 1260 + Vitamin E) on Basal and LH Stimulated Lipid Peroxidation and Hydroxyl Radical of Leydig Cells *in vitro*. Each bar denotes mean±S.E.M. Significance at $p < 0.01$, a: compare to control, b: compare to PCB-exposed dose.

Aroclor 1260 exposed Leydig cells. When A1260 alone doses were compared with Aroclor 1260 + vitamin E doses, a significant increase occurred at 6, 12 and 24 hours in vitamin E added doses in terms of amount of 3β-HSD and 17β-HSD enzymes.

Lipid Peroxidation and Hydroxyl Radical

Figure 26.4 shows levels of lipid peroxidation and hydroxyl radical in the groups treated with A1260 ve A1260 + vitamin E. Lipid peroxidation significantly increased in the groups treated with A1260 compared to the control groups. It was observed that lipid peroxidation was significantly reduced at the end of 6 hours but there was no variation at the end of 12 and 24 hours when the groups treated with A1260 + vitamin E and treated with only A1260 were compared. Hydroxyl radical levels significantly increased in the A1260-exposed groups compared to the control group while it

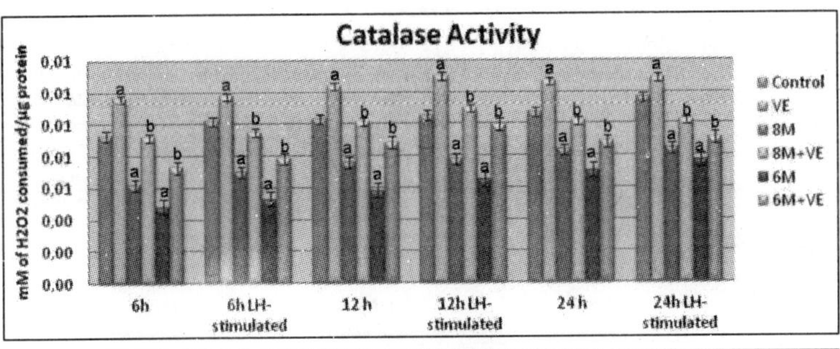

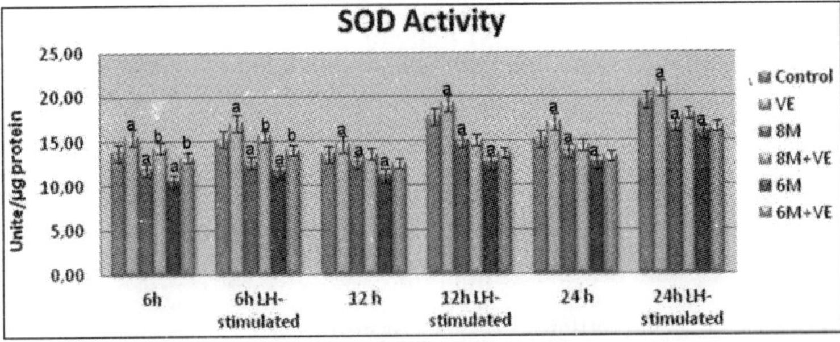

Figure 26.5– Dose- and Time-Dependent Effects of PCB (Aroclor 1260) and PCB + Vitamin E (Aroclor 1260 + Vitamin E) on Basal and LH Stimulated Antioxidant Enzymes such as SOD, and CAT Activities of Leydig Cells *in vitro*. Each bar denotes mean±S.E.M. Significance at $p < 0.05$, a: compare to control, b: compare to PCB-exposed dose.

was significantly reduced in the groups treated with vitamin E in addition to A1260 at all doses compared with the groups treated with only A1260.

Leydig Cellular Antioxidant Enzymes

Figuress 26.5 and 26.6 provide data on cellular levels of antioxidant enzymes, such as SOD, CAT, GPx, and GST in A1260 and A1260 + vitamin E-exposed Leydig cells under basal and LH-stimulated conditions. The antioxidant enzymes, such as SOD, CAT, GPx and GST, were significantly diminished in a dose-dependent manner under basal and LH-stimulated conditions when compared with respective control. Comparison of the Vitamin E-exposed groups and the groups treated with only A1260 showed that CAT and GPx enzyme amounts significantly increased at all doses while SOD

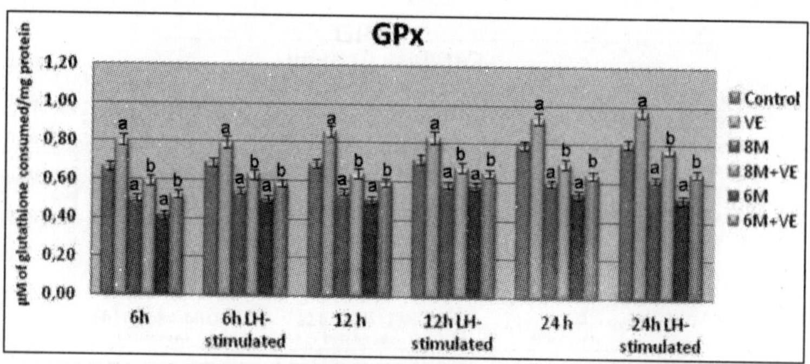

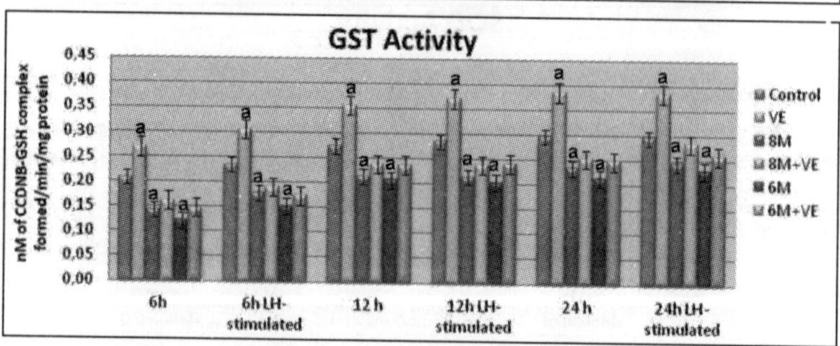

Figure 26.6– Dose- and Time-Dependent Effects of PCB (Aroclor 1260) and PCB + Vitamin E (Aroclor 1260 + Vitamin E) on Basal and LH Stimulated Antioxidant Enzymes such as GPx, and GST Activities of Leydig Cells *in vitro*. Each bar denotes mean±S.E.M. Significance at $p < 0.05$, a: compare to control, b: compare to PCB-exposed dose.

enzyme amount increased significantly in the 6-hour tests for the Vitamin E-exposed groups. On the other hand, any significant variation was not observed for the amount of GST enzyme.

Discussion

This study displays that PCBs inhibit testosterone synthesis in Leydig cells under basal and LH-stimulated conditions and Vitamin E's antioxidant role against toxicity caused by PCB. Disruption of androgen biosynthesis in Leydig cells has a direct relation with the reduction in activities of steroidogenic enzymes and enzymatic antioxidants. PCBs reduce viability of cells, depending on dose, at higher doses and depending on time (Johansson, 1989; Kovacevic *et al.*, 1995; Murugesan *et al.*, 2007b; Murugesan *et al.*, 2008).

3β-HSD and 17β-HSD enzymes are important enzymes taking place in the testosterone synthesis pathway (Payne, 2007). Many studies have been conducted to understand whether or not the administered PCBs affect testosterone synthesis by investigating activities of these two enzymes (Murugesan *et al.*, 2005a; 2005b; 2007b; 2008). A1260, like other PCBs, reduces activities of 3 β-HSD and 17β-HSD enzymes depending on dose and time. However, vitamin E has a protective effect on activities of 3β-HSD and 17β-HSD enzymes in the groups exposed to A1260 for 6 and 12 hours.

It is known that oxidative damage affects important molecules in cells like lipids, proteins or DNA (Halliwell and Gutteridge, 1990; Dalle-Donne *et al.*, 2003). This damage in the cells is caused by reactive oxygen species (ROS) (Cai *et al.*, 1997). ROS create lipid peroxidation products resulting from the oxidation of lipids on the membrane; thus, integrity of the membrane is distorted (Halliwell and Gutteridge, 1990). PCBs, especially high chlorinated PCBs, promote cytochrome P450s, which are possibly a source for ROS, and as a result, lipid peroxidation occurs due to the interaction between the produced ROS and cell membrane (Parkinson *et al.*, 1983; Peltola *et al.*, 1994; Banudevi *et al.*, 2006; Venkataraman *et al.*, 2008). Lipid peroxidation is suppressed by vitamin E simultaneously administered in addition to PCBs because formation of lipid peroxidation products is inhibited (Oda *et al.*, 1987). Lipid peroxidation increases in Leydig cells exposed to A1260, as it has occurred in case A1254 (Murugesan *et al.*, 2008). Damage is significantly reduced in the groups supplemented with vitamin E.

ROS may harm StAR protein and important elements including cytochrome P450 enzymes in Leydig cells in steroidogenic pathways (Georgiou *et al.*, 1987; Diemer *et al.*, 2003). Moreover, ROS cause reduction in activities of antioxidant enzymes, inhibition of protein synthesis, DNA damage and finally, death of cells (Halliwell and Gutteridge, 1990). PCBs may interact with hydroxyl radical to produce hydrogen peroxide (Kamohara *et al.*, 1984). The significant increase in number of hydroxyl radicals in the groups exposed to A1260, as it has occurred in case of A1254 (Murugesan *et al.*, 2007b; 2008) and the decrease of the number in the groups supplemented with vitamin E evidence vitamin E's protective role.

Enzymatic and non-enzymatic antioxidants are the natural defense system of the body against tissue damage caused by free radical intermediates (Peltola *et al.*, 1994). Superoxide dismutase (SOD) is an antioxidant enzyme which takes part in this defense system and reduces superoxide anion to hydrogen peroxide and oxygen (Murugesan *et al.*, 2008). The amount of the SOD enzyme significantly decreases in the A1260 exposed groups while it increases in the groups treated with vitamin E in addition to A1260 compared with those treated with A1260 alone. Catalase, an enzyme which takes place in intracellular antioxidant a defense system is produced by activities of free radicals inside the cell and converts hydrogen peroxide, which is very toxic to the cell, into water and molecular oxygen (Goldman and Blobel, 1978; Murugesan *et al.*, 2008). Catalase enzyme activity is inhibited by the superoxide anion radical (Kono ve Fridovich, 1982). A1254 reduces the amount of catalase enzymes in Leydig and Sertoli cells (Senthil kumar *et al.*, 2004; Murugesan *et al.*, 2007b; 2008). GPx enzyme is a selenium-binding enzyme, which catalyzes reduction of hydrogen peroxide or lipid hydro peroxide (LOOH) by glutathione. The GPx enzyme uses glutathione as a hydrogen donor to ensure formation of GSSG, which is an oxidized form of glutathione, and water in removing hydrogen peroxide (Delibas and Özcankaya, 1995; Simsek, 1999; Burçak and Andican, 2004; Murugesan *et al.*, 2008). It was observed that the decrease in the glutathione level and the increase in peroxide levels, upon administration of PCB, reduced GPx enzyme activity (Twaroski *et al.*, 2001). GST enzyme, which is another intracellular antioxidant enzyme, catalyzes conjugation of free radicals with glutathione (Banudevi *et al.*, 2006). It is known that glutathione has direct interaction with free radicals and suppresses them

(Chausseud, 1979). A1260 administration reduces GST enzyme activity depending on time and dose. The lack of vitamin E's protective effect against this reduction displays that vitamin E does not have an effect on glutathione levels.

Vitamin E is the most active chain breaking antioxidant taking place in the cell membrane and has a great role in keeping the integrity of cell membrane. Vitamin E is effective in removing free radicals in cells (Scott, 1969; Hsu *et al.*, 1998). It reduces free radicals' level and increases enzymatic and non-enzymatic antioxidants taking place in the antioxidant system (Banudevi *et al.*, 2006). This study evidences that vitamin E protects cells in the groups treated with Aroclor 1260 against lipid peroxidation and oxidative damages and it reduces number of free radicals and thus, promotes activities of enzymes taking place in the intracellular antioxidant system.

Finally, it was displayed that Aroclor 1260, which is a commercial PCB mixture, reduces activities of 3β-HSD and 17β-HSD enzymes, which have a significant role in testosterone synthesis in TM3 Leydig cells and thus, it reduces testosterone amount. It was also discovered that it promotes toxicity in TM3 Leydig cells while vitamin E has a protective role against oxidative damage.

Acknowledgements

This work was supported by Istanbul University Scientific Research Projects (T-2857).

References

Ahlborg, U.G., Becking, G.C., Birnbaum, L.S., Brouwer, A., Derks, H.J.G.M., Feeley, M., Golor, G., Hanberg, A., Larsen, J.C., Liem, A.K.D., Safe, S.H., Schlatter, C., Woem, F., Younes, M. and Yrjanheikki, E., 1994. Toxic equivalency factors for dioxin–like PCB's. *Chemosphere,* 28: 1049–1067.

Andric, S.A., Kostic, T.S., Dragisic, S.M., Andric, N.L., Stanko, S.S. and Kovacevic, R.Z., 2000a. Acute effects of polychlorinated biphenyl–containing and free transformer fluids on rat testicular steroidogenesis. *Environmental Health Perspectives,* 108: , 955–959.

Andric, S.A., Stojilkovic, S.S., Stanko, S.S. and Kovacevic, R.Z., 2000b. Inhibition of rat testicular androgenesis by a polychlorinated biphenyl mixture Aroclor 1248. *Biology of Reproduction*, 62: 1882–1888.

Apostoli, P., Mangili, A., Carasi, S. amd Manno, M., 2003. Relationship between PCBs in blood and D–glucaric acid in urine. *Toxicology Letters*, 144: 17–26.

Amaral Mendes, J. J., 2002. The endocrine disrupters: a major medical challenge. *Food and Chemical Toxicology*, 40 (6): 781–788.

Atsdr, 1997. Toxicological Profile for Polychlorinated Biphenyls (update). Atlanta: US Department of Health and Human Services.

Atsdr, 2000. Toxicological Profile for Polychlorinated Biphenyls (PCBs). Atlanta: US Department of Health and Human Services.

Banudevi, S., Krishnamoorthy, G., Venkataraman, P., Vignesh, C., Aruldhas, M.M. and Arunakaran, J., 2006. Role of α–tocopherol on antioxidant status in liver, lung and kidney of PCB exposed male albino rats. *Food and Chemical Toxicology*, 44: 2040–2046.

Bergmeyer, M.U., 1974. Steroid dehidrogenase. In: Bergmeyer H.U. editor, *Methods of Enzymatic Analysis*. New York: Academic Press, 476–477.

Burçak, G. and Andican, G., 2004. Oksidatif DNA hasari ve yaþlanma, *Cerrahpasa. Journal of Medicine*, 35: 159–169.

Cai, Q., Rahn, R.O. and Zhang, R., 1997. Dietary flavonoids, quercetin, luteolin and genistein, reduce oxidative DNA damage and lipid peroxidation and quench free radicals. *Cancer Letters*. 119(1): 99–107.

Chausseud, L.F., 1979. The role of glutathione and glutathione–S–transferase in the metabolism of chemical carcinogens and other electrophilic agents. *Advances of Cancer Research*, 29: 175–274.

Coskun, T., 2005. Fonksiyonel besinlerin sagligimiz üzerine etkileri. *Çocuk Sagligi ve Hastaliklari Dergisi*, 48: 69–84.

Dalle–Donne, I., Giustarini, D., Colombo, R., Rossi, R. and Milzani, A., 2003. Protein carbonylation in human diseases. *Trends Molecular Medicine*, 9: 169–176.

De Voogt, P. and Brinkman, U.A.Th., 1989. *Production, properties and usage of polychlorinated biphenyls*, In: Kimbourgh, R, Jensen, A.A. (Eds),: Halogenated Biphenyls, Terphenyls, Naphthalenes,

Dibenzodioxins and Related Products. Elsevier, Amsterdam, 3–45.

De Vos, S., Maervoet, J., Schepens, P. and De Schrijver, R., 2003. Polichlorinated biphenyls in broiler diets: their digestibility and incorporation in body tissues. *Chemosphere,* 51: , 7–11.

Delibas, N. and Özcankaya, R., 1995. Serbest radikaller, SDÜ Tip Fakültesi Dergisi 2(3): 11–17.

Devasagayam, T.P. and Tarachand, U., 1987. Decreased lipid peroxidation in the rat kidney during gestation. *Biochemical and Biophysical Research Communications,* 198: 273–80.

Diemer, T., Allen, J.A., Hales, K.H. and Hales D.B., 2003. Reactive Oxygen Disrupts Mitochondria in MA–10 Tumor Leydig Cells and Inhibits Steroidogenic Acute Regulatory (StAR) Protein and Steroidogenesis. *Endocrinology,* 144(7): 2882–2891.

Eisler, R. and Belisle, A.A., 1996. *Planar PCB Hazards to Fish, Wildlife, and Invertebrates: A Synoptic Review,* Contaminant Hazard Reviews. *Biological Report,* 31: 1081–2911

Georgiou, M., Prekins, L.M. and Payne, A.H., 1987. Steroid synthesis dependent oxygen–mediated damage of mitochondrial and microsomal cytochrome P–450 enzymes in rat Leydig cell cultures. *Endocrinology,* 121: 1390–1399.

Goldman, B.M. and Blobel, G., 1978. Biogenesis of peroxisomes: Intracellular site of synthesis of catalase and uricase. *Proceedings in National Academic Science,* 75(10): 5066–5070.

Habig, W.H., Pabst, M.J. and Jakoby, W.B., 1974. Glutathione–S–transferase. the first enzymatic step in mercapturic acid formation. *The Journal of Biological Chemistry,* 249: 7130–7139.

Hafeman, D.G., Sunde, R.A. and Hoekstra, W.G., 1974. Effect of dietary selenium on erythrocyte and liver glutathione peroxidase in the rat. *Journal of Nutrition,* 104: 580–587.

Halliwell, B. and Gutteridge, J.M.C., 1990. Role of free radicals and catalytic metal ions in human disease: an overview. *Methods in Enzymology,* 186: 1–85.

Hansen, L.G., 1999. *The Ortho Side of PCB's Occurrence and Disposition.* Kluwer Academic Publishers, Boston, London.

Hsu, P.C., Liu, M.Y., Hsu, C.C., Chen, L.Y. and Guo, Y.L., 1998. Effects of vitamin E and/or C on reactive oxygen species–related lead toxicity in the rat sperm. *Toxicology*, 128: 169–179.

Hutzinger, O., Safe, S. and Zitko, V., 1974. *The Chemistry of PCB's.* CRC Press, Cleveland.

Johansson, B., 1989. Effects of polychlorinated biphenyls (PCBs) on *in vitro* biosynthesis of testosterone and cell viability in mouse leydig cells. *Bulletin of Environmental Contamination and Toxicology*, 42(1): 9–14.

Kamohara, K., Yagi, N. and Itokawa, Y., 1984. Mechanism of lipid peroxide formation in polychlorinated biphenyls (PCB) and dichlorodiphenyltrichloroethane (DDT)–poisoned rats. *Environmantal Research*, 34: 18–23.

Kono, Y. and Fridovich, I., 1982. Superoxide radicals inhibit catalase. *Journal of Biological Chemistry*, 257: 5751–5754.

Kosowski, A. and Clouatre, D.L., 2009. *Vitamin E: Natural vs. Synthetic*, Tocotrienols Vitamin E Beyond Tocopherols, CRC Press, New york, 61–75.

Kovacevic, R., Vojinovic–Miloradov, M., Teodorovic, I. and Andric, S., 1995. Effect of PCBs on androgen production by suspension of adult rat leydig cells *in vitro*. *Journal Steroid of Biochemistry Molecular Biology*, 52: 595–597.

Lilienthal, H., Fastabend, A. and Hany, J., 2000. Reduced levels of 1,25–Dihydroxy vitamin D3 in rat dams and offspring after exposure to a reconstituted PCB mixture. *Toxicological Sciences*, 57: 292–301.

Lowry, O.H., Rosebrough, N.J., Farr, A.L. and Randall, R.J., 1951. Protein measurement with folin phenol reagent. *Journal Of Biological Chemistry*, 193: 265–275.

Marklund, S. and Marklund, G., 1974. Involvement of the superoxide anion radical in the autooxidation of pyrogallol and a convenient assay for superoxide dismutase. *European Journal of Biochemistry*, 47: 469–474.

Mossman, T., 1983. Rapid colourimetric assay for cellular growth and survival: application to proliferation and cytotoxicity assays. *Journal of Immunological Methods*, 65: 55–63.

Murugesan, P., Kanagaraj, P., Yuvaraj, S., Balasubramanian, K., Aruldhas, M.M. and Arunakaran, J., 2005a. The inhibitory effects of polychlorinated biphenyl (Aroclor 1254) on Leydig cell LH receptors, steroidogenic enzymes, and antioxidant enzymes in adult rats. *Reproductive Toxicology,* 20: 117–126.

Murugesan, P., Muthusamy T., Balasubramanian, K. and Arunakaran, J., 2005b. Studies on the protective role of vitamin C and E against polychlorinated biphenyl (aroclor 1254)–induced oxidative damage in leydis cells. *Free Radical Research,* 39(11): 1259–1272.

Murugesan, P., Muthusamy T., Balasubramanian, K. and Arunakaran, J., 2007a. Effects of vitamins C and E on steroidogenic enzymes mRNA expression in polychlorinated bipenyls (Aroclor 1254) exposed adult rat Leydig cells. *Toxicology,* 232: 170–182.

Murugesan, P., Balaganesh, M., Balasubramanian, K. and Arunakaran, J., 2007b. Effects of polychlorinated biphenyl (Aroclor 1254) on steroidogenesis and antioxidant system in cultured adult rat Leydig cells. *Journal of Endocrinology,* 192, 325–338.

Murugesan, P., Muthusamy T., Balasubramanian, K. and Arunakaran, J., 2008. Polychlorinated biphenyl (Aroclor 1254) inhibits testosterone biosynthesis and antioxidant enzymes in cultured rat Leydig cells. *Reproductive Toxicology,* 25: 447–454.

Oakley, G.G., Devanaboyina, U., Robertson, L.W. and Gupta, R.C., 1996. Oxidative DNA damage induced by activation of polychlorinated biphenyls (PCBs): implications for PCB induced oxidative stress in breast cancer. *Chemical Research in Toxicology,* 9: 1285–1292.

Oda, H., Yamashita, K., Sasaki, S., Horio, F. and Yoshida, A., 1987. Long–term effects of dietary polychlorinated biphenyl and high level of vitamin E on ascorbic acid and lipid metabolism in rats. *Journal of Nutrition,* 117: 1217–1223.

Parkinson, A., Safe, S.H., Robertson, L.W., Thomas, P.E., Ryan, D.E., Reik, L.M. and Levin, W., 1983. Immunochemical quantitation of cytochrome P–450 isoenzymes and epoxide hydrolase in liver microsomes from polychlorinated or polybrominated biphenyls–treated rats. A study of structure–activity relationships, *Journal of Biological Chemistry*, 258: 5967–5976.

Payne, A.H., 2007. *Steroidogenic Enzymes in Leydig Cells*, The Leydig Cell in Health and Disease. Humana Press Inc., Totowa, New Jersey, 3–30.

Peltola, V., Mantyla, E., Huhtaniemi, I. and Ahotupa, M., 1994. Lipid peroxidation and antioxidant enzyme activities in the rat testis after cigarette smoke inhalation or administration of polychlorinated biphenyls or polychlorinated naphthalenes. *Jounral of Andrology*, 15: 353–361.

Pflieger–Bruss, S., Schuppe, H. C. and Schill, W. B., 2004. The male reproductive system and its susceptibility to endocrine disrupting chemicals. *Andrologia*, 36 (6): 337–345.

Puntarulo, S. and Cederbaum, A.I., 1988. Effects of oxygen concentration on microsomal oxidation of ethanol and generation of oxygen radicals. *Biochemical Journal*, 251: 787–794.

Rogan, W.J. and Gladen, B.C., 1992. Neurotoxicology of PCBs and related compounds. *Neurological Toxicology*, 13: 27–36.

Safe, S., 1994. Polychlorinated bipheyls (PCBs): Environmental impact, biochemical and toxic responses and implications for risk assessment. *Critical Reviews Toxicology*, 24: 87– 149.

Scott, M.L., 1969. *Studies on vitamin E and related factors in nutrition and metabolism*. In: Deluca, The fat soluble vitamins. University of Wisconsin Press, Madison, pp.355–368.

Senthilkumar, J., Banudevi, S., Sharmila, M., Murugesan, P., Srinivasan, N., Balasubramanian, K., Aruldhas, M.M. and Arunakaran, J., 2004. Effects of vitamin C and E on PCB (Aroclor 1254) induced oxidative stress, androgen binding protein and lactate in rat Sertoli cells. *Reprodtive Toxicology*, 19: 201–208.

Seyran, A. and Erisis, M., 2008. Poliklorlu bifeniller ve saglik üzerine etkileri. *Firat Üniversitesi Saglik Bilimleri Dergisi*, 22 (1): 33–40.

Shain, W., Overmann, S.R., Wilson, L.R., Kostas J. and Bush B., 1986. A congener analysis of polychlorinated biphenyls accumulating in rat pups after perinatal exposure. *Archives of Environmental Contamination and Toxicology*, 15: 687–707.

Sharpe, R.M. and Skakkebaek, N.E., 1993. Are oestrogens involved in falling sperm counts and disorders of the male reproductive tract. *Lancet*, 341: 1392–1395.

Simone, R. and Palozza, P., 2008. *Antioxidant activity of tocotrienols in cells and serum.* Tocotrienols: Vitamin E beyond tocopherols, CRC Press, 99–106.

Sinha, A.K., 1972. Colourimetric assay of catalase. *Analytical Biochemistry*, 47: 389–394.

Sridhar, M., Venkataraman, P., Siva, D., Arun Kumar, A., Aruldhas, M.M. and Srinivasan, N., 2004. Impact of polychlorinated biphenyl (Aroclor 1254) and vitamin C on antioxidant system of rat ventral prostate. *Asian Journal of Andrology*, 6: 19–22.

Simsek, F., 1999. Serbest Oksijen Radikalleri, Antioksidanlar Velipit Peroksidasyonu. *Pediatri Dergisi*, 8(1): 42–47.

Takacs, P., Kauma, S. W., Sholley, M. M., Walsh, S. W., Dinsmoor, M. J. and Green, K., 2001. Increased circulating lipid peroxides in severe preeclampsia activate NF–kappaB and upregulate ICAM–1 in vascular endothelial cells. *FASEB J* 15: 279–281.

Twaroski, L.W., O'brien, M.L., Larmonier, N., Glauret, H.P. and Robertson, L.W., 2001. Polychlorinated biphenyl (PCB) congeners on hepatic glutathione, glutathione–related enzymes and selenium status: implications for oxidative stress. *Biochemical Pharmacology*, 62: 273–278.

Venkataraman, P., Krishnamoorthy, G., Vengatesh, G., Srinivasan, N., Aruldhas, M.M. and Arunakaran, J., 2008. Protective role of melatonin on PCB (Aroclor 1254) induced oxidative stress and changes in acetylcholine estrease and membrane bound ATPases in cerebellum, cerebral cortex and hippocampus of adult rat brain. *International Journal of Developmental Neuroscience*, 26: 585–591.

WHO., 1993. Polychlorinated biphenyls and terphenyls (Second Edition). *Environmental Health Criteria*, Geneva, 140.

Previous Volumes

— Volume 1 —

2002, xiii+308p., figs., tabls., ind., 23 cm Rs. 650

ISBN 978-81-7035-272-3

Section I: Animal Ecology

Section II: Animal Reproduction

— Volume 2 —

2004, xi+387p., figs., tabls., ind., 23 cm **Rs. 850**

ISBN 978-81-7035-322-5

Section I: Animal Ecology

— Volume 3 —

2006, xvii+443p., figs., tabls., ind., 23 cm Rs. 1150

ISBN 978-81-7035-424-6

Section I: Animal Ecology

Section II: Animal Reproduction

— Volume 4 —

2007, xiv+550p., figs., tabls., ind., 23 cm **Rs. 1400**

ISBN 978-81-7035-459-8

Section I: Animal Ecology

Section II: Animal Reproduction

— Volume 5 —

2008, xiv+476p., col. plts., figs., tabls., ind., 23 cm **Rs. 1600**

ISBN 978-81-7035-563-2

Section I: Animal Ecology

Section II: Animal Reproduction

— Volume 6 —

2010, xiv+497p., col. plts., figs., tabls., ind., 23 cm Rs. 1500

ISBN 978-81-7035-635-6

Section I: Animal Ecology

Section II: Animal Reproduction

— Volume 7 —

2011, xiv+494p., col. plts., figs., tabls., ind., 23 cm Rs. 1700

ISBN 978-81-7035-746-9

Section I: Animal Ecology

Section II: Animal Reproduction

— Volume 8 —

2011, xviii+533p., col. plts., figs., tabls., ind., 23 cm Rs. 1800

ISBN 978-81-7035-747-6

Section I: Animal Ecology

Section II: Animal Reproduction

Index

U

V